Molecular Biology of Drug Addiction

Molecular Biology of Drug Addiction

Edited by

Rafael Maldonado

Laboratory of Neuropharmacology
Health and Life Sciences School
Universitat Pompeu Fabra
Barcelona, Spain

Humana Press  **Totowa, New Jersey**

999 Riverview Drive, Suite 208
Totowa, New Jersey 07512
www.humanapress.com

This publication is printed on acid-free paper. ∞

ANSI Z39.48-1984 (American National Standards Institute) Permanence of Paper for Printed Library Materials.

Cover design by Patricia F. Cleary.

Cover illustration: Disruption of the CREB gene in brain by Cre/loxP-mediated recombination (Chapter 2, Fig. 5). *See* complete illustration on p. 33 and discussion on p. 30.

For additional copies, pricing for bulk purchases, and/or information about other Humana titles, contact Humana at the above address or at any of the following numbers: Tel: 973-256-1699; Fax: 973-256-8341; E-mail: humana@humanapr.com or visit our website at http://humanapress.com

Printed in the United States of America. 10 9 8 7 6 5 4 3 2 1

Library of Congress Cataloging-in-Publication Data

Molecular biology of drug addiction / edited by Rafael Maldonado

p. cm.

Includes bibliographical references and index.

ISBN 1-58829-060-3 (alk. paper)

1. Drug abuse--Molecular aspects. I. Maldonado, Rafael.

RC564 .M638 2003

616.86--dc21

2002024057

Preface

The neurobiological mechanisms involved in drug addiction have been investigated for several decades with a variety of pharmacological and biochemical approaches. These studies have associated several neuroanatomical and neurochemical mechanisms with different components of drug-addictive processes, and this has led to the identification of possible targets for new treatment strategies. Progress has been accelerated dramatically in the last few years by novel research tools that selectively remove or enhance the expression of specific genes encoding proteins responsible for the biological responses of these drugs. These new models, most of them obtained from the recent advances in molecular biology's technology, have provided definitive advances in our understanding of the neurobiological mechanisms of drug addiction. Classical behavioral, biochemical, and anatomical techniques have been adapted to take a maximum advantage of these new molecular tools. These recent studies have clarified the different molecular and intracellular mechanisms involved in addictive processes, as well as the interactions among these endogenous neurobiological mechanisms; and they have provided new insights toward identifying other genetic bases of drug addiction.

The main purpose of *Molecular Biology of Drug Addiction* is to offer an extensive survey of the recent advances in molecular biology and complementary techniques used in the study of the neurobiological basis of drug dependence and addiction. Ours is a multidisciplinary review of the most relevant molecular, genetic, and behavioral approaches used in this field. The definitive advances given by the new molecular and behavioral tools now available provide a unique opportunity for such an approach. Each chapter in this book is not simply a review of the research activities of the author's laboratory, but rather provides a critical review of the main advances in the corresponding topic. Sixteen different chapters organized in four parts have been included in the book. The first part is devoted to the advances in the knowledge of the neurobiological mechanisms of opioid addiction provided in the last few years using the new available techniques, and some of the new therapeutic perspectives now opening up in this field. The second part addresses the most recent findings on the molecular, genetic, and neurochemical mechanisms involved in psychostimulant addiction, which have changed some of the basic knowledge of the neurobiological substrate of these processes. The third part of the book is focused on cannabinoid addiction. New molecular tools have also been used recently to elucidate the biological substrate of cannabinoid dependence. The behavioral models now available, which allow evaluation of the different components of cannabinoid dependence, have optimized results in this particular field. The last part addresses several molecular, genetic, and behavioral aspects of alcohol and nicotine addiction, which have provided decisive progress in our understanding of these addictive processes.

Molecular Biology of Drug Addiction addresses the main advances in understanding the molecular mechanisms involved in the complex physiological and behavioral processes underlying drug addiction and will, we hope, serve as a useful reference guide for a wide range of neuroscientists. This book also provides basic information of interest for scientists and clinicians interested in the new therapeutic approaches to drug addiction. The different sections of the book are presented by the most relevant scientific personalities for each area. I deeply thank the authors for their effort and expert contribution in the different chapters, and Elyse O'Grady at Humana Press for offering this rewarding opportunity. Finally, I thank Raquel Martín especially for help in manuscript preparation and administrative assistance and Dr. Patricia Robledo and Dr. Olga Valverde for scientific assistance and help in library research.

Rafael Maldonado

Contents

Preface .... *v*
Contributors .... *ix*

PART I. OPIOID ADDICTION

1 Molecular Mechanisms of Opioid Dependence by Using Knockout Mice
Brigitte L. Kieffer and Frédéric Simonin ***3***

2 Molecular Genetic Approaches
Theo Mantamadiotis, Günther Schütz, and Rafael Maldonado ***27***

3 Opiate Addiction: *Role of the cAMP Pathway and CREB*
Lisa M. Monteggia and Eric J. Nestler ***37***

4 Different Intracellular Signaling Systems Involved in Opioid Tolerance/Dependence
Thomas Koch, Stefan Schulz, and Volker Höllt ***45***

5 Inhibitors of Enkephalin Catabolism: *New Therapeutic Tool in Opioid Dependence*
Florence Noble and Bernard P. Roques ***61***

PART II. PSYCHOSTIMULANT ADDICTION

6 Recent Advances in the Molecular Mechanisms of Psychostimulant Abuse Using Knockout Mice
Cécile Spielewoy and Bruno Giros ***79***

7 Opioid Modulation of Psychomotor Stimulant Effects
Toni S. Shippenberg and Vladimir I. Chefer ***107***

8 Influence of Environmental and Hormonal Factors in Sensitivity to Psychostimulants
Michela Marinelli and Pier Vincenzo Piazza ***133***

9 Development and Expression of Behavioral Sensitization: *Temporal Profile of Changes in Gene Expression*
Peter W. Kalivas ***161***

PART III. CANNABINOID ADDICTION

10 New Advances in the Identification and Physiological Roles of the Components of the Endogenous Cannabinoid System
Ester Fride and Raphael Mechoulam ***173***

11 Integration of Molecular and Behavioral Approaches to Evaluate Cannabinoid Dependence
Dana E. Selley, Aron H. Lichtman, and Billy R. Martin ***199***

12 Opioid System Involvement in Cannabinoid Tolerance and Dependence
Rafael Maldonado ***221***

PART IV. ALCOHOL AND NICOTINE ADDICTION

13 Current Strategies for Identifying Genes for Alcohol Sensitivity
John C. Crabbe ***249***

14 Genetic Basis of Ethanol Reward
Christopher L. Cunningham and Tamara J. Phillips ***263***

15 Behavioral and Molecular Aspects of Alcohol Craving and Relapse
Rainer Spanagel ***295***

16 Molecular and Behavioral Aspects of Nicotine Dependence and Reward
Emilio Merlo Pich, Christian Heidbreder, Manolo Mugnaini, and Vincenzo Teneggi ***315***

Index ***339***

Contributors

VLADIMIR I. CHEFER • *Integrative Neuroscience Section, National Institute on Drug Abuse Intramural Research Program, Baltimore, MD*
JOHN C. CRABBE • *Portland Alcohol Research Center, VA Medical Center and Department of Behavioral Neuroscience, Oregon Health & Science University, Portland, OR*
CHRISTOPHER L. CUNNINGHAM • *Department of Behavioral Neuroscience and Portland Alcohol Research Center, Oregon Health & Science University, Portland, OR*
ESTER FRIDE • *Department of Behavioral Sciences, College of Judea and Samaria, Ariel, Israel*
BRUNO GIROS • *INSERM U513, Faculté de Médecine de Créteil, Créteil, France*
CHRISTIAN HEIDBREDER • *Biology Department, GlaxoSmithKline Psychiatry CEDD Research Centre, Verona, Italy*
VOLKER HÖLLT • *Department of Pharmacology, Otto-von-Guericke University, Magdeburg, Germany*
PETER W. KALIVAS • *Department of Physiology, Medical University of South Carolina, Charleston, SC*
BRIGITTE L. KIEFFER • *Institut de Génétique et de Biologie Moléculaire et Cellulaire, CNRS/INSERM/ULP, Illkirch, France*
THOMAS KOCH • *Department of Pharmacology, Otto-von-Guericke University, Magdeburg, Germany*
ARON H. LICHTMAN • *Department of Pharmacology and Toxicology, Virginia Commonwealth University Medical College of Virginia, Richmond, VA*
RAFAEL MALDONADO • *Laboratory of Neuropharmacology, Health and Life Sciences School, Universitat Pompeu Fabra, Barcelona, Spain*
THEO MANTAMADIOTIS • *Differentiation and Transcription Laboratory, Peter MacCallum Cancer Institute, Vic, Australia*
MICHELA MARINELLI • *INSERM U259, Université de Bordeaux II, Bordeaux, France*
BILLY R. MARTIN • *Department of Pharmacology and Toxicology, Virginia Commonwealth University, Medical College of Virginia, Richmond, VA*
RAPHAEL MECHOULAM • *Department of Medicinal Chemistry and Natural Products, Medical Faculty, Hebrew University of Jerusalem, Jerusalem, Israel*
LISA M. MONTEGGIA • *Department of Psychiatry, UT Southwestern Medical Center, Dallas, TX*

MANOLO MUGNAINI • *Biology Department, GlaxoSmithKline Psychiatry CEDD Research Centre, Verona, Italy*
ERIC J. NESTLER • *Department of Psychiatry, University of Texas Southwestern Medical Center, Dallas, TX*
FLORENCE NOBLE • *INSERM U266–CNRS UMR 8600, Université René Descartes (Paris V), Paris, France*
TAMARA J. PHILLIPS • *VA Medical Center, Department of Behavioral Neuroscience and Portland Alcohol Research Center, Oregon Health & Science University, Portland, OR*
PIER VINCENZO PIAZZA • *INSERM U 259, Université de Bordeaux II, Bordeaux, France*
EMILIO MERLO PICH • *Investigative Medicine, GlaxoSmithKline Psychiatry CEDD Research Centre, Verona, Italy*
BERNARD P. ROQUES • *INSERM U266-CNRS UMR 8600, Université René Descartes (Paris V), Paris, France*
STEFAN SCHULZ • *Department of Pharmacology, Otto-von-Guericke University, Magdeburg, Germany*
GÜNTHER SCHÜTZ • *Division Molecular Biology of the Cell I, German Cancer Research Center, Heidelberg, Germany*
DANA E. SELLEY • *Department of Pharmacology and Toxicology, Virginia Commonwealth University Medical College of Virginia, Richmond, VA*
TONI S. SHIPPENBERG • *Integrative Neuroscience Section, National Institute on Drug Abuse Intramural Research Program, Baltimore, MD*
FRÉDÉRIC SIMONIN • *Institut de Génétique et de Biologie Moléculaire et Cellulaire, CNRS/INSERM/ULP, Illkirch, France*
RAINER SPANAGEL • *Department of Psychopharmacology, Central Institute of Mental Health, Mannheim, Germany*
CÉCILE SPIELEWOY • *The Scripps Research Institute, La Jolla, CA*
VINCENZO TENEGGI • *Investigative Medicine, GlaxoSmithKline Psychiatry CEDD Research Centre, Verona, Italy*

Part I

Opioid Addiction

1

Molecular Mechanisms of Opioid Dependence by Using Knockout Mice

Brigitte L. Kieffer and Frédéric Simonin

1. Introduction

Opium, extracted from the seed of the poppy *Papaver somniferum*, has been used and abused for several thousand years. This substance is highly efficient to relieve pain or treat dysentery, and also shows strong euphoric and addictive properties. Due to their exceptional therapeutic potential, the active ingredients of opium have been the subject of intense investigations. Morphine, named after Morpheus, the Greek god of dreams, was isolated in 1806 *(1)* and is considered the prototypic opioid compound. This compound retains both analgesic and addictive properties of opium. Despite numerous adverse effects *(2)*, morphine remains the best painkiller in contemporary medicine, and its clinical use is under tight regulation. In 1898 heroin was chemically synthesized by morphine diacetylation, in an attempt to obtain a drug with lower abuse liability. In fact, this morphine derivative showed even higher addictive potential due to its distinct pharmacokinetic properties. Heroin is being illegally abused worldwide and represents a major public health problem. Attempts to dissociate opioid analgesia from opioid addiction have been unsuccessful so far.

Opioids have been classified as narcotic drugs (from the Greek word for stupor), due to their pharmacological profile very distinct from that of other drugs of abuse, such as pyschostimulants (cocaine, amphetamine), cannabinoids, nicotine, or alcohol *(3)*. As for other substances of abuse, though, opioid addiction typically develops in four stages *(4)*: (a) the initiation phase, in which drug exposure produces positive subjective effects (euphoria); (b) the maintenance phase, in which drug-taking becomes compulsive, indicating that dependence has developed; (c) withdrawal, which develops when drug levels decrease in the body and is recurrently experienced by drug abusers; and (d) craving—or the intense desire to use the drug—and relapse, which are most critical from a therapeutic standpoint. Not every individual exposed to opioids will develop addiction, depending on social, contextual, or perhaps genetic factors *(5)*. However, opioids are considered strongly addictive, and it has been proposed that incremental—perhaps irreversible—neuroadaptations profoundly modify the central nervous system (CNS) following repeated opioid exposure, and contribute to the establishment of opioid dependence *(6)*.

Opioid addiction is a complex phenomenon. Opioid drugs act by activating opioid receptors distributed throughout the CNS and stimulate a number of pathways, among

From: *Molecular Biology of Drug Addiction*
Edited by: R. Maldonado © Humana Press Inc., Totowa, NJ

which the so-called reward pathways located in the limbic system *(7)* are particularly relevant to the addictive process. Repeated opioid stimulation will modify and dysregulate opioid receptor activity and, consequently, interfere with a tightly regulated endogenous opioid system *(8)*, which is critically involved in the control of natural rewards and motivation *(7,9)*, as well as responses to stress *(10)* and pain *(11)*. The endogenous opioid system itself interacts with other neurotransmitter systems, and long-term exposure to exogenous opioids may ultimately remodel associated neuronal networks within brain circuits *(12)* and activate antiopioid systems that counteract opioid effects *(13–15)*, thereby modifying hedonic homeostasis *(16)*.

Recent research aims at clarifying the molecular mechanisms of neuroadaptations to chronic opioids. Cellular models have highlighted regulatory processes, which occur at the level of opioid receptors and their associated signaling proteins, and are believed to contribute to the development of opioid tolerance and withdrawal. Receptor uncoupling from second messenger systems, receptor downregulation, and adenylyl cyclase upregulation were largely shown in neuroblastoma cells expressing opioid receptors endogenously *(17)*. Agonist-induced receptor phosphorylation, desensitization, internalisation, and trafficking were demonstrated more recently using recombinant opioid receptors (e.g., *18–24*). These studies, however, addressed a limited aspect of opioid adaptations, and the link between early agonist-induced events and integrated behavioral responses remains to be established. In vivo, biochemical studies have confirmed upregulation of the cAMP pathway in several brain areas, shown modifications of tyrosine hydroxylase, glutamate receptor subunits, or cytoskeleton protein levels, and proposed a role for growth and transcription factors in the establishment of opioid addiction *(6,25)*.

Gene manipulation in rodents provides a unique mean to correlate molecular events with complex behavior, and is now used to study substance abuse. Possible approaches include (a) targeted gene inactivation using homologous recombination in embryonic stem cells (knockouts), (b) gene overexpression by egg microinjection (transgenics), (c) gene overexpression by viral-mediated gene transfer in adult mice, and (d) gene downregulation by antisense oligonucleotides (26). In this chapter we will focus on gene knockout models, in an attempt to analyze what these unique genetic tools have taught us about opioid addiction. Recently, a number of null mutant mice have been subjected to chronic morphine treatments and their responses found to differ from their wild-type controls (*see* Tables 1–3). These observations have highlighted a role for a number of known genes in behavioral responses to opioids, and allow us to establish a connection between the activity of these genes and molecular neuroadaptations subsequent to chronic opioid treatments in vivo.

2. The Behavioral Models

The manifestations of opioid addiction and dependence can be evaluated in mice using a large panel of behavioral models *(27)*. The reinforcing properties of opioids are currently investigated using conditioned place preference (CPP) or self-administration (SA) procedures. The development of tolerance is observed at the level of opioid analgesia. Typically, tail withdrawal latencies are measured in response to thermal or mechanical pain (tail flick, tail immersion, tail pinch, or hot plate). Latencies are prolonged following acute treatment (analgesia) and gradually return to control values under

Table 1
Effects of Morphine in Knockout Mice of the Opioid System[a]

Gene knockout	Acute morphine	Tolerance to analgesia	Morphine reward	Morphine withdrawal	Reference
MOR	Analgesia abolished		CPP abolished	Somatic and vegetative signs absent	*28*
	Hyperlocomotion abolished				*29*
			SA below saline		*30*
DOR	Analgesia unchanged	Abolished (TF)			*31*
				Somatic signs unchanged	Pintar J., personal communication
KOR	Analgesia unchanged		CPP unchanged	Somatic signs reduced	*32*
PreproENK		Abolished (TF)			*31*
				Somatic signs unchanged	Pintar J., personal communication

[a]CPP, conditioned place preference; SA, self-administration; TF, tail-flick.

Table 2
Effects of Morphine in Knockout Mice for Neuropeptides and Receptors[a]

Gene knockout	Acute morphine	Tolerance to analgesia	Sensitization to hyperlocomotion	Morphine reward	Morphine withdrawal	Reference
CB1	Analgesia unchanged	Unchanged (HP, TI)		SA abolished	Somatic signs reduced	*33*
				DA increase in Nuc Acc abolished		*34*
	Hyperlocomotion unchanged		Abolished	CPP abolished	Withdrawal CPA unchanged	*35*
				SA abolished		*36*
					Somatic signs reduced	*37*
D2R	Analgesia unchanged; hyperlocomotion unchanged			CPP abolished	Somatic signs unchanged	*38*
				CPP maintained in naive but absent during withdrawal	Somatic signs unchanged; withdrawal CPA abolished	*39*
DAT	Analgesia unchanged; hyperlocomotion abolished			CPP enhanced	Some somatic signs reduced (but not jump)	*40*

NK1	Analgesia unchanged; hyperlocomotion abolished			CPP abolished	Jump abolished; other somatic signs unchanged; withdrawal CPA reduced	*41*
GluR-A and GluR-A(R/R)[b]	Analgesia unchanged; hyperlocomotion unchanged	Abolished in GluR-A; unchanged in GluR-A(R/R) (TF)	Context-independent sensitization abolished		Somatic signs reduced in GluR-A, unchanged in GluR-A(R/R))	*42*
OFQ/N	Analgesia unchanged	Unchanged (TI)			Jump increased	*43*
ORL-1	Analgesia unchanged	Reduced (TP)				*44*
	Analgesia unchanged	Reduced (TP and TF)			Jump reduced; other somatic signs abolished	*45*
αCGRP	Analgesia reduced; SIA abolished	Unchanged (TF)		Heroin SA unchanged	Somatic signs reduced	*46*
IL6	Analgesia reduced; SIA abolished	Faster (HP)				*47*

[a]CPA, conditioned place aversion; CPP, conditioned place preference; DA, dopamine; HP, hot plate; Nuc Acc, nucleus accumbens; SA, self-administration; SIA, stress-induced analgesia; TF, tail flick; TI, tail immersion; TP, tail pinch.

[b]In GluR-A(R/R) mutant mice the Q582 residue of the GluR-A subunit is replaced by an arginine residue, which reduces the calcium permeability and channel conductance of receptors containing this subunit.

Table 3
Effects of Morphine in Knockout Mice for Signaling Proteins[a]

Gene knockout	Acute morphine	Tolerance to analgesia	Morphine reward	Morphine withdrawal	Reference
$G_{z\alpha}$	Analgesia unchanged	Higher and faster (TI, HP)			*48*
	Analgesia reduced				*49*
β-Arr2	Analgesia prolonged				*50*
		Acute and chronic abolished (HP)		Somatic and vegetative signs unchanged	*51*
PKCγ			CPP abolished		*52*
CREB	Analgesia unchanged; hyperlocomotion unchanged	Reduced (HP)		Somatic signs reduced	*53*

[a]CPP: conditioned place preference; HP, hot plate; TI, tail immersion.

repeated exposure (tolerance). As for psychostimulants, sensitization to opioids occurs at the level of locomotor activity and is expressed as an enhancement of morphine-induced hyperactivity with repeated administrations. Physical dependence to opioids is revealed by withdrawal, which can be precipitated in chronically treated animals by a single naloxone (opioid antagonist) injection and scored using somatic (jump, sniffing, teeth chattering, tremors, ptosis, diarrhea) or vegetative (temperature drop and weight loss) signs. Also, the negative emotional aspect of opioid withdrawal is detectable using conditioned place aversion, under conditions of mild withdrawal where physical signs are not present. These behavioral paradigms have been used to study the knockout models reviewed here.

3. Knockout Mice Under Scrutiny

Seventeen different knockout mice have been subjected to morphine treatments. Obvious models to test were mice mutated within the opioid system (*see* Table 1) and include mice deficient in μ (MOR) *(28–30)*, δ (DOR) *(31)* or κ (KOR) *(32)* receptors, as well as mice lacking the preproenkephalin gene (preproENK, *31*). Mice deficient in neuropeptides or receptors from other neurotransmitter systems were investigated, on the basis of their postulated functional interactions with the opioid system (*see* Table 2). Thus, morphine activity was examined in mice lacking genes for the CB1 cannabinoid receptor (CB1) (33–37), the dopamine D2 receptor (D2R) *(38,39)*, the dopamine transporter (DAT) *(40)*, the substance P receptor (NK1) *(41)*, the GluR-A subunit of AMPA-type glutamate receptor (GluR-A) *(42)*, the orphaninFQ/nociceptin peptide (OFQ/N) *(43)* or receptor (ORL1) *(44,45)*, the neuropeptide αCGRP *(46)* and interleukin-6 (IL6, *47*). Finally null mutations of genes involved in opioid signaling were analyzed in the context of addictive behaviors, and mice lacking the α subunit of the G protein Gz (Gzα) *(48,49)*, the G-protein receptor modulator β-arrestin 2 (β-Arr 2) *(50,51)*, the protein kinase C isoform γ (PKCγ) *(52)*, and the transcription factor CREB *(53)* were exposed to morphine (*see* Table 3). Phenotypic modifications of these knockout mice in response to chronic morphine are summarized in Tables 1–3 and main features are discussed below. In addition, some of the knockout mice described here have also been treated with other drugs of abuse *(33,35–37,41,49,54–63)*. The results are summarized in Table 4 and are discussed together with data from morphine administration.

4. Mu Receptors: *The Gate for Opioid Addiction*

Three opioid receptor genes, MOR, DOR, and KOR, have been cloned and characterized at the molecular level *(64)*. The genes encode μ, δ, and κ receptors respectively, previously identified by the pharmacology *(65)* and hypothesized to mediate either the euphoric (μ, δ) or dysphoric (κ) properties of opioids *(66)*. Responses to morphine were investigated in mice lacking the MOR gene *(67)*. Morphine conditioned place preference and withdrawal were abolished in mutant mice *(28)*, indicating that μ receptor is essential for morphine reward and physical dependence. In addition, morphine self-administration was below saline controls, suggesting the possibility of an aversive κ receptor-mediated morphine activity in those animals *(30)*. Of note is the finding that reward and withdrawal produced by deltorphin were also abolished in these mice, suggesting that the addictive properties of this prototypic δ agonist are, in fact, mediated by μ receptors *(68)*. Many other pharmacological actions of morphine, unrelated to

Table 4
Effects of Other Drugs in Knockout Mice Studied for Opiate Addiction[a]

Gene knockout	Drug treatment	Acute treatment	Tolerance to analgesia/ Sensitization to hyperlocomotion	Reward	Withdrawal	Ref.
MOR	Ethanol			SA reduced		*54*
	Ethanol			SA and CPP reduced		*55*
	Cannabinoid				Somatic signs reduced	*37*
	Cannabinoid	Analgesia, hypolocomotion, hypothermia unchanged	Tolerance unchanged (TI)	CPP abolished	Unchanged	*56*
KOR	Cannabinoid	Analgesia, hypolocomotion, hypothermia unchanged	Tolerance slightly reduced (TI)	CPP unchanged	Unchanged	*56*
DOR	Ethanol			SA enhanced		*57*
PreproENK	Cannabinoid	Reduced analgesia; hypolocomotion, catalepsy, and hypothermia unchanged	Tolerance reduced (TI)		Somatic signs reduced	*58*

CB1	Cannabinoid	Analgesia, hypolocomotion, hypothermia, hypotension abolished		SA abolished	Somatic signs absent	*33*
	Cocaine	Hyperlocomotion unchanged	Sensitization unchanged	CPP unchanged		*35*
	Cocaine D-Amphetamine Nicotine			SA unchanged		*36*
D2R	Ethanol			SA reduced		*59*
	Ethanol			SA reduced		*60*
	Ethanol			CPP reduced		*61*
DAT	Cocaine	Hyperlocomotion abolished		SA unchanged		*62*
	Cocaine			CPP unchanged		*63*
NK1	Cocaine			CPP unchanged		*41*
$G_{z\alpha}$	Cocaine	Hyperlocomotion enhanced				*49*
	Catecholamine uptake inhibitors	Decrease immobility in the forced swim test abolished				

[a]CPP, conditioned place preference; SA, self-administration; TI, tail immersion.

addiction, were also tested. All of them, including analgesia (see ref. *67*), respiratory depression *(69)*, immunosuppression *(70,71)*, and constipation *(72)*, were absent. Altogether, these data demonstrate unambiguously that μ receptor is essential in mediating the main biological activities of the prototypic opioid. Importantly, these findings further indicate that activation of μ receptor necessarily and concomitantly triggers both the therapeutic and adverse—including addictive—effects of morphine. These conclusions most probably apply to other noted opioids, including heroin and the clinically useful opioids such as fentanyl and methadone. The latter drugs indeed bind preferably at μ receptor *(73)*, and heroin analgesia was also shown to be absent in MOR-deficient mice *(74)*. In conclusion, μ receptor is considered a key receptor for morphine activity in vivo and represents the molecular gate for opioid addiction. This receptor seems also to be implicated in mediating the effect of other drugs of abuse. Ethanol reward measured either by CPP or SA was strongly reduced, cannabinoid CPP was abolished, and somatic signs of cannabinoid withdrawal were reduced in MOR-deficient mice (*see* Table 4). The μ receptors are not direct molecular targets for these drugs, therefore implication of this receptor presumably occurs indirectly, through activation of the endogenous opioid system. Because MOR-deficient mice do not respond to morphine at all, discussion of this knockout model will not be pursued further.

5. Acute Morphine in Knockout Models

Before investigating the long-term effects of opioids, acute responses to morphine were tested in almost all the mutant mice reviewed here (Tables 1–3). Morphine analgesia was unchanged in mice lacking δ and κ receptors, suggesting lack of morphine crossreactivity at these two μ-highly homologous opioid receptors. The antinociceptive activity of morphine was also maintained in mice lacking other receptors and transporters, including D2R, DAT, NK1, CB1, ORL1, and GluR-A, suggesting that none of these neurotransmitter systems is required to produce opioid analgesia. As for ORL-1 knockout mice, morphine analgesia was unchanged in the OFQ/N null mutants, indicating that the ORL-1/OFQ/N system does not contribute to this acute effect of morphine. Finally, the transcription factor CREB does not seem to be implicated in morphine analgesia. Conflicting results have been reported for the signaling protein $G_{z\alpha}$, suggesting that this protein is either not *(48)* or slightly *(49)* implicated in acute morphine analgesia.

In contrast, αCGRP knockout mice showed reduced morphine analgesia, reflecting the previously reported role of this peptide in nociception and its complex interaction with opioids *(46)*. β-Arr2 knockout mice showed prolonged nociception following a single morphine administration, an effect that could be related to the hypothesized role of β-Arr2 in promoting μ receptor desensitization (discussed further in Section 6.1.). Morphine analgesia was decreased in IL6-deficient mice, at a low dose only (1.25 mg/kg). In these mice, stress-induced analgesia was absent and, together, changes in morphine potency and stress effect could be correlated to marked modifications of the endogenous opioid system, which most probably result from the absence of IL6 during development rather than a direct IL6–opioid interaction in the adult.

Another acute effect of morphine was investigated in some of the mutants. The morphine-induced stimulation of locomotor activity was unchanged in CB1, D2R, GluR-A, and CREB knockout mice. However, it was abolished in NK1 null mutant mice. This suggests that a substance P-mediated mechanism is involved in morphine-evoked hyperlocomotion, a phenomenon that seems to be—in some cases—related to morphine reward (*see* Section 6.2.). A similar phenotype was reported in DAT knockout mice, characterized by a constitutive elevation of dopamine mesolimbic transmission *(75)*. This result was intriguing because morphine was still able to elevate extracellular dopamine levels in the nucleus accumbens, to diminish vertical locomotor activity, and to produce CCP in mutant animals. This shows that morphine locomotor stimulation can be dissociated from morphine reward (*see* Section 6.2.).

In summary, many of the targeted genes discussed here show no role in short-term opioid transmission in vivo. A few genes, however, could be directly implicated in physiological responses triggered by a time-limited interaction of morphine with μ receptors. The analgesic and locomotor-stimulating activities of morphine may require release of αCGRP and substance P, respectively, and β-Arr2 seems to hamper rapid μ-receptor signaling in vivo.

6. Chronic Morphine in Knockout Models

6.1. Tolerance and Sensitization

Biological responses to drugs of abuse are generally modified with repeated exposure. The drug potency may gradually decrease, as is the case for opioid analgesia (tolerance), or increase, which is observed for morphine hyperlocomotion (sensitization). These processes could participate to escalating drug use (tolerance) or relapse (sensitization) *(6,76)*. Although molecular mechanisms of tolerance have been investigated extensively *(77)*, little is known about sensitization. Whether common biochemical adaptations participate to the establishment of both phenomena is unknown. Tolerance to morphine analgesia has been evaluated in 11 knockout mouse strains (including MOR; *see* Section 4.) of the 17 described in this review, and sensitization to hyperlocomotion has been measured for only two of the mutant strains (Tables 1–3).

Morphine tolerance was unchanged in CB1 knockout mice *(33)*, while sensitization to hyperlocomotion was abolished *(35)*. This result indicates that the two processes can be dissociated, at least at the level of the cannabinoid system. In addition, this effect seemed specific to morphine, because sensitization to cocaine was unchanged (Table 4). Tolerance was also unchanged in αCGRP and OFQ/N-deficient mice.

Tolerance was reduced in ORL1 knockout mice, confirming the proposed role for this receptor as a component of antiopioid systems *(78)*. This result seems to contradict the observation that morphine tolerance is not modified in OFQ/N mutant mice, since the peptide OFQ/N was identified as the endogenous ligand for the ORL1 *(79,80)* receptor. These conflicting results could indicate the presence of an additional, yet uncharacterized, endogenous ligand for ORL1 or could reflect the absence in OFQ/N knockout mice of two other biologically active peptides originating from the same precursor protein *(81)*, nocistatin and OFQ/NII. Tolerance was also reduced in mice deficient for α and δ isoforms of CREB. This result suggests that the development of tolerance, like physical dependence (*see* Section 6.3.) needs—at least in part—modification of gene expression.

Morphine tolerance was abolished in mice lacking the GluR-A subunit of AMPA receptors. Glutamate-mediated neurotransmission, as well as NMDA and AMPA glutamate receptor levels have been shown altered following chronic drug treatments *(82–86)*. AMPA receptors have been essentially implicated in sensitization to drugs of abuse *(87,88)*. Therefore the results obtain with GluR-A-deficient mice further extend the role of AMPA receptors to the development of morphine tolerance. Context-independent sensitization was also abolished in GluR-A mutant mice, thus confirming the role of AMPA receptors in one aspect of morphine sensitization. A second strain of mice bearing a mutation in the GluR-A subunit (GluR-A[R/R]), which reduces the calcium permeability and channel conductance of the receptors, was also tested. In this case mutant mice were affected for context-independent sensitization but not tolerance. This suggests that context-independent sensitization is related to a modification of Ca^{2+} conductance in the GluR-A subunit-containing AMPA receptor channels, while the mechanism by which AMPA receptors are involved in tolerance is unknown *(42)*.

Morphine tolerance was also abolished in mice lacking β-Arr2. Numerous in vitro studies indicate that arrestins are involved in desensitization of G-protein-coupled receptors *(89)*, and it had long been proposed that this family of proteins could be involved in the development of tolerance in vivo. Since desensitization of G-protein-coupled receptors in cell lines is a short-term phenomenon (minutes), while tolerance to morphine or other drugs develops over continued use for several days or weeks, this hypothesis needed experimental confirmation. The results obtained with β-Arr-2 knockout mice nicely demonstrate that this protein is a key element for both the development of short-term tolerance *(50)*, also called acute tolerance, and the development of long-term tolerance *(51)*.

Abolition of tolerance to morphine was also observed in DOR- and preproENK-deficient mice. This result is in good agreement with the current notion that DOR mediates the effects of enkephalins *(90)* and emphasizes the critical role of this ligand-receptor system in the development of morphine tolerance. The complete absence of tolerance in these two strains of mice appeared fairly dramatic *(31)*, since previous studies using either a selective DOR antagonist *(91)* or DOR antisense oligonucleotides *(92)* showed only partial blockade in the development of morphine tolerance. PreproENK mutants also showed a reduced tolerance to cannabinoids (Table 4), indicating that the endogenous enkephalinergic system participates in the development of tolerance to other drugs of abuse.

Finally, morphine tolerance developed faster in IL6 knockout mice. As for the acute morphine treatment (*see* Section 5.), the reduction of morphine potency following chronic administration most likely reflects the lower density of μ-opioid receptors in the brain of those mice. Tolerance also developed at a faster rate in mice deficient for $G_{z\alpha}$ morphine . In addition, these mutants displayed a greater degree of tolerance than wild-type mice. These results suggest that G_z could play a role in counteracting signaling pathways associated with tolerance. One should note, however, that this effect was observed in the hot-plate test but not in the tail-flick test, suggesting that G_i may compensate for G_z function in spinal opioid tolerance but not in supraspinal tolerance *(48)*.

In summary, studies with knockout mice demonstrate an important role of signaling molecules downstream of the μ receptor, particularly β-Arr2, in the development of tolerance to morphine. They also reveal a role of AMPA receptors in this adaptative phenomenon, beyond NMDA glutamatergic neurotransmission *(93)*, and highlight a role of the δ-receptor–enkephalin system. They finally confirm the fact that ORL1 receptor may be part of an antiopioid system. In contrast to tolerance, morphine sensitization has been little investigated in knockout mice. At present, data confirm the key role of AMPA receptors in this process and point to possible μ-CB1 receptors interactions for the development of morphine sensitization. Data from the two knockout strains in which both tolerance and sensitization were studied (CB1 and GluR-A) highlight distinct rather than common mechanisms to these adaptations to chronic morphine. Studies on sensitization in genetically modified mice will probably expand greatly in the future.

6.2. Reward

A prominent common action of all drugs of abuse, including opioids, is the activation of reward pathways. The rewarding action of most drugs is thought to be mediated mainly by the mesolimbic dopamine system *(6,16,94,95)*. For opioids, it has been proposed that activation of dopaminergic neurons in the ventral tegmental area increases dopamine release in the nucleus accumbens and other regions of the limbic forebrain, and that high dopamine levels in those areas form the neural basis for morphine reinforcement. Mapping of sites for morphine reward throughout the brain has also suggested that morphine reward could be mediated by dopamine-independent mechanisms in other brain regions, including the nucleus accumbens, the hypothalamus, and the amygdala *(4,96)*. Rewarding properties of morphine have been evaluated in eight knockout mice (including MOR; *see* Section 4.) of the 17 described in this review, using either conditioned place preference (CPP) or self-administration (SA) paradigms or both (Tables 1–3).

Opioid reward was unchanged in KOR and αCGRP-deficient mice. The fact that morphine CPP was not altered in KOR-deficient mice was surprising. The reported opposing action of μ- and κ-agonists in modulating the endogenous tone of mesolimbic dopaminergic neurons *(97)* and the ability of κ-agonists to block morphine reward *(98)* suggest that μ-mediated reward could be potentiated in the absence of κ receptors. The expected increase of morphine CPP was not observed in the KOR knockout mice, either due to the experimental conditions or to compensatory modifications in the mutant mice. This should be further investigated using other experimental conditions or behavioral paradigms (SA, for example).

The rewarding action of morphine was abolished in CB1-deficient mice, in both CPP and SA paradigms (Table 2). In accord, increase of dopamine release in the nucleus accumbens following morphine administration was not observed *(34)*. These results suggest a critical role of the cannabinoid, neurotransmitter system in mediating the reinforcing properties of morphine. In the same way, the opioid system seems important in mediating rewarding effects of cannabinoids since THC did not induce place preference in MOR knockout mice (Table 4). Together, the data from knockout mice support the notion of bi-directional interactions between the opioid and cannabinoid systems suggested previously by the pharmacology for other responses *(99)*. In addition, the close interaction between cannabinoids and opioids seems specific to these systems, since the rewarding action and sensitization to hyperlocomotion of psychostimulants are unchanged in mice lacking the CB1 receptor (see Table 4).

Morphine rewarding effects were also absent in D2R mutant mice. This result is in agreement with the well-documented role of dopaminergic activity in opioid reward and indicates that the D2 receptor represents a critical dopamine receptor subtype for motivational response to opioids *(38)*. Motor and reward responses to opioids have been reported to be closely related *(100)*. However, morphine was still able to stimulate the locomotor response of D2R-deficient mice, indicating that these two phenomena can be dissociated. Food reward was unaffected in mice lacking D2R *(38)*, suggesting a specific role of this receptor in opioid, but not in natural forms of reward. Different results have been obtained by Dockaster and co-workers *(39)* using another strain of D2R knockout mice. In these mice, morphine reward was lost only in opioid-dependent and withdrawn mice but not in naive animals. The authors suggested that this difference could result from the distinct genetic backgrounds of the two strains. Finally, ethanol reward was reduced in D2R mutant mice, extending the involvement of dopamine D2 receptor in motivational response to non-opioid drugs of abuse (Table 4).

Morphine reward was also abolished in NK1 receptor-deficient animals. As is the case for CB1 receptor, this loss of rewarding properties was opioid-specific, since both cocaine- and food-induced CPP were maintained in these mice. In addition, the hyperlocomotor effect of morphine (but not of cocaine) was abolished in NK1 mutant mice, indicating that, unlike CB1 and D2 receptors, NK1 receptor is important for both motor and motivational responses to opioids.

A loss of morphine rewarding properties was also shown for PKCγ-deficient mice. This result is rather intriguing, since Narita et al. *(52,101)* have shown that several functions of μ-opioid receptor are enhanced in PKCγ knockout animals. The μ-opioid receptor-mediated analgesia and activation of G-proteins in the spinal cord were increased following both acute and chronic DAMGO treatments, suggesting that the absence of PKCγ protects μ receptors from a phosphorylation-induced degradation mechanism. Therefore, the absence of morphine reward in those mice would result, rather, from an indirect action of the kinase. PKCγ is involved in synaptic plasticity *(102,103)*, and the phosphorylation of NMDA receptors, also shown to be players in morphine reward, may be required *(104,105)*.

Finally, morphine reward was enhanced in only one of the knockout models, the DAT mutant mice. Although basal extracellular level of dopamine is high in those mice, morphine was still able to increase dopamine further, allowing the expression of morphine reward. The authors suggested that increased morphine reward in DAT-deficient mice could be associated with enhanced drug sensitivity or faster acquisition of conditioned place preference. Morphine hyperlocomotor effect was abolished in these animals, indicating that, as for mice lacking D2R, the ability of morphine to produce reward can be dissociated from its capacity to increase locomotion.

In summary, studies using knockout mice have identified key molecular components of opioid reward. Results obtained from the analysis of D2R- and DAT-deficient animals confirm the important role of the dopamine system in mediating the reinforcing properties of morphine. Interestingly, a previously unsuspected role of CB1 and NK1 receptors in morphine reward was uncovered. It is tempting to propose that NK1 and CB1 receptors may account for both dopamine- and non-dopamine-mediated mechanisms of opioid reward. These receptors may represent new potential targets for the treatment of opioid addiction.

6.3. Withdrawal

Drug abstinence following chronic drug exposure results in the development of a withdrawal syndrome that comprises two components, a negative emotional state and physical (somatic and vegetative) signs of withdrawal *(4,95)*. The negative emotional state results in dysphoria, anxiety, and irritability, is considered a source of negative reinforcement, and is produced by most drugs of abuse, including opioids. It has been proposed that both the positive reinforcing effect of the drug (reward) and the negative reinforcing effect of withdrawal participate to the establishment of drug addiction *(95)* but little is known about the key proteins involved in the aversive aspect of opioid withdrawal. Physical signs of withdrawal develop differently depending on the substance. They barely develop for most nonopioid drugs of abuse, while they are particularly severe following morphine abstinence. Because of their dramatic manifestation, the physical signs of morphine withdrawal have been examined extensively at the pharmacological and biochemical levels. Numerous molecular events have been shown to occur, involving transcriptional regulations and posttranscriptional mechanisms *(6)*. In the knockout mouse models, as for previous studies, opioid physical withdrawal was broadly examined (14 knockout mouse strains, including MOR; *see* Section 4.), while the negative emotional aspect of withdrawal has been measured in a few strains only (three; Tables 1–3).

Morphine withdrawal was unchanged in both DOR and preproENK-deficient mice (Table 1). This result is consistent with the notion that DOR and enkephalins are part of the same neurotransmitter–receptor system (*see* Section 6.2.). The absence of implication of preproENK in physical withdrawal is not general to all drugs of abuse, since cannabinoid withdrawal was reduced in mice lacking Pre-proenk gene (Table 4). Somatic signs of morphine withdrawal were also unchanged in D2R knockout mice. In these mice, however, abstinence-induced CPA was abolished, showing an implication of D2 receptors in the negative emotional component of morphine withdrawal. This result, together with the total suppression of morphine rewarding properties in D2R-deficient animals (*see* above), underscores the importance of D2 receptors mainly in the motivational aspect of opioid addiction. Physical signs of morphine withdrawal were also fully present in mice lacking β-Arr2. This result suggests that, although β-Arr2 plays a role in tolerance following acute and chronic morphine treatment *(50,51)*, this regulatory protein is not necessarily involved in every adaptive response to morphine. It would be interesting to investigate the two other major actions of the drug (reward and sensitization) before definitely concluding on the specific implication of β-Arr2 in morphine tolerance.

Morphine withdrawal was reduced in KOR-deficient mice. Therefore, the KOR gene product, although not essential, seems to participate in the expression of morphine abstinence. This result is in agreement with a previous study showing partial precipitation of morphine withdrawal with a κ-selective antagonist *(106)*. In mice lacking CB1 receptors, several signs of withdrawal were strongly reduced, including jumping, wet dog shakes, and body tremor, resulting in a 50% decrease of the global withdrawal score. This indicates that, in addition to a critical role in morphine reward, the CB1 receptor also participates in the behavioral expression of morphine withdrawal. Several signs of withdrawal were also strongly reduced in DAT knockout mice, in particular tremor, sniffing, and ptosis, supporting the participation of dopamine in these behaviors. Since morphine withdrawal was unchanged in D2R-deficient mice it is likely that other dopamine receptor subtypes are involved in this phenomenon. In NK1-deficient mice, only the jumping behavior was absent following precipitation of the withdrawal syndrome, indicating a limited implication of this receptor in the somatic manifestations of morphine withdrawal. Interestingly, morphine abstinence-induced CPA was strongly reduced in the NK1 knockout mutants. This result, together with the absence of morphine CPP in these animals, indicates that NK1 receptor plays an important role in mediating both positive and negative motivational responses to opioids. Morphine withdrawal was also reduced in mice lacking the GluR-A subunit of AMPA receptors, with a decrease of about 40% in the number of withdrawal symptoms. This result, together with reduced morphine tolerance in those mice, suggests that GluR-A subunit-expressing neurons, possibly the interneurons known to express GluR-A but not GluR-B subunits *(107)*, are involved in neurochemical pathways normally affected by morphine. Withdrawal syndrome, like tolerance, was not affected in the GluR-A(R/R) knockin mice, indicating that this adaptation to chronic morphine is not related to a modification of Ca^{2+} conductance in these AMPA receptor channels *(42)*.

Morphine withdrawal was strongly reduced in ORL1-deficient mice, with an important reduction in all the somatic signs of withdrawal, including jumping. As for tolerance, a different result was found in mice lacking the OFQ/N precusor protein, with an increase in jumping behavior following precipitation of morphine withdrawal syndrome. Again, this could be due to the presence of a yet unknown endogenous ligand for ORL1 or the absence of nocistatin and OFQ/NII in the OFQ/N knockout mice. Behavioral manifestations of morphine withdrawal syndrome were also strongly reduced in αCGRP-deficient mice, with a decrease of up to 70% in the number of withdrawal signs. This result suggests that some behavioral signs of withdrawal may be mediated by the peripheral nervous system involved in neurogenic inflammatory responses. Finally, a strong reduction of somatic signs of withdrawal was also observed in CREB-deficient mice and, very recently, mutant animals with all CREB isoforms deleted specifically in the CNS showed a similar phenotype (G. Schutz, personal communication). These results confirm the role of CREB as a key transriptional factor mediating adaptations to chronic morphine treatment *(6)*. As CREB is an ubiquitous protein and is expressed early in the development, the phenotype observed in CREB-deficient animals might be an indirect consequence of profound developmental modifications. In this case, conditional gene deletion *(108)* would be extremely useful to confirm this phenotype.

In summary, morphine withdrawal was reduced in most knockout mouse models examined so far, that is, 11 mutant strains including those for seven different receptors for neurotransmitters. This agrees with the notion that the morphine-dependent state is a complex phenomenon and implicates a large number of genes. The knockout models have revealed a role for several gene products that were not expected from previous studies, including CB1, NK1, AMPA GluR-A, ORL1 receptors, dopamine transporter, and αCGRP. Specific neuronal pathways in which these proteins modulate morphine withdrawal are unknown. An intriguing role of the peripheral nervous system in the behavioral manifestation of morphine abstinence has been suggested from the observation of αCGRP knockout mice. The role of dopamine in drug reward has been extended to morphine physical dependence with DAT-deficient mice. Finally, the important role of CREB in adaptation to chronic morphine treatment is confirmed in vivo. Obviously, the future examination of other gene-deleted mice will allow the identification of additional receptor–neurotransmitter systems or other cellular proteins involved in the expression of morphine withdrawal.

6. Conclusion and Perspectives

The use of knockout mice to study molecular mechanisms of opioid dependence has highlighted several important features. From the data, the various adaptative processes to chronic opioids can be dissociated, with distinct sets of genes being involved in tolerance and sensitization, or reward and withdrawal. An observation is that morphine reward is either intact or fully ablated throughout the knockout models that have been tested. This may be due to the behavioral paradigms used (CPP and SA), which may not allow easy detection of subtle modifications. Alternatively, the genes identified from knockout models may have a permissive activity in drug reinforcement, which would require sequential steps within the reward circuitry *(7)*. In contrast, withdrawal and tolerance are generally attenuated but not abolished, suggesting that several genes participate concomitantly in the development and expression of these behavioral adaptations, which take place gradually over time and are most probably more widespread throughout the central nervous system *(109)*.

Some phenotypes described in this chapter may be explained by developmental modifications, as was clear for IL6-deficient mice. Also, null mutations in those knockout models lead to protein deletion throughout the entire animal and provide no indication on the neural sites of opioid adaptations. In the future, the analysis of inducible and site-specific knockout animals will be more appropriate to delineate the molecular components and neurons responsible for the behavioral manifestations of opioid dependence *(108)*.

Finally, a number of knockout models have been tested for cocaine or nicotine (for example D1, D3, D4, nAChR-β2 knockout mice, reviewed in ref. *110*), but not morphine. It is likely that the genes inactivated in those mice are also implicated in responses to chronic opioid exposure, and this should be tested in future studies. More generally, a systematic screen of mice genetically modified in the central nervous system could be considered, because many genes involved in opioid addiction remain to be discovered.

References

1. Sertürner, F. W. A. (1806) Darstellung der reinen Mohnsäure (Opiumsäure), nebst einer chemischen Untersuchung des Opium. *J. Pharm. f. Artze. Apoth. Chem.* **14,** 47–93.
2. Shug, S. A., Zech, D., and Grond, S. (1992) Adverse effects of systemic opioid analgesics. *Drug Safety* **7,** 200–213.
3. Goldstein, M. D. (1994) *From biology to drug policy*. Freeman, New York.
4. van Ree, J. M., Gerrits, M. A., and Vanderschuren, L. J. (1999) Opioids, reward and addiction: an encounter of biology, psychology, and medicine. *Pharmacol. Rev.* **51,** 341–396.
5. LaForge, K. S., Yuferov, V., and Kreek, M. J. (2000) Opioid receptor and peptide gene polymorphisms: potential implications for addictions. *Eur. J. Pharmacol.* **410,** 249–268.
6. Nestler, E. J. (2001) Molecular basis of long-term plasticity underlying addiction. *Nat. Rev. Neurosci.* **2,** 119–128.
7. Koob, G. F. (1992) Drug of abuse: anatomy, pharmacology and function of reward pathways. *Trends Pharmacol. Sci.* **13,** 177–184.
8. Akil, H., Watson, S. J., Young, E., Lewis, M. E., Khachaturian, H., and Walker, J. J. (1984) Endogenous opioids: Biology and function. *Ann. Rev. Neurosci.* **7,** 223–255.
9. Di Chiara, G., and North, A. (1992) Neurobiology of opiate abuse. *Trends Pharm. Sci.* **13,** 185–193.

10. Przewlocki, R. (1993) Opioid systems and stress. In *Handbook of Experimental Pharmacology: Opioids II* (A. Herz, ed.), pp. 293-324, Springer Verlag, Berlin.
11. Ossipov, M. H., Malan, T. P., Lai, J. J., and Porreca, F. (1997) Opioid Pharmacology of acute and chronic pain. In *Handbook of Experimental Pharmacology* (A. Dickenson and J. M. Besson, eds.), pp. 305-327, Springer Verlag, Berlin Heidelberg.
12. Robinson, T. E., and Kolb, B. (1999) Morphine alters the structure of neurons in the nucleus accumbens and neocortex of rats. *Synapse* **33,** 160–162.
13. Cesselin, F. (1995) Opioid and anti-opioid peptides. *Fundam. Clin. Pharmacol.* **9,** 409–433.
14. Mogil, J. S., Grisel, J. E., Reinscheid, R. K., Civelli, O., Belknap, J. K., and Grandy, D. K. (1996) Orphanin FQ is a functional anti-opioid peptide. *Neurosci.* **75,** 333–337.
15. Rothman, R. B. (1992) A review of the role of anti-opioid peptides in morphine tolerance and dependence. *Synapse* **12,** 129–138.
16. Koob, G. F., and Moal, M. L. (1997) Drug abuse: hedonic homeostatic dysregulation. *Science* **278,** 52–58.
17. Childers, S. R. (1991) Opioid receptor-coupled second messenger systems. *Life Sci.* **48,** 1991–2003.
18. Pei, G., Kieffer, B. L., Lefkowitz, R. J., and Freedman, N. J. (1995) Agonist-dependent phosphorylation of the mouse delta-opioid receptor: involvement of G protein-coupled receptor kinases but not protein kinase C. *Mol. Pharmacol.* **48,** 173–177.
19. Yu, Y., Zhang, L., Yin, X., Sun, H., Uhl, G. R., and Wang, J. B. (1997) Mu-opioid receptor phosphorylation, desensitization and ligand efficacy. *J. Biol. Chem.* **272,** 28,869–28,874.
20. Arden, J. R., Segredo, V., Wang, Z., Lameh, J., and Sadée, W. (1995) Phosphorylation and agonist specific intracellular trafficking of an epitope-tagged m-opioid receptor expressed in HEK 293 cells. *J. Neurochem.* **65,** 1636–1645.
21. Keith, D. E., Murray, S. R., Zaki, P. A., Chu, P. C., Lissin, D. V., Kang, L., Evans, C. J., and Von Zastrow, M. (1996) Morphine activates opioid receptors without causing their rapid internalization. *J. Biol. Chem.* **277,** 19,021–19,024.
22. Cvejic, S., Trapaidze, N., Cyr, C., and Devi, L. A. (1996) Thr353, located within the COOH-terminal tail of the δ opiate receptor, is involved in receptor down-regulation. *J. Biol. Chem.* **271,** 4073–4076.
23. Trapaidze, N., Keith, D. E., Cvejic, S., Evans, C. J., and Devi, L. A. (1996) Sequestration of the δ-opioid receptor: role of the C terminus in agonist-mediated internalization. *J. Biol. Chem.* **271,** 29,279–29,285.
24. Whistler, J. L., Chuang, H. H., Chu, P., Jan, L. Y., and von Zastrow, M. (1999) Functional dissociation of mu opioid receptor signaling and endocytosis: implications for the biology of opiate tolerance and addiction. *Neuron* **23,** 737–746.
25. Nestler, E. J. (1996) Under siege: The brain on opiates. *Neuron* **16,** 897–900.
26. Nestler, E. J. (2000) Genes and addiction. *Nat. Genet.* **26,** 277–281.
27. Crawley, J. N. (2000) *What's wrong with my mouse*. Wiley-Liss, New York.
28. Matthes, H. W. D., Maldonado, R., Simonin, F., Valverde, O., Slowe, S., Kitchen, I., Befort, K., Dierich, A., LeMeur, M., Dollé, P., Tzavara, E., Hanoune, J., Roques, B. P., and Kieffer, B. L. (1996) Loss of morphine-induced analgesia, reward effect and withdrawal symptoms in mice lacking the μ-opioid receptor gene. *Nature* **383,** 819–823.
29. Tian, M., Broxmeyer, H. E., Fan, Y., Lai, Z., Zhang, S., Aronica, S., Cooper, S., Bigsby, R. M., Steinmetz, R., Engle, S. J., Mestek, A., Pollock, J. D., Lehman, M. N., Jansen, H. T., Ying, M., Stambrook, P. J., Tischfield, J. A., and Yu, L. (1997) Altered hematopoiesis, behavior, and sexual function in μ opioid receptor-deficient mice. *J. Exp. Med.* **185,** 1517–1522.
30. Becker, A., Grecksch, G., Brodemann, R., Kraus, J., Peters, B., Schroeder, H., Thiemann, W., Loh, H. H., and Hollt, V. (2000) Morphine self-administration in mu-opioid receptor-deficient mice. *Naunyn Schmiedebergs Arch. Pharmacol.* **361,** 584–589.

31. Zhu, Y., King, M. A., Schuller, A. G. P., Nitsche, J. F., Riedl, M., Elde, R. P., Unterwald, E., Pasternak, G. W., and Pintar, J. E. (1999) Retention of supraspinal delta-like analgesia and loss of morphine tolerance in delta opioid receptor knockout mice. *Neuron* **24,** 243–252.
32. Simonin, F., Valverde, O., Smadja , S., Slowe, S., Kitchen, I., Dierich, A., Le Meur, M., Roques, B. P., Maldonado, R., and Kieffer, B. L. (1998) Disruption of the κ-opioid receptor gene in mice enhances sensitivity to chemical visceral pain, impairs pharmacological actions of the selective κ-agonist U-50,488H and attenuates morphine withdrawal. *EMBO J.* **17,** 886–897.
33. Ledent, C., Valverde, O., Cossu, G., Petitet, F., Aubert, J. F., Beslot, F., Bohme, G. A., Imperato, A., Pedrazzini, T., Roques, B. P., Vassart, G., Fratta, W., and Parmentier, M. (1999) Unresponsiveness to cannabinoids and reduced addictive effects of opiates in CB1 receptor knockout mice. *Science* **283,** 401–404.
34. Mascia, M. S., Obinu, M. C., Ledent, C., Parmentier, M., Bohme, G. A., Imperato, A., and Fratta, W. (1999) Lack of morphine-induced dopamine release in the nucleus accumbens of cannabinoid CB(1) receptor knockout mice. *Eur. J. Pharmacol.* **383,** R1–R2.
35. Martin, M., Ledent, C., Parmentier, M., Maldonado, R., and Valverde, O. (2000) Cocaine, but not morphine, induces conditioned place preference and sensitization to locomotor responses in CB1 knockout mice. *Eur. J. Neurosci.* **12,** 4038–4046.
36. Cossu, G., Ledent, C., Fattore, L., Imperato, A., Bohme, G. A., Parmentier, M., and Fratta, W. (2001) Cannabinoid CB1 receptor knockout mice fail to self-administer morphine but not other drugs of abuse. *Behav. Brain Res.* **118,** 61–65.
37. Lichtman, A. H., Sheikh, S. M., Loh, H. H., and Martin, B. R. (2001) Opioid and cannabinoid modulation of precipitated withdrawal in delta(9)-tetrahydrocannabinol and morphine-dependent mice. *J. Pharmacol. Exp. Ther.* **298,** 1007–1014.
38. Maldonado, R., Saiardi, A., Valverde, O., Samad, T. A., Roques, B. P., and Borrelli, E. (1997) Absence of opiate rewarding effects in mice lacking dopamine D2 receptors. *Nature* **388,** 586–589.
39. Dockstader, C. L., Rubinstein, M., Grandy, D. K., Low, M. J., and van der Kooy, D. (2001) The D2 receptor is critical in mediating opiate motivation only in opiate-dependent and withdrawn mice. *Eur. J. Neurosci.* **13,** 995–1001.
40. Spielewoy, C., Gonon, F., Roubert, C., Fauchey, V., Jaber, M., Caron, M. G., Roques, B. P., Hamon, M., Betancur, C., Maldonado, R., and Giros, B. (2000) Increased rewarding properties of morphine in dopamine-transporter knockout mice. *Eur. J. Neurosci.* **12,** 1827–1837.
41. Murtra, P., Sheasby, A. M., Hunt, S. P., and De Felipe, C. (2000) Rewarding effects of opiates are absent in mice lacking the receptor for substance P. *Nature* **405,** 180–183.
42. Vekovischeva, O. Y., Zamanillo, D., Echenko, O., Seppala, T., Uusi-Oukari, M., Honkanen, A., Seeburg, P. H., Sprengel, R., and Korpi, E. R. (2001) Morphine-induced dependence and sensitization are altered in mice deficient in AMPA-type glutamate receptor-A subunits. *J. Neurosci.* **21,** 4451–4459.
43. Kest, B., Hopkins, E., Palmese, C. A., Chen, Z. P., Mogil, J. S., and Pintar, J. E. (2001) Morphine tolerance and dependence in nociceptin/orphanin FQ transgenic knock-out mice. *Neuroscience* **104,** 217–222.
44. Ueda, H., Yamaguchi, T., Tokuyama, S., Inoue, M., Nishi, M., and Takeshima, H. (1997) Partial loss of tolerance liability to morphine analgesia in mice lacking the nociceptin receptor gene. *Neurosci. Lett.* **237,** 136–138.
45. Ueda, H., Inoue, M., Takeshima, H., and Iwasawa, Y. (2000) Enhanced spinal nociceptin receptor expression develops morphine tolerance and dependence. *J. Neurosci.* **20,** 7640–7647.
46. Salmon, A. M., Damaj, M. I., Marubio, L. M., Epping-Jordan, M. P., Merlo-Pich, E., and Changeux, J. P. (2001) Altered neuroadaptation in opiate dependence and neurogenic inflammatory nociception in alpha CGRP-deficient mice. *Nat. Neurosci.* **4,** 357–358.
47. Bianchi, M., Maggi, R., Pimpinelli, F., Rubino, T., Parolaro, D., Poli, V., Ciliberto, G., Panerai, A. E., and Sacerdote, P. (1999) Presence of a reduced opioid response in interleukin-6 knock out mice. *Eur. J. Neurosci.* **11,** 1501–1507.

48. Hendry, I. A., Kelleher, K. L., Bartlett, S. E., Leck, K. J., Reynolds, A. J., Heydon, K., Mellick, A., Megirian, D., and Matthaei, K. I. (2000) Hypertolerance to morphine in G(z alpha)-deficient mice. *Brain. Res.* **870,** 10–19.
49. Yang, J., Wu, J., Kowalska, M. A., Dalvi, A., Prevost, N., O'Brien, P. J., Manning, D., Poncz, M., Lucki, I., Blendy, J. A., and Brass, L. F. (2000) Loss of signaling through the G protein, Gz, results in abnormal platelet activation and altered responses to psychoactive drugs. *Proc. Natl. Acad. Sci. USA* **97,** 9984–9989.
50. Bohn, L. M., Lefkowitz, R. J., Gainetdinov, R. R., Peppel, K., Caron, M. G., and Lin, F. T. (1999) Enhanced morphine analgesia in mice lacking beta-arrestin 2. *Science* **286,** 2495–2498.
51. Bohn, L. M., Gainetdinov, R. R., Lin, F. T., Lefkowitz, R. J., and Caron, M. G. (2000) Mu-opioid receptor desensitization by beta-arrestin-2 determines morphine tolerance but not dependence. *Nature* **408,** 720–723.
52. Narita, M., Aoki, T., Ozaki, S., Yajima, Y., and Suzuki, T. (2001) Involvement of protein kinase Cgamma isoform in morphine-induced reinforcing effects. *Neuroscience* **103,** 309–314.
53. Maldonado, R., Blendy, J. A., Tzavara, E., Gass, P., Roques, B. P., Hanoune, J., and Schutz, G. (1996) Reduction of morphine abstinence in mice with a mutation in the gene encoding CREB. *Science* **273,** 657–659.
54. Roberts, A., Mcdonald, J. S., Heyser, C. J., Kieffer, B. L., Matthes, H. W. D., Koob, G. F., and Gold, L. H. (2000) Mu-opioid receptor knockout mice do not self-administer alcohol. *J. Pharm. Exp. Ther.* **293,** 1002–1008.
55. Hall, F. S., Sora, I., and Uhl, G. R. (2001) Ethanol consumption and reward are decreased in mu-opiate receptor knockout mice. *Psychopharmacology* (Berl) **154,** 43–49.
56. Ghozland, S., Matthes, H. W., Simonin, F., Filliol, D., Kieffer, B. L., and Maldonado, R. (2001) Motivational effects of cannabinoids are mediated by mu- and kappa-opioid receptors. *J. Neurosci.* in press.
57. Roberts, A. J., Gold, L. H., Polis, I., McDonald, J. S., Filliol, D., Kieffer, B. L., and Koob, G. F. (2001) Increased ethanol self-administration in delta opioid receptor knockout mice. *Alcohol. Clin. Exp. Res.* **25,** 1249–1256
58. Valverde, O., Maldonado, R., Valjent, E., Zimmer, A. M., and Zimmer, A. (2000) Cannabinoid withdrawal syndrome is reduced in pre-proenkephalin knock-out mice. *J. Neurosci.* **20,** 9284–9289.
59. Phillips, T. J., Brown, K. J., Burkhart-Kasch, S., Wenger, C. D., Kelly, M. A., Rubinstein, M., Grandy, D. K., and Low, M. J. (1998) Alcohol preference and sensitivity are markedly reduced in mice lacking dopamine D2 receptors. *Nat. Neurosci.* **1,** 610–615.
60. Risinger, F. O., Freeman, P. A., Rubinstein, M., Low, M. J., and Grandy, D. K. (2000) Lack of operant ethanol self-administration in dopamine D2 receptor knockout mice. *Psychopharmacology* (Berl) **152,** 343–350.
61. Cunningham, C. L., Howard, M. A., Gill, S. J., Rubinstein, M., Low, M. J., and Grandy, D. K. (2000) Ethanol-conditioned place preference is reduced in dopamine D2 receptor-deficient mice. *Pharmacol. Biochem. Behav.* **67,** 693–699.
62. Rocha, B. A., Fumagalli, F., Gainetdinov, R. R., Jones, S. R., Ator, R., Giros, B., Miller, G. W., and Caron, M. G. (1998) Cocaine self-administration in dopamine-transporter knockout mice. *Nat. Neurosci.* **1,** 132–137.
63. Sora, I., Wichems, C., Takahashi, N., Li, X. F., Zeng, Z., Revay, R., Lesch, K. P., Murphy, D. L., and Uhl, G. R. (1998) Cocaine reward models: conditioned place preference can be established in dopamine- and in serotonin-transporter knockout mice. *Proc. Natl. Acad. Sci. USA* **95,** 7699–7704.
64. Kieffer, B. L. (1995) Recent advances in molecular recognition and signal transduction of active peptides: receptors for opioid peptides. *Cell. Mol. Neurobiol.* **15,** 615–635.

65. Goldstein, A. and Naidu, A. (1989) Multiple opioid receptors: ligand selectivity profiles and binding site signatures. *Mol. Pharmacol.* **36,** 265–272.
66. Pan, Z. Z. (1998) Mu-opposing actions of the kappa-opioid receptor. *Trends Pharmacol. Sci.* **19,** 94–98.
67. Kieffer, B. L. (1999) Opioids: first lessons from knock-out mice. *Trends Pharmacol. Sci.* **20,** 537–544.
68. Hutcheson, D. M., Matthes, H. W. D., Valjent, E., Sanchez-Blazquez, P., Rodriguez-Diaz, M., Garzon, J., Kieffer, B. L., and Maldonado, R. (2001) Lack of dependence and rewarding effects of deltorphin II in MOR-deficient mice. *Eur. J. Neurosci.* **13,** 153–161.
69. Matthes, H. W. D., Smadja, C., Valverde, O., Vonesch, J.-L., Foutz, A. S., Boudinot, E., Denavit-Saubier, M., Severini, C., Negri, L., Roques, B. P., Maldonado, R., and Kieffer, B. L. (1998) Activity of the δ-opioid receptor is partially reduced while activity of the κ-receptor is maintained in mice lacking the μ-receptor. *J. Neurosci.* **18,** 7285–7295.
70. Gavériaux-Ruff, C., Matthes, H. W. D., Peluso, J., and Kieffer, B. L. (1998) Abolition of morphine-immunosuppression in mice lacking the μ-opioid receptor gene. *Proc. Natl. Acad. Sci. USA* **95,** 6326–6330.
71. Roy, S., Barke, R. A., and Loh, H. H. (1998) Mu opioid receptor knockout mice: role of mu opioid receptor in morphine mediated immune functions. *Mol. Brain Res.* **61,** 190–194.
72. Roy, S., Liu, H. C., and Loh, H. H. (1998) Mu-Opioid receptor-knockout mice: the role of mu-opioid receptor in gastrointestinal transit. *Mol. Brain Res.* **56,** 281–283.
73. Corbett, A. D., Paterson, S. J., and Kosterlitz, H. W. (1993) Selectivity of ligands for opioid receptors, in: *Opioids I* (A. Herz, ed.), Springer-Verlag, Berlin, pp. 645–673.
74. Kitanaka, N., Sora, I., Kinsey, S., Zeng, Z., and Uhl, G. R. (1998) No heroin or morphine 6-beta-glucuronide analgesia in mu-opioid receptor knockout mice. *Eur. J. Pharmacol.* **355,** R1–R3.
75. Giros, B., Jaber, M., Jones, S. R., Wightman, R. M., and Caron, M. G. (1996) Hyperlocomotion and indifference to cocaine and amphetamine in mice lacking the dopamine transporter. *Nature* **379,** 606–612.
76. Berke, J. D., and Hyman, S. E. (2000) Addiction, dopamine, and the molecular mechanisms of memory. *Neuron* **25,** 515–532.
77. Taylor, D. A., and Fleming, W. W. (2001) Unifying perspectives of the mechanisms underlying the development of tolerance and physical dependence to opioids. *J. Pharmacol. Exp. Ther.* **297,** 11–18.
78. Harrison, L. M., Kastin, A. J., and Zadina, J. E. (1998) Opiate tolerance and dependence: receptors, G-proteins, and antiopiates. *Peptides* **19,** 1603–1630.
79. Meunier, J.-C., Mollereau, C., Toll, L., Suaudeau, C., Moisand, C., Alvinerie, P., Butour, J.-L., Guillemot, J.-C., Ferrara, P., Monsrrat, B., Mazarguil, H., Vassart, G., Parmentier, M., and Costentin, J. (1995) Isolation and structure of the endogenous agonist of opioid receptor-like ORL_1 receptor. *Nature* **377,** 532–535.
80. Reinscheid, R. K., Nothacker, H.-P., Bourson, A., Ardati, A., Henningsen, R. A., Bunzow, J. R., Grandy, D. K., Langen, H., Monsma, F. J., and Civelli, O. (1995) Orphanin FQ: A neuropeptide that activates an opioid like G protein-coupled receptor. *Science* **270,** 792–794.
81. Reinscheid, R. K., Nothacker, H., and Civelli, O. (2000) The orphanin FQ/nociceptin gene: structure, tissue distribution of expression and functional implications obtained from knockout mice. *Peptides* **21,** 901–906.
82. White, F. J., Hu, X.-T., Henry, D. J., and Zhang, X.-F. (1995) Neurophysiological alterations in the mesocorticolimbic dopamine system with repeated cocaine administration, in: *The Neurobiology of Cocaine* (R. P. Hammer, ed.), CRC Press, New York, pp. 99–119.
83. Fitzgerald, L. W., Ortiz, J., Hamedani, A. G., and Nestler, E. J. (1996) Drugs of abuse and stress increase the expression of GluR1 and NMDAR1 glutamate receptor subunits in the rat ventral tegmental area: common adaptations among cross-sensitizing agents. *J. Neurosci.* **16,** 274–282.

84. Churchill, L., Swanson, C. J., Urbina, M., and Kalivas, P. W. (1999) Repeated cocaine alters glutamate receptor subunit levels in the nucleus accumbens and ventral tegmental area of rats that develop behavioral sensitization. *J. Neurochem.* **72,** 2397–2403.
85. Lu, W. and Wolf, M. E. (1999) Repeated amphetamine administration alters AMPA receptor subunit expression in rat nucleus accumbens and medial prefrontal cortex. *Synapse* **32,** 119–131.
86. Jang, C. G., Rockhold, R. W., and Ho, I. K. (2000) An autoradiographic study of [3H]AMPA receptor binding and in situ hybridization of AMPA sensitive glutamate receptor A (GluR-A) subunits following morphine withdrawal in the rat brain. *Brain Res. Bull.* **52,** 217–221.
87. Carlezon, W. A., Jr., Boundy, V. A., Haile, C. N., Lane, S. B., Kalb, R. G., Neve, R. L., and Nestler, E. J. (1997) Sensitization to morphine induced by viral-mediated gene transfer. *Science* **277,** 812–814.
88. Carlezon, W. A., Jr., Haile, C. N., Coppersmith, R., Hayashi, Y., Malinow, R., Neve, R. L., and Nestler, E. J. (2000) Distinct sites of opiate reward and aversion within the midbrain identified using a herpes simplex virus vector expressing GluR1. *J. Neurosci.* **20,** RC62.
89. Ferguson, S. S. G., Downey III, W. E., Colapietro, A.-M., Barak, L. S., Menard, L., and Caron, M. G. (1996) Role of β-arrestin in mediating agonist-promoted G protein-coupled receptor internalization. *Science* **271,** 363–366.
90. Simon, E. J. and Hiller, J. M. (1994) Opioid peptides and opioids receptors, in: *Basic NeuroChemistry: Molecular, Cellular and Medical Aspects* (Siegel, G. J., et al., eds.), Raven Press, Ltd., New York, pp. 321–339.
91. Hepburn, M. J., Little, P. J., Gingras, J., and Kuhn, C. M. (1997) Differential effects of naltrindole on morphine-induced tolerance and physical dependence in rats. *J. Pharmacol. Exp. Ther.* **281,** 1350–1356.
92. Kest, B., Lee, C. E., McLemore, G. L., and Inturrisi, C. E. (1996) An antisense oligodeoxynucleotide to the delta opioid receptor (DOR-1) inhibits morphine tolerance and acute dependence in mice. *Brain Res. Bull.* **39,** 185–188.
93. Mao, J. (1999) NMDA and opioid receptors: their interactions in antinociception, tolerance and neuroplasticity. *Brain Res. Rev.* **30,** 289–304.
94. Wise, R. A. and Bozarth, M. A. (1987) A psychomotor stimulant theory of addiction. *Psychol. Rev.* **94,** 469–492.
95. Koob, G. F., Sanna, P. P., and Bloom, F. E. (1998) Neuroscience of addiction. *Neuron* **21,** 467–476.
96. Leshner, A. I. and Koob, G. F. (1999) Drugs of abuse and the brain. *Proc. Assoc. Am. Physicians* **111,** 99–108.
97. Shippenberg, T. S., Bals-Kubik, R., and Herz, A. (1987) Motivational properties of opioids: evidence that an activation of δ-receptors mediates reinforcement processes. *Brain Res.* **436,** 234–239.
98. Funada, M., Suzuki, T., Narita, M., Misiwa, M., and Nagase, H. (1993) Blockade of morphine reward through the activation of κ-opioid receptors in mice. *Neuropharmacol.* **32,** 1315–1323.
99. Manzanares, J., Corchero, J., Romero, J., Fernandez-Ruiz, J. J., Ramos, J. A., and Fuentes, J. A. (1999) Pharmacological and biochemical interactions between opioids and cannabinoids. *Trends Pharmacol. Sci.* **20,** 287–294.
100. Salamone, J. D. (1994) The involvement of nucleus accumbens dopamine in appetitive and aversive motivation. *Behav. Brain Res.* **61,** 117–133.
101. Narita, M., Mizoguchi, H., Suzuki, T., Dun, N. J., Imai, S., Yajima, Y., Nagase, H., and Tseng, L. F. (2001) Enhanced mu-opioid responses in the spinal cord of mice lacking protein kinase Cgamma isoform. *J. Biol. Chem.* **276,** 15,409–15,414.
102. Abeliovich, A., Paylor, R., Chen, C., Kim, J. J., Wehner, J. M., and Tonegawa, S. (1993) PKC gamma mutant mice exhibit mild deficits in spatial and contextual learning. *Cell* **75,** 1263–1271.

103. Abeliovich, A., Chen, C., Goda, Y., Silva, A. J., Stevens, C. F., and Tonegawa, S. (1993) Modified hippocampal long-term potentiation in PKC gamma-mutant mice. *Cell* **75,** 1253–1262.
104. Suzuki, T., Aoki, T., Kato, H., Yamazaki, M., and Misawa, M. (1999) Effects of the 5-HT(3) receptor antagonist ondansetron on the ketamine- and dizocilpine-induced place preferences in mice. *Eur. J. Pharmacol.* **385,** 99–102.
105. Suzuki, T., Kato, H., Aoki, T., Tsuda, M., Narita, M., and Misawa, M. (2000) Effects of the non-competitive NMDA receptor antagonist ketamine on morphine-induced place preference in mice. *Life Sci.* **67,** 383–389.
106. Maldonado, R., Negus, S., and Koob, G. F. (1992) Precipitation of morphine withdrawal syndrome in rats by administration of mu-, delta- and kappa-selective opioid antagonists. *Neuropharmacol.* **31,** 1231–1241.
107. Geiger, J. R., Melcher, T., Koh, D. S., Sakmann, B., Seeburg, P. H., Jonas, P., and Monyer, H. (1995) Relative abundance of subunit mRNAs determines gating and Ca^{2+} permeability of AMPA receptors in principal neurons and interneurons in rat CNS. *Neuron* **15,** 193–204.
108. Metzger, D. and Feil, R. (1999) Engineering the mouse genome by site-specific recombination. *Curr. Op. Biotech.* **10,** 470–476.
109. Christie, M. J., Williams, J. T., Osborne, P. B., and Bellchambers, C. E. (1997) Where is the locus of opioid withdrawal? *Trends Pharm. Sci.* **18,** 134–140.
110. Pich, E. M. and Epping-Jordan, M. P. (1998) Transgenic mice in drug dependence research. *Ann. Med.* **30,** 390–396.

2

Molecular Genetic Approaches

Theo Mantamadiotis, Günther Schütz, and Rafael Maldonado

1. Introduction

Genetic influences are pivotal in determining the sensitivity to drugs of abuse. The spectrum of genes involved in the behavioral manifestation of drug dependence or withdrawal has not been fully determined, but there are a number of candidate genes that appear to be important. The complexity of the underlying molecular mechanisms governing the adaptation of the neuronal system has prevented the straightforward study of the genetic influences involved. Animal models have allowed the identification of genes involved in drug-related behaviors and have created tools with which to pursue the pharmacogenetic research necessary for the molecular dissection of biochemical pathways involved. A great leap forward in the development of molecular genetic animal models came with the progress in the field of stem cell research. Mouse embryonic stem (ES) cell technology in the late 1980s became amenable to routine research applications *(1,2)*. The gene of choice could be silenced in the mouse and the consequences of this analyzed in the living organism. More sophisticated techniques allowing for the conditional deletion of genes both temporally and tissue specifically have become available, bypassing either pleiotropic or developmental effects of gene loss *(3)*. These advances will be discussed in detail in this chapter.

Our work has focused on the transcription factor cAMP response element-binding protein (CREB) and the related members CRE response element modulation protein (CREM) and activating transcription factor 1(ATF1). This is of particular significance to the study of drug addiction, because the cAMP signal transduction cascade has been implicated in drug-induced cellular responses *(4,5)* (Fig. 1). CREB activity has previously been shown to be altered in response to a number of drugs, including opiates, both in cells and in vivo *(5–7)*. Here we discuss the use of a number of previously described and novel mouse models, using both the classical and conditional gene knockout approaches, in which CREB protein is either reduced or completely absent, to study the role of this important transcription factor in substance abuse.

2. CREB Function in Brain

CREB is expressed in almost all mammalian cells and is a transcription factor with important functions in many tissues, including brain. It harbors an N-terminal activation domain and C-terminal DNA-binding dimerization domain (Fig. 2) and is a member of the basic leucine zipper (bZIP) protein superfamily. CREB is able to either

From: *Molecular Biology of Drug Addiction*
Edited by: R. Maldonado © Humana Press Inc., Totowa, NJ

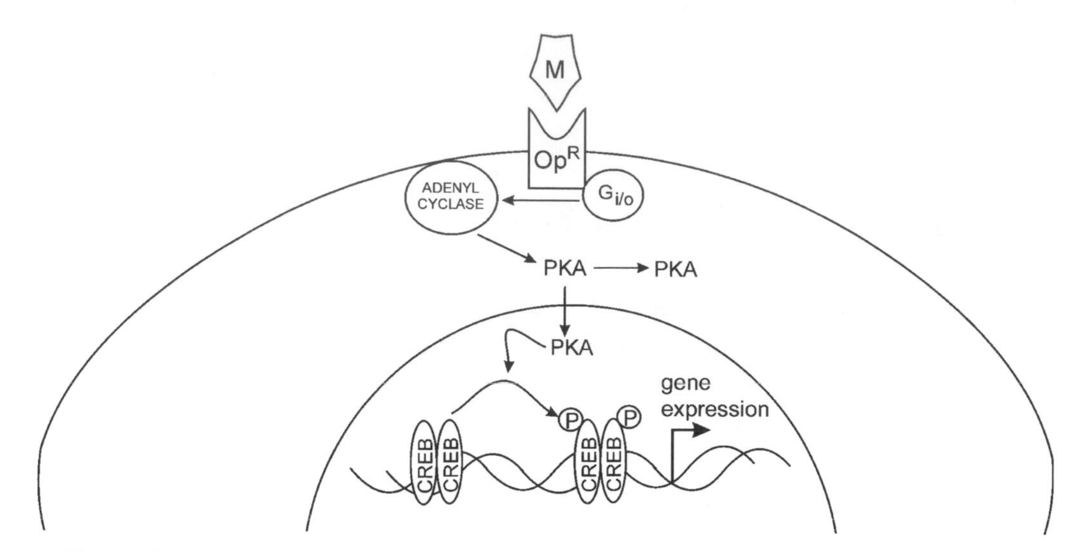

Fig. 1. The opiate signal transduction pathway. Opiates such as morphine bind to G_i- or G_o-coupled opioid receptors. Acute opiate exposure results in adenyl cyclase inhibition, reduction in cAMP levels and cAMP-dependent protein kinase activity and the phosphorylation of both cytoplasmic and nuclear targets, including CREB. On the other hand, chronic opiate exposure increases the levels of these factors.

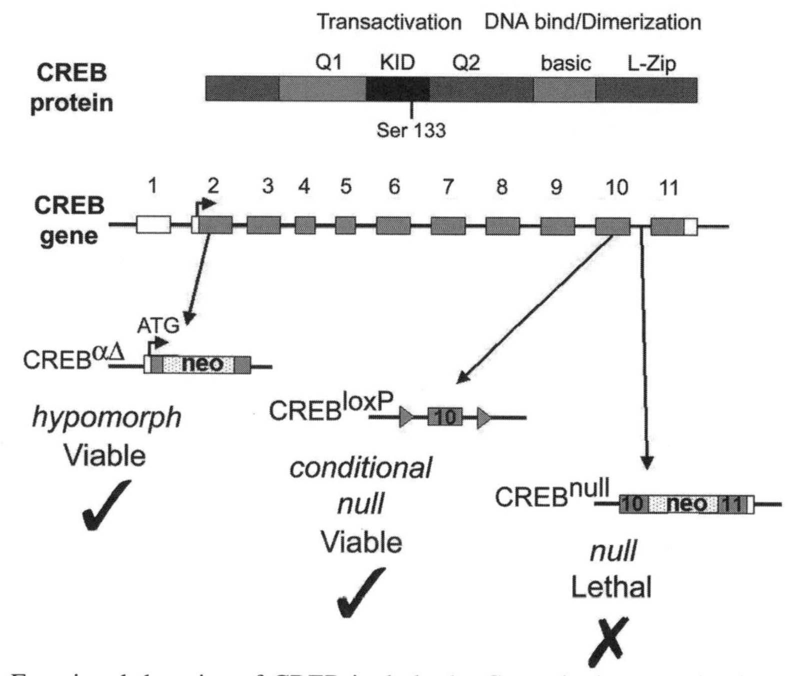

Fig. 2. Functional domains of CREB include the C-terminal transactivation composed of two glutamine-rich domains (Q1 and Q2) flanking the kinase inducible domain (KID), which harbors serine residue 133 which is phosphorylated upon cellular stimulation via the cAMP pathway as well as other signaling pathways. The C-terminal domain harbors the basic leucine zipper (L-Zip) domains, which are involved in DNA binding and dimerization. The CREB gene is comprised by at least 11 exons shown as rectangles (white for untranslated and gray for translated). The three genetically modified CREB mutant mice generated in our laboratory are indicated with the disrupted exons shown. The only genetically modified CREB mutant mice that are viable are the CREBaD and the conditional CREB^{loxP} mice. These have been used in drug studies described elsewhere *(8,9)* and herein.

homodimerize or heterodimerize with its closely related factors, CREM and ATF1 *(10)*. Upon phosphorylation on a critical serine-133 residue, it can bind to cAMP-responsive elements (CREs) and recruit the CREB-binding protein (CBP) and other transcriptional cofactors to transactivate a large number of target genes important for cellular function (Fig. 1). Apart from CREB's role in the cellular responses triggered by drugs of abuse, specific functions attributed to CREB in brain include neuronal survival *(11,12)*, hypothalamic/pituitary growth axis *(13)*, circadian rhythm *(14–16)*, and learning and memory *(17,18)*.

Several molecular changes have been described during exposure to opioids *(19–22)*. Acute opioid administration inhibits adenylyl cyclase activity, whereas chronic opioid treatment leads to a dramatic upregulation of the cAMP pathway at every major step of the cascade between receptor activation and physiological response *(23)* (Fig. 1). This upregulation occurs in discrete brain areas including the locus coeruleus (LC) and the nucleus accumbens (NAc), providing a neuroanatomical link for opioid physical dependence and rewarding effects, respectively *(24–26)* Upregulation of the cAMP pathway also seems to be involved in the addictive mechanisms of other drugs of abuse, such as cocaine *(23,27)*. The phosphorylation state of CREB was shown to be decreased in the LC after acute morphine administration, whereas chronic morphine produces an increase in the phosphorylation and expression of CREB in this structure *(5–7)*. We have previously demonstrated that CREB is an important factor involved in the onset of behavioral manifestations of opiate withdrawal, where the major signs of morphine abstinence were strongly attenuated in CREB$^{\alpha\Delta}$ mutant mice, which lack the major transactivating CREB α and Δ isoforms *(8)* (*see* Section 3.). Work utilizing antisense oligonucleotides has also implicated decreased CREB expression in the LC with attenuated withdrawal and electrophysiological responses *(7)*. CREB has also been implicated in the motivational properties of morphine and cocaine *(27,28)*, although the functional relevance in molecular genetic animal models has not yet been determined.

3. Genetically Altered CREB Mutant Mice

To investigate the role of CREB in drug dependence and motivational responses, we made use of two independent genetically modified CREB mutant mice. On the one hand, we have used CREB mice that lack the two major α and Δ CREB isoforms *(8)* These mice were generated by targeting the second CREB exon, which harbors the first translated ATG codon (Fig. 2) *(29)*. Although the major CREB α and Δ isoforms were ablated, this mutation allowed for the translation of a novel and previously unidentified CREB β isoform, at levels higher than in wild-type mice *(30)*. In essence, these mice carry a hypomorphic CREB allele and are termed CREB$^{\alpha\Delta}$ mice.

To study the consequences of complete CREB loss, a second mouse was generated by targeting the region encoding the entire DNA binding and dimerization domain *(31)* (Fig. 2). Homozygous CREB null mice die at birth, due to the failure of the lungs to inflate. Therefore, until recently the CREB$^{\alpha\Delta}$ mice represented the only viable mouse model with a genetically modified CREB gene.

4. Attenuated Naloxone-Precipitated Withdrawal Response in CREB$^{\alpha\Delta}$ Mice

Various manifestations of somatic signs of naloxone-precipitated withdrawal were evaluated in CREB$^{\alpha\Delta}$ mice. Opioid dependence was induced by repeated morphine injection. The morphine dose (ip) was progressively increased from 20 to 100 mg/kg over a period of 5 d. Morphine withdrawal syndrome was precipitated by naloxone (1 mg/kg, sc) 2 h after the last morphine administration. Mouse behavior was observed immediately after naloxone administration. Opiate withdrawal syndrome is characterized by a number of behavioral and physiological signs. Some of these responses, such as jumping and teeth chattering, are dependent on the central nervous system, while other responses are mediated by the peripheral nervous system, including diarrhea, weight loss, ptosis, and lacrimation. CREB$^{\alpha\Delta}$ mice exhibited significant attenuation in nine classical withdrawal responses immediately following naloxone injections. All nine responses were significantly attenuated in the mutant animals compared with the control group (Fig. 3). Importantly, the reduction in withdrawal symptoms was due to the reduced CREB levels and not a result of altered opioid receptors, as receptor studies showed that neither the affinity nor number of receptors was changed in mutant mice *(8)*.

Acute administration of morphine was also evaluated in CREB$^{\alpha\Delta}$ mice by assessing the analgesic effects using the hot-plate test. Mutant and wild-type mice exhibited similar nociceptive threshold and analgesic responses to 3, 9, and 20 mg/kg morphine, manifested by increased licking and jumping latencies. As the development of physical dependence is also associated with tolerance, which is the diminishing response to a given drug dose over time, the development of opioid tolerance was examined in CREB$^{\alpha\Delta}$ mice by monitoring antinociceptive responses during chronic morphine treatment (5 mg/kg ip, twice daily for 5 d). There was no difference in antinociceptive responses between mutant and wild-type mice in the hot-plate test upon acute morphine administration (3 and 9 mg/kg ip). However, though both morphine doses generated significant antinociception in mutant mice, the effect is slightly attenuated in naive CREB$^{\alpha\Delta}$ mice for licks and jumps latency (Fig. 4). In summary, CREB$^{\alpha\Delta}$ mice do develop tolerance to morphine analgesia, but to a lesser degree than wild-type mice.

5. Brain-Specific CREB Loss in Mice

As mentioned in Section 3, the only viable genetically modified CREB mutant mouse model available for drug studies to date has been the CREB$^{\alpha\Delta}$ hypomorph mouse. To study the brain-specific loss of CREB in adult mice, free of the complications inherent in classical knockout models, such as pleiotropic effects during embryonic development and postnatal physiology, we employed the Cre/loxP recombination system to conditionally eliminate CREB only in brain, leaving a normal intact CREB gene in all other tissues. To generate the nervous system-specific CREB mutant mice, we used homologous recombination in ES cells to generate a modified CREB allele in which CREB exon 10, encoding the first part of the bZIP domain, was flanked with loxP sites (Fig. 2). Mice harboring the CREB^{loxP} allele were crossed with transgenic mice possessing a transgene for Cre recombinase under the control of the *nestin* promoter and enhancer *(32)* (Fig. 5A). The mice lacking CREB in brain are referred to as CREB^{NesCre} mice.

CREB^{NesCre} mice lacked CREB immunoreactivity in almost all neurons and glia, probing with either of three antibodies recognizing CREB epitopes from the N-terminal half to the C-terminal end, indicating that no CREB protein, including truncated forms, were present (Fig. 5B). Phenotypically, CREBNesCre mice are essentially normal except for a reduction in body size due to a deficiency in growth hormone (T. M., unpublished data).

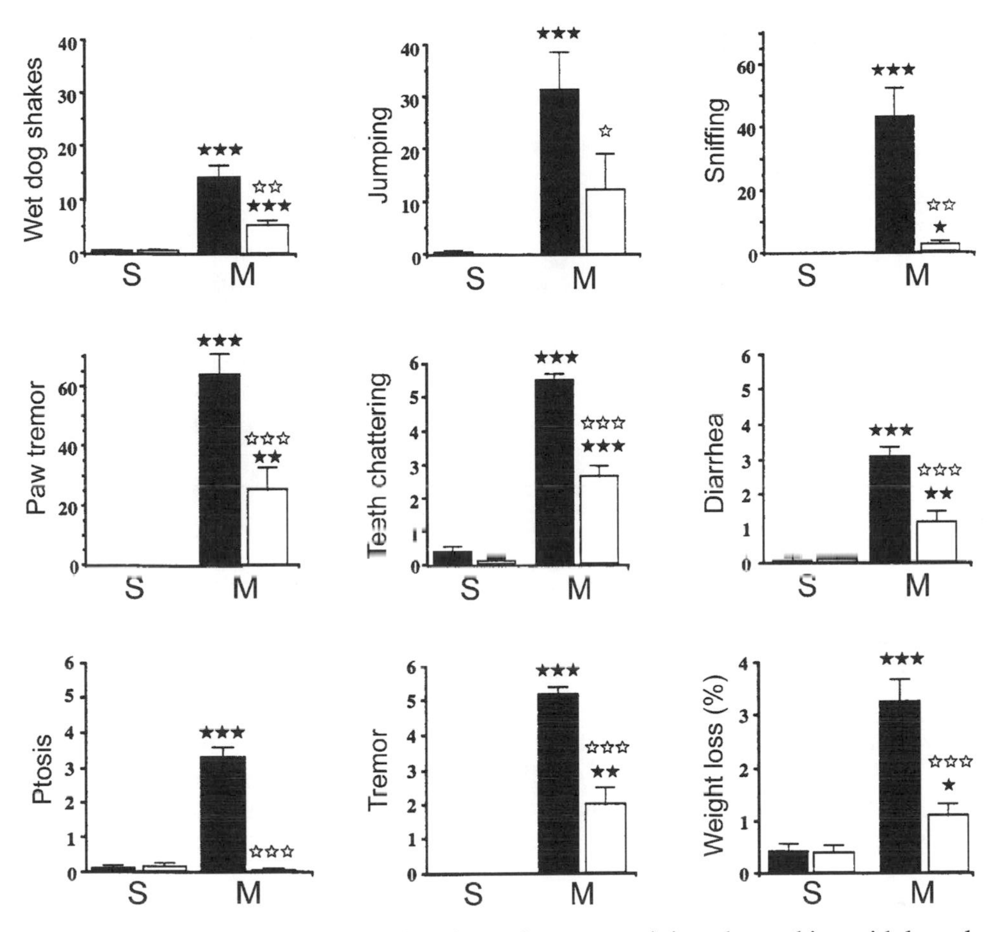

Fig. 3. Behavioral signs measured during naloxone-precipitated morphine withdrawal syndrome in CREB hypomorph mice (white columns), and their wild-type controls (black columns). Opiate dependence was induced by repeated ip injections of morphine-HCl (increasing dose) every 8 h during 3 d. Withdrawal was precipitated once in each mouse by naloxone-HCl injection (1 mg/kg sc) 2 h after the last morphine injection. The mice were placed individually into test chambers 30 min before naloxone injection, and the behavioral signs of withdrawal were evaluated after injection for 30 min. Data were subjected to two-way analysis of variance between animals. The number of animals per group was 12–16. Black stars, comparison between morphine-treated mice (M) and saline-treated mice (S); white stars, comparisons between wild-type and mutant groups receiving the same treatment; one star, $p < 0.05$; two stars, $p < 0.01$; three stars, $p < 0.001$.

6. Ongoing Analyses of CREB Hypomorph and Conditional Mutant Mice

The previous work on $CREB^{\alpha\Delta}$ mice has recently been extended, in parallel with novel studies on $CREB^{NesCre}$ mice. $CREB^{\alpha\Delta}$ mice used in this work were backcrossed for seven generations into a C57/BL6 strain, to determine the contribution of genetic background on the withdrawal behavior. The backcrossed $CREB^{\alpha\Delta}$ mice exhibited almost identical withdrawal responses to those described previously, showing that the attenuated withdrawal syndrome in these mice is a robust phenotype apparently independent of genetic background.

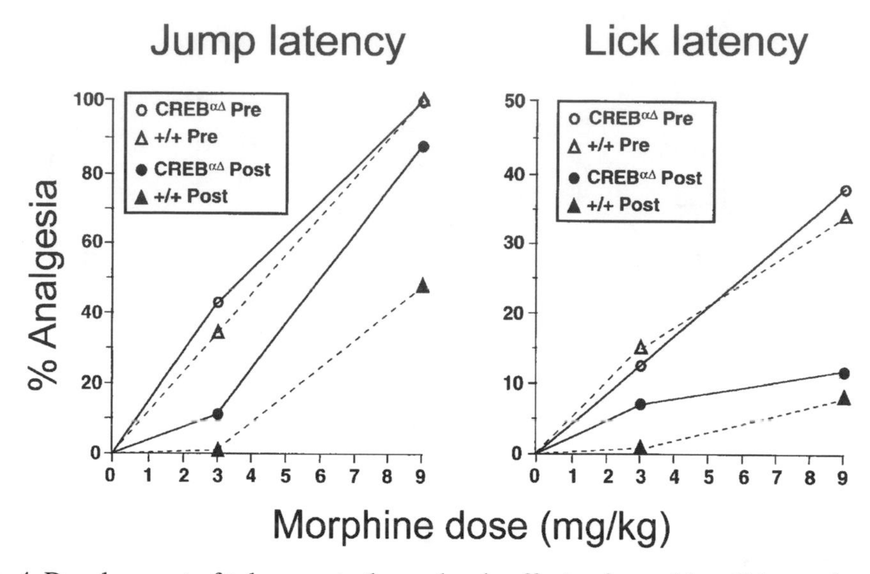

Fig. 4. Development of tolerance to the analgesic effects of morphine. Prior to chronic morphine treatment, mice were examined in the hot-plate test. Fifteen minutes after acute morphine administration (3 or 9 mg/kg ip) the percentage of analgesia was calculated as (test latency minus control latency) divided by (cutoff time minus control latency) × 100. Test latency is the time it takes for the animal to jump off the hot plate after saline injection. Cutoff time is 120 s. Mice were treated with morphine (5 mg/kg ip) for 4 d and reexamined on the hot-plate test 9 h after the last morphine injection. Circles, $CREB^{\alpha\Delta}$ mutant mice; triangles, wild-type mice; open symbols, percentage analgesia before chronic morphine; filled symbols, percentage analgesia after chronic morphine.

Preliminary results indicate that $CREB^{NesCre}$ mice exhibit significantly attenuated withdrawal responses, similar to $CREB^{\alpha\Delta}$ mice, supporting the notion that the phenotype observed in $CREB^{\alpha\Delta}$ mice is primarily a consequence of CREB loss in the nervous system. Furthermore, the CREB α and Δ are probably the major CREB isoforms involved in the expression of morphine withdrawal syndrome. A more detailed analysis of the conditional CREB mutants will allow for the distinction between CNS and peripheral CREB-dependent mechanisms.

An elevated activity of LC neurons has been postulated to contribute to the expression of opiate withdrawal in morphine-dependent rats. Controversial data have been previously reported on the role played by the LC in the expression of morphine abstinence. The firing rate of LC neurons was strongly increased during spontaneous and antagonist-precipitated morphine withdrawal, which seems to contribute to the behavioral expression of the somatic signs of abstinence. Moreover, the LC was the most sensitive brain structure to precipitate the somatic signs of morphine withdrawal by microinjection of opioid antagonists, and its electrolytic lesion strongly inhibited opioid abstinence. Other studies, however, found that morphine treated rats failed to exhibit opiate withdrawal hyperactivity in the LC or that lesions of the noradrenergic brain pathways emanating from the LC failed to attenuate the somatic signs of opioid withdrawal. To examine whether CREB plays a role in this withdrawal-induced hyperactiv-

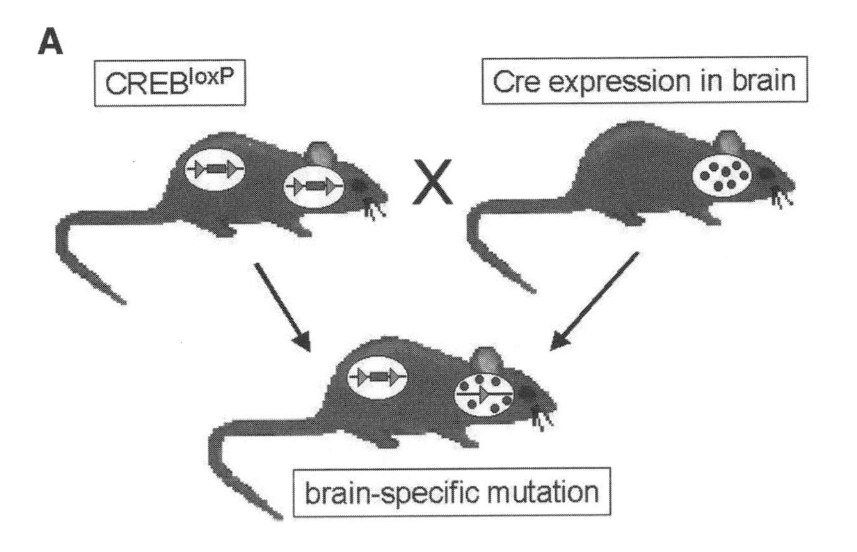

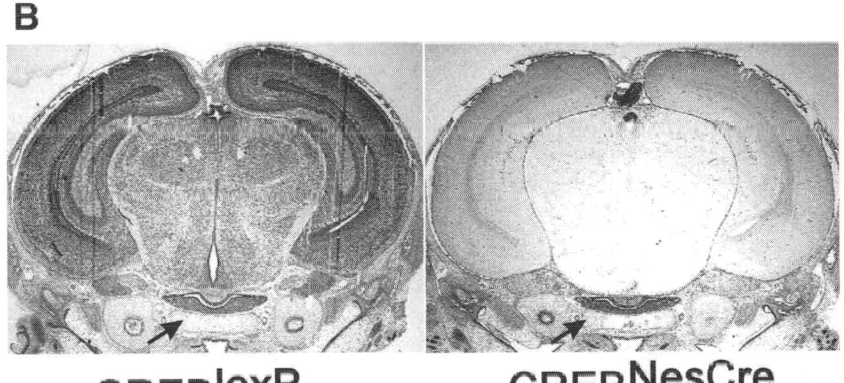

Fig. 5. Disruption of the CREB gene in brain by Cre/loxP-mediated recombination. **(A)** Once mice homozygous for the CREB^{loxP} allele are crossed with mice expressing Cre recombinase specifically in brain, the result is CREB loss restricted to the nervous system. Use of various Cre transgenic lines would result in distinct anatomical and temporal patterns of CREB gene ablation. **(B)** Cre-recombinase expression under the control of the nestin promoter and enhancer results in almost complete CREB loss in brain. CREB^{loxP} brains show normal widespread nuclear protein expression revealed by using anti-CREB antibodies, while CREB^{NesCre} mutant mice exhibit almost complete loss of CREB protein. The anatomical specificity of CREB loss is highlighted by the failure of CREB recombination in the pituitary cells of CREB^{NesCre} mice.

ity, single-unit extracellular recordings of LC neurons in brain slices from wild-type, CREB^{NesCre}, and CREB$^{\alpha\Delta}$-deficient mice will be performed following chronic morphine treatment.

Interesting studies focussing on the role of CREB in rewarding behavior have recently been reported. Using rats in the conditioned place preference paradigm, where a herpes simplex virus vector expressing dominant-negative CREB was injected into the NAc of rat brain, a significant enhancement in cocaine rewarding effects was seen, while overexpression of wild-type CREB had an aversive effect *(27)*. More recently, studies using CREB$^{\alpha\Delta}$ mice suggest that there may be differences in the way CREB modulates downstream target genes, depending on whether morphine or cocaine is used to induce reward. In this study, CREB$^{\alpha\Delta}$-deficient mice do not respond to the reinforcing properties of morphine but do show an enhanced response to cocaine *(9)*. We are currently using both our hypomorph and conditional knockout CREB mutant models to investigate these reward responses. In contrast to this last study, our preliminary data suggest that both CREB$^{\alpha\Delta}$ mice and CREB^{NesCre} mice show a reward response to morphine.

The conditional CREB mutant mice will prove to be useful in further studies as more Cre transgenic mice become available, allowing for more precise anatomical and temporal control over CREB ablation. For example, we now have Cre transgenic mice that will allow for the selective postnatal loss of CREB in either all neurons or dopamine D1 receptor-positive neurons, further refining the neuroanatomical and developmental molecular dissection of CREB function in mouse behavioral studies. The conditional disruption of CREB in either the peripheral or central nervous system will also allow us to distinguish between effects dependent on either or both the central or peripheral nervous system.

7. Conclusions

Neuroadapatations arising during prolonged exposure to opioids and the development of addiction are complex. Using established and emerging techniques in the manipulation of the mouse genome, we have been able to disrupt the CREB gene in the whole organism or specifically in the nervous system. These evolving technologies will bring forward the understanding of the molecular mechanisms involved in the development of drug addiction. As CREB plays a pivotal role in drug addiction, the ongoing studies described here may provide a handle on how to intervene pharmacologically in the biochemical pathways involved in opioid withdrawal syndrome and drug addiction in general.

References

1. Doetschman, T., Gregg, R. G., Maeda, N., Hooper, M. L., Melton, D. W., Thompson, S., and Smithies, O. (1987) Targeted correction of a mutant HPRT gene in mouse embryonic stem cells. *Nature* **330,** 576–578.
2. Mansour, S. L., Thomas, K. R., and Capecchi, M. R. (1988) Disruption of the proto-oncogene int-2 in mouse embryo-derived stem cells: a general strategy for targeting mutations to non-selectable genes. *Nature* **336,** 348–352.
3. Gu, H., Marth, J. D., Orban, P. C., Mossmann, H., and Rajewsky, K. (1994) Deletion of a DNA polymerase beta gene segment in T cells using cell type-specific gene targeting. *Science* **265,** 103–106.
4. Chakrabarti, S., Wang, L., Tang, W. J., and Gintzler, A. R. (1998) Chronic morphine augments adenylyl cyclase phosphorylation: relevance to altered signaling during tolerance/dependence. *Mol. Pharmacol.* **54,** 949–953.
5. Nestler, E. J. (1993) Cellular responses to chronic treatment with drugs of abuse. *Crit. Rev. Neurobiol.* **7,** 23–39.
6. Guitart, X., Thompson, M. A., Mirante, C. K., Greenberg, M. E., and Nestler, E. J. (1992)

Regulation of cyclic AMP response element-binding protein (CREB) phosphorylation by acute and chronic morphine in the rat locus coeruleus. *J. Neurochem.* **58,** 1168–1171.
7. Lane-Ladd, S. B., Pineda, J., Boundy, V. A., Pfeuffer, T., Krupinski, J., Aghajanian, G. K., and Nestler, E. J. (1997) CREB (cAMP response element-binding protein) in the locus coeruleus: biochemical, physiological, and behavioral evidence for a role in opiate dependence. *J. Neurosci.* **17,** 7890–7901.
8. Maldonado, R., Blendy, J. A., Tzavara, E., Gass, P., Roques, B. P., Hanoune, J., and Schutz, G. (1996) Reduction of morphine abstinence in mice with a mutation in the gene encoding CREB [see comments]. *Science* **273,** 657–659.
9. Walters, C. L., Blendy, J. A. (2001) Different requirements for cAMP response element binding protein in positive and negative reinforcing properties of drugs of abuse. *J. Neurosci.* **21,** 9438–9444.
10. Mayr, B. and Montminy, M. (2001) Transcriptional regulation by the phosphorylation-dependent factor CREB. *Nat. Rev. Mol. Cell Biol.* **2,** 599–609.
11. Riccio, A., Ahn, S., Davenport, C. M., Blendy, J. A., and Ginty, D. D. (1999) Mediation by a CREB family transcription factor of NGF-dependent survival of sympathetic neurons. *Science* **286,** 2358–2361.
12. Mantamadiotis, T., Lemberger, T., Bleckmann, S. C., Kern, H., Kretz, O., Villalba, A. M., et al. (2002) Disruption of CREB function in brain leads to neurodegeneration. *Nat. Genet.* **31,** 47–54.
13. Struthers, R. S., Vale, W. W., Arias, C., Sawchenko, P. E., and Montminy, M. R. (1991) Somatotroph hypoplasia and dwarfism in transgenic mice expressing a non-phosphorylatable CREB mutant. *Nature* **350,** 622–624.
14. Belvin, M. P., Zhou, H., and Yin, J. C. (1999) The Drosophila dCREB2 gene affects the circadian clock. *Neuron* **22,** 777–787.
15. Ginty, D. D., Kornhauser, J. M., Thompson, M. A., Bading, H., Mayo, K. E., Takahashi, J. S., and Greenberg, M. E. (1993) Regulation of CREB phosphorylation in the suprachiasmatic nucleus by light and a circadian clock. *Science* **260,** 238–241.
16. Obrietan, K., Impey, S., Smith, D., Athos, J., and Storm, D.R. (1999) Circadian regulation of cAMP response element-mediated gene expression in the suprachiasmatic nuclei. *J. Biol. Chem.* **274,** 17748–1756.
17. Bourtchuladze, R., Frenguelli, B., Blendy, J., Cioffi, D., Schutz, G., Silva, A. J. (1994) Deficient long-term memory in mice with a targeted mutation of the cAMP-responsive element-binding protein. *Cell* **79,** 59–68.
18. Gass, P., Wolfer, D. P, Balschun, D., Rudolph, D., Frey, U., Lipp, H. P., Schutz G. (1998) Deficits in memory tasks of mice with CREB mutations depend on gene dosage. *Learn. Mem.* **5,** 274–288.
19. Hyman, S. E. (1996) Addiction to cocaine and amphetamine. *Neuron* **16,** 901–904.
20. Koob, G. F., Sanna, P. P., and Bloom, F. E. (1998) Neuroscience of addiction. *Neuron* **21,** 467–476.
21. Nestler, E. J. and Aghajanian, G. K. (1997) Molecular and cellular basis of addiction. *Science* **278,** 58–63.
22. Spanagel, R. and Weiss, F. (1999) The dopamine hypothesis of reward: past and current status. *Trends Neurosci.* **22,** 521–527.
23. Nestler, E. J. (1997) Molecular mechanisms of opiate and cocaine addiction. *Curr. Opin. Neurobiol.* **7,** 713–719.
24. Di Chiara, G. and North, R. A. (1992) Neurobiology of opiate abuse. *Trends Pharmacol. Sci.* **13,** 185–193.
25. Koob, G. F. and Le Moal, M. (1997) Drug abuse: hedonic homeostatic dysregulation. *Science* **278,** 52–58.
26. Maldonado, R., Stinus, L., Gold, L. H., and Koob, G. F. (1992) Role of different brain struc-

tures in the expression of the physical morphine withdrawal syndrome. *J. Pharmacol. Exp. Ther.* **261,** 669–677.
27. Carlezon, W. A., Thome, J., Jr., Olson, V. G., Lane-Ladd, S. B., Brodkin, E. S., Hiroi, N., Duman, R. S., Neve, R. L., and Nestler, E. J. (1998) Regulation of cocaine reward by CREB. *Science* **282,** 2272–2275.
28. Widnell, K. L., Self, D. W., Lane, S. B., Russell, D. S., Vaidya, V. A., Miserendino, M. J., Rubin, C. S., Duman, R. S., and Nestler, E. J. (1996) Regulation of CREB expression: in vivo evidence for a functional role in morphine action in the nucleus accumbens. *J. Pharmacol. Exp. Ther.* **276,** 306–315.
29. Hummler, E., Cole, T. J.,. Blendy, J. A, Ganss, R., Aguzzi, A., Schmid, W., Beermann F., and Schutz, G. (1994) Targeted mutation of the CREB gene: compensation within the CREB/ATF family of transcription factors. *Proc. Natl. Acad. Sci. USA* **91,** 5647–5651.
30. Blendy, J. A., Kaestner, K. H., Schmid, W., Gass, P., and Schutz, G. (1996) Targeting of the CREB gene leads to up-regulation of a novel CREB mRNA isoform. *EMBO J.* **15,** 1098–106.
31. Rudolph, D., Tafuri, A., Gass, P., Hammerling, G. J., Arnold, B., and Schutz, G. (1998) Impaired fetal T cell development and perinatal lethality in mice lacking the cAMP response element binding protein, *Proc. Natl. Acad. Sci. USA* **95,** 4481–4486.
32. Tronche, F., Kellendonk, C., Kretz, O., Gass, P., Anlag, K., Orban, P. C., Bock, R., Klein, R., and Schutz, G. (1999) Disruption of the glucocorticoid receptor gene in the nervous system results in reduced anxiety. *Nat. Genet.* **23,** 99–103.
33. Kogan, J. H., Nestler, E. J., and Aghajanian, G. K. (1992) Elevated basal firing rates and enhanced responses to 8-Br-cAMP in locus coeruleus neurons in brain slices from opiate-dependent rats. *Eur. J. Pharmacol.* **211,** 47–53.
34. Rasmussen, K., Beitner-Johnson, D. B., Krystal, J. H., Aghajanian, G. K., and Nestler, E. J. (1990) Opiate withdrawl and the rat locus coeruleus: behavioral, electrophysiolgical, and biochemical correlates. *J. Neurosci.* **10,** 2308–2317.
35. Bell, J. A. and Grant, S. J. (1998) Locus coeruleus neurons from morphine-treated rats do not show opiate-withdrawal hyperactivity in vitro. *Brain Res.* **788,** 237–244.
36. Britton, K. T., Svensson, T., Schwartz, J., Bloom, F. E., and Koob, G. F. (1984) Dorsal noradrenergic bundle lesions fail to alter opiate withdrawal or suppression of opiate withdrawal by clonidine. *Life Sci.* **34,** 133–139.
37. Caille, S., Espejo, E. F., Reneric, J. P., Cador, M., Koob, G. F., and Stinus, L. (1999) Total neurochemical lesion of noradrenergic neurons of the locus ceruleus does not alter either naloxone-precipitated or spontaneous opiate withdrawal nor does it influence ability of clonidine to reverse opiate withdrawal. *J. Pharmacol. Exp. Ther.* **290,** 881–892.
38. Delfs, J. M., Zhu, Y., Druhan, J. P., and Aston-Jones, G. (2000) Noradrenaline in the ventral forebrain is critical for opiate withdrawl-induced aversion. *Nature* **403,** 430–437.

3

Opiate Addiction

Role of cAMP Pathway and CREB

Lisa M. Monteggia and Eric J. Nestler

Addiction is characterized by compulsive drug seeking and taking despite adverse consequences. Addiction is a gradual process, which occurs in most individuals only after repeated drug exposure. Thus, it appears that individuals addicted to drugs undergo time-dependent alterations or neuroadaptations in the brain that occur during a course of repeated drug administration. Chronic drug intake produces these neuroadaptations in neuronal cell types located within specific brain regions. These neuroadaptations alter the function of these individual neurons and ultimately the neural circuitry that mediates the abnormalities in complex behaviors, which characterize aspects of addiction.

Opiates, a class of drugs with high abuse potential, produce several types of behavioral abnormalities that are mediated by neuroadaptations within specific brain regions. Chronic opiate administration produces dependence, a functional alteration that is described as the need for continued drug administration to avoid a withdrawal syndrome. The opiate withdrawal syndrome is characterized by physical and psychological symptoms; the latter include restlessness, extreme anxiety, and depressed mood. These negative emotional symptoms of withdrawal are thought to contribute to addiction by inducing relapse to drug use during the withdrawal phase. The physical dependence associated with opiates has been ascribed to numerous brain regions as well as spinal cord. The psychological symptoms of withdrawal have been mapped partly to the mesolimbic dopamine system *(1)*, which arises from dopaminergic neurons in the ventral tegmental area (VTA) of the midbrain and projects to the nucleus accumbens (NAc) and other forebrain limbic structures. Tolerance, another type of functional alteration seen after chronic opiate administration, describes the need for an increasing drug dose to achieve the same effect. Tolerance occurs to many opiate actions, and is thought to be mediated by neuroadaptations in the brain and spinal cord regions that mediate those actions. Tolerance may contribute to addiction by causing escalation of drug doses. Opiates can also produce sensitization, or reverse tolerance, in which the same dose of the drug elicits progressively greater responses. Of particular importance is sensitization that occurs to the rewarding effects of opiates, which is believed to be mediated by neuroadaptations in the mesolimbic dopamine system. Sensitization may contribute to addiction by promoting relapse to drug use long after withdrawal periods subside.

From: *Molecular Biology of Drug Addiction*
Edited by: R. Maldonado © Humana Press Inc., Totowa, NJ

Table 1
Upregulation of the cAMP Pathway and Opiate Addiction

Site of upregulation	Functional consequence
Locus coeruleus[a]	Physical dependence and withdrawal
Ventral tegmental area[b]	Dysphoria during early withdrawal periods
Periaqueductal gray[b]	Dysphoria during early withdrawal periods, and physical dependence and withdrawal
Nucleus accumbens	Dysphoria during early withdrawal periods
Amygdala	Conditioned aspects of addiction
Dorsal horn of spinal cord	Tolerance to opiate-induced analgesia
Myenteric plexus of gut	Tolerance to opiate-induced reductions in intestinal motility and increases in motility during withdrawal

[a]The cAMP pathway is up-regulated within the principal noradrenergic neurons located in this region.

[b]Indirect evidence suggests that the cAMP pathway is up-regulated within GABAergic neurons that innervate the dopaminergic (ventral tegmental area) and serotonergic (periaqueductal gray) neurons located in these regions. During withdrawal, the upregulated cAMP pathway would become fully functional and could contribute to a state of dysphoria by increasing the activity of the GABAergic neurons, which would then inhibit the dopaminergic and serotonergic neurons *(2–4)*.

To date, numerous neuroadaptations have been described in the brain and spinal cord after chronic administration of opiates, and it has been increasingly possible to relate certain of these changes to the behavioral abnormalities associated with opiate addiction. This review focuses on one particular neuroadaptation, upregulation of the cAMP second messenger pathway, which is one of the best-established molecular and cellular mechanisms of opiate tolerance and dependence. Upregulation of the cAMP pathway has been demonstrated in several regions of the nervous system, and related to different aspects of opiate action within those regions (*see* Table 1) *(5)*. Here, we focus on two such brain regions: the locus coeruleus (LC) and the nucleus accumbens (NAc).

1. Upregulation of the cAMP Pathway in the Locus Coeruleus

The LC, located on the floor of the fourth ventricle in the anterior pons, is the major noradrenergic nucleus in brain. The LC is a densely packed homogeneous nucleus that regulates attention, arousal, and autonomic tone. The LC's noradrenergic neurons have extremely widespread projections and innervate virtually all regions of the brain and spinal cord. Activation of the LC, which occurs upon precipitation of opiate withdrawal, increases noradrenergic transmission throughout the central nervous system and mediates some of the physical symptoms of withdrawal. Indeed, excitation of LC neurons is both necessary and sufficient to elicit certain behavioral signs of opiate withdrawal. Although physical dependence is not thought to be an important contributor to addiction *per se*, studies of opiate action in the LC have provided important insight into the types of neuroadaptations that opiates elicit in responsive neurons.

Acutely, opiates inhibit the firing rate of LC neurons by binding to μ opioid receptors that are coupled to the $G_{i/o}$ family of G proteins (Fig. 1) *(3,5)* $G_{i/o}$ proteins couple directly to an inward rectifier K^+ channel to mediate an increase in outward K^+ current. The coupling to $G_{i/o}$ also causes inhibition of adenylyl cyclase to decrease cAMP levels and reduce activity of cAMP-dependent protein kinase (PKA). This leads to a decrease in the conductance of a nonspecific cation channel, which has not yet been isolated at the molecular level *(6)*. These two opiate actions, activation of an inward-rectifying K^+ channel and inhibition of a nonspecific cation channel, diminish LC neuronal excitability.

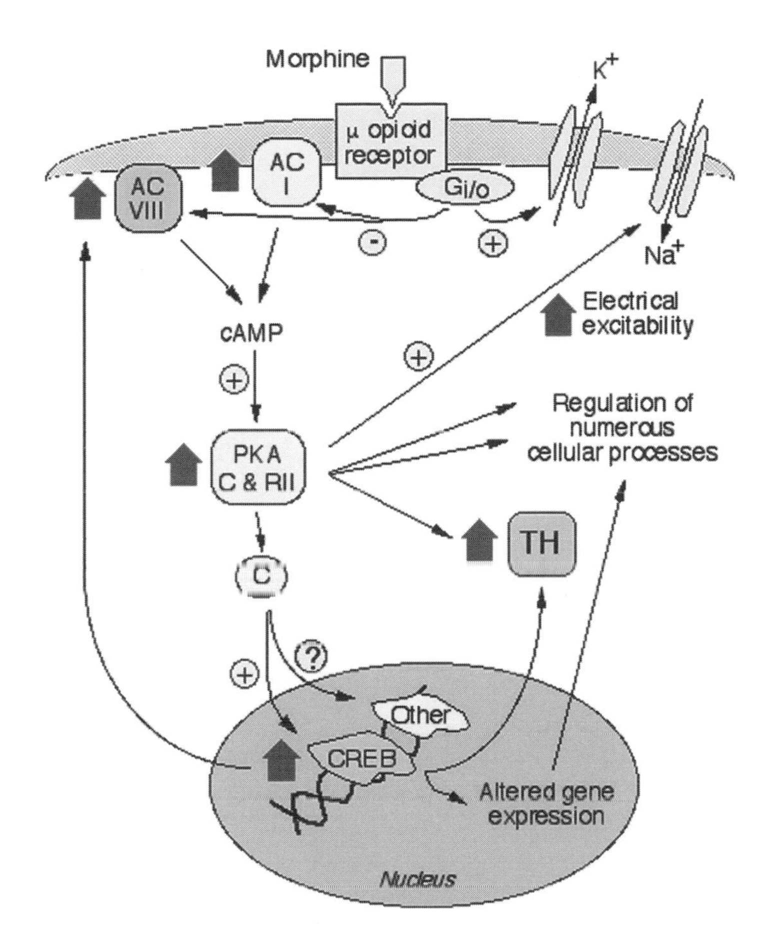

Fig. 1. Scheme illustrating opiate actions in the locus coeruleus. Opiates acutely inhibit LC neurons by increasing the conductance of an inwardly rectifying K^+ channel via coupling with subtypes of $G_{i/o}$, and by decreasing a Na^+-dependent inward current via coupling with $G_{i/o}$ and the consequent inhibition of adenylyl cyclase. Reduced levels of cAMP decrease protein kinase A (PKA) activity and the phosphorylation of the responsible channel or pump. Inhibition of the cAMP pathway also decreases phosphorylation of numerous other proteins and thereby affects many additional processes in the neuron. For example, it reduces the phosphorylation state of CREB, which may initiate some of the longer-term changes in locus coeruleus function. Upward bold arrows summarize effects of chronic morphine in the locus coeruleus. Chronic morphine increases levels of types I (ACI) and VIII (ACVIII) adenylyl cyclase, PKA catalytic (C) and regulatory type II (RII) subunits, and several phosphoproteins, including CREB and tyrosine hydroxylase (TH), the rate-limiting enzyme in norepinephrine biosynthesis. These changes contribute to the altered phenotype of the drug-addicted state. For example, the intrinsic excitability of LC neurons is increased via enhanced activity of the cAMP pathway and Na^+-dependent inward current, which contributes to the tolerance, dependence, and withdrawal exhibited by these neurons. Upregulation of ACVIII and TH is mediated via CREB, whereas upregulation of ACI and of the PKA subunits appears to occur via a CREB-independent mechanism not yet identified *(5)*.

In contrast, chronic opiate exposure increases the intrinsic excitability of LC neurons via upregulation of the cAMP pathway *(5)*. In particular, chronic opiate exposure increases levels of expression of type I and type VIII adenylyl cyclase and of the catalytic and type II regulatory subunits of PKA *(7)* This upregulation has been shown to increase the excitability of LC neurons partly via activation of the nonspecific cation channel. Upregulation of the cAMP pathway can be viewed as a homeostatic response of LC neurons to opiate inhibition and, therefore, as a mechanism of opiate tolerance and dependence in these neurons. The continued presence of morphine and the up-regulated cAMP pathway results in LC neurons firing at near-normal levels. When opiates are removed, the up-regulated cAMP pathway is unopposed and leads to overactivation of these neurons, as seen during withdrawal. Interestingly, the time course for the recovery of the cAMP pathway to normal, over a course of hours to days, parallels that of the diminution of the withdrawal syndrome.

2. Role for CREB in the Locus Coeruleus

Many long-lasting, activity-dependent changes in neuronal function are mediated by changes in gene transcription. This occurs via activity-dependent regulation of transcription factors, proteins that recognize and bind to specific DNA sites within the genes they control. These transcription factors then influence gene expression and contribute to the long-lasting phenotypic changes in neurons. One of the best-characterized transcription factors in the nervous system is CREB (cAMP response element-binding protein). Considerable work has shown that CREB is an important mediator of the upregulation of the cAMP pathway seen in the LC during chronic opiate administration.

CREB, and its family members, ATF1 (activating transcription factor 1) and CREM (cAMP response-element modulator), are expressed ubiquitously and are highly regulated by extracellular stimuli. CREB binds as a dimer to CRE (cAMP response element) consensus sites within the regulatory region of many genes. CREB is activated by its phosphorylation on a specific serine residue, Ser 133. Upon phosphorylation, CREB interacts with the adaptor proteins, CBP (CREB-binding protein) and p300, to stimulate (and in some cases inhibit) gene transcription. CREB can be phosphorylated on Ser 133 by many different protein kinases, including PKA (protein kinase A) *(8)*, Ca^{2+}-calmodulin kinase II and IV *(9–11)*, Rsk2 (pp90 ribosomal S6 kinase) *(12,13)*, MAPKAP kinase 2 (MAPK-activated protein kinase 2) *(14)*, Akt/PKB (protein kinase B) *(15)*, p70 S6K (p70/S6 kinase) *(16)*, and MSK1 (mitogen and stress-activated protein kinase 1) *(17)*. These findings suggest that diverse types of extracellular stimuli and intracellular pathways converge at the level of CREB (and related proteins) to regulate the expression of CRE-containing genes *(18,19)*.

Addiction researchers focused attention on CREB based on the upregulation of the cAMP pathway that occurs in response to drugs of abuse. Thus, according to the scheme shown in Fig. 1, the up-regulated cAMP pathway in the LC would be expected to cause increased activation of CREB and induction of several CRE-containing genes. Indeed, CREB is highly regulated by opiates in this brain region. Acute morphine has no effect on CREB immunoreactivity, but, it decreases the level of CREB phosphorylation (activation) in the LC *(4)*. After chronic opiate administration, levels of phosphoCREB return to control levels, while there is an increase in CREB expression. Levels of phosphoCREB increase dramatically upon the onset of opiate withdrawal. These data can be explained in the following way. Opiate inhibition of the cAMP pathway leads to decreased PKA activity and lower levels of phosphoCREB, which feeds back to increase transcription of CREB via a CRE site within the CREB gene. As the level of CREB rises it eventually is phosphorylated either by the now normal levels of PKA (which is itself induced by chronic opiates independently of CREB; see below) or by other protein kinases. Higher levels of phosphoCREB then increase expression of many genes that contain CRE sites, such as adenylyl cyclase type VIII and tyrosine hydroxylase *(7,20)*.

Further evidence for a role of CREB in physical opiate dependence comes from mice with a targeted mutation in the CREB gene, which show a reduction in opiate physical withdrawal *(21)*. However, this mouse has several shortcomings for investigating detailed mechanisms of opiate dependence. First, the CREB mutation occurs throughout the mouse, not just in particular brain regions involved in opiate dependence such as the LC. Second, CREB expression is not completely lost in this mouse, since the mutation results in increased expression of an alternative isoform of CREB. Third, since the mutation occurs at the earliest stages of development, any phenotype seen could reflect abnormalities during development. Future generations of genetically engineered mice that allow inducible CREB mutations targeted to specific brain regions of adult animals are necessary to define the role of CREB in opiate dependence more clearly *(22)*.

Although opiate regulation of CREB mediates many aspects of the up-regulated cAMP pathway in the LC, other mechanisms also exist. Two examples are the upregulation of adenylyl cyclase type I and PKA subunits, genes that do not contain CRE sites. The ability of chronic morphine to increase these proteins is not affected by blockade of CREB in the LC *(7)*. Although the mechanism underlying induction of adenylyl cyclase type I is unknown, upregulation of PKA subunits appears to occur via a posttranscriptional mechanism. Sustained PKA activation in cultured cells leads to decreased levels of the catalytic and type II regulatory subunits, without affecting their mRNA levels *(23)*. The ubiquitin-dependent protein degradation pathway apparently mediates the proteolysis of the PKA subunits after their activation by cAMP. Our hypothesis is that opiates, by inhibiting cAMP formation and PKA activation, lead to an accumulation of PKA subunits due to reduced degradation *(4)*. Work is now needed to study this scheme within the LC in vivo.

3. Role of the cAMP Pathway and CREB in the Nucleus Accumbens

As stated earlier, the mesolimbic dopamine system is a major neural substrate for the motivational and rewarding effects of opiates. This occurs via two mechanisms *(3)*. Opiates increase dopaminergic transmission to the NAc by activating VTA dopamine neurons. This occurs indirectly through opiate inhibition of GABAergic interneurons within the VTA that inhibit the dopamine neurons. Opiates also act directly on opioid receptors expressed by NAc neurons. The rewarding effects of other drugs of abuse are mediated via similar actions in the VTA–NAc pathway, although each drug produces these effects via drug-specific mechanisms *(1,24)*. In addition, the mesolimbic dopamine system appears to be play a similar role in mediating the actions of natural reinforcers, such as food, drink, sex, and social interactions.

It is interesting, therefore, that chronic administration of opiates or several other drugs of abuse up-regulates the cAMP pathway in the NAc *(4)*. The upregulation, similar to that observed in the LC, involves increased levels of adenylyl cyclase and PKA, although the specific isoforms of adenylyl cyclase that are induced have not yet been identified. In addition, the drugs decrease levels of $G_{\alpha i/o}$, which would further augment

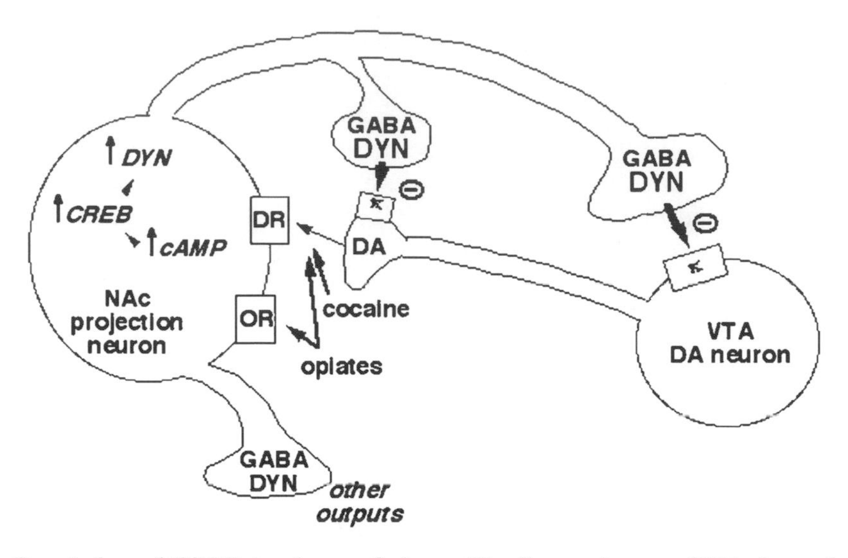

Fig. 2. Regulation of CREB by drugs of abuse. The figure shows a VTA dopamine (DA) neuron innervating a class of NAc GABAergic projection neuron that expressed dynorphin (dyn). Dynorphin serves a negative feedback mechanism in this circuit: dynorphin, released from terminals of the NAc neurons, acts on κ-opioid receptors located on nerve terminals and cell bodies of the DA neurons to inhibit their functioning. Chronic exposure to cocaine or opiates up-regulates the activity of this negative feedback loop via upregulation of the cAMP pathway, activation of CREB, and induction of dynorphin *(4)*

activity of the cAMP pathway. As would be expected, chronic opiate or cocaine administration also causes a sustained increase in the functional activity of CREB in the NAc *(25,26)*.

Insight into the functional consequences of the up-regulated cAMP pathway in the NAc has come from several studies. Infusion of inhibitors of the cAMP pathway (e.g., PKA inhibitors) directly into the NAc makes animals less responsive to the rewarding effects of cocaine and, possibly, opiates *(27,28)*. Conversely, infusion of activators of the cAMP pathway (e.g., adenylyl cyclase or PKA activators) have the opposite effects. Consistent with these findings are observations obtained from direct manipulation of CREB within the NAc by use of viral-mediated gene transfer. Overexpressing CREB specifically within the NAc decreases the rewarding effects of cocaine and opiates, whereas expressing a dominant negative inhibitor of CREB (termed mCREB) increases the drugs' rewarding effects *(29)*.

This action of CREB appears to be mediated in part via regulation of dynorphin gene transcription (Fig. 2). Dynorphin is an opioid peptide expressed in a subset of NAc neurons. Dynorphin acts on κ-opioid receptors to inhibit dopamine neuron cell bodies in the VTA and their terminals in the NAc, and thereby inhibits drug reward *(30)*. The dynorphin gene has been shown to be a target for CREB both in vitro and within the NAc in vivo *(25,29)*. Moreover, the effect of increased CREB activity in the NAc on drug reward can be completely blocked by a κ-receptor antagonist *(29)*. Thus, it appears that CREB negatively regulates drug reward by increasing the gain of the dynorphin feedback loop.

Recent work has provided a more complete understanding of the behavioral phenotype mediated by CREB activity in the NAc. Not only does CREB activity decrease sensitivity to drug reward, it also renders an animal less sensitive to a range of external stimuli, including both natural rewards (e.g., sucrose preference) and aversive stimuli (e.g., anxiogenic, aversive, and nociceptive stimuli) *(31,32)*. Based on these data, our hypothesis is that upregulation of the cAMP pathway induced in the NAc by chronic drug exposure numbs an animal's emotional responsiveness. This state presumably contributes to part of the negative emotional state that occurs during drug withdrawal and helps drive relapse to drug use.

4. Conclusions

Great excitement surrounds the progress that is being made in understanding the alterations in specific neurons that occur following chronic opiate administration, and relating these adaptations to complex behavior. This review focused on one alteration, namely, upregulation of the cAMP pathway and CREB and the very different roles it subserves in two different brain regions. Upregulation of the cAMP pathway is, however, just one of a large number of changes at the molecular and cellular levels that have been documented after chronic drug administration. A current challenge is to understand how these many changes interact with one another to produce the complex abnormalities in behavior that characterize drug addiction.

References

1. Koob, G. F., Sanna P. P., and Bloom F. E. (1998) Neuroscience of addiction. *Neuron* **21,** 467–476.
2. Jolas, T., Nestler, E. J., and Aghajanian, G. K. (2000) Chronic morphine increases GABA tone on serotonergic neurons of the dorsal raphe nucleus: association with an upregulation of the cyclic AMP pathway. *Neuroscience* **95,** 433–443.
3. Williams, J. T., Christie, M. J., and Manzoni, O. (2001) Cellular and synaptic adaptations mediating opioid dependence. *Physiol. Rev.* **81,** 299–343.
4. Nestler, E. J. (2001) Molecular basis of neural plasticity underlying addiction. *Nature Rev. Neurosci.* 2,119–128.
5. Nestler, E. J. and Aghajanian, G. K. (1997) Molecular and cellular basis of addiction. *Science* **278,** 58–63.
6. Alreja, M. and Aghajanian, G. K. (1994) QX-314 blocks the potassium but not the sodium-dependent component of the opiate response in locus coeruleus neurons. *Brain Res.* **639,** 320–324.
7. Lane-Ladd, S. B., Pineda, J., Boundy, V., Pfeuffer, T., Krupinski, J., Aghajanian, G. K., and Nestler, E. J. (1997) CREB in the locus coeruleus: biochemical, physiological, and behavioral evidence for a role in opiate dependence. *J. Neurosci.* **17,** 7890–7901.
8. Gonzalez, G. A. and Montminy, M. R. (1989) Cyclic AMP stimulates somatostatin gene transcription by phosphorylation of CREB at serine 133. *Cell* **59,** 675–680.
9. Sheng, M., Thompson, M. A., Greenberg, M. E. (1991) CREB: a Ca^{2+}-regulated transcription factor phosphorylated by calmodulin-dependent kinases. *Science* **252,** 1427–1430.
10. Dash, P. K., Karl, K. A., Colicos, M. A., Prywes, R., and Kandel, E. R. (1991) cAMP response element-binding protein is activated by Ca^{2+}/calmodulin- as well as cAMP-dependent protein kinase. *Proc. Natl. Acad. Sci. USA* **88,** 5061–5065.
11. Sun, P., Enslen, H., Myung, P. S., Maurer, R. A. (1994) Differential activation of CREB by Ca^{2+}/calmodulin-dependent protein kinases type II and type IV involves phosphorylation of a site that negatively regulates activity. *Genes Dev.* **8,** 2527–2539.

12. Xing, J., Ginty, D. D., and Greenberg, M. E. (1996) Coupling of the RAS-MAPK pathway to gene activation by RSK2, a growth factor-regulated CREB kinase. *Science* **273,** 959–963.
13. Impey, S., Obrietan, K., Wong, S. T., Poser, S., Yano, S., Wayman, G., Deloulme, J. C., Chan, G., Storm, D. R. (1998) Cross talk between ERK and PKA is required for Ca^{2+} stimulation of CREB-dependent transcription and ERK nuclear translocation. *Neuron* **21,** 869–883.
14. Tan, Y., Rouse, J., Zhang, A., Cariati, S., Cohen, P., and Comb, M. J. (1996) FGF and stress regulate CREB and ATF-1 via a pathway involving p38 MAP kinase and MAPKAP kinase-2. *EMBO J.* **15,** 4629–4642.
15. Du, K. and Montminy, M. (1998) CREB is a regulatory target for the protein kinase Akt/PKB. *J. Biol. Chem.* **273,** 32,377–32,379.
16. de Groot, R. P., Ballou, L. M., and Sassone-Corsi, P. (1994) Positive regulation of the cAMP-responsive activator CREM by the p70 S6 kinase: an alternative route to mitogen-induced gene expression. *Cell* **79,** 81–91.
17. Deak, M., Clifton, A. D., Lucocq, L. M., and Alessi, D. R. (1998) Mitogen- and stress-activated protein kinase-1 (MSK1) is directly activated by MAPK and SAPK2/p38, and may mediate activation of CREB. *EMBO J.* **17,** 4426–4441.
18. Shaywitz, A. J. and Greenberg, M. E. (1999) CREB: a stimulus-induced transcription factor activated by a diverse array of extracellular signals. *Annu. Rev. Biochem.* **68,** 821–861.
19. De Cesare, D., and Sassone-Corsi, P. (2000) Transcriptional regulation by cyclic AMP-responsive factors. *Prog. Nucleic Acid. Res. Mol. Biol.* **64,** 343–369.
20. Boundy, V. A., Gold, S. J., Messer, C. J., Chen, J., Son, J. H., Joh, T. H., Nestler, E. J. (1998) Regulation of tyrosine hydroxylase promoter activity by chronic morphine in TH9.0-LacZ transgenic mice. *J. Neurosci.* **18,** 9989–9995.
21. Maldonado, R., Blendy, J. A., Tzavara, E., Gass, P., Roques, B. P., Hanoune, J., and Schutz, G. (1996) Reduction of morphine abstinence in mice with a mutation in the gene encoding CREB. *Science* **273,** 657–659.
22. Chen, J. S., Kelz, M. B., Zeng, G. Q., Sakai, N., Steffen, C., Shockett, P. E., Picciotto, M., Duman, R. S., Nestler, E. J. (1998) Transgenic animals with inducible, targeted gene expression in brain. *Mol. Pharmacol.* **54,** 595–503.
23. Boundy, V.A., Chen, J. S., and Nestler, E. J. (1998) Regulation of cAMP-dependent protein kinase subunit expression in CATH.a and SH-SY5Y cells. *J. Pharmacol. Exp. Ther.* **286,** 1058–1065.
24. Wise, R. A. (1998) Drug-activation of brain reward pathways. *Drug Alcohol Dependence* **51,** 13–22.
25. Hyman, S. (1996) Addiction to cocaine and amphetamine. *Neuron* **16,** 901–904.
26. Shaw-Lutchman, T. Z., Barrot, M., Wallace, T., Gilden, L., Zachariou, V., Impey, S., Duman, R. S., Storm, D., and Nestler, E. J. (2002) Regional and cellular mapping of CRE-mediated transcription during naltrexone-precipitated morphine withdrawal. *J. Neurosci.* **22,** 3663–3672.
27. Self, D. W. and Nestler, E. J. (1995) Molecular mechanisms of drug reinforcement and addiction. *Annu. Rev. Neurosci.* **18,** 463–495.
28. Self, D. W., Genova, L. M., Hope, B. T., Barnhart, W. J., Spencer, J. J., and Nestler, E. J. (1998) Involvement of cAMP-dependent protein kinase in the nucleus accumbens in cocaine self-administration and relapse of cocaine-seeking behavior. *J. Neurosci.* **18,** 1848–1859.
29. Carlezon, W. A., Jr., Thome, J., Olson, V. G., Lane-Ladd, S. B., Brodkin, E. S., Hiroi, N., Duman, R. S., Neve, R. L., and Nestler, E. J. (1998) Regulation of cocaine reward by CREB. *Science* **282,** 2272–2275.
30. Shippenberg, T. S. and Rea, W. (1997) Sensitization to the behavioral effects of cocaine: modulation by dynorphin and kappa-opioid receptor agonists. *Pharmacol. Biochem. Behav.* **57,** 449–455.
31. Pliakas A. M., Carlson R. R., Neve R. L., Konradi C., Nestler E .J., Carlezon W. A. Jr. (2001) Altered responsiveness to cocaine and increased immobility in the forced swim test associated with elevated CREB expression in the nucleus accumbens. *J. Neurosci.* **21,** 7397–7403.
32. Barrot, M., Olivier, J. D. A., Perrotti, L. I., Impey, S., Storm, D. R., Neve, R. L., Zachariou, V., and Nestler, E. J (2001) CREB activity in the nucleus accumbens shell gates behavioral responses to emotional stimuli. *Soc. Neurosci.* **Abstr. 420.6.**

4

Different Intracellular Signaling Systems Involved in Opioid Tolerance/Dependence

Thomas Koch, Stefan Schulz, and Volker Höllt

Although opioids are highly effective for the treatment of pain, they are also known to be intensively addictive. After chronic opioid intake, the drug becomes less effective, so that higher doses are needed to produce the same effect as before—a phenomenon that is called tolerance. At the same time, a situation develops in which the interruption of taking the drugs results in withdrawal sickness, unmasking a state called dependence *(1)*. Both tolerance and dependence result from biochemical changes in the brain.

Pharmakokinetic effects, such as altered liver metabolism contribute very little, if any, to the tolerance in response to chronic opioid treatment. Thus, tolerance and dependence occur predominantly at the cellular level. Critical for opioid tolerance and dependence, therefore, is the regulation of multiple second messenger systems associated with receptor–effector coupling, receptor trafficking, and nuclear signaling. This review centers on the adaptive changes in intracellular signaling systems in response to chronic opioid treatment.

1. Adaptations Involved in the Development of Opioid Tolerance

Opioid receptors (μ, δ, κ) couple via heterotrimeric G proteins to a variety of downstream effectors including adenylate cyclase *(2)*, phospholipase C *(3–5)*, and mitogen-activated protein kinase *(6–9)*. The opioid receptor coupling with these multiple signaling systems occurs via GTP-binding proteins G_i/G_o and involves inhibition of intracellular cAMP production, mobilization of intracellular Ca^{2+} through phosphoinositide metabolism, activation of K^+ channels *(10,11)*, and inhibition of voltage-dependent Ca^{2+} channels *(12)*.

Early adaptive processes such as acute receptor desensitization and internalization, as well as long-term adaptations such as receptor downregulation, or counterregulatory processes such as adenylate cyclase superactivation, have been suggested to be crucial to the development of opioid tolerance.

1.1. Agonist-Induced Opioid Receptor Desensitization

Like other G-protein-coupled receptors (GPCRs), the activities of opioid receptors are attenuated after chronic agonist treatment. This attenuation of opioid receptor activities is associated with receptor desensitization and downregulation. A primary

From: *Molecular Biology of Drug Addiction*
Edited by: R. Maldonado © Humana Press Inc., Totowa, NJ

mechanism of desensitization involves agonist-induced opioid receptor phosphorylation resulting in the uncoupling of the receptor from the G-proteins.

1.1.1. Agonist-Induced Opioid Receptor Phosphorylation

Agonist-induced opioid receptor phosphorylation was demonstrated first by Pei et al. *(13)* for the δ-opioid receptor, by Arden et al. *(14)* with the μ-opioid receptor, and by Appleyard et al. *(15)* with the κ-opioid receptor.

Further studies suggested that the agonist-induced phosphorylation is mediated by G-protein-coupled receptor kinases (GRKs) *(16)* or second messenger-regulated protein kinases, such as Ca^{2+}/calmodulin-dependent kinase II *(17,18)*, but not by protein kinase A *(10)* and protein kinase C *(19)*.

1.1.1.1. G-Protein-Coupled Receptor Kinase-Mediated Phosphorylation

G-protein-coupled receptor kinases (GRKs) are a family of serine/threonine protein kinases that specifically recognize agonist-activated G-protein-coupled receptor proteins as substrates. Phosphorylation of an activated receptor by a GRK terminates signaling by initiating the binding of arrestin and consequently by uncoupling of the receptor from heterotrimeric G proteins *(20,21)*. Six distinct mammalian GRKs are known, which differ in tissue distribution and regulatory properties *(22)*. Several reports indicate the important roles of GRK2, GRK3, GRK5, and GRK6 in the short- and long-term adaptive changes in opioid receptor activity *(23–25)*.

Specific GRK2 phosphorylation sites involved in the agonist-induced receptor desensitization could be identified for the μ- and δ-opioid receptors. A cluster of serine/threonine residues (T354/S355/S356/T357) in the C terminus of the μ-opioid receptor has been shown to play an important role in GRK2-mediated receptor desensitization *(26)*. Another important phosphorylation site (T394) affecting the agonist-induced desensitization of the μ-opioid receptor *(27)* is suggested to be a GRK2 substrate *(28,29)*. It was further demonstrated that agonist-induced δ-opioid receptor phosphorylation occurs exclusively at two phosphate acceptor sites (T358 and S363) of GRK2 at the receptor carboxyl terminus *(30)*. Threonine-180 in the second intracellular loop was shown to play an important role in the GRK3-mediated desensitization of the μ-opioid receptor, whereas serine-369 appears to be necessary for the GRK3 mediated desensitization of the κ-opioid receptor *(31)*. The role of the phosphorylation of serine 369 for the desensitization of the κ-opioid receptor was also shown in mouse brain in vivo using a specific antibody which recognizes the phosphorylated kappa receptor only. The phosphorylation signal obtained with this antibody increase by 110% in the brain of mice made behaviorally tolerant with the κ agonist U50,488 for 5 d *(32)*.

Overexpression of GRK5 and β-arrestin 2 (β-Arr2)together with the μ- or κ-opioid receptor in *Xenopus* oocytes led to an increase in the rate of receptor desensitization *(25,31)*, but up to now no specific GRK5 phosphorylation sites for the various opioid receptor types could be identified. Furthermore, chronic treatment of rats with sufentanil induced analgesic tolerance associated with an upregulation of GRK6 and β-Arr2 in the rat brain indicating a physiological role for GRK6 in the development of opioid tolerance *(23)*.

1.1.1.2. CaM Kinase II-Mediated Phosphorylation

The modulation of μ-opioid receptor desensitization by culmodulin-dependent protein kinase II (CaM kinase II) was first described by Mestek et al. *(18)*. It was further observed that the basal phosphorylation rate of the μ-opioid receptor is Ca^{2+} dependent *(33)*. Direct evidence for the involvement of CaM kinase II in the desensitization of the rat μ-opioid receptor was provided by the identification of serine-266 in the intracellular loop as the critical phosphorylation site for the CaM kinase II-mediated receptor desensitization *(17,34)*. It was further demonstrated that CaM kinase II is collocalized with the μ-opioid receptor in a specific region of the rat brain and thus in a position to phosphorylate the μ-opioid receptor and to contribute to the development of tolerance to opioid analgesics *(35)*. In vivo studies revealed that morphine treatment significantly increased activities of CaM kinase II in the hippocampus of rats (36) and that administration of CaM kinase II antisense oligonucleotides into the hippocampus, which decreases the expression of CaM kinase II specifically, also attenuated morphine tolerance and dependence *(37)*. Taken together, these results suggest that CaM kinase II is critically involved in the development of opioid tolerance and that inhibition of this kinase may be of therapeutic benefit in the treatment of opioid tolerance and dependence.

The mechanisms by which opioid receptors regulate CaM kinase II activity are not yet clear. It has been shown that stimulation of opioid receptors can elevate phosphatidyl hydrolysis with subsequent release from Ca2+ from intracellular stores *(38)*. The increased free Ca^{2+} concentrations, in turn, may result in the activation of CaM kinase II.

1.1.1.3. Protein Kinase A (PKA)-Mediated Phosphorylation

Based on the majority of studies performed in mice or in *Xenopus* oocyte systems it can be supposed that PKA plays no important role in the process of direct opioid receptor phosphorylation and desensitization *(10,39–41)*. However, another study performed in hypothalamic slices of guinea pig revealed PKA dependency in the development of tolerance to μ-opioid agonists *(42)*. Although the involvement of the cAMP-PKA signal transduction pathway in the opioid receptor desensitization is not completely clear, its implication in the development of opioid dependence is well documented (*see* Section 1.4.1.).

1.1.1.4. Protein Kinase C (PKC)-Mediated Phosphorylation

There are many reports characterizing the role of PKC in the desensitization of opioid receptors. Desensitization of μ- and δ-opioid receptors by PKC has been observed in the *Xenopus* oocyte expression system *(10,18)* and in in vivo experiments in mice *(40,43,44)*. However, cell culture studies with the μ-opioid receptor *(19)* and with the δ-opioid receptor *(13)* suggested that agonist-induced opioid receptor phosphorylation is not mediated by PKC. It can be speculated that PKC activation after opioid receptor-mediated PLC stimulation might not be an obligatory event for opioid receptor desensitization but may be involved in the phosphorylation of other signaling proteins.

1.1.1.5 Mitogen Activated Protein (MAP) Kinase-Mediated Phosphorylation

Evidence has been provided for an involvement of the MAP kinase pathway in the homologous desensitization of the μ-opioid receptor *(7,45)*. Specific inhibitors of the MAP kinase diminish the agonist-induced desensitization and phosphorylation of the μ-opioid receptor in a dose-dependent manner *(45)*.

A recent report shows that the μ-opioid receptor-mediated activation of the MAP kinase involves a transactivation of the EGF receptor which is calmodulin dependent *(46)*.

The μ-opioid receptor signaling via the MAP kinase cascade is also desensitized upon prolonged agonist exposure in cultured cells. The MAP kinase cascade also appears to undergo neuroadaptation during chronic opioid exposure in vivo. By monitoring the activation state of the MAP kinase using phosphospecific antibodies, neuronal MAP kinase activity in the rat brain was potently repressed after repeated morphine administration *(6)*.

1.1.2. Agonist-Induced Opioid Receptor Internalization

Agonist-induced internalization of G-protein-coupled receptors causes a rapid spatial uncoupling of the endocytosed receptor from the cell surface, providing an additional mechanism for desensitization *(47)*. The binding of β-arrestins to the phosphorylated receptors is a crucial step in the internalization of G-protein-coupled receptors.

The agonist-dependent phosphorylation of G-protein coupled receptors by GRKs increases the affinity of the receptors for cytosolic arrestin proteins, which bind preferentially to agonist-activated phosphorylated receptors and further uncouple the receptors from interacting G proteins *(22)*. From the six members of the arrestin family, only β-arrestin 1 and β-arrestin 2 are well characterized and exhibit ubiquitous patterns of expression *(21)*.

Co-expression of β-arrestin 2 with various GRKs, such as GRK2, GRK3, GRK5, and GRK6, was shown to increase the rate of agonist-induced receptor desensitization *(31)*. Furthermore, in knockout mice lacking β-arrestin 2 (β-Arr2 −/−) no desensitization of the μ-opioid receptor occurs after chronic morphine treatment *(48)*.

For continued G-protein-receptor signaling, desensitized receptors need to reestablish their responsiveness to extracellular signals through a receptor resensitization process. One potential mechanism proposed to play an important role in receptor resensitization is internalization. Desensitized G-protein-coupled receptors are internalized via clathrin-coated pits in early endosomes, where they are thought to be dephosphorylated and recycled back to the cell surface *(21,49)*. Interestingly, β-arrestin proteins play a dual role in regulating G-protein-coupled receptors responsiveness by contributing to both receptor desensitization and clathrin-mediated receptor internalization. Because internalization of G-protein-coupled receptors enhances receptor resensitization and recycling *(50–52)*, β-arrestins may also be critical determinants for G-protein-coupled receptor resensitization *(53)*.

The β-arrestins interact with at least two main classes of signaling proteins *(54)*. First, an interaction with molecules such as clathrin, phosphoinosite-AP-2 adapter protein *(55)*, and N-ethylmaleimide-sensitive fusion protein (NSF), an ATPase essential for many intracellular transport reactions *(56)*, directs the clathrin-mediated internalization of G-protein-coupled receptors. Second, an interaction with molecules with Src, Raf, Erk, apoptosis signal regulating kinase 1 (ASK1) and c-Jun amino-terminal kinase 3 (JNK3) appears to regulate several pathways that result in the activation of MAP kinases *(54,57)*. These recent discoveries indicate that the β-arrestins play widespread roles as scaffolds and/or adapter molecules that organize a variety of complex signaling pathways emanating from heptahelical receptors.

1.2. Long-Term Uncoupling of Opioid Receptors from G-Proteins

Using ^{35}S-GTPγS binding, the opioid receptor-coupled G-protein activity was studied after chronic treatment with opioids. ^{35}S-GTPγS binding, which reflects GTPase activity, was found to be decreased in several brain areas of rats chronically (12 d) with morphine or in rats treated self-administering heroine for 39 d *(58)*. However, no change in ^{35}S-GTPγS binding was observed in the brain of rats treated acutely with opioids. Similar effects observed in cultured SH-SY5Y cells were accompanied by a downregulation of μ-opioid receptors determined by a decrease in the maximal binding capacity for the opioid antagonist [^{3}H]-diprenorphine of −66%, but with no change in binding affinity *(59)*.

1.3. Downregulation of Opioid Receptors

Downregulation of opioid receptors in response to chronic opioid treatment is a long-term adaptive process that can contribute to opioid tolerance. Using the mouse neuroblastoma cell line N4TG, Chang et al. reported first a decrease in B_{max} after exposure of the δ-receptor agonist DADLE *(60)*. Also, chronic treatment of rats with the selective μ-agonist PL0 17 was shown to downregulate μ-opioid receptors *(61,62)*. In the SH-SY5Y human neuroblastoma cell line, which contains both μ- and δ-opioid receptors, morphine was able to downregulate both types of opioid receptors *(63,64)*. In another study, using antibodies directed against μ opioid receptors, it was shown that chronic treatment with morphine decreases immunoreactive μ-receptor proteins in mice *(65)*. These findings indicate no differences in the downregulation of μ-opioid receptors under morphine compared to other opioid receptor agonists.

In contrast to these findings, binding experiments in NG108-15 cells suggested that only full agonist can downregulate the opioid receptors. In these cells, etorphine and opioid peptides, but not morphine, which is not a full agonist, were able to downregulate opioid receptors *(66)*. Further, in vitro binding studies in membranes of C6 glial cells expressing the μ-opioid receptor showed that chronic treatment with agonists reduced [3H]DAMGO binding with the rank order etorphine > DAMGO = β-endorphin > morphine > butorphanol *(67)*. These results were supported by in vivo binding studies showing only a small reduction in the total number of binding sites in brain of guinea pigs treated chronically with morphine *(68)*. Further, in vivo downregulation of both μ and δ receptors has been observed in certain brain areas of rat treated chronically with etorphine, whereas morphine was shown to cause an upregulation *(69)*. Mutation analysis revealed that serine-355 and serine-363 were required for this etorphine-induced downregulation of the μ-opioid receptor *(70)*. Together, these data suggest that the rate of morphine-induced downregulation of opioid receptors depends on species, brain regions, and cell types tested.

1.3.1. Downregulation of Opioid Receptors at Transcriptional Level

Downregulation of opioid receptors may result from the degradation of internalized receptors and/or from a decrease of receptor resynthesis. Therefore, receptor regulation during chronic drug exposure has also been investigated at the level of the opioid receptor mRNA. The observed decrease in the mRNA level of the μ- and δ-opioid receptors after chronic agonist treatment is suggested to be due to changes in the intracellular cAMP level *(71,72)*. camp-dependent PKA is known to phosphorylate CRE-binding protein (CREB) and thereby to stimulate the activity of a transcritional activation domain of several promotors. Promoter analysis of the rat μ-opioid receptor revealed that an increase of the intracellular cAMP level by forskolin enhances promoter activity in transfected SH SY5Y cells, whereas DAMGO, by inhibiting cAMP formation, decreases transcription driven by the μ-receptor promoter *(71)*. On the other hand, elevation of cAMP levels by forskolin in NG108-15 cells has been shown to reduce levels of δ-opioid receptor mRNA, an effect that was not explained by a decrease in mRNA stability *(73)*. Conversely, in another study it was reported that in NG108-15 cells forskolin causes a transient decrease in δ-opioid receptor mRNA levels within 5 h, followed by an increase after 48 h *(74)*.

However, to date, neither for the μ- nor for the δ-opioid receptor could CREB elements be detected in the receptor promotor sequences. An alternative mechanism for PKA-induced stimulation of transcription involves transcription factor AP2 or might be indirectly mediated by transcription factor AP1. In fact, reporter gene assays and electrophoretic mobility shift assays indicated that transcriptional regulation of the δ-opioid receptors involves the transcription factor AP2 *(74)*.

1.4. Adaptations in the Signaling Systems Involved in Opioid Tolerance/Dependence

Opioid tolerance after chronic treatment is due not only to a rapid decrease in receptor activity following phosphorylation and G-protein uncoupling, but also to a compensatory mechanism counteracting receptor function. In addition, to maintain normal function, the presence of opioid agonists is required. This dependence on opioids results in withdrawal signs upon their removal.

1.4.1. Activation of Adenylate Cyclase (AC) and Protein Kinase A (PKA) After Chronic Opioid Treatment

Early studies revealed that opioids acutely inhibit AC activity in NG108-15 cells, but in the continued presence of morphine, there was an upregulation of AC activity *(75)*. This phenomenon, by which chronic activation of inhibitory $G_{i/o}$-coupled receptors lead to an increase in cAMP signaling in the cells, has also been termed AC superactivation *(76)*. The authors observed that acute activation of the μ-opioid receptor inhibited the activity of adenylate cyclase isozymes I, V, VI and VIII, but stimulated types II, IV, and VII and did not affect type III activity. Conversely, chronic receptor activation led to a superactivation of AC types I, V, VI, and VIII and to a reduction in the activities of types II, III, IV, and VII. The upregulation was sensitive to pertussis toxin and to agents that scavenged free $G_{\beta\gamma}$ subunits *(77,78)*. These results suggest that isozyme-specific AC superactivation may represent a general means of cellular adaptation to the activation of inhibitory receptors and that the presence/absence and intensity of the AC response in different brain areas (or cell types) could be explained by the expression of different AC isozyme types in these areas.

However, to date, the mechanism by which AC activity is regulated by chronic exposure to inhibitory agonists remains largely unclear. Opioid receptors are predominantly coupled to AC via $G_{\alpha 1}$ subunits of G proteins. Some recent reports show that there is also an enhanced opioid receptor signaling via $G_{\beta\gamma}$ subunits during the development of opioid tolerance *(79,80)*. Chronic morphine exposure also induces a shift in the relative preponderance of opioid receptor $G_{1\alpha}$ inhibition to $G_{\beta\gamma}$ stimulation of AC activity. Furthermore, an increasing body of results suggests that opioid receptors can also couple to G_s proteins after chronic opioid exposure, leading to an increase in AC activity. It has further been shown that a direct phosphorylation of AC isoforms (type II family) can significantly increase their stimulatory responsiveness to $G_{s\alpha}$ and $G_{\beta\gamma}$; this mechanism could underlie, in part, the predominance of opioid AC stimulatory signaling observed in opioid tolerant/dependent tissue *(81)*.

Although the upregulation of AC in cell lines is a clear phenomenon, studies on the effects of chronic morphine treatment on AC activity in brain have produced mixed results *(81–83)*. Chronic (in vivo) administration of morphine pellets for 5 d, treatment known to induce opiate tolerance and dependence, increased basal, GTP- and forskolin-stimulated adenylate cyclase in the locus coeruleus (LC) *(82)*. It has been reported that whereas most regions showed no regulation in response to chronic morphine, nucleus accumbens (NAc) and amygdala did show increases in adenylate cyclase and cyclic AMP-dependent protein kinase activity, and thalamus showed an increase in cyclic AMP-dependent protein kinase activity only *(83)*. Thus, opioid tolerance involves an alteration in the activity of AC and of cyclic AMP-dependent protein kinase *(42,84)*. The activation of cAMP-dependent PKA seems to be important for the development of tolerance, because drugs that inhibit PKA reverse the antinociceptive tolerance to morphine *(85)*.

But what is the mechanism by which PKA enhances opioid tolerance? Chronic administration of opiates has been observed to upregulate the cAMP pathway and to activate PKA in locus coeruleus cells. Stimuli that upregulate the cAMP pathway after chronic administration (e.g., stress or opiates) increase the excitability of locus coeruleus neurons, whereas stimuli that downregulate the cAMP pathway (e.g., antidepressants) exert the opposite effect. In these cells PKA activates a nonspecific cation current that modulates pacemaker activities of these cells *(86)*. PKA has also been shown to modulate the G_s/G_i protein selectivity of the β-adrenergic receptor *(87,88)*. Thus, activation of PKA during chronic opioid exposure might also result in AC stimulation by switching the opioid receptor from G_i to G_s coupling, leading to an increase of the intracellular cAMP level.

1.4.2 Activation of Mitogen-Activated Protein (MAP) Kinase After Chronic Opioid Exposure

The activation state of MAP kinase was further determined in morphine-dependent rats using phospho-specific antibodies. Whereas neuronal MAP kinase activity was potently repressed after repeated morphine administration, after withdrawal by naloxone, MAP kinase was strongly activated in many brain regions including the locus coeruleus, which has been proposed to play a primary causal role in the expression of the withdrawal syndrome *(6)*. A possible explanation would be that increased PKA activity during withdrawal stimulates activation of the MAP kinase pathway, since PKA has been shown to activate Rap1 and Raf directly leading to MAP kinase activation. Alternatively, PKA-mediated phosphorylation of a K^+ channel would result in Ca^{2+} influx, leading to activation of the Ras/MAP kinase cascade. In contrast, opposite results have been very recently reported in μ-opioid receptor-expressing C6 glioma cells in vitro. The precipitated withdrawal by naloxone in these cells chronically pretreated with opioids produces a decrease of phospho-MAP kinase levels to near-undetectable levels. This effect was blocked by PKA inhibitors *(84)*. The reason for the discrepancy is not clear, but it indicates that the effect may be cell type-specific.

1.4.3. Activation of Phospholipase C (PLC) and Protein Kinase C (PKC) After Chronic Opioid Treatment

Phospholipid pathways are also altered in vivo during morphine tolerance. Chronic morphine treatment of rats upregulates phospholipase Cγ1 in the ventral tegmental areas (VTA) of rats *(89)*. Moreover, injection of phosphatidylinositol-specific phospholpase C inhibitors or protein kinase C inhibitors significantly reversed tolerance in mice, indicating a potential role for inositol 1,4,5-trisphosphate (IP3) and PKC in opioid tolerance *(3)*. In addition, inhibitors of PKC have been shown to inhibit μ-opioid receptor internalization and to block acute tolerance to the μ-agonist DAMGO in mice *(90)*. Thus, chronic opioid exposure might lead to changes in the phospholipid metabolism and protein kinase C activity that have a direct role in maintaining the state of tolerance.

1.4.4. Involvement of Nitric Oxide (NO) in the Development of Opioid Dependence/Tolerance

Nitric oxide (NO) has been postulated to contribute significantly to analgesic effects of opiates as well as to the development of tolerance and physical dependence to morphine. Several studies have shown that nitric oxide synthase (NOS) inhibitors can reduce signs of opioid withdrawal and attenuate tolerance to morphine in mice and rats *(91–94)*.

Repeated morphine administration has been shown to increase NOS biosynthesis in the rat spinal cord, which may reflect adaptive changes accounting for development of opiate tolerance and dependence *(95)*. Immunohistochemical studies revealed that morphine dependence produced an increase in the number of neuronal NOS (nNOS)-positive cells in the main and accessory olfactory bulb, olfactory nuclei, cerebellum, locus coeruleus, medulla oblongata (nucleus of the solitary tract and prepositus hypoglossal nucleus), and a decrease in nNOS immunoreactivity in hypothalamus of mice *(96)*. The administration of naloxone to morphine-dependent mice to induce abstinence increased nNOS immunoreactivity in the hypothalamus and locus coeruleus *(96)*. The increased NO synthesis during chronic morphine treatment potentiates morphine analgesia and enhances the development of morphine tolerance in mice *(97)*. The NMDA receptor has been implicated in the development of opioid-induced tolerance and dependence because several NMDA antagonists, such as MK-801, inhibited morphine tolerance and dependence *(98–100)*. NO production has been linked to the NMDA complex. The activation of NMDA receptors has enhanced the entry of extracellular Ca^{2+}, thus stimulating enzymatic production of NO, which in turn increases the formation of cyclic GMP by activating guanylylcyclase *(101)*.

1.4.5. Changes in the Firing Rate of Locus Coeruleus Neurons After Chronic Opioid Exposure

Acute morphine causes an inhibition of spontaneous firing in the locus coeruleus of rats. During continous morphine treatment of the rats the firing rate returned to control levels, indicating tolerance. After naloxone there was an increased firing rate, indicating a cellular sign of withdrawal *(102)*. Later studies showed that the increase in firing was a result from an augmented glutamatergic input to the locus coeruleus *(103)* and not due to compensatory action on either potassium or calcium conductances *(104)*.

In addition, an increased cation conductance has been reported to be responsible for the increased excitability in periaqueductal gray neurons in vitro *(105)*. This cation conductance is proposed to be regulated by the cAMP cascade.

1.5. Stimulatory Signaling in Response to Chronic Opioid Exposure

In addition to the AC signalling, other signaling pathways of the opioid receptor mediated by G_s and/or $G_{\beta\gamma}$ subunits are stimulated after chronic opioid exposure. Excitatory effects mediated by G_s proteins have been observed for most opioids *(106)*. After chronic opioid exposure, neurons become supersensitive to the excitatory effects of opioid agonists *(107)*.

In addition, $G_{\beta\gamma}$ subunits derived from heterotrimeric G proteins, upon activation, can activate certain AC isoforms (type II family). This increase in the stimulatory responsiveness of signaling proteins to G_s and $G_{\beta\gamma}$ subunits might be regulated by phosphorylation of signaling proteins after chronic opioid treatment.

1.5.1. Phosphorylation of Signaling Proteins After Chronic Opioid Exposure

The finding that phosphorylation of adenylate cyclase isoforms (type II family) can significantly increase their stimulatory responsiveness to $G_{s\alpha}$ and $G_{\beta\gamma}$ indicates that such a mechanism could underlie, in part, the predominance of opioid AC stimulatory signaling observed in opioid tolerant/dependent tissue *(79,81)*.

In fact, in addition to agonist-induced receptor phosphorylation, chronic opioid exposure leads also to phosphorylation of kinases and other signaling proteins, such as GRK2/3, β-arrestin, adenylate cyclase isoforms (type II family), and G-protein subunits, modulating their ability to associate *(80,81)*. $G_{\beta\gamma}$ subunits derived from heterotrimeric G proteins, upon activation of GPCRs, play a complex role in signal transduction *(108)*. Potassium channels as well as many signal transduction enzymes, for example, phospholipase A2, phospholipase C_β, AC isoforms, and MAP kinase, are regulated by $G_{\beta\gamma}$ *(109)*. Phosphorylation of GRK2/3 increases its association with $G_{\beta\gamma}$ subunit and its kinase activity *(80)*. Src tyrosine kinase as well as PKC have been shown to phosphorylate GRK2 directly *(110,111)*. Recent studies revealed that activation of PKC selectively phosphorylates $G\alpha_{(i\text{-}1)}$ and $G\alpha_{(i\text{-}2)}$, but not $G\alpha_{(i\text{-}3)}$ or $G\alpha_{(o)}$, and blocked inhibition of AC mediated by somatostatin receptors coupled to $G_{(i\text{-}1)}$ and opioid receptors coupled to $G_{(i\text{-}2)}$, but not by muscarinic $M_{(2)}$ and adenosine $A_{(1)}$ receptors coupled to $G_{(i\text{-}3)}$ *(112)*. Phosphorylation of G_α subunits by PKC also affected responses mediated by $G_{\beta\gamma}$-subunits via a decrease in reassociation and availability of heterotrimeric G proteins.

PKC-mediated phosphorylation of PLCβ3 has been demonstrated to rapidly attenuate opioid-induced phosphoinositide turnover in NG108-15 cells *(113)*. It is likely that this feedback mechanisms limits the involvement of PLCβ in the chronic action of opioids.

1.6. Sensitization During Chronic Opioid Exposure

It is very likely that the described intracellular changes in signal transduction contribute to the phenomenon of sensitization observed as an increased locomotoric and psychomotoric activation in response to repetitive treatment of opioids. This long-lasting sensitization has been proposed to be a primary factor for the drug-induced addiction *(114,115)*. In fact, persistent marked elevation of c-fos expression was observed in the mesolimbic structures in brain of rats treated chronically with morphine *(116,117)*. One component may be the activation of the cAMP cascade through D1 dopamine in the ventral tegmental area, since blockade of D1 dopamine receptors and inhibitors of PKA prevented the induction of sensitization to stimulants. The *c-fos* gene activated by the stimulated cAMP could be responsible for the altered gene expression underlying certain aspects of opioid addiction.

References

1. Goldstein, A. (1994) Addiction, in: *From Biology to Drug Policy*. Freeman, New York.
2. Johnson, P. S., Wang, J. B., Wang, W. F., and Uhl, G. R. (1994) Expressed mu opiate receptor couples to adenylate cyclase and phosphatidyl inositol turnover. *Neuroreport* **5,** 507–509.
3. Smith, F. L., Lohmann, A. B., and Dewey, W. L. (1999) Involvement of phospholipid signal transduction pathways in morphine tolerance in mice. *Br. J. Pharmacol.* **128,** 220–226.
4. Smart, D., Smith, G., and Lambert, D. G. (1995) Mu-opioids activate phospholipase C in SH-SY5Y human neuroblastoma cells via calcium-channel opening. *Biochem. J.* **305,** 577–581.
5. Smart, D., Smith, G., and Lambert, D. G. (1994) Mu-opioid receptor stimulation of inositol (1,4,5)trisphosphate formation via a pertussis toxin-sensitive G protein. *J. Neurochem.* **62,** 1009--1014.
6. Schulz, S. and Hollt , V. (1998) Opioid withdrawal activates MAP kinase in locus coeruleus neurons in morphine-dependent rats in vivo. *Eur. J. Neurosci.* **10,** 1196–1201.
7. Schmidt, H., Schulz, S., Klutzny, M., Koch, T., Handel, M., and Hollt, V. (2000) Involvement of mitogen-activated protein kinase in agonist-induced phosphorylation of the mu-opioid receptor in HEK 293 cells. *J. Neurochem.* **74,** 414–422.
8. Li, L. Y. and Chang, K. J. (1996)The stimulatory effect of opioids on mitogen-activated protein kinase in Chinese hamster ovary cells transfected to express mu-opioid receptors. *Mol. Pharmacol.* **50,** 599–602.
9. Burt, R. P., Chapple, C. R., and Marshall, I. (1996) The role of diacylglycerol and activation of protein kinase C in alpha 1A-adrenoceptor-mediated contraction to noradrenaline of rat isolated epididymal vas deferens. *Br. J. Pharmacol.* **117,** 224–230.
10. Chen, Y. and Yu, L. (1994) Differential regulation by cAMP-dependent protein kinase and protein kinase C of the mu opioid receptor coupling to a G protein-activated K^+ channel. *J. Biol. Chem.* **269,**7839–7842.
11. Ikeda, K., Kobayashi, T., Ichikawa, T., Usui, H., and Kumanishi, T. (1995) Functional couplings of the delta- and the kappa-opioid receptors with the G-protein-activated K^+ channel. *Biochem. Biophys. Res. Commun.* **208,** 302–308.
12. Kaneko, S. (1995) [Analysis of receptor-ion channel functions in *Xenopus* oocyte translation system]. *Nippon Yakurigaku Zasshi* **106,** 243–253.
13. Pei, G., Kieffer, B. L., Lefkowitz, R. J., and Freedman, N. J. (1995) Agonist-dependent phosphorylation of the mouse delta-opioid receptor: involvement of G protein-coupled receptor kinases but not protein kinase C. *Mol. Pharmacol.* **48,** 173–177.
14. Arden, J. R., Segredo, V., Wang, Z., Lameh, J., and Sadee, W. (1995) Phosphorylation and agonist-specific intracellular trafficking of an epitope-tagged mu-opioid receptor expressed in HEK 293 cells. *J. Neurochem.* **65,** 1636–1645.
15. Appleyard, S. M., Patterson, T. A., Jin, W., and Chavkin, C. (1997) Agonist-induced phosphorylation of the kappa-opioid receptor. *J. Neurochem.* **69,** 2405–2412.
16. Kovoor, A., Nappey, V., Kieffer, B. L., and Chavkin. C. (1997) Mu and delta opioid receptors are differentially desensitized by the coexpression of beta-adrenergic receptor kinase 2 and beta-arrestin 2 in xenopus oocytes. *J. Biol. Chem.* **272,** 27605–27611.
17. Koch, T., Kroslak, T., Mayer, P., Raulf, E., and Hollt, V. (1997) Site mutation in the rat mu-opioid receptor demonstrates the involvement of calcium/calmodulin-dependent protein kinase II in agonist-mediated desensitization. *J. Neurochem.* **69,** 1767–1770.
18. Mestek, A., Hurley, J. H., Bye, L. S., Campbell, A. D., Chen, Y., Tian, M., Liu, J., Schulman, H., and Yu, L. (1995) The human mu opioid receptor: modulation of functional desensitization by calcium/calmodulin-dependent protein kinase and protein kinase C. *J. Neurosci.* **15,** 2396–2406.
19. Zhang, L., Yu, Y., Mackin, S., Weight, F. F., Uhl, G. R., and Wang, J. B. (1996) Differential mu opiate receptor phosphorylation and desensitization induced by agonists and phorbol esters. *J. Biol. Chem.* **271,** 11,449–11,454.

20. Lohse, M. J., Krasel, C., Winstel, R., and Mayor, F., Jr. (1996) G-protein-coupled receptor kinases. *Kidney Int.* **49,** 1047–1052.
21. Ferguson, S. S., Barak, L. S., Zhang, J., and Caron, M. G. (1996) G-protein-coupled receptor regulation: role of G-protein-coupled receptor kinases and arrestins. *Can. J. Physiol. Pharmacol.* **74,** 1095–1110.
22. Premont, R. T., Inglese, J., and Lefkowitz, R. J. (1995) Protein kinases that phosphorylate activated G protein-coupled receptors. *FASEB J.* **9,**175–182.
23. Hurle, M. A. (2001) Changes in the expression of G protein-coupled receptor kinases and beta-arrestin 2 in rat brain during opioid tolerance and supersensitivity. *J. Neurochem.* **77,**486–492.
24. Li, A. H. and Wang, H. L. (2001) G protein-coupled receptor kinase 2 mediates mu-opioid receptor desensitization in GABAergic neurons of the nucleus raphe magnus. *J. Neurochem.* **77,** 435–444.
25. Kovoor, A., Celver, J. P., Wu, A., and Chavkin, C. (1998) Agonist induced homologous desensitization of mu-opioid receptors mediated by G protein-coupled receptor kinases is dependent on agonist efficacy. *Mol. Pharmacol.* **54,** 704–711.
26. Wang, H. L. (2000) A cluster of Ser/Thr residues at the C-terminus of mu-opioid receptor is required for G protein-coupled receptor kinase 2-mediated desensitization. *Neuropharmacology* **39,** 353–363.
27. Wolf, R., Koch, T., Schulz, S., Klutzny, M., Schroder, H., Raulf, E., Buhling, F., and Hollt, V. (1999) Replacement of threonine 394 by alanine facilitates internalization and resensitization of the rat mu opioid receptor. *Mol. Pharmacol.* **55,** 263–268.
28. Deng, H. B., Yu, Y., Pak, Y., O'Dowd, B. F., George, S. R., Surratt, C. K., Uhl, G. R., and Wang, J. B. (2000) Role for the C-terminus in agonist-induced mu opioid receptor phosphorylation and desensitization. *Biochemistry* **39,** 5492–5499.
29. Pak, Y., O'Dowd, B. F., and George, S. R. (1997) Agonist-induced desensitization of the mu opioid receptor is determined by threonine 394 preceded by acidic amino acids in the COOH-terminal tail. *J. Biol. Chem.* **272,** 24961–24965.
30. Guo, J., Wu, Y., Zhang, W., Zhao, J., Devi, L. A., Pei, G., and Ma, L. (2000) Identification of G protein-coupled receptor kinase 2 phosphorylation sites responsible for agonist-stimulated delta-opioid receptor phosphorylation. *Mol. Pharmacol.* **58,** 1050–1056.
31. Appleyard, S. M., Celver, J., Pineda, V., Kovoor, A., Wayman, G. A., and Chavkin, C. (1999) Agonist-dependent desensitization of the kappa opioid receptor by G protein receptor kinase and beta-arrestin. *J. Biol. Chem.* **274,** 23802–23807.
32. McLaughlin, J. P. (2001) Phosphospecific antibody recognizes the desensitized form of the kappa opioid receptor (KOR). 32nd International Narcotics Research Conference, Helsinki, Finland.
33. Wang, Z., Arden, J., and Sadee, W. (1996) Basal phosphorylation of mu opioid receptor is agonist modulated and Ca^{2+}-dependent. *FEBS Lett.* **387,** 53–57.
34. Koch, T., Kroslak, T., Averbeck, M., Mayer, P., Schroder, H., Raulf, E., and Hollt, V. (2000) Allelic variation S268P of the human mu-opioid receptor affects both desensitization and G protein coupling. *Mol. Pharmacol.* **58,** 328–334.
35. Bruggemann, I., Schulz, S., Wiborny, D., and Hollt, V. (2000) Colocalization of the mu-opioid receptor and calcium/calmodulin- dependent kinase II in distinct pain-processing brain regions. *Brain Res. Mol. Brain Res.* **85,** 239–250.
36. Lou, L., Zhou, T., Wang, P., and Pei., G. (1999) Modulation of Ca2+/calmodulin-dependent protein kinase II activity by acute and chronic morphine administration in rat hippocampus: differential regulation of alpha and beta isoforms. *Mol. Pharmacol.* **55,** 557–563.
37. Fan, G. H., Wang, L. Z., Qiu, H. C., Ma, L., and Pei, G. (1999) Inhibition of calcium/calmodulin-dependent protein kinase II in rat hippocampus attenuates morphine tolerance and dependence. *Mol. Pharmacol.* **56,** 39–45.
38. Zimprich, A., Simon, T., and Hollt, V. (1995) Transfected rat mu opioid receptors (rMOR1 and rMOR1B) stimulate phospholipase C and Ca^{2+} mobilization. *Neuroreport* **7,** 54–56.

39. Narita, M., Mizoguchi, H., and Tseng, L. F. (1995) Inhibition of protein kinase C, but not of protein kinase A, blocks the development of acute antinociceptive tolerance to an intrathecally administered mu-opioid receptor agonist in the mouse. *Eur. J. Pharmacol.* **280,** R1–R3.
40. Narita, M., Mizoguchi, H., Kampine, J. P., and Tseng, L. F. (1996) Role of protein kinase C in desensitization of spinal delta-opioid- mediated antinociception in the mouse. *Br. J. Pharmacol.* **118,** 1829–1835.
41. Wang, Z. and Sadee, W. (2000) Tolerance to morphine at the mu-opioid receptor differentially induced by cAMP-dependent protein kinase activation and morphine. *Eur. J. Pharmacol.* **389,**165–171.
42. Wagner, E. J., Ronnekleiv, O. K., and Kelly, M. J. (1998) Protein kinase A maintains cellular tolerance to mu opioid receptor agonists in hypothalamic neurosecretory cells with chronic morphine treatment: convergence on a common pathway with estrogen in modulating mu opioid receptor/effector coupling. *J. Pharmacol. Exp. Ther.* **285,** 1266–1273.
43. Narita, M., Ohsawa, M., Mizoguchi, H., Kamei, J., and Tseng. L. F. (1997) Pretreatment with protein kinase C activator phorbol 12,13-dibutyrate attenuates the antinociception induced by mu- but not epsilon-opioid receptor agonist in the mouse. *Neuroscience* **76,**291–298.
44. Narita, M., Mizoguchi, H., Nagase, H., Suzuki, T., and Tseng, L. F. (2001) Involvement of spinal protein kinase Cgamma in the attenuation of opioid mu-receptor-mediated G-protein activation after chronic intrathecal administration of [D-Ala2,N-MePhe4,Gly-Ol(5)]-enkephalin. *J. Neurosci.* **21,** 3715–3720.
45. Polakiewicz, R. D., Schieferl, S. M., Dorner, L. F., Kansra, V., and Comb. M. J. (1998) A mitogen-activated protein kinase pathway is required for mu-opioid receptor desensitization. *J. Biol. Chem.* **273,** 12402–12406.
46. Belcheva, M. M., Szucs, M., Wang, D., Sadee, W., and Coscia, C. J. (2001) Mu opiod receptor-mediated ERK-activation involves calmodulin-dependent EGF receptor transactivation. *J. Biol. Chem.* **16,** 16.
47. Koenig, J. A. and Edwardson, J. M. (1997) Endocytosis and recycling of G protein-coupled receptors. *Trends Pharmacol. Sci.* **18,** 276–287.
48. Bohn, L. M., Gainetdinov, R. R., Lin, F. T., Lefkowitz, R. J., and Caron, M. G. (2000) Mu-opioid receptor desensitization by beta-arrestin-2 determines morphine tolerance but not dependence. *Nature* **408,** 720–723.
49. Lefkowitz, R. J., Pitcher, J., Krueger, K., and Daaka, Y. (1998) Mechanisms of beta-adrenergic receptor desensitization and resensitization. *Adv. Pharmacol.* **42,** 416–420.
50. Koch, T., Schulz, S., Schroder, H., Wolf, R., Raulf, E., and Hollt, V. (1998) Carboxyl-terminal splicing of the rat mu opioid receptor modulates agonist-mediated internalization and receptor resensitization. *J. Biol. Chem.* **273,** 13652–13657.
51. Law, P. Y., Erickson, L. J., El-Kouhen, R., Dicker, L., Solberg, J., Wang, W., Miller, E., Burd, A. L., and Loh, H. H. (2000) Receptor density and recycling affect the rate of agonist-induced desensitization of mu-opioid receptor. *Mol. Pharmacol.* **58,** 388–398.
52. Koch, T., Schulz, S., Pfeiffer, M., Klutzny, M., Schroder, H., Kahl, E., and Hollt, V. (2001) C-terminal splice variants of the mouse mu-opioid receptor differ in morphine-induced internalization and receptor resensitization. *J. Biol. Chem.* **276,** 31,408–31,414..
53. Ferguson, S. S., Zhang, J., Barak, L. S., and Caron, M. G. (1998) Molecular mechanisms of G protein-coupled receptor desensitization and resensitization. *Life Sci.* **62,** 1561–1565
54. Miller, W. E. and Lefkowitz, R. J. (2001) Expanding roles for beta-arrestins as scaffolds and adapters in GPCR signaling and trafficking. *Curr. Opin. Cell. Biol.* **13,**139–145.
55. Gaidarov, I. and Keen, J. H. (1999) Phosphoinositide-AP-2 interactions required for targeting to plasma membrane clathrin-coated pits. *J. Cell. Biol.* **146,** 755–764.
56. McDonald, P. H., Cote, N. L., Lin, F. T., Premont, R. T., Pitcher, J. A., and Lefkowitz, R. J. (1999) Identification of NSF as a beta-arrestin1-binding protein. Implications for beta2-adrenergic receptor regulation. *J. Biol. Chem.* **274,** 10,677–10,680.

57. Luttrell, L. M., Ferguson, S. S., Daaka, Y., Miller, W. E., Maudsley, S., Della Rocca, G. J., Lin, F., Kawakatsu, H., Owada, K., Luttrell, D. K., Caron, M. G., and Lefkowitz, R. J. (1999) Beta-arrestin-dependent formation of beta2 adrenergic receptor-Src protein kinase complexes. *Science* **283,** 655–661.
58. Sim-Selley, L. J., Selley, D. E., Vogt, L. J., Childers, S. R., and Martin, T. J. (2000) Chronic heroin self-administration desensitizes mu opioid receptor- activated G-proteins in specific regions of rat brain. *J. Neurosci.* **20,** 4555–4562.
59. Elliott, J., Guo, L., and Traynor, J. R. (1997) Tolerance to mu-opioid agonists in human neuroblastoma SH-SY5Y cells as determined by changes in guanosine-5'-O-(3-[35S]-thio)triphosphate binding. *Br. J. Pharmacol.* **121,**1422–1428.
60. Chang, K. J., Eckel, R. W., and Blanchard, S. G. (1982) Opioid peptides induce reduction of enkephalin receptors in cultured neuroblastoma cells. *Nature* **296,** 446–448.
61. Tao, P. L., Lee, H. Y., Chang, L. R., and Loh, H. H. (1990) Decrease in mu-opioid receptor binding capacity in rat brain after chronic PL017 treatment. *Brain Res.* **526,** 270–275.
62. Tao, P. L., Han, K. F., Wang, S. D., Luc, W. M., Eldc, R., Law, P. Y., and Loh, II. II. (1998) Immunohistochemical evidence of down-regulation of mu-opioid receptor after chronic PL-017 in rats. *Eur. J. Pharmacol.* **344,** 137–142.
63. Zadina, J. E., Chang, S. L., Ge, L. J., and Kastin, A. J. (1993) Mu opiate receptor down-regulation by morphine and up-regulation by naloxone in SH-SY5Y human neuroblastoma cells. *J. Pharmacol. Exp. Ther.* **265,** 254–262.
64. Zadina, J. E., Harrison, L. M., Ge, L. J., Kastin, A. J., and Chang, S. L. (1994) Differential regulation of mu and delta opiate receptors by morphine, selective agonists and antagonists and differentiating agents in SH- SY5Y human neuroblastoma cells. *J. Pharmacol. Exp. Ther.* **270,** 1086–1096.
65. Bernstein, M. A. and Welch, S. P. (1998) Mu-opioid receptor down-regulation and cAMP-dependent protein kinase phosphorylation in a mouse model of chronic morphine tolerance. *Brain Res. Mol. Brain Res.* **55,** 237–242.
66. Law, P. Y., Hom, D. S., and Loh, H. H. (1983) Opiate receptor down-regulation and desensitization in neuroblastoma X glioma NG108-15 hybrid cells are two separate cellular adaptation processes. *Mol. Pharmacol.* **24,** 413–424.
67. Yabaluri, N. and Medzihradsky, F. (1997) Down-regulation of mu-opioid receptor by full but not partial agonists is independent of G protein coupling. *Mol. Pharmacol.* **52,** 896–902.
68. Werling, L. L., McMahon, P. N., and Cox, B. M. (1989) Selective changes in mu opioid receptor properties induced by chronic morphine exposure. *Proc. Natl. Acad. Sci. USA* **86,** 6393–6397.
69. Tao, P. L., Law, P. Y., and Loh, H. H. (1987)Decrease in delta and mu opioid receptor binding capacity in rat brain after chronic etorphine treatment. *J. Pharmacol. Exp. Ther.* **240,** 809–816.
70. Burd, A. L., El-Kouhen, R., Erickson, L. J., Loh, H. H., and Law, P. Y. (1998) Identification of serine 356 and serine 363 as the amino acids involved in etorphine-induced down-regulation of the mu-opioid receptor. *J. Biol. Chem.* **273,** 34,488–34,495.
71. Kraus, J., Horn, G., Zimprich, A., Simon, T., Mayer, P., and Hollt, V. (1995) Molecular cloning and functional analysis of the rat mu opioid receptor gene promoter. *Biochem. Biophys. Res. Commun.* **215,** 591–597.
72. Kim, D. S., Chin, H., and Klee, W. A. (1995) Agonist regulation of the expression of the delta opioid receptor in NG108-15 cells. *FEBS Lett.* **376,** 11–14.
73. Buzas, B., Rosenberger, J., and Cox, B. M. (1997) Regulation of delta-opioid receptor mRNA levels by receptor-mediated and direct activation of the adenylyl cyclase-protein kinase A pathway. *J. Neurochem.* **68,** 610–615.
74. Woltje, M., Kraus, J., and Hollt, V. (2000) Regulation of mouse delta-opioid receptor gene transcription: involvement of the transcription factors AP-1 and AP-2. *J. Neurochem.* **74,** 1355–1362.

75. Sharma, S. K., Klee, W. A., and Nirenberg, M. (1975) Dual regulation of adenylate cyclase accounts for narcotic dependence and tolerance. *Proc. Natl. Acad. Sci. USA* **72,** 3092–3096.
76. Nevo, I., Avidor-Reiss, T., Levy, R., Bayewitch, M., Heldman, E., and Vogel, Z. (1998) Regulation of adenylyl cyclase isozymes on acute and chronic activation of inhibitory receptors. *Mol. Pharmacol.* **54,** 419–426.
77. Avidor-Reiss, T., Bayewitch, M., Levy, R., Matus-Leibovitch, N., Nevo, I., and Vogel, Z. (1995) Adenylylcyclase supersensitization in mu-opioid receptor-transfected Chinese hamster ovary cells following chronic opioid treatment. *J. Biol. Chem.* **270,** 29,732–29,738.
78. Avidor-Reiss, T., Nevo, I., Levy, R., Pfeuffer, T., and Vogel, Z. (1996) Chronic opioid treatment induces adenylyl cyclase V superactivation. Involvement of Gbetagamma. *J. Biol. Chem.* **271,** 21,309–21,315.
79. Chakrabarti, S., Rivera, M., Yan, S. Z., Tang, W. J., and Gintzler, A. R. (1998) Chronic morphine augments G(beta)(gamma)/Gs(alpha) stimulation of adenylyl cyclase: relevance to opioid tolerance. *Mol. Pharmacol.* **54,** 655–662.
80. Chakrabarti, S., Oppermann, M., and Gintzler, A. R. (2001) Chronic morphine induces the concomitant phosphorylation and altered association of multiple signaling proteins: a novel mechanism for modulating cell signaling. *Proc. Natl. Acad. Sci. USA* **98,** 4209–4214.
81. Chakrabarti, S., Wang, L., Tang, W. J., and Gintzler, A. R. (1998) Chronic morphine augments adenylyl cyclase phosphorylation: relevance to altered signaling during tolerance/dependence. *Mol. Pharmacol.* **54,** 949–953.
82. Duman, R. S., Tallman, J. F., and Nestler, E. J. (1988) Acute and chronic opiate-regulation of adenylate cyclase in brain: specific effects in locus coeruleus. *J. Pharmacol. Exp. Ther.* **246,** 1033–1039.
83. Terwilliger, R. Z., Beitner-Johnson, D., Sevarino, K. A., Crain, S. M., and Nestler, E. J. (1991) A general role for adaptations in G-proteins and the cyclic AMP system in mediating the chronic actions of morphine and cocaine on neuronal function. *Brain Res.* **548,** 100–110.
84. Haddad, L. B., Hiller, J. M., Simon, E. J., and Kramer, H. K. Opioid withdrawal decreases basal MAP kinase levels through protein kinase A (PKA). 32nd International Narcotics Research Conference, Helsinki, Finland.
85. Narita, M., Feng, Y., Makimura, M., Hoskins, B., and Ho, I. K. (1994) A protein kinase inhibitor, H-7, inhibits the development of tolerance to opioid antinociception. *Eur. J. Pharmacol.* **271,** 543–545.
86. Nestler, E. J., Alreja, M., and Aghajanian, G. K. (1999) Molecular control of locus coeruleus neurotransmission. *Biol. Psychiatr.* **46,** 1131–1139.
87. Daaka, Y., Luttrell, L. M., and Lefkowitz, R. J. (1997) Switching of the coupling of the beta2-adrenergic receptor to different G proteins by protein kinase A. *Nature* **390,** 88–91.
88. Luo, X., Zeng, W., Xu, X., Popov, S., Davignon, I., Wilkie, T. M., Mumby, S. M., and Muallem, S. (1999) Alternate coupling of receptors to Gs and Gi in pancreatic and submandibular gland cells. *J. Biol. Chem.* **274,** 17,684–17,690.
89. Wolf, D. H., Numan, S., Nestler, E. J., and Russell, D. S. (1999) Regulation of phospholipase Cgamma in the mesolimbic dopamine system by chronic morphine administration. *J. Neurochem.* **73,** 1520–1528.
90. Ueda, H., Inoue, M., and Matsumoto, T. (2001) Protein kinase C-mediated inhibition of mu-opioid receptor internalization and its involvement in the development of acute tolerance to peripheral mu-agonist analgesia. *J. Neurosci.* **21,** 2967–2973.
91. Kimes, A. S., Vaupel, D. B., and London, E. D. (1993) Attenuation of some signs of opioid withdrawal by inhibitors of nitric oxide synthase. *Psychopharmacology* **112,** 521–524.
92. London, E. D., Kimes, A. S., and Vaupel, D. B. (1995) Inhibitors of nitric oxide synthase and the opioid withdrawal syndrome. *NIDA Res. Monogr.* **147,** 170–181.

93. Adams, M. L., Kalicki, J. M., Meyer, E. R., and Cicero, T. J. (1993) Inhibition of the morphine withdrawal syndrome by a nitric oxide synthase inhibitor, NG-nitro-L-arginine methyl ester. *Life Sci.* **52,** L245–L249.
94. Xu, J. Y., Hill, K. P., and Bidlack, J. M. (1998) The nitric oxide/cyclic GMP system at the supraspinal site is involved in the development of acute morphine antinociceptive tolerance. *J. Pharmacol. Exp. Ther.* **284,** 196–201.
95. Machelska, H., Ziolkowska, B., Mika, J., Przewlocka, B., and Przewlocki, R. (1997) Chronic morphine increases biosynthesis of nitric oxide synthase in the rat spinal cord. *Neuroreport* **8,** 2743–2747.
96. Cuellar, B., Fernandez, A. P., Lizasoain, I., Moro, M. A., Lorenzo, P., Bentura, M. L., Rodrigo, J., and Leza, J. C. (2000) Up-regulation of neuronal NO synthase immunoreactivity in opiate dependence and withdrawal. *Psychopharmacology* (Berl.) **148,** 66–73.
97. Pataki, I. and Telegdy, G. (1998) Further evidence that nitric oxide modifies acute and chronic morphine actions in mice. *Eur. J. Pharmacol.* **357,** 157–162.
98. Trujillo, K. A. and Akil, H. (1991) Inhibition of morphine tolerance and dependence by the NMDA receptor antagonist MK-801. *Science* **251,** 85–87.
99. Marek, P., Ben-Eliyahu, S., Vaccarino, A. L., and Liebeskind, J. C. (1991) Delayed application of MK-801 attenuates development of morphine tolerance in rats. *Brain Res.* **558,** 163–165.
100. Tiseo, P. J. and Inturrisi, C. E. (1993) Attenuation and reversal of morphine tolerance by the competitive N-methyl-D-aspartate receptor antagonist, LY274614. *J. Pharmacol. Exp. Ther.* **264,** 1090–1096.
101. Bredt, D. S. and Snyder, S. H. (1992) Nitric oxide, a novel neuronal messenger. *Neuron* **8,** 3–11.
102. Aghajanian, G. K. (1978) Tolerance of locus coeruleus neurones to morphine and suppression of withdrawal response by clonidine. *Nature* **276,** 186–188.
103. Akaoka, H. and Aston-Jones, G. (1991) Opiate withdrawal-induced hyperactivity of locus coeruleus neurons is substantially mediated by augmented excitatory amino acid input. *J. Neurosci.* **11,** 3830–3839..
104. Williams, J. T., Christie, M. J., and Manzoni, O. (2001) Cellular and synaptic adaptations mediating opioid dependence. *Physiol. Rev.* **81,** 299–343.
105. Chieng, B. and Christie, M. D. (1996) Local opioid withdrawal in rat single periaqueductal gray neurons in vitro. *J. Neurosci.* **16,** 7128–7136.
106. Crain, S. M. and Shen, K. F. (1990) Opioids can evoke direct receptor-mediated excitatory effects on sensory neurons. *Trends Pharmacol. Sci.* **11,** 77–81.
107. Crain, S. M. and Shen, K. F. (1992) After chronic opioid exposure sensory neurons become supersensitive to the excitatory effects of opioid agonists and antagonists as occurs after acute elevation of GM1 ganglioside. *Brain Res.* **575,** 13–24.
108. Clapham, D. E. and Neer, E. J. (1993) New roles for G-protein beta gamma-dimers in transmembrane signalling. *Nature* **365,** 403–406..
109. Daaka, Y., Pitcher, J. A., Richardson, M., Stoffel, R. H., Robishaw, J. D., and Lefkowitz, R. J. (1997) Receptor and G betagamma isoform-specific interactions with G protein- coupled receptor kinases. *Proc. Natl. Acad. Sci. USA* **94,** 2180–2185.
110. Sarnago, S., Elorza, A., and Mayor, F., Jr. (1999) Agonist-dependent phosphorylation of the G protein-coupled receptor kinase 2 (GRK2) by Src tyrosine kinase. *J. Biol. Chem.* **274,** 34,411–34,416.
111. Chuang, T. T., LeVine, H., 3rd, and De Blasi, A. (1995) Phosphorylation and activation of beta-adrenergic receptor kinase by protein kinase C. *J. Biol. Chem.* **270,** 18,660–18,665.
112. Murthy, K. S., Grider, J. R., and Makhlouf, G. M. (2000) Heterologous desensitization of response mediated by selective PKC- dependent phosphorylation of G(i-1) and G(i-2). *Am. J. Physiol. Cell. Physiol.* **279,** C925–C934.
113. Strassheim, D., Law, P. Y., and Loh, H. H. (1998) Contribution of phospholipase C-beta3 phosphorylation to the rapid attenuation of opioid-activated phosphoinositide response. *Mol. Pharmacol.* **53,**1047–1053.

114. Robinson, T. E. and Berridge, K. C. (1993) The neural basis of drug craving: an incentive-sensitization theory of addiction. *Brain Res. Brain Res. Rev.* **18,** 247–291.
115. Robinson, T. E. and Berridge, K. C. (2001) Incentive-sensitization and addiction. *Addiction* **96,** 103–114.
116. Erdtmann-Vourliotis, M., Mayer, P., Riechert, U., Grecksch, G., and Hollt, V. (1998) Identification of brain regions that are markedly activated by morphine in tolerant but not in naive rats. *Brain Res. Mol. Brain Res.* **61,** 51–61.
117. Erdtmann-Vourliotis, M., Mayer, P., Linke, R., Riechert, U., and Hollt, V. (1999) Long-lasting sensitization towards morphine in motoric and limbic areas as determined by c-fos expression in rat brain. *Brain Res. Mol. Brain Res.* **72,** 1–16.

5

Inhibitors of Enkephalin Catabolism

New Therapeutic Tool in Opioid Dependence

Florence Noble and Bernard P. Roques

Typically, long-term administration of opiates leads to a reduction in the magnitude and duration of effects produced by a given dose (tolerance) and to physical dependence, a state manifested by withdrawal symptoms when drug taking is terminated or significantly reduced. Addiction or psychological dependence implies the compulsive self-administration of the drug, caused by both its reinforcing or rewarding effects and the unpleasant experience (i.e., abstinence syndrome) produced by the sudden interruption of its consumption. In contrast to tolerance or physical dependence, addiction is not simply a biological phenomenon. Psychological, environmental, and social factors can strongly influence its development. All these factors could explain the important variability of the clinical syndrome of addiction, and the various approaches that have been proposed to help addicts to stop their drug use. Nevertheless, these methods do not work equally well for all types of addicts.

Clinicians recognize that addiction is fundamentally a behavioral syndrome, and a chronic, relapsing disease. The major clinical problems produced by addiction are not tolerance and withdrawal symptoms. These phenomena can easily be managed and, in fact, they occur with no serious consequences in patients who receive chronic opiate medications for medical reasons such as pain. Indeed, addiction is relatively rare among patients receiving opiate analgesics for long-term treatment, despite the occurrence of tolerance, dependence, and a withdrawal syndrome if the medication is abruptly interrupted or if the dose is significantly decreased. In the treatment of addiction, the most difficult aspect is the strong proneness to relapse, which continues long after the drug has been cleared from the body, despite the knowledge by addicts of the negative consequences of this decision on their social life.

Numerous pharmacological agents have begun to be explored in morphine abstinence syndrome in animals, such as N-methyl-D-aspartate (NMDA) antagonists, benzodiazepine-, adenosine-, α_2-agonists, partial dopamine agonists, or opioid agonists.

Among the pharmacological treatments, the most commonly used in heroin addicts is administration of long-acting agonists such as methadone and buprenorphine. Methadone or buprenorphine maintenance reduces the "rush" produced by heroin, but not the sensation of "well-being" induced by the opioid compound. This results in significant reduction in heroin use and thus is expected to facilitate heroin abstinence. Neverthe-

From: *Molecular Biology of Drug Addiction*
Edited by: R. Maldonado © Humana Press Inc., Totowa, NJ

less, various problems are associated with the use of these substitutes. Like heroin, they are classified as drugs, and even if the methadone withdrawal syndrome is less severe than that observed with heroin, it is much longer-lasting. This constitutes a serious problem for long-term methadone-maintained patients who wish to end the maintenance phase of treatment. Moreover, chronic methadone use can lead to biochemical and physiological alterations of important functions *(1)*. Finally, methadone and, even more, buprenorphine can be misused.

Thus, the use of a more "physiological" maintenance treatment by increasing the level of endogenous opioid peptides could be an interesting new approach in the treatment of drug abuse.

1. The Endogenous Opioid System

The endogenous opioid system is consists of three distinct neuronal pathways that are widely distributed throughout the central nervous system (CNS). The three opioid precursors of the opioid peptides are proopiomelanocortin *(2)*, proenkephalin *(3)*, and prodynorphin *(4)*, each generating biologically active peptides that are released by specific opioidergic neurons. These peptides exert their physiological actions by interacting with various classes of opioid receptor types (μ, δ, κ) *(5–7)*, present on both pre- and postsynaptic membranes of opioid and opioid-target neurons *(8)*. The most important opioid peptides seems to be the pentapeptides enkephalins *(9)* which interact with both μ- and δ-receptors, their affinities being significantly better for the latter. This has triggerred intensive studies to elucidate the role of enkephalins in the brain and to develop putative novel effective treatments in analgesia *(10)* and CNS disorders, including drug addiction.

Therapeutic approaches to increase endogenous opioid levels, such as electrostimulation or acupuncture, have been proposed in the treatment of addiction. Nevertheless, although it now is well established that such stimuli indeed increase the levels of endogenous opioid peptides, there is not yet clear evidence that they constitute successful treatments in opioid addiction. This could be related to the unsufficient increase in extracellular levels of endogenous opioid peptides induced by these techniques. This limitation could be overcome with the use of enkephalin-degrading enzyme inhibitors. Indeed, early studies on the enkephalins showed that they have a very short half-life in both in vitro and in vivo conditions. In contrast to catecholamines and amino acid transmitters, which are essentially cleared from the extracellular space by reuptake mechanisms, neuropeptides appear to be inactivated by peptidases cleaving the biologically active peptides into inactive fragments *(11)*. This process was clearly demonstrated for the enkephalins, which are degraded by the concomitant action of two peptidases: neutral endopeptidase (neprilysin, NEP) and aminopeptidase N (APN) (Fig. 1).

This well-admitted interruption by the two peptidases of the messages conveyed by enkephalins is in good agreement with the demonstration of a collocalization of NEP and APN in brain areas where opioid peptides and receptors are present *(12–14)*.

2. Development of Complete Inhibitors of Enkephalin-Degrading Enzymes

Based on these results, it was tempting to inhibit the two enkephalin-inactivating enzymes to study the physiological role of the endogenous opioid system *(10)*. With

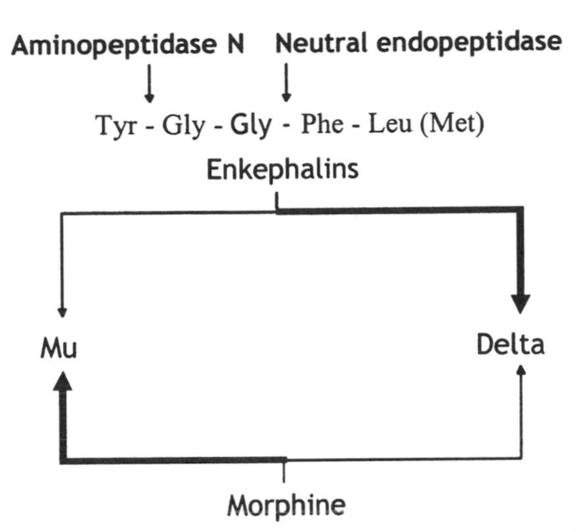

Fig. 1. Enkephalins may act on μ- and δ-opioid receptors as morphine, and are degraded by two enzymes, aminopeptidase N and neutral endopeptidase.

this aim and accounting for the fact that NEP and APN belong to the same class of zinc metallopeptidases, the concept of mixed inhibitors was developed *(15)*.

At first, mixed-inhibitor prodrugs were designed, by linking through a disulfide bond, two very efficient inhibitors with nanomolar affinities for APN and NEP, respectively. Among the various compounds synthesized, RB 101 is a systemically active prodrug that completely inhibits both enzymes *(16)* and increases the extracellular concentrations of Met-enkephalin in brains of freely moving rats *(17)*. Following intraveinous, intraperitoneal, or oral administration (at high doses), RB 101 or derivative compounds induce naloxone-reversible antinociceptive responses in all animals models of pain in which morphine is active, including neuropathic pain in which opiates exhibit weak potency *(10,11,18)*.

Very recently, true dual inhibitors able to recognize with nanomolar affinities the two peptidases and belonging to the family of aminophosphinic compounds have been obtained and were shown to posses a longer duration of action than RB 101 *(19)* (Fig. 2).

3. Chronic Treatments and Side Effects of Mixed Inhibitors of NEP and APN

The main advantage of modifying the concentration of endogenous peptides by use of peptidase inhibitors is that pharmacological effects are induced only at receptors tonically or phasically stimulated by the natural effectors. Moreover, in contrast to exogenous agonists or antagonists, chronic administration of mixed enkephalin-degrading enzymes inhibitors does not induce changes in the synthesis of its peptide precursors, as well as in the secretion of the active peptide *(20)*.

As expected from their mechanims of action, the mixed inhibitors are devoid of the main drawbacks of morphine (i.e., respiratory depression *(21)*, constipation and physical and psychic dependence *[22,23]* [Fig. 3]). This is due mainly to the weaker, but more specific, stimulation of the opioid-binding sites by the tonically or phasically released endogenous opioids, thus minimizing receptor desensitization or down-

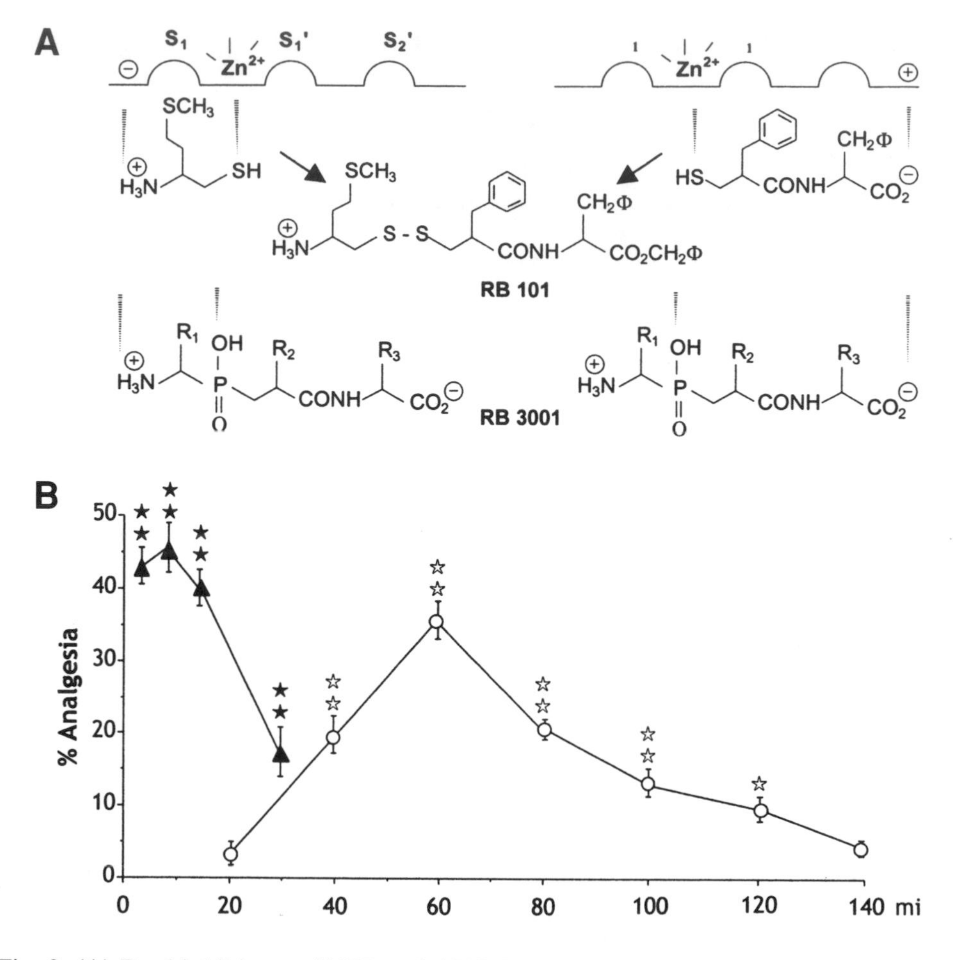

Fig. 2. **(A)** Dual inhibitors of NEP and APN. Two strategies have been followed: RB 101, association by a disulfide bond of a potent APN inhibitor with a potent NEP inhibitor, and RB 3001, a true dual NEP/APN inhibitor. **(B)** Differences in the duration of antinociceptive responses (hot-plate test in mice) provided by RB 101 ▲ (10 mg/kg iv) and RB 3001 ○ (25 mg/kg iv). ⋆p <0.05 and ⋆⋆p <0.01 compared with control.

regulation that usually occurs after ubiquitous activation of opioid receptors by exogenous agonists. This limited opioid receptor occupation by the endogenous peptides is in agreement with in vivo binding studies, demonstrating that the increase in tonically released endogenous enkephalins is too low to saturate opioid receptors *(24)*.

Moreover, chronic morphine induces a hypersensitivity of noradrenaline-containing neurons in the locus coeruleus, considered one of the main causes of the withdrawal syndrome *(25–27)*. In this brain region a very low tonic release of enkephalins was observed *(28)*, and this is probably one of the main reasons why the withdrawal syndrome is almost absent after chronic treatment by peptidase inhibitors as compared to exogenous opioids. Moreover, because of their higher intrinsic efficacy, enkephalins need to occupy fewer opioid receptors than does morphine to generate the same pharmacological responses *(29)*.

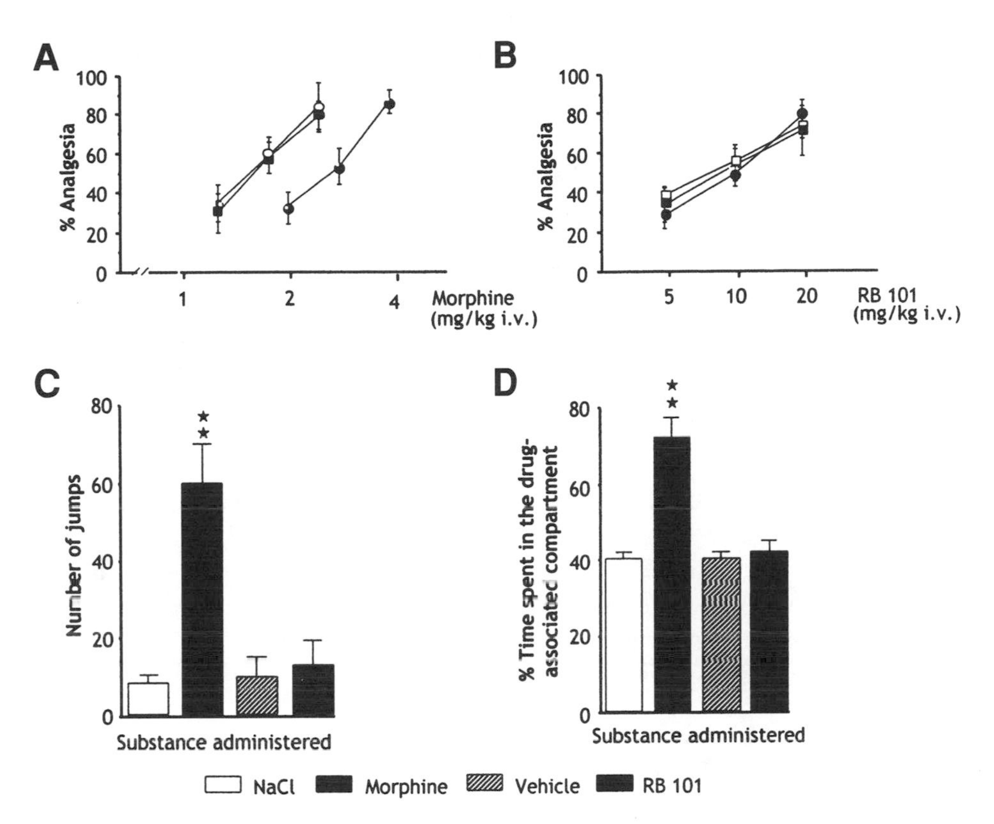

Fig. 3. **(A)** Antinociceptive dose–response curves recorded in the hot-plate test (i.e., jump response) 10 min after iv administration of morphine to mice chronically pretreated with saline ○, RB 101 ■ (80 mg/kg), or morphine ● (3 mg/kg), ip twice daily, for 4 d. **(B)** Antinociceptive dose–response curves recorded in the hot-plate test (jump response) 10 min after iv administration of RB 101 to mice chronically pretreated with vehicle □, RB 101 ■ (80 mg/kg), or morphine ● (3 mg/kg), ip, twice daily for 4 d. **(C)** Comparison of the withdrawal symptoms induced by naloxone after chronic treatment with morphine (6 mg/kg) or RB 101 (160 mg/kg), injected ip, twice daily for 5 d. **(D)** Comparison of the psychic dependence induced by chronic morphine (6 mg/kg) or RB 101 (160 mg/kg), injected ip, in the place preference test. ★★$p < 0.01$, as compared with other groups.

Several studies have used the expression of immediate early genes as markers for neuronal activity in an attempt to differentiate the effects of exogenously administered opioids from tonically released endogenous opioid peptides. By measuring c-Fos expression generated by either noxious thermal stimuli or chronic pain, it was shown that dual inhibitors of NEP and APN reduced this expression to a maximum of 60%, as compared to 90% for morphine *(30–32)*, and a lack of tolerance to repeated administration of RB 101 was observed *(33)*.

The moderate degree or lack of tolerance or physical dependence observed following chronic treatment with mixed inhibitors could also be due to weaker intracellular modifications. Indeed, opioid side effects are complex processes involving multiple cellular targets including receptors involved at central or peripheral levels on respiration, cardiovascular control, intestinal functioning, and so on *(34)*. The lack of undesirable effects might also be related to differences in the mechanisms of receptor endocytosis following receptor activation by enkephalins or by alcaloids or synthetic heterocyclic agonists *(35)*. Similarly, different desensitization of opioid receptors may occur, which could be caused by an alteration in adenylyl cyclase functioning or expression, phosphorylation and/or dimerization of receptors, or changes in G-protein subunit concentrations.

The biochemical mechanisms involved in psychic dependence that could result in craving remain still unclear. However, it is well established that dopaminergic neurons that project from the ventral tegmental area (VTA) to the nucleus accumbens play a major role in this process *(36)*. The failure of RB 101 to induce psychic dependence in various animal models used to predict the addictive properties of a given compound *(23)* probably results from a lower recruitment of opioid receptors and weaker modifications of intracellular events for endogenous enkephalins than for morphine. This hypothesis is supported by the weaker changes in dopamine release in the nucleus accumbens after administration in the VTA of a dual NEP/APN inhibitor, compared with a μ-opioid agonist *(37)*, and by the apparent absence of effects on the levels of dopamine and metabolites in the nucleus accumbens following administration of a selective NEP inhibitor *(38)*.

Accordingly, when injected in the VTA of rats, mixed inhibitors, unlike morphine, have no effect or only slightly decrease the rate of intracranial self-stimulation, a response strongly triggered by addictive drugs *(39)*.

4. Clinical Interest in Enkephalin-Degrading Enzyme Inhibitors in Opioid Addiction Treatment

Addiction involves the compulsive seeking of the drug (craving), caused both by its reinforcing or rewarding effects and by the unpleasant experience (i.e., abstinence syndrome) produced by the sudden interruption of its consumption. It is important to emphasize that, in contrast to tolerance or physical dependence, which could be explained by events occurring at the receptor level, addiction is more complicated. Thus, psychological, environmental, and social factors can strongly influence its development. All these factors could explain the important variability among addicts to reach drug abstinence. Many approaches have been developed to help addicts interrupt their drug consumption. It seems that a combination of psychological assistance and pharmacotherapy has been so far the most efficient treatment, the most difficult problem remaining the strong susceptibility to relapse.

It has been suggested that the craving and self-administration of drugs could be explained either by a preexisting deficit in the endogenous opioid system or by a deficit that could occur after chronic administration of opiates. This hypothesis is in agreement with recent results *(40)* showing an important increase of Met-enkephalin outflow in morphine-dependent rats as compared to control animals in the periaqueductal gray (PAG), which contains high levels of μ-receptors as well as NEP and APN *(12,14)*, and could be an important site of action for the development of physical dependence *(27,41)*. This large increase in synaptic levels of enkephalin tone leads to a new state of the enkephalinergic neural circuitry that is not modified following naloxone-precipitated withdrawal syndrome. This lack of compensatory increase in extracellular amounts of endogenous enkephalins could participate in the withdrawal syndrome. Consistent with this hypothesis, an increase in endogenous enkephalins induced by peptidase inhibitors was shown to reduce the severity of the withdrawal syndrome in rats *(27,42)*, and direct injection of enkephalins in rodent brain reduced morphine withdrawal *(43)*.

Nevertheless, although early abstinence syndrome may be an important clinical problem, the most difficult aspect of the treatment of addiction is the protracted abstinence syndrome, one of the main factors contributing to relapse. Indeed, in the first days after cessation of prolonged drug use an acute withdrawal syndrome is observed, which consists of physiological changes (i.e., agitation, hyperalgesia, tachycardia, hypertension, diarrhea, and vomiting) and a variety of phenomena, i.e., cardiovascular, visceral, thermoregulatory, and subjective changes including a depressive state that may persist for months or more after the last dose of opiate. Thus, the main challenge in the management of opioid addiction is to develop a pharmacotherapy to minimize the short-term withdrawal syndrome and protracted opiate abstinence syndrome. Compounds used classically for clinical treatment of opiate withdrawal, such as clonidine and methadone, were investigated using a spontaneous abstinence, and their effects were compared with those of RB 101. As previously mentioned, methadone is the opiate agonist most currently used for maintenance treatment, and the α_2-adrenoceptor agonist, clonidine, is the most effective nonopioid drug for improving some aspects of opiate withdrawal *(44)*.

As expected, the effect of clonidine was limited to spontaneous withdrawal, while methadone was effective in reducing the withdrawal syndrome and appeared to be an efficient treatment in the maintenance period *(45)*. The responses induced by RB 101 in this model of spontaneous withdrawal were similar to those induced by methadone *(45)* (Fig. 4). Thus, RB 101 could be particularly interesting as a therapeutic alternative in the maintenance of opiate addicts, especially to avoid methadone-dependence risk. Moreover, interestingly, it has been suggested that defects in the endogenous opioid systems might be involved in the etiology of depression. Accordingly, the behavioral responses triggered by forced swimming, conditioned suppression of motility, and learned helplessness, which are currently used as animal tests to screen antidepressants, were attenuated by treatment with dual inhibitors *(46–50)*. These experiments have demonstrated that the inhibitors could modulate the functioning of the mesocorticolimbic and nigrostriatal dopaminergic systems, which are implicated in mood control and connected with enkephalin pathways. In line with this finding, the increased levels of endogenous enkephalins induced by RB 101 produce antidepressant-like effects, which were suppressed by both the δ-opioid antagonist naltrindole and the dopamine D1 antagonist SCH23390 *(47)* (Fig. 5).

Furthermore, in a recent study we have shown that the Met-enkephalin outflow in the nucleus accumbens is modified by the induction of psychic dependence using the conditioned place preference paradigm, which is a test considered closer to addictive situations in humans than the self-administration procedure *(51)*. In this experiment, rats with cannulae implanted in the nucleus accumbens for microdialysis were confined alternatively in distinct compartments under reinforced (morphine) or nonreinforced (saline) treatments. Using this model, opposite changes in the Met-enkephalin outflow were observed after a conditioning period of 6 d. Thus, Met-enkephalin level was found to be enhanced in the drug-paired compartment and reduced in the saline-paired one *(40)* (Fig. 6). The transient increase in enkephalin efflux observed in the nucleus accumbens when the animals were placed in the drug-paired compartment during the

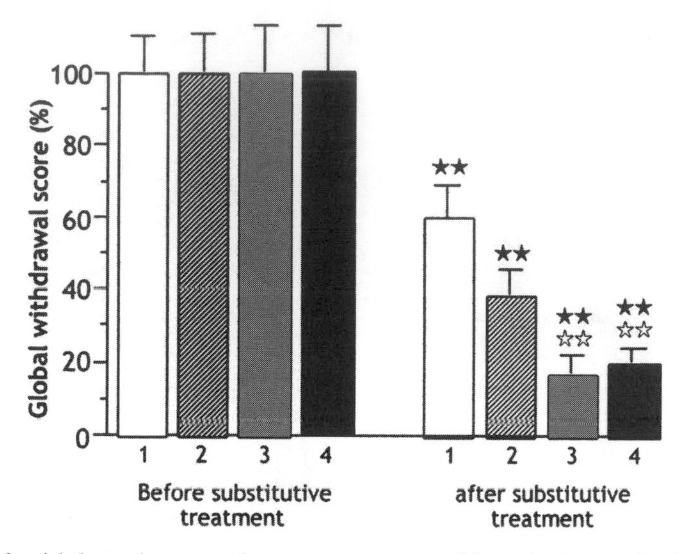

Fig. 4. Global withdrawal score of spontaneous morphine abstinence before substitutive treatments evaluated 36 h after the last injection of morphine, or after substitutive treatments (twice daily for 4 d) with (1) saline; (2) clonidine (0.025 mg/kg); (3) methadone (2 mg/kg); (4) RB 101 (40 mg/kg). ★★$p < 0.01$ vs value of the same group before substitutive treatment, ☆☆$p < 0.01$ vs value of saline in the same session.

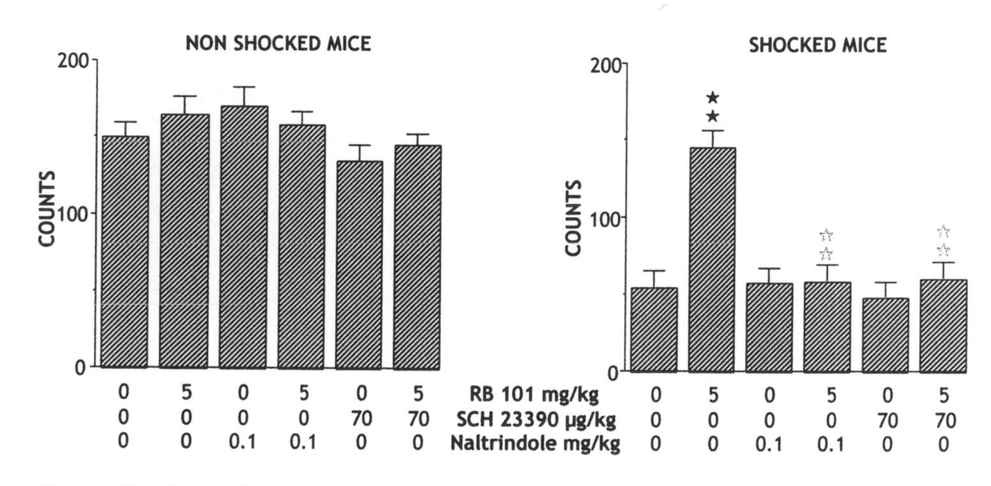

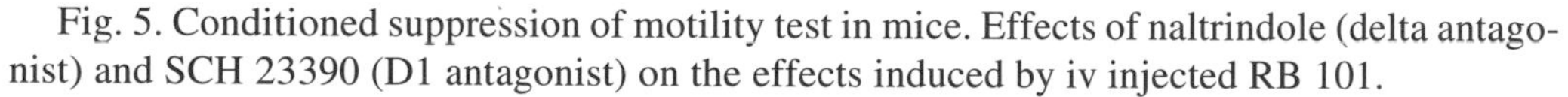

0	5	0	5	0	5	RB 101 mg/kg	0	5	0	5	0	5
0	0	0	0	70	70	SCH 23390 µg/kg	0	0	0	0	70	70
0	0	0.1	0.1	0	0	Naltrindole mg/kg	0	0	0.1	0.1	0	0

Fig. 5. Conditioned suppression of motility test in mice. Effects of naltrindole (delta antagonist) and SCH 23390 (D1 antagonist) on the effects induced by iv injected RB 101.

microdialysis experiment may reflect an anticipation of the rewarding effect, associated with the memory of the reinforcing effects obtained with morphine in this compartment during the conditioning phase. In contrast, when the rats were placed in the saline-paired compartment, a decrease in the extracellular level of Met-enkephalin was observed, which may be related to an aversive effect. Consistent with these results, several studies have clearly demonstrated a role for endogenous opioid peptides in the perception of reward and in the mediation of behavioral reinforcement. Thus, it has been shown that water deprivation induced a reduction in opioid release, and that this effect was reversed in animals receiving water *(52)*. On the other hand, rats isolated

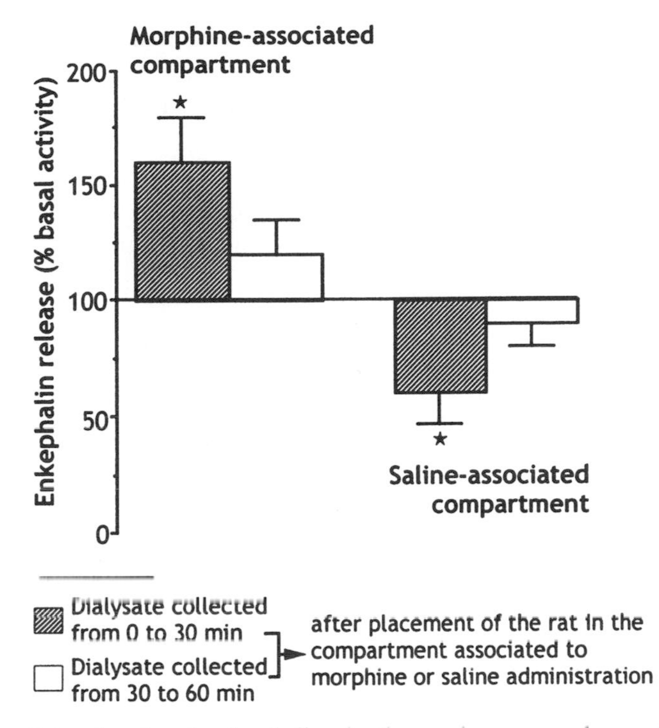

Fig. 6. Extracellular levels of enkephalins in the nucleus accumbens of rats chronically treated with morphine in the place preference test to induce a psychic dependence. Briefly, the conditioning apparatus used in this experiment consisted of a rectangular plexiglas box divided into two square compartments of the same size (45 × 45 × 30 cm). Two distinctive sensory cues differentiated the compartments: the wall coloring (black or stripes), and the floor texture (grid or smooth). The protocole consisted of three phases. (1) Habituation (preconditioning) phase (1 d): drug-naive rats had free access to both compartments of the conditioning apparatus. (2) Conditioning phase (6 d). on alternate days each animal was injected with morphine (5 mg/kg ip) and confined in one compartment (d 1, 3, and 5) for 30 min, or injected with saline and confined in the other compartment (d 2, 4, and 6) for 30 min. Control groups were injected with saline every day and placed alternatively in both compartments. At the end of the conditioning phase (on d 7) rats were connected to the microdialysis pumps and 2 h after the beginning of the perfusion, two samples were collected to determine the basal efflux of neuropeptides. Then animals were transferred in conditioning apparatus and two microdialysis samples were collected.

and submitted to repetitive mild stress and then placed in presence of a congener (a situation considered fearful) have shown a lack of enkephalin release in the nucleus accumbens as compared to controls *(53)*.

The increase of opioids observed when rats were placed in the morphine-paired compartment may contribute to the perception of reward. Thus, injection of enkephalin in the nucleus accumbens may serve as reinforcement for self-administration behavior *(54,55)*, and an increased level of endogenous opioid peptides was observed when the compulsive seeking for drugs of abuse is high in the self-administration procedure *(56,57)*.

The regulation of endogenous opioid peptides observed suggests that enkephalins may be a neural substrate for reward expectation, as already suggested for dopamine

(58,59). A signal may be deliver when the animal is placed in the drug-paired compartment, increasing the release of enkephalins, which may influence the processing of predictions and the choice of reward-maximizing action. This process may be either (a) dopamine-independent, involving only opioid receptors localized postsynaptically in the nucleus accumbens, as previously demonstrated in the maintenance of heroin self-administration *(60)*, or (b) dopamine-dependent, consistent with the demonstration that activation of μ- or δ-opioid receptors in the nucleus accumbens increased dopamine release *(61,62)*. Thus, activation of μ- and/or δ-opioid receptors by high levels of enkephalins may lead to an increased release of dopamine, while reduction of their stimulation by a lower amounts of opioid peptides may reduce the dopamine tone, resulting in a subsequent decrease in the activity of D_1 receptors present in the nucleus accumbens. This regulation of extracellular dopamine efflux may be important in the expression of morphine-induced psychic dependence, as opioid reward measured by the conditioned place preference paradigm depends on midbrain dopamine-related mechanisms *(63,64)*.

The social contexts in which drug addiction occurred and withdrawal syndrome was undergone are critically important. It is well established that reexposure to the conditioned environmental cues that had initially been associated with drug use can be a major factor causing persistent or recurrent drug cravings and relapses.

Thus, it could be speculated that when the drug abuser is reexposed to few of behavioral and social-context cue components associated with the use of drugs, a cascade of biochemical modifications occurs in the brain, with an increase of enkephalins in the limbic system, as observed by microdialysis when rats were placed in the drug-paired compartment. These modifications are assumed to participate in the control of emotions and feeling of pleasure. The increase of endogenous opioid peptides, which may reflect reward expectation, is short-lasting, and may be very quickly followed by a "distress" state. To avoid this negative effect the addict will self-administer drug to maintain activation of the hedonic pathway and even to increase it. Based on this hypothesis, the dual inhibitors of enkephalin-degrading enzymes may be proposed as a new therapy in opioid addiction, increasing the half-life of the endogenous pentapeptides (Fig. 7). This seems to be confirmed by preliminary results showing that RB 101 is able to strongly reduce the number of heroin administrations in the model of self-administrations in rats (E. Ambrosio, personnal communication).

5. Conclusion

The main advantages of modifying the concentrations of endogenous opioid peptides by use of peptidase inhibitors is that pharmacological effects are induced only at receptors tonically or phasically stimulated by the natural effectors. Moreover, in contrast to exogenous agonists or antagonists, chronic administration of dual enkephalin-degrading enzyme inhibitors does not induce changes in the synthesis of the clearing peptidases and in the synthesis of its target peptide precursors, as well as in the secretion of the active peptides *(20)*. They could represent more efficient compounds than methadone in the treatment of opioid addiction, both because they seem to be unable to trigger dependence and because they are less suceptible to toxicological problems than the long-lasting agonist methadone, for instance. The protracted abstinence syndrome also could be reduced owing to the antidepressant-like properties of the dual inhibitors,

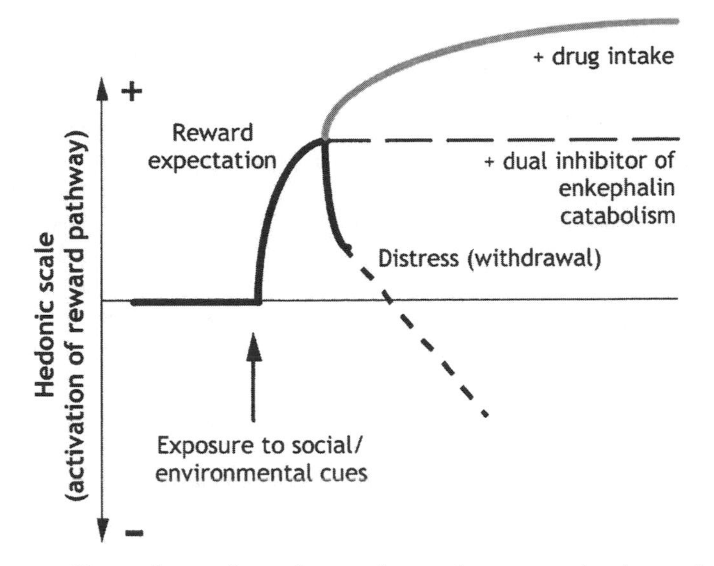

Fig. 7. Diagram illustrating various phases of reward system activation and expected role of dual enkephalin-degrading enzymes inhibitors.

thus reducing the risks of relapse, the most important problem in the management of opioid addiction.

References

1. Kreek, M. J. and Hartman, N. (1982) Chronic use of opioids and antipsychotic drugs: side effects, effects on endogenous opioids, and toxicity. *Ann. N.Y. Acad. Sci.* **398,** 151–172.
2. Young, E., Bronstein, D., and Akil, H. (1993) Proopiomelanocortin biosynthesis, processing and secretion: functional implications, in *Handbook of Experimental Pharmacology, Opioids I* (Herz, A., Akil, H., and Simon, E. J., eds.), Springer-Verlag, Berlin, Heidelberg, pp. 393–421.
3. Rossier, J. (1993) Biosynthesis of enkephalins and proenkephalin-derived peptides, in *Handbook of Experimental Pharmacology, Opioids I* (Herz, A., Akil, H., and Simon, E. J., eds.), Springer-Verlag, Berlin, Heidelberg, pp. 423–447.
4. Day, R., Trujillo, K. A., and Akil, H. (1993) Prodynorphin biosynthesis and posttranslational processing, in *Handbook of Experimental Pharmacology, Opioids I* (Herz, A., Akil, H., and Simon, E. J., eds.), Springer-Verlag, Berlin, Heidelberg, pp. 449–470.
5. Simon, E. J., Hiller, J. M., and Edelman, I. (1973) Stereospecific binding of the potent narcotic analgesic ^{3}H-etorphin to rat brain homogenate. *Proc. Natl. Acad. Sci. USA* **70,** 1947–1949.
6. Pert, C. B. and Snyder, S. H. (1973) Opiate receptor: demonstration in nervous tissue. *Science* **179,** 1011–1014.
7. Terenius, L. (1973) Stereospecific interaction between narcotic analgesics and a synaptic plasma membrane fraction of rat brain cortex. *Acta Pharmacol. Toxicol.* **32,** 317–320.
8. Besse, D., Lombard, M. C., Zajac, J. M., and Roques, B. P. (1990) Pre- and post-synaptic distribution of μ and δ opioid receptors in the superficial layers of the cervical dorsal horn of the rat spinal cord. *Brain Res.* **521,** 15–22.
9. Hughes, J., Smith, T. W., Kosterlitz, H. W., Fotherghill, L. A., Morgan, B. A., and Morris, H. R. (1975) Identification of two related pentapeptides from the brain with potent opiate agonist activity. *Nature* **258,** 577–579.
10. Roques, B. P. (2000) Novel approaches to targeting neuropeptide systems. *Trends Pharmacol. Sci.* **21,** 475–483.

11. Roques, B. P., Noble, F., Daugé, V., Fournié-Zaluski, M. C., and Beaumont, A. (1993) Neutral endopeptidase 24.11: structure, inhibition, and experimental and clinical pharmacology. *Pharmacol. Rev.* **45,** 87–146.
12. Waksman, G., Hamel, E., Fournié-Zaluski, M. C., and Roques, B. P. (1986) Autoradiographic comparison of the distribution of the neutral endopeptidase "enkephalinase" and of mu and delta opioid receptors in rat brain. *Proc. Natl. Acad. Sci. USA* **83,** 1523–1527.
13. Mansour, A., and Watson, S. J. (1993) Anatomical distribution of opioid receptors in mammalians: an overview, in *Handbook of Experimental Pharmacology. Opioids I* (Herz, A., Akil, H., and Simon, A. J., eds.), pp. 79–105, Springer-Verlag, Berlin, Germany.
14. Noble, F., Banisadr, G., Jardinaud, F., Popovici, T., Lai-Kuen, R., Chen, H., Bischoff, L., Melik Parsadaniantz, S., Fournié-Zaluski, M. C., and Roques, B. P. (2001) First discrete autoradiographic distribution of aminopeptidase N in various structures of rat brain and spinal cord using the selective iodinated inhibitor [^{125}I]RB 129. *Neuroscience* **105,** 479–488.
15. Fournié-Zaluski, M. C., Chaillet, P., Bouboutou, R., Coulaud, A., Chérot, P., Waksman, G., Costentin, J., and Roques, B. P. (1984) Analgesic effects of kelatorphan, a new highly potent inhibitor of multiple enkephalin degrading enzymes. *Eur. J. Pharmacol.* **102,** 525–528.
16. Noble, F., Soleilhac, J. M., Soroca-Lucas, E., Turcaud, S., Fournié-Zaluski, M. C., and Roques, B. P. (1992) Inhibition of the enkephalin-metabolizing enzymes by the first systemically active mixed inhibitor prodrug RB 101 induces potent analgesic responses in mice and rats. *J. Pharmacol. Exp. Ther.* **261,** 181–190.
17. Daugé, V., Mauborgne, A., Cesselin, F., Fournié-Zaluski, M. C., and Roques, B. P. (1996) The dual peptidase inhibitor RB 101 induces a long lasting increase in the extracellular level of Met-enkephalin in the nucleus accumbens of freely moving rats. *J. Neurochem.* **67,** 1301–1308.
18. Coudoré-Civiale, M. A., Meen, M., Fournié-Zaluski, M. C., Boucher, M., Roques, B. P., and Eschalier, A. (2001) Enhancement of the effects of a complete inhibitor of enkephalin-catabolizing enzymes, RB 101, by a cholecystokinin-B receptor antagonist in diabetic rats. *Br. J. Pharmacol.* **133,** 179–185.
19. Chen, H., Noble, F., Roques, B. P., and Fournié-Zaluski, M. C. (2001) Long lasting antinociceptive properties of enkephalin degrading enzyme (NEP and APN) inhibitor prodrugs. *J. Med. Chem.* **44,** 3523–3530.
20. Delay-Goyet, P., Zajac, J. M., and Roques, B. P. (1989) Effects of repeated treatment with haloperidol on rat striatal neutral endopeptidase EC 3.4.24.11, and on μ and δ opioid binding sites: comparison with chronic morphine and chronic kelatorphan. *Neurosci. Lett.* **103,** 197–202.
21. Boudinot, E., Morin-Surun, M. P., Foutz, A. S., Fournié-Zaluski, M. C., Roques, B. P., and Denavit-Saubié, M. (2001) Effects of the potent analgesic enkephalin-catabolizing enzyme inhibitors RB 101 and kelatorphan on respiration. *Pain* **90,** 7–13.
22. Noble, F., Coric, P., Fournié-Zaluski, M. C., and Roques, B. P. (1992) Lack of physical dependence in mice after repeated systemic administration of the mixed inhibitor prodrug of enkephalin-degrading enzymes, RB 101. *Eur. J. Pharmacol.* **223,** 91–96.
23. Hutcheson, D. M., Subhan, F., Pache, D. M., Maldonado, R., Fournié-Zaluski, M. C., Roques, B. P., and Sewell, R. D. E. (2000) Analgesic doses of the enkephalin degrading enzyme inhibtor RB 120 do not have discriminative stimulus properties. *Eur. J. Pharmacol.* **401,** 197–204.
24. Ruiz-Gayo, M., Baamonde, A., Turcaud, S., Fournié-Zaluski, M. C., and Roques, B. P. (1992) *in vivo* occupation of mouse brain opioid receptors by endogenous enkephalins: blockade of enkephalin degrading enzymes by RB 101 inhibits [^{3}H]diprenorphine binding. *Brain Res.* **571,** 306–312.
25. Aghajanian, G. K. (1978) Tolerance of locus coeruleus neurons to morphine and suppression of withdrawal response by clonidine. *Nature* (Lond.) **276,** 186–188.
26. Duman, R. S., Tallman, J. F., and Nestler, E. J. (1988) Acute and chronic opiate-regulation of adenylate cyclase in brain: specific effects in locus coeruleus. *J. Pharmacol. Exp. Ther.* **246,** 1033–1039.

27. Maldonado, R., Stinus, L., Gold, L., and Koob, G. F. (1992) Role of different brain structures in the expression of the physical morphine withdrawal syndrome. *J. Pharmacol. Exp. Ther.* **261,** 669–677.
28. Williams, J. T., Christie, M. J., North, R. A., and Roques, B. P. (1987) Potentiation of enkephalin action by peptidase inhibitors in rat locus coeruleus. *J. Pharmacol. Exp. Ther.* **243,** 397–401.
29. Noble, F. and Roques, B. P. (1995) Assessment of endogenous enkephalins efficacy in the hot plate test in mice: comparative study with morphine. *Neurosci. Lett.* **185,** 75–78.
30. Abbadie, C., Honoré, P., Fournié-Zaluski, M. C., Roques, B. P., and Besson, J. M. (1994) Effects of opioids and non opioids on c-Fos-like immunoreactivity induced in rat lumbar spinal cord neurons by noxious heat stimulation. *Eur. J. Pharmacol.* **258,** 215–227.
31. Tölle, T. R., Schadrack, J., Castro-Lopes, J., Evan, G., Roques, B. P., and Zieglgänsberger, W. (1994) Effects of kelatorphan and morphine before and after noxious stimulation on immediate-early gene expression in rat spinal cord neurons. *Pain* **56,** 103–112.
32. Honoré, P., Buritova, J., Fournié-Zaluski, M. C., Roques, B. P., and Besson, J. M. (1997) Antinociceptive effects of RB 101, a complete inhibitor of enkephalin-catabolizing enzymes, are enhanced by a CCK-B receptor antagonist as revealed by noxiously-evoked spinal c-fos expression in the rat. *J. Pharmacol. Exp. Ther.* **281,** 208–217.
33. le Guen, S., Noble, F., Fournié-Zaluski, M. C., Roques, B. P., Besson, J. M., and Buritova, J. (2002) RB101(S), a dual inhibitor of enkephalinases, does not induce antinociceptive tolerance, or cross-tolerance with morphine: a c-Fos study at the spinal cord level. *Eur. J. Pharmacol.* **441,** 141–150.
34. Nestler, E. J. and Aghajanian, G. K. (1997) Molecular and cellular basis of addiction. *Science* **278,** 58–63.
35. Whistler, J. L., Chuang, H. H., Chu, P., Jan, L. Y., and von Zastrow, M. (1999) Functional dissociation of mu opioid receptor signaling and endocytosis: implications for the biology of opiate tolerance and addiction. *Neuron* **23,** 737–746.
36. Di Chiara, G. (1999) Drug addiction as dopamine-dependent associative learning disorder. *Eur. J. Pharmacol.* **375,** 13–30.
37. Daugé, V., Kalivas, P. W., Duffy, T., and Roques, B. P. (1992) Effect of inhibiting enkephalin catabolism in the VTA on motor activity and extracellular dopamine. *Brain Res.* **599,** 209–214.
38. Dourmap, N., Michael-Titus, A., and Costentin, J. (1990) Acetorphan, an enkephalinase inhibitor, modulates dopaminergic transmission in rat olfactory tubercle, but not in the nucleus accumbens and striatum. *Eur. J. Neurosci.* **2,** 783–787.
39. Heidbreder, C., Gewiss, M., Lallemand, S., Roques, B. P., and De Witte, P. (1992) Inhibition of enkephalin metabolism and activation of mu- or delta-opioid receptors elicit opposite effects on reward and motility in the ventral mesencephalon. *Neuropharmacology* **31,** 293–298.
40. Mas Nieto, M., Wilson, J., Cupo, A., Roques, B. P., and Noble, F. (2002) Chronic morphine treatment modulates the extracellular levels of endogenous enkephalins in rat brain structures involved in opiate dependence. A microdialysis study. *J. Neurosci.* **22,** 1034–1041.
41. Laschka, E., Teschemacher, H., Mehraein, P., and Herz, A. (1976) Sites of action of morphine involved in the development of physical dependence in rats. II. Morphine withdrawal precipitated by application of morphine antagonists into restricted parts of the ventricular system and by microinjection into various brain areas. *Psychopharmacologia* **46,** 141–147.
42. Haffmans, J. and Dzoljic, M. R. (1987) Inhibition of enkephalinase activity attenuates naloxone-precipitated withdrawal symptoms. *Gen. Pharmacol.* **18,** 103–105.
43. Bhargava, H. N. (1977) Opiate-like action of methionine-enkephalin in inhibiting morphine abstinence syndrome. *Eur. J. Pharmacol.* **41,** 81–84.
44. O'Brien, C. P. (1993) Opioid addiction, in *Handbook of Pharmacology. Opioids II* (Herz, A., Akil, H. and Simon, E. J., eds.), Springer-Verlag, Berlin, pp. 803–823.

45. Ruiz, F., Fournié-Zaluski, M. C., Roques, B. P., and Maldonado, R. (1996) Similar decrease in spontaneous morphine abstinence by methadone and RB 101, an inhibitor of enkephalin catabolism. *Br. J. Pharmacol.* **119,** 174–182.
46. Gibert-Rahola, J., Tejedor, P., Chover, A. J., Payana, M., Rodriguez, M. M., Leonsegui, I., Mellado, M., Mico, J. A., Maldonado, R., and Roques, B. P. (1990) RB 38B, a selective endopeptidase inhibitor, induced several of escape deficits cause by inescapable shocks pretreatment in rats. *Eur. J. Pharmacol.* **183,** 2317–2325.
47. Baamonde, A., Daugé, V., Ruiz-Gayo, M., Fulga, I. G., Turcaud, S., Fournié-Zaluski, M. C., and Roques, B. P. (1992) Antidepressant-type effects of endogenous enkephalins protected by systemic RB 101 are mediated by δ opioid and dopamine D_1 receptor stimulation. *Eur. J. Pharmacol.* **216,** 157–166.
48. Tejedor-Real, R., Mico, T. P., Maldonado, R., Roques, B. P., and Gibert-Rahola, J. (1995) Implication of endogenous opioid system in the learned helplessness model of depression. *Pharmacol. Biochem. Behav.* **52,** 145–152.
49. Tejedor-Real, R., Mico, J. A., Maldonado, R., Roques, B. P., and Gibert-Rahola, J. (1993) Effect of a mixed (RB 38A) and selective (RB 38B) inhibitors of enkephalin-degrading enzymes on a model of depression in the rat. *Biol. Psychiatr.* **34,** 100–107.
50. Smadja, C., Maldonado, R., Turcaud, S., Fournié-Zaluski, M. C., and Roques, B. P. (1995) Opposite role of CCK-A and CCK-B receptors in the modulation of endogenous enkephalins antidepressant-like effects. *Psychopharmacology* **128,** 400–408.
51. Van Ree, J. M., Gerrits, M. A., and Vanderschuren, L. J. (1999) Opioids, reward and addiction: an encounter of biology, psychology, and medicine. *Pharmacol. Rev.* **51,** 341–396.
52. Blake, M. J., Stein, E. A., and Czech, D. A. (1987) Drinking-induced alterations in reward pathways: an *in vivo* autoradiographic analysis. *Brain Res.* **413,** 111–119.
53. Bertrand, E., Smadja, C., Mauborgne, A., Roques, B. P., and Daugé, V. (1997) Social interaction increases the extracellular levels of Met-enkephalin in the nucleus accumbens of control but not of chronic mild stressed rats. *Neuroscience* **80,** 17–20.
54. Stein, L. and Belluzzi, J. D. (1979) Brain endorphins. possible role in reward and memory formation. *Fed. Proc.* **38,** 2468–2472.
55. Goeders, N. E., Lane, J. D., and Smith, J. E. (1984) Self-administration of methionine enkephalin into the nucleus accumbens. *Pharmacol. Biochem. Behav.* **20,** 451–455.
56. Cappendijk, S. L. T., Hurd, Y. L., Nylander, I., van Ree, J. M., and Terenius, L. (1999) A heroin-, but not cocaine-expecting, self-administration state preferentially alters endogenous brain peptides. *Eur. J. Pharmacol.* **365,** 175–182.
57. Gerrits, M., Wiegant, V.M., and van Ree, J. M. (1999) Endogenous opioids implicated in the dynamics of experimental drug addictions: an *in vivo* autoradiographic analysis. *Neuroscience* **89,** 1219–1227.
58. Schultz, W., Dayan, P., and Montague, P. R. (1997) A neural substrate of prediction reward. *Science* **275,** 1593–1599.
59. Garris, P. A., Kilpatrick, M., Bunin, M. A., Michael, D., Walker, Q. D., and Wightman, R. M. (1999) Dissociation of dopamine release in the nucleus accumbens from intracranial self-stimulation. *Nature* **398,** 67–69.
60. Koob, G. F. (1992) Drugs of abuse: anatomy, pharmacology and function of reward pathways. *Trends Pharmacol. Sci.* **13,** 177–184.
61. Spanagel, R., Herz, A., and Shippenberg, T. S. (1990) The effects of opioid peptides on dopamine release in the nucleus accumbens: an in vivo microdialysis study. *J. Neurochem.* **55,** 1734–1740.
62. Yokoo, H., Yamada, S., Yoshida, M., Tanaka, T., Mizoguchi, K., Emoto, H., Koga, C., Ishii, H., Ishikawa, M., Kurasaki, N., Matsui, M., and Tanaka, M. (1994) Effect of opioid peptides on dopamine release from nucleus accumbens after repeated treatment with methamphetamine. *Eur. J. Pharmacol.* **256,** 335–338.

63. Bozarth, M. A. (1987) neuroanatomical boundaries of the reward-relevant opiate-receptor field in the ventral tegmental area as mapped by the conditioned place preference method in rats. *Brain Res.* **414,** 77–84.
64. Bals-Kubik, R., Ableitner, A., Herz, A., and Shippenberg, T. S. (1993) Neuroanatomical sites mediating the motivational effects of opioids as mapped by the conditioned place preference paradigm in rats. *J. Pharmacol. Exp. Ther.* **264,** 489–495.

Part II

Psychostimulant Addiction

6

Recent Advances in the Molecular Mechanisms of Psychostimulant Abuse Using Knockout Mice

Cécile Spielewoy and Bruno Giros

In the last years, generalization of homologous recombination approaches in the mouse has been extremely fruitful for our general understanding of the molecular mechanisms underlying the acute and chronic effects of psychostimulant drugs. Indeed, this technique based on genetic manipulations provides advantages that surpass inconveniences from pharmacological tools and offer unexpected insights into the field of action of psychostimulant drugs. For example, the extremely sharp precision of gene invalidation surpasses the specificity of drugs that can be available for in vivo experiments, and this is especially important in the study of psychostimulants given that these drugs are not specific within the family of their target proteins. Also, even though the technique may be hampered by genetic heterogeneity and developmental adaptations bias, the life-long consequences of the gene deletion together with the possibility of a reliable reproducibility in animal groups present a strong and exciting basis of investigation.

In this chapter, we will successively review studies performed on mice strains that were developed from genetic disruption of psychostimulant primary targets (transporters), secondary targets (receptors), and upstream (presynaptic) or downstream (postsynaptic) significant proteins.

1. Amine Transporters

The amine transporters, first described in the 1960s *(1)*, are presynaptic proteins terminating a transmission event by emptying the synaptic cleft from their cognate neurotransmitter. They act via an active reuptake of the released neurotransmitter from the extracellular space back into the presynaptic element. Thus, these transporters directly regulate the temporal and spatial action of the neurotransmitter inside and beyond the synapse. Consequently, their blockade triggers very important changes in the brain homeostasis, with mainly a potentiation of the transmission to an extent that cannot be reached under regular physiological functioning. Cocaine and amphetamine, the two psychostimulant drugs to which we refer in this review, are potent blockers of all three classes of transporters, the dopamine (DAT), norepinephrine (NET), and serotonin (SERT) transporters. However, even though these psychostimulants share common targets (*see* Table 1), they do not have the same mechanism of action. Indeed, both cocaine and amphetamine the reuptake of the neurotransmitter by blocking the transporter, but amphetamine is in addition able to trigger reverse transport of the transmit-

From: *Molecular Biology of Drug Addiction*
Edited by: R. Maldonado © Humana Press Inc., Totowa, NJ

Table 1
K_i Values for Prototypical Transporter Inhibitors

K_i (n*M*)	hDAT	hSERT	hNET
Psychostimulants			
Cocaine	260	180	91
S(+) D-amphetamine	180	54,000	66
R(−) L-amphetamine	760	266,000	35
DAT > (SERT, NET)			
Methylphenidate	24	44,000	234
GBR12935	27	940	310
Bupropion	630	15,600	2,300
SERT > (NET , DAT)			
Paroxetine	490	0.1	40
Citalopram	28,100	1.2	4,070
Fluoxetine	1,600	12	280
NET > (SERT, DAT)			
Mazindol	8.1	39	0.5
Desipramine	9,250	17.6	1.2
Reboxetine	>10,000	720	11

ter from the cytoplasm to the synapse by activating its release from neuronal storage vesicles. In comparison to cocaine, this additional property of amphetamine amplifies the elevation in extracellular concentration of the transmitter and allows the possibility of a release of the transmitter under extreme conditions when the neuron is electrically silent.

Because of the very close K_i values of psychostimulants for DAT, NET, and SERT and in the absence of selective transporter inhibitors, psychopharmacological studies were unable to assign directly a precise role for one of these transporters in the mechanisms of action of cocaine and amphetamine. However, indirect approaches slowly highlighted the importance of the DAT in being the main target for the reinforcing properties of psychostimulants. For example, the observation that antidepressant drugs, such as the tricyclics or the serotonin selective reuptake inhibitors (SSRI), poorly inhibit DAT and are devoid of addictive or reinforcing properties.

More systematically, Kuhar and collaborators were able to correlate the potency of various drugs to elicit a positive reinforcement in monkeys or rodents with their respective affinities to inhibit the DAT on in vitro preparations *(2)*. At the same time, the group of Di Chiara showed that dopamine (DA) is released in the nucleus accumbens (NAC) of rats that have been administered any class of drugs known to be addictive in humans *(3)*. These two observations established a conceptual framework for the action of psychostimulant drugs in which DA and the DAT were almost self-sufficient to square the circle.

In the early 1990s, all amine transporters were cloned and identified in various species including rodents and humans *(4–6)*. They were classified as part of the large family of Na^+/Cl^--dependent transporters, described to be single-subtype representative for each amine and in the brain, to be exclusively expressed in amine-containing neurons, at a presynaptic level. This molecular knowledge allowed the production of specific tools for the study of the amine transporters, as probes for *in situ* hybridization, antibodies for protein localization, eukaryotic cell lines for the study of their expression and functional characterization. Regarding their key role in neurotransmission, it appeared essential to create specific gene disruption of these transporters in order to investigate in more depth their functional role in homeostasis processes, as well as to develop useful models for the study of integrated brain functions.

1.1. Dopamine Transporter Null Mice

Mice with a genetic disruption of the DAT gene are viable, but are dwarf and display a decrease in survival *(7)*. Overall, the DAT-deficient mice present many physiological modifications, such as an anterior pituitary hypoplasia and an inability to lactate *(8)* that has dramatic consequences on their maternal behavior *(7,9)*, as well as altered distal colonic motility that may affect their eating behavior *(10)*. These findings revealed an important role for endogenous DA in modulating these peripheric functions. Regarding basic behaviors, DAT-deficient mice exhibit disturbed sensorimotor gating as measured by a deficit in the prepulse inhibition of the startle response *(11)*, important baseline sleep–wake abnormalities despite normal circadian patterns of inactivity and activity *(12,13)*, and cognitive dysfunctions *(14)*. On the contrary, these mutant mice show normal social behavior and no changes in aggressiveness when compared to their wild-type littermates *(13)*. Taken together, these observations offer a large image of the physiological as well as integrated consequences of the DAT gene removal and indicate clear differences in the implication of DA in various behaviors. In this chapter, however, we will focus on the effects of the DAT deletion on the action of psychostimulants, providing unpredicted mechanisms in the absence of the supposed main target of the drugs.

As expected from our knowledge about the DAT function, its removal dramatically prolongs the lifetime of extracellular DA (by at least a persistence in the synaptic space of 100 times longer), leading to five-times greater extracellular DA concentrations in the mutant mice compared to controls, that is, a hyperdopaminergia phenotype *(7)*. Consequently, adaptive mechanisms take place in order to compensate the phenotype, establishing a biochemical situation that can be described as a DA-deficient, but functionally hyperactive, mode of DA neurotransmission. Indeed, Giros et al. *(7)* and Jones et al. *(15)* showed that although there is this dramatic increase in extracellular DA concentrations, the tissue contents of the transmitter is reduced by 95% in DAT-deficient mice. Accordingly, the protein level of tyrosine hydroxylase (TH), the rate-limiting DA synthesizing enzyme, is decreased by 90% in the striatum (STR) and 35% in the NA, and its maximal activity as measured in brain homogenates is 80% lower in mutant than in wild-type mice *(16)*. However, the DA synthesis rates are doubled because of the decrease of inhibitory control normally driven by newly transported DA in the cytoplasm and the DA D2 autoreceptors. Thus, the few TH molecules present in the DAT-deficient mice are very efficient in converting tyrosine to L-DOPA *(15)*. Importantly, these studies reported no changes in the anatomy of the TH projections *(16)*. Examination of the electrically-stimulated release of DA indicated a reduction of 75% in the amplitude of the transmitter released, due to a disturbed burst firing activity of the DA neurons and to the low concentrations of intracellular DA *(15,17)*. These two laboratories *(15,17)* further showed that no other processes besides diffusion can compensate for the elimination of the extracellular DA and that in the DAT mutant mice, the DA diffusion distance is 10 times higher than in wild-type animals. Finally, Jones et al. *(18)* demonstrated that in the absence of DAT, the important inhibitory feedback mechanism provided by DA D2 autoreceptors on impulse-, synthesis-, and release-regulating functions of DA neurons is dramatically affected, as these autoreceptors show a 90% loss of activity. Thus, it is interesting to note that the DAT mutant mice reveal that the DAT is not only regulating the lifetime (duration and intensity) of extracellular DA, but is also critically involved in the regulation of presynaptic events as the balance among DA synthesis, release, and degradation.

Furthermore, disruption of the DAT gene also deeply affects postsynaptic DA homeostasis. First reported *(7)* and since largely confirmed *(19,20)* is the dramatic downregulation of the main DA D1 and D2 receptors, both at their mRNA (postsynaptically in the striatum and NA and for the D2 receptor, also presynaptically in the ventral tegmental area [VTA] and the substantia nigra, [SN]) and protein (STR and NA) levels. This 50% decrease in receptor density has never been reached before using any other kind of genetic, pharmacological, lesion, or behavioral manipulation. Interestingly, the mRNA expression of the DA D3 receptor is slightly increased in the NA *(19)*, while the DA D4 and D5 receptors have not been investigated so far. Moreover, it was recently reported that in these mice the remaining DA D1 receptors are mostly internalized in the rough endoplasmic reticulum and Golgi apparatus *(21)*, an effect that is shown to be related directly to the enhanced DA transmission and hereby highlights the fact that the abundance and availability of DA receptors on the extracellular membrane are dependent on the neurotransmitter tone. As DA D1 and D2 receptors are collocalized with neuropeptides, a modulation of these peptide's expression was observed in the DAT-deficient mice *(7)*. In the STR and NA, preproenkephalin mRNA is decreased in D2-containing GABAergic neurons and preprodynorphin mRNA increases in D1-containing GABAergic neurons. However, substance P mRNA expression is unchanged in D1-containing neurons.

This unprecedented plasticity of the DA system to compensate the hyperdopaminergia phenotype is not sufficient to prevent a huge rise in spontaneous locomotor activity in the DAT-deficient mice *(7)*. When introduced in a novel environment, the mutant mice exhibit a four- to five-times increase in locomotor activity. This hyperactivity is long-lasting, requiring 5–6 h to slowly decrease to the level of locomotion of normal mice *(9)* and is characterized by a nonfocal perseverative pattern of locomotion *(11)*. Moreover, no locomotor habituation is observed when the DAT-deficient mice are repeatedly tested in the same environment *(9)*. Overall, it has been reported that the motor hyperactivity of the DAT-deficient mice disturbs the expression of normal locomotor habituation, exploratory behavior, and response to inescapable stress *(13)*.

Interestingly, cocaine and amphetamine are unable to induce an increase in horizontal or vertical motor activity in DAT-deficient mice administered with the drugs after 2 h of habituation to the testing chamber *(7)*. Paradoxically, when the psychostimulants are administered just before, or 30 min after, introduction of the mice in the chamber, the animals exhibit a dramatic hypolocomotor behavior *(22,23)*. This hypolocomotion can also be induced by agents acting on different levels of the serotonin (5-HT) neurotransmission to increase the extracellular concentration of 5-HT *(22,23)*, unmasking a negative counterbalanced participation of the 5-HT system in the hyperlocomotor property of psychostimulants in normal mice. This contribution of the 5-HT neurotransmission is, however, independent of any changes in the density of two important regulators of the 5-HT system activity, the SERT and 5-HT_{1A} receptors *(23)*. The calming action of psychostimulants on locomotion of DAT-deficient mice does not change upon chronic administration *(23)*, either toward a decrease (tolerance) or an increase (sensitization), suggesting a ceiling effect of the drugs on the behavior.

Even more puzzling than the hypolocomotor effect of psychostimulant in DAT-deficient mice, is the observation that these mutant mice are still responsive to the reinforcing effects of cocaine to an extent similar to that of their wild-type littermates. The remaining property of cocaine was observed in two different strains of DAT-deficient mice generated in two independent laboratories and was measured by using either a conditioned place preference paradigm *(24)* or a self-administration protocol *(25)*. These observations totally unravel the framework of the cocaine-reinforcing effects being mediated through a blockage of the DAT!

Obviously, cocaine is still able to elicit its reinforcing properties through an inhibition of alternative targets, possibly the NET or SERT (Table 1). Another unexpected finding is the absence of induction of DA release in the STR after cocaine administration despite the clear reinforcing properties of the drug *(25)*. This *a priori* impossible situation is tricky to understand and can hardly offer a connection between how cocaine can still trigger rewarding effects in the absence of the DAT, which was supposed to be its primary target, and in the absence of DA release in the midbrain which was supposed to be a *sine qua non* obligation for long-term dependency to drugs of abuse *(3)*. However, one has to keep in mind that the DA system is profoundly perturbed in DAT-deficient mice and that, for example, their long-term hyperdopaminergia phenotype may induce adaptive changes in the neuronal network used by drugs to activate their properties. In support, Rocha et al. *(25)* observed that 1 h after administration of a single dose of cocaine, the expression of the immediate early gene *c-fos* mRNA is increased in the STR, NA, and olfactory tubercle of wild-type mice, but not in DAT-deficient mice, suggesting that one of the first molecular events thought to participate in the early steps of cocaine action is not also obligatory for self-administration in these mutant mice. Fascinatingly, however, the authors reported that after the cocaine treatment, both DAT-deficient and wild-type mice showed an increased in *c-fos* mRNA in the anterior olfactory nuclei and piriform cortex, and to a lesser extend the orbital cortex. These brain regions are predominantly innervated by serotonin (5-HT) neurons and thus the results suggest a participation of this circuitry in the reinforcing effects of cocaine. This possibility is sustained by the fact that the cocaine congener [125]RTI-55 binds to the SERT in DAT-deficient mice, indicating a positive contribution of the 5-HT system in the maintenance of the rewarding properties of cocaine in DAT-deficient mice *(25)*. Finally, Sora et al. *(26)* recently reported that the place preference for cocaine is totally abolished in double knockout mice with no DAT gene copy and either no or one copy of the SERT (DAT/SERT knockout mice vs DAT knockout/SERT heterozygous mice), supporting the idea of a 5-HT dependence of cocaine reward in DAT-deficient mice. However, participation of the NET and the norepinephrine (NE) system in the reinforcing action of cocaine in DAT-deficient mice cannot be ruled out. Indeed, although *(22)* reported that amphetamine and fluoxetine, a SERT inhibitor, are unable to induce a release of DA in the STR of DAT-deficient mice, we recently demonstrated in collaboration with the group of Di Chiara that amphetamine and cocaine dose-dependently retain their capacity to increase DA release in the NA *(27)*. Thus, even in the absence of DAT, this mechanism seems still mandatory for the expression of the rewarding properties of psychostimulant drugs. Moreover, using reboxetine, a specific NET inhibitor, we are able to show a release of DA in the NA of DAT−/− mice, which is not the case in wild-type mice or by using fluoxetine. Together, these results indicate that the remaining positive-reinforcing effect of cocaine in the absence of DAT can be mediated by an inhibition of the NET that unusually and specifically activates a release of DA in the mutant mice. Our hypothesis for a contribution of the NE system is further supported by the observation that reboxetine is able to induce by its own a place preference behavior in the DAT-deficient mice only (Marika Nosten-Bertrand and Bruno Giros, unpublished data).

In summary, these findings seem to indicate that in the particular situation of a long-term hyperdopaminergia, the DAT is not required for the mediation of the reinforcing effects of cocaine and amphetamine even though this is still the case for its psychostimulant locomotor property. However, an increase of extracellular DA levels in the NA induced by the drugs remains a prerequisite event for the expression of the reinforcing effect, besides that in the DAT-deficient mice the release of DA is mediated by the NET, unraveling an unusual function of this transporter. On the other hand, the SERT appears extremely important in the paradoxical calming action of cocaine and amphetamine on the locomotor behavior of DAT mutant mice.

1.2. Serotonin Transporter Null Mice

Mice with a deletion of the SERT gene are viable, and present no obvious decrease in fertility, litter size, weight gain, and survival *(28)*. These mice show decreased responses to SSRI, and to the hallucinogenic drug, MDMA. However, there is no change at all in basal locomotion and the acute effect of psychostimulants in the SERT-deficient mice compared to their control littermates *(24,26,28)*.

1.3. Norepinephrine Transporter Null Mice

The NET-deficient mice are the most recent knockouts produced of the three amine transporters *(29)*. As already observed with DAT and SERT, deletion of the NET induces changes at different levels of the NE homeostasis, indicating that the transporter is not only regulating the extracellular lifetime of NE but is also critically involved in presynaptic and postsynaptic neuronal functions. Thus, directly related to the disruption of NET, Xu et al. *(29)* reported a twofold increase of extracellular NE levels in brain tissues from mutant mice compared to control littermates, an effect that is associated to a decrease in α1-adrenoreceptor binding. Furthermore, the authors observed a 55–70% decrease in tissue concentrations of NE in the NET-deficient mice, while the synthesis rates of the transmitter is increased, indicating that the NE storage levels in neurons are determined primarily by the transmitter reuptake through the NET rather than by synthesis rates *(29)*. Finally, Xu et al. *(29)* showed a reduction by 60% of the NE release in response to electrical stimulation and rates of clearance that are at least six-fold slower in brain tissues from mutants than from wild-type littermates.

Using simple experimental paradigms designed to evaluate the efficacy of antidepressant drugs, *(29)* found that the mutant mice are less immobile in the forced-swim test and the tail-suspension test because of a higher struggling activity when compared to control littermates. The NET-deficient mice actually act like normal mice treated with antidepressant drugs. This effect is not biased by locomotor disturbances related to the deletion, since the mutant mice present an overall lower locomotion level than the wild-type mice. Consequently, the NET-deficient mice are insensitive to the action of antidepressant drugs, a result consistent with the proposed role of NE in the action of these agents.

Administration of cocaine induces a significantly higher dose-dependent increase in locomotion in NET-deficient mice compared to wild-type controls *(29)*. A similar enhanced response is observed with a single dose of amphetamine. However, when treated for 5 d with cocaine according to a sensitization protocol, the mutant mice do not show a further augmentation in locomotion, which is the case for the wild-type mice *(29)*. This result suggests that the NET-deficient mice are already sensitized to the action of cocaine and that a repeated treatment with the higher dose of cocaine used (20 mg/kg) is not able to surpass a ceiling effect on locomotion already reached acutely with this dose of the drug. However, by using a conditioned place preference paradigm, the authors were able to further demonstrate the supersensitivity of these mutant mice to cocaine, since the NET-deficient mice presented a greater preference for the chamber previously associated with the injection of 20 mg/kg of cocaine than did control mice *(29)*. Once again, these results show that the locomotor and reinforcing effects of psychostimulants are dissociable for the same dose of drug. In order to explain the supersensitivity to psychostimulants observed here, Xu et al. *(29)* examined the DA neurotransmission that is viewed as the important mediator of drug effects. It is thus observed that the presynaptic DA activity is downregulated in NET-deficient mice and that the percent changes in extracellular DA concentrations in the STR in response to cocaine are unchanged in the mutant mice compared to their control littermates, two effects that do not sustain supersensitivity to the drug. Fortunately, further investigations showed an increased locomotor response to DA D2/D3 receptor agonists in the NET-deficient mice than in the wild-type animals, associated with an higher coupling of the DA D2/D3 receptors to their G proteins, in the absence of changes in their density. These results suggest that the D2/D3 supersensitivity may be responsible for the enhanced responses to psychostimulant drugs in NET-deficient mice. Overall, these findings also suggest that the NET deletion may have unraveled an inhibitory control of the NE transmission over DA transmission.

In summary, the generating of DAT-, SERT-, and NET-deficient mice teaches us about the implication of these three amine transporters in the direct mechanisms of action of cocaine and amphetamine to mediate their locomotor and reinforcing properties. Regarding locomotion, there is a total abolition of the psychostimulant effect of the drug in the absence of DAT that cannot be compensated by the inhibition of SERT or NET, suggesting a principal positive participation of the DAT target in the drug-induced behavior. On the contrary, removal of SERT does not affect this behavior, indicating no implication of SERT in the psycholocomotor action of cocaine and amphetamine and the maintenance of the behavior through the DAT. In the case of NET, there is an enhancement of the locomotor responses to the psychostimulant drugs, which can be related to a synergistic action of the drugs on DAT associated to the DA D2/D3 receptors supersensitivity, together indicating a negative influence of NET on this property of cocaine and amphetamine. Concerning the reinforcing properties of these drugs, it appears here that the NET (and, to be demonstrated, the SERT) can compensate the absence of DAT, suggesting that this supposed main target of psychostimulant drugs is not that important. However, removal of both the SERT and NET enhanced the reinforcing effects of cocaine and amphetamine, clearly showing that these two transporters normally play an inhibitory role in DAT function. In conclusion, all of these observations about amine transporter knockout mice point out the dramatic extended role of the transporter on presynaptic and postsynaptic events that deeply modify the homeostasis of the neurotransmitter system as well as its responses to psychostimulant drugs. In the following paragraphs, we will present how knockout mice for indirect targets of cocaine and amphetamine give us further insights into the cellular and molecular mechanisms of action of these drugs.

1.4. Vesicular Monoamine Transporter (VMAT) Null Mice

The vesicular monoamine transporter (VMAT) accumulates amines (dopamine, serotonin, noradrenaline, adrenaline, and histamine) in vesicles against their gradient. This process uses the proton and potential gradient which is established by the V-ATPase proton pump; amines are exchanged (antiport mechanism) against protons and can reach a concentration that is 100,000 times their cytoplasmic concentration. Packaging in the vesicles plays a dual role, protecting the amines from oxidation in an acidic compartment and filling the vesicles for the next release. Vesicular transporters are therefore mandatory proteins to establish the phenotype and functionality of their cognate neurons.

Administration of reserpine or tetrabenazine, which are very potent vesicular monoamine transporter inhibitors *(30,31)*, results in complete neurotransmitter depletion from amine-containing neurons. When given chronically, treatment with these drugs will have anti-hypertensive properties by depleting the pool of blood circulating monoamines; reserpine is actually the cheapest hypotensive drug, still widely used in countries that cannot afford expensive treatments *(32)*. Depletion in the central pool of monoamines by these inhibitors may produce depressive-like effects, as opposed to those that are observed in antidepressant treatments. Such observations in the late 1950s were actually the basis for establishing the role of aminergic transmission in psychiatric disorders *(33)*. In the long term, high doses of reserpine will eventually be lethal. However, it was surmised that such lethality is a consequence of both central and peripheral effects on monoamines. The molecular cloning of two different subtypes *(34,35)*, which were not suspected from pharmacological studies, show that the vesicular monoamine type 1 (VMAT-1) was mostly present in the adrenal, whereas VMAT-2 was expressed in the central nervous system. These two transporters are closely related, with primary sequence and a pharmacological pattern almost analogous. It was therefore extremely tempting to remove only the brain-specific subtype VMAT-2, in order to understand the physiological consequences of a central dopamine, noradrenaline, serotonin, adrenaline and histamine depletion. Again, from experiments with inhibitory drugs, it was not totally surprising to observe that the VMAT-2 knockout mice cannot survive long after birth . All studies reporting the VMAT-2 knockout found that $-/-$ mice all died shortly after birth; none of them survived after the second postnatal week *(36–38)*. Death was probably a consequence of the complete disappearance of the main monoamines, dopamine, serotonin, and noradrenaline *(37,38)*. However, indicating that the synthesis and metabolism pathways of monoamines were only little affected, all metabolites, DOPAC, HVA, and 5-HIAA, were at the same level in knockout and wild type mice. In primary midbrain cultures from VMAT2 knockout mice *(37)*, K^+-induced depolarization cannot release any dopamine, whereas amphetamine can release some, probably by reversal of the plasmic dopamine transporter *(39,40)*, again indicating that the synthesis pathway is functional, even if dopamine cannot accumulate at all. Interestingly, taking advantage of this observation, Fon et al. were able to increase the VMAT-2 knockout mice's survival up to 2 wk, by subcutaneous administration of amphetamine to the pups *(37)*. This amphetamine administration actually increased the feeding behavior of the VMAT-2 knockout mice, and this was reversible by nomifensine, again indicating that DAT is recruited for dopamine release (otherwise, nomifensine would have potentiated the amphetamine effects).

However, it was possible to study the VMAT-2 heterozygous mice, and to compare their behavior and physiology to the wild-type animals. The idea was to consider that the VMAT-2 density in vesicles is not in excess, something that was actually confirmed later vesicles from manipulated synapses *(41,42)*. Therefore, it was observed that amphetamine *(36,38)*, but also cocaine, the D1/D2 agoniste apomorphine, and alcohol *(38)*, all produced a higher motor response in VMAT-2 heterozygous mice as compared to their control littermates. This supersensitivity to dopamine agonists is presently not understand, but is probably of the same kind as was observed on reserpinized rats *(33)*. This presensitized state is also evidenced using a locomotor sensitization protocol, where the VMAT-2 heterozygous mice do not sensitize to chronic administration of cocaine for 1 wk. This lack of sensitization actually happened because the heterozygous animals have a locomotor activity on the first day of the protocol that is already as elevated as the activity of wild type animals after a 1-wk sensitization *(38)*.

Using a conditioned place preference paradigm, Takahashi et al. *(36)* were able to assess that the heterozygous VMAT-2 mice would actually spend more time in the amphetamine-paired compartment. Even if the heterozygous mice spend less time than their control littermates, they can still retain the appetitive effects of the psychostimulant drug. Interestingly, this is again an experimental condition that involves aminergic transmission, in which a lack of drug sensitization (or, alternatively a presensitized state) is not paralleled by the lack of appetitive effects of the drug as evidenced by conditioned place preference. It is, however, difficult to go further in the explanation of this difference, because VMAT-2 is not specifically from a given aminergic pathway. Conditional knockout of VMAT-2, in more homogeneous neuronal populations or after development, would be necessary for a better understanding of these mechanisms.

2. Dopamine Receptors (*see* Table 2)

So far, five distinct DA receptor subtypes have been isolated and characterized within the central nervous system (CNS). They all belong to the family of the seven-transmembrane-domain G-protein-coupled receptors and have been divided into two subfamilies, the D1- and D2-like receptors, on the basis of their interaction with the G proteins and their transductional properties *(44)*. Classically, the D1-like receptor subfamily consists of two members, the so-called D1 and D5 receptors, which are positive regulators of the cAMP pathway by coupling to a stimulatory G_s protein, which induces

Table 2
Principal In Vitro Characteristics of Dopamine Receptors[a]

	D1-like		D2-like		
	D1	D5	D2	D3	D4
Structure	446 AA no intron	475 AA no intron	444 AA 6 introns	446 AA 5 introns	385 AA 3 introns
Distribution	Substantia nigra pars reticulata Striatum Nucleus accumbens Olfactory tubercles	Hippocampus Thalamus	Substantia nigra pars compacta Ventral tegmental area Striatum Nucleus accumbens Olfactory tubercles Amygdala Cortex	Substantia nigra pars compacta Ventral tegmental area Nucleus accumbens shell Olfactory tubercles Island of Calleja Cortex	Striatum Nucleus accumbens Amygdala Hippocampus Cortex
Expression	High	Low	High	Low	Low
Co-expression	Weak with D2/D3 receptors	Weak with D2 receptors	Weak with D1/D3/D5 receptors	Weak with D1/D2 receptors	ND
Colocalization	Dynorphin Substance P	ND	Enkephalin Neurotensin	Substance P Neurotensin	ND
Function	Postsynaptic receptor	Postsynaptic receptor	Autoreceptor Postsynaptic receptor	Autoreceptor Postsynaptic receptor	Autoreceptor? Postsynaptic receptor
Coupling to cAMP pathway	Positive	Positive	Negative	Negative	Negative
Affinity for dopamine	+/–	+	+	++	++
Agonist	SKF-38393	SKF-38393	Apomorphine	7-OHDPAT	Apomorphine
Antagonist	SCH-23390	SCH-23390	Haloperidol	Nafadotride	Spiperone

[a]ND, not determined; 7-OHDPAT, 7-hydroxy-dipropylaminotetralin.

adenylate cyclase (AC) activity. In contrast, modulation of ion channels and/or inhibition of AC activity via an interaction with an inhibitory $G_{i/o}$-protein is a common property of the D2, D3, and D4 receptors that compose the D2-like receptor subfamily. Besides this distinct biochemical property, the two subfamilies of DA receptors are differentiated also structurally, physiologically, pharmacologically, and in their anatomical distribution *(45,46)*. Differences remain also within each subfamily of DA receptors. Discriminating the role of every subtype has been complicated by the absence of selective pharmacological tools able to distinguish specifically between D1 and D5 receptors, or among the three members of the D2-like receptors subfamily.

Therefore, pharmacological studies investigating the contribution of dopamine receptors in the action of psychostimulant drugs have focused on D1 and D2 receptors. Blockage of these receptors induces an inhibition of well-characterized properties of cocaine and amphetamine, including locomotor hyperactivity and stereotyped behaviors *(47)*, rewarding and reinforcing properties assessed by self-administration and place-preference paradigms *(48–50)*, and the development of psychostimulant-induced behavioral sensitization *(51)*. Moreover, there is some evidence that D1 and D2 receptors contribute in opposite ways to mediate the action of psychostimulant drugs on downstream targets, such as the expression of the immediate early genes *c-fos* and *zif*268 *(52,53)* and that of neuropeptide genes coexpressed with these receptors, including substance P, dynorphin, enkephalin, and neurotensin *(54)*. Recently, development of more selective ligands for the D3 receptor provides the opportunity to investigate the role of this receptor in the action of psychostimulant drugs. It was found that D3 receptors have opposite effects to the D2 subtype on cocaine self-administration behavior *(55–57)* and gene expression of neurotensin *(58)*. However, the selectivity of the ligands used and therefore the results obtained have been discussed. In addition, as reported before, the absence of ligands selective for the D4 and D5 receptors has limited the study of these receptors in the biochemical and behavioral properties of psychostimulant drugs.

In this context, genetically manipulated mice represent an original alternative to study more precisely the intrinsic roles of the individual subtypes of DA receptors and to give more insights about their specific participation in the mechanisms of action of psychostimulant drugs of abuse.

2.1. D1-Like Receptors

2.1.1. D1 Receptor Null Mice

Two independent laboratories have generated mice lacking functional D1 receptors using gene targeting by homologous recombination *(59,60)*. In both case, the D1 mutant mice are growth-retarded and show a general failure to thrive after weaning, characterized by impairment in motivated-driven behaviors toward food and water. Analyses of motor behaviors in an open field environment reveal some inconsistencies, as Xu et al. *(59)* originally reported a small hyperactivity in D1-deficient mice associated with a decrease in rearing, while Drago et al. *(60)* showed that the mutant mice displayed hypoactivity and fewer rearing events. Finally, it has been suggested that D1-deficient mice exhibit a prominent shift between individual elements of behaviors that is characterized by a reduction in some forms of rearing, and a moderate increase in locomotion and grooming *(61)*. Further examinations of these mutant mice revealed a significant reduction in substance P and dynorphin striatal levels and a disorganization of the striosome-matrix cellular complex of the striatum, which might lead to an abnormal development of this structure in the absence of D1 receptors *(60,62)*. Consistent with the lack of D1 receptor, no stimulation of D1-mediated cAMP production is observed in the CNS *(62)*. These mice also exhibit higher levels of DA and DA metabolites restricted to the midbrain, whereas D2-like receptor binding, dopamine transporter binding, and enkephalin levels are unaltered *(59,60)*.

The first studies investigating the effects of psychostimulant drugs in D1-deficient mice showed a failure of acute cocaine administration to activate locomotion in the mutant mice and the abolition of certain stereotyped behaviors, such as rearing, sniffing, and grooming *(60,62)*. If anything, high doses of the drug induced hypoactivity. The complete absence of cocaine-induced locomotor activity and stereotyped behaviors in D1 mutant mice is consistent with pharmacological data reporting in rats and mice the full prevention of these effects of cocaine by D1 antagonists *(63,64)*. Thus, the present finding underscores the fundamental role of D1 receptors in some actions of cocaine. However, Xu et al. *(62)* showed that NA neurons were still poorly sensitive to cocaine in D1 mutant mice; a remaining effect of the drug that might be sustained by an action on the SERT and thus implicating an inhibition of the accumbal neurons via a serotoninergic mechanism. The induction of hypoactivity at high doses of cocaine is a parameter that can support the participation of serotonin, as it has been described that limbic serotonergic projection from the raphe nucleus has opposing influences on cocaine-induced locomotor activity *(65)*.

In contrast to cocaine, acute administration of moderate doses of amphetamine tends to decrease the locomotion of mutant mice, whereas a higher dose generates significant hyperactivity, even though the intensity of the effect is attenuated compared to wild-type mice *(66,67)*. However, amphetamine also decreases the number of rearing and grooming behaviors in these mutant mice. The difference between the locomotor effects of acute cocaine and amphetamine might be related to the higher potential of amphetamine to increase extracellular DA levels in the STR and NA *(40)*, thereby overlapping the negative influence of serotonin on locomotion in the absence of D1 receptors.

A similar divergence is also revealed when both drugs are administered subchronically, using a psychostimulant-induced locomotor sensitization design. Indeed, whereas repeated intermittent administrations of cocaine totally lack to develop a progressive increment in the level of locomotion in D1-deficient mice, the same treatment with amphetamine generates a moderate locomotor sensitization in the mutants as compared to drug-treated wild-type mice *(66,67)*. This result is in contrast to an earlier study reporting the absence of a progressive increase in locomotor activity across days in D1-deficient mice treated with amphetamine *(68)*. However, there were differences in the experimental design used and the dose administered *(66,68)*, which actually support the idea that amphetamine is dose-dependently able to maintain its locomotor effects in the D1 mutant mice. More striking is the observation that although the drug schedule used by *(68)* did not develop an increase of locomotion across days, D1-deficient mice exhibited behavioral sensitization when retested with the drug after 3 d of abstinence. This is in accordance with the idea that different neuronal mechanisms sustain the development and expression of locomotor sensitization *(69)*. The study also confirmed the importance of D1 receptors in the development of sensitization, whereas its role in the expression of the phenomena remains unclear *(51,70)*. The retained action of the drug might be more mediated by other DAergic and/or heterolog neuronal mechanisms. Consistently, *(68)* reports a relation between the absence of progressive increase in locomotion during the treatment with amphetamine and the absence of drug-induced cAMP protein kinase (PKA) activity.

Moreover, the presence of neuronal processes able to preserve some action of psychostimulant drugs in the absence of D1 receptors is supported by the persistence of conditioned place preference to cocaine in the mutant mice *(71)*. Thus, the present result challenges the presumed important role of D1 receptors in the rewarding and reinforcing effect of cocaine and argues for an independent implication from alternative neuronal mechanisms in the expression of environmental cues-associated learning.

Finally, it was shown that cocaine and amphetamine are deficient to activate the immediate early genes *c-fos* or *Jun*B/*zif*268 and to regulate genes encoding the neuropeptide dynorphin that is co-expressed in D1 striatal neurons *(72,73)*. Thus, induction of these genes is dependent of D1 receptors, and their insensitivity to psychostimulant drugs might explain the loss or attenuation of some locomotor effects of these drugs. In contrast, cocaine-treatment increased substance P expression is not blocked in D1-deficient mice but produces an abnormal elevation in brain areas where the neuropeptide was not activated in wild-type mice, suggesting the involvement of mechanisms other than D1 receptors *(73)*. Futhermore, a D2-receptor antagonist (haloperidol) or agonist (quinelorane) remains effective in regulating the immediate early genes *c-Fos* and *Jun*B in D1-deficient mice *(72)*, demonstrating that some functions of the D2 receptors are still available in the absence of D1 receptors.

In conclusion, the D1 mutant mice reveal that the locomotor-activating action of psychostimulant drugs is strongly related to D1 receptor stimulation, whereas some other properties of these drugs, such as their rewarding effect, can be mediated by alternative homolog or heterolog neuronal mechanisms.

2.1.2. D5 Receptor Null Mice

A generation of D5-deficient mice has been recently obtained using homologous recombination techniques. The mice survive, do not present gross developmental deficits, and present normal fertility. Preliminary studies showed that these mutant mice exhibit a significantly general higher score of motor activity in an open field environment that is not correlated to a reduced level of anxiety *(74)*. This result suggests that D5 receptors do normally decrease locomotion, in opposition to the facilitating action of its related D1-like receptor subtype.

Unfortunately, no experiment has been published so far investigating the action of cocaine and amphetamine in D5-deficient mice, still questioning the potential role of this receptor in the mechanisms of action of psychostimulants drugs.

2.2. D2-Like Receptors

2.2.1. D2 Receptor Null Mice

Two independent groups of investigators have generated D2-deficient mice and reported discrepancies in the behavioral phenotypes observed. Indeed, Baik et el. *(75)* claimed a Parkinsonian-like phenotype, as their D2 mutant mice exhibit severe neurological impairments, with abnormal posture and gait, and slow movements. These mutant mice are also cataleptic and impaired on the rotarod test, a measure of locomotor coordination and learning. In contrast, Kelly et al. *(76)* did not report the same severe impairments in their D2-deficient mice, since these mutant mice are not cataleptic, exhibit normal posture and gait, and no tremor or ataxia. Also, even though D2-deficient mice are impaired on the rotarod test, these investigators reported that the mutant mice are able to learn and perform the task. Discrepancies between the two groups also concern the growth, development, fertility, breeding, and body weight of these deficient mice *(75,76)*. The divergences in the results obtained by the two laboratories might be determined by dissimilarities in the gene dosage and genetic background of the two strains of D2-deficient mice.

Biochemically, D2-deficient mice are characterized by a loss of autoreceptor function in DA neurons, thereby supporting the important role that D2 receptors play at this level *(77)*. This lack of function is not associated with changes in electrophysiological properties of DA neurons or expression of tyrosine hydroxylase gene expression, suggesting that the DA pathway is unaffected by the absence of D2 receptors and excludes a role for D3 autoreceptors in regulating DA neuronal activity *(77,78)*. No changes in the expression of others members of the DA receptor family (D1, D3, and D4 receptors) are observed in the mutant mice, whereas there is an increase in the gene expression of enkephalin, a neuropeptide contained in D2-positive neurons, and unexpectedly, a minor upregulation of substance P expression, a neuropeptide collocalized with D1 receptors *(75,76,78)*. Interestingly, *(79)* reported that D2-deficient mice exhibit decrease DAT function (in the absence of modification in the density of the transporter) with no consequences on the level of DA release in the dorsal STR, thereby supporting the regulation of DAT activity by D2 receptors. However, the absence of changes in DA synthesis and release remain unclear and might be due to the development of compensatory mechanisms.

Although D2-deficient mice have been shown to exhibit no rewarding effects of morphine *(80)* and a reduced sensitivity to ethanol *(81)*, no study has thus far reported behavioral or biochemical experiments with psychostimulant drugs. Only a comment from unpublished data suggested that D2-deficient mice appeared to have a lowered locomotor response to cocaine *(46)*. Recently, a meeting abstract reported an increased sensitivity to the rewarding effect of cocaine, as indicated by a higher self-administration of cocaine in the mutant mice compared to controls *(82)*.

2.2.2. D3 Receptor Null Mice

Two laboratories developed D3-deficient mice at the same time and reported no gross developmental deficits. Both studies showed that the mutant animals are hyperactive in a novel environment *(83,84)*. The idea that reduced anxiety-related behaviors in these mice would explain the elevation in locomotion remains controversial *(84,85)*. However, the locomotor phenotype of these mutant mice is consistent with pharmacological results showing that D3 receptor stimulation has an inhibitory action on locomotion *(86,87)*, whereas the opposite is observed with D3 receptor antagonist *(88)*. Otherwise, D3-deficient mice present no alteration in D1 and D2 receptor binding or function, no changes in DA transporter density, and no abnormality in tyrosine hydroxylase immunoreactivity in the STR, suggesting that general DA neuronal function are normal in the absence of D3 receptors and that these receptors do not have an important role as autoreceptor *(84,89)*. In addition, there is no change in basal expression of the immediate early gene *c-fos*, or in the expression of genes encoding for neuropeptides co-expressed with D1 receptors such as dynorphin, or D2 receptors such as enkephalin and neurotensin *(90)*. However, higher basal extracellular levels of dopamine have been observed in the ventral STR of the mutant mice *(89)*.

D3-deficient mice are supersensitive to the locomotor and rewarding effects of amphetamine and cocaine *(84)*. The phenotype is characterized by an increase in locomotion and expression of stereotyped behaviors at lower doses than for wild-type mice, and by a stronger and earlier expression of place preference to amphetamine. This is in accordance with pharmacological studies reporting that D3-receptor ligands modulate both cocaine self-administration *(56,57,91)* and psychostimulant-induced hyperlocomotion *(51)*. The facilitating action of psychostimulants in D3-deficient mice seems to be strongly related to a desinhibition of the synergism existing between D1 and D2 receptors. Indeed, Xu et al. *(84)* demonstrated that D3-deficient mice are more active when both receptors are simultaneously stimulated with selective agonists or cocaine than when activated alone. The same difference between mutant and wild-type mice persists when both genotype are DA-depleted, suggesting that the dampening effect of D3 receptors on the cooperative action of D1 and D2 receptors is likely to be a postsynaptic phenomena *(84)*. As the firing activity of single cells stimulated with D1 and D2 agonists is not different between mutant and wild-type mice, these investigators argued for an inhibitory effect of D3 receptors located on distinct neurons and acting at a system level. Further examinations revealed that in D3-deficient mice, cocaine induces strong cellular responses associated to D1 receptors, as indicated by a high increase in *c-fos* and dynorphin gene expression, whereas there is no modification in the pattern of gene expression of enkephalin, a neuropeptide co-expressed in D2-receptor neurons *(90)*. This increase in gene expression related to D1-receptor activation is not correlated to brain areas where D3 and D1 receptors are co-expressed, sustaining the idea *(84)* that D3 receptors modulate D1 receptor activity via an indirect neuronal network. The difference in sensitivity to psychostimulant drugs between mutant and wild-type mice disappear at high doses of the drugs, indicating that the inhibiting action of D3 receptors can be overcome when D1 receptors are overstimulated, leading to the reestablishment of its cooperation with D2 receptors. For example, the analysis in wild-type and D3 mutant mice of the effect of cocaine on neurotensin gene expression *(92)* demonstrate that whereas acute administration of the drug induced no difference between the genotypes, a chronic treatment with cocaine revealed a higher expression of neurotensin in D3-deficient mice compared to their wild-type littermates. This observation suggests a role for D3-receptors in the modulation by D1 receptors of the activity of D2 receptors during chronic treatment with cocaine.

In conclusion, the D3 mutant mice confirm that the motor and rewarding action of psychostimulant drugs is attenuated by D3 receptors but reveal that this dampening effect is mostly directed against D1 receptor activity.

2.2.3. D4 Receptor Null Mice

So far, one laboratory has generated D4-deficient mice and reported that the mice have no gross structural developmental abnormalities and are fertile *(93)*. Their main phenotype is a hypoactivity in a novel and familiar environment, suggesting a positive contribution of D4 receptors in the expression of some motor behaviors. Further examinations of the phenotype revealed that the reduced response to novelty is related to a diminution in approach and explorating behaviors *(94)*. This is in accordance with recent human studies of association between some alleles of the highly polymorphic D4 receptor gene and the risk for novelty-seeking personality *(95,96)*. Under basal conditions, biochemical analysis of the D4 mutant mice show no alteration in D1 and D2 receptor binding, but an increase in both the biosynthesis and turnover of DA in the dorsal STR *(93)*. It has been proposed that in D4-deficient mice, glutamatergic corticonigral and corticostriatal projections are no longer under the inhibitory tone normally provided by mesocortical DA via D4 receptors and thus affect DA synthesis and turnover in the dorsal STR. This phenotype is, however, in clear contrast to the reduced locomotion behavior exhibited by D4-deficient mice, as it is generally admitted that extracellular DA levels in the STR are positively correlated to the level of motor behavior *(47)*. However, the modification in DA turnover can intervene at a qualitative level, seeing that the mutant mice exhibit enhanced motor coordination performance in the rotarod test, supporting the importance of DA in the realization of complex motor behavior *(93)*.

D4-deficient mice show a supersensitivity to the locomotor-stimulating effect of cocaine and amphetamine characterized by a greater intensity when compared to the effect observed in the wild-type littermates *(93)*. First, this result reveals that D4 receptors normally have an inhibitory action on psychostimulant locomotor action. Second, it indicates that enhanced basal DA activity is associated with enhanced sensitivity to the locomotor stimulant effects. Conversely, the increased locomotor sensitivity to psychostimulant in these mutant mice does not support the assumption that the level of locomotor responsivity to novelty is predictive of the degree of response to a rewarding stimulus *(47)*.

In conclusion, the D4 mutant mice reveal a dissociative contribution on locomotion between the basal level and psychostimulant drug-induced conditions, which may or may not be linked to basal elevated extracellular DA levels.

3. Dopaminoceptive Downstream Proteins (*see* Table 3)

Recently, remarkable work has been undertaken to unravel the effects of psychostimulant drugs on the molecular actions of dopamine on its target cells. Taking advantage of the precision of the knockout technique, these studies provide further insights into the complex mechanisms mediating and underlying the long-term changes in gene expression associated with acute or repeated drug exposure. As it is now well established that an intriguing property of psychostimulants is the activation of the DA–cAMP pathway through an action via the D1 receptor, focus has been given to the investigation of this molecular cascade *(97–99)*.

3.1. $G_{olf\alpha}$ *Protein Null Mice*

It is generally assumed that the coupling of D1 receptors to AC is mediated by the stimulatory GTP-binding protein $G_{s\alpha}$. However, the $G_{olf\alpha}$ subunit, which is highly expressed in the striatum, may represent the preferred D1-coupled G_{α} subunit *(100)*. The generation of $G_{olf\alpha}$ null mice *(101)* showed that this protein, which shares 88% homology with $G_{s\alpha}$, is actually mediating D1 signaling in the DA nigrostriatal path-

Table 3
Principal Observations Concerning Dopamine Receptor Null Mice After Psychostimulant Drug Treatment[a]

	Basal characteristics	Acute exposure to psychostimulants	Chronic exposures to psychostimulants
D1 receptor null mice	Hyperactivity *(59)*	Cocaine: no induction of locomotion and stereotypes	Cocaine: no locomotor sensitization and no conditioned place preference
	Hypoactivity *(60)*	Amphetamine: attenuated induction of locomotion	Amphetamine: no locomotor sensitization *(68)* or moderate locomotor sensitization *(29)*
	Disorganization of the striatal complex	ND	Cocaine: elevation of substance P levels No cAMP pathway stimulation
	No cAMP pathway stimulation Elevated midbrain dopamine levels		
D5 receptor null mice	Hyperactivity Reduced level of anxiety	ND	ND
D2 receptor null mice	Severe neurological impairments *(75)*	Cocaine: Lowered locomotor function (unpublished data)	ND
	Modest neurological impairments *(76)* Loss of autoreceptor function Elevated enkephalin levels Decrease in DAT function	ND	ND
D3 receptor null mice	Hyperactivity	Cocaine and amphetamine: supersensitivity to locomotor and stereotyped effects	Cocaine and amphetamine: facilitated place preference
	Elevated striatal dopamine levels	Cocaine: heightened c-fos and dynorphin levels	Cocaine: heightened expression of neurotensin
D4 receptor null mice	Hypoactivity	Cocaine and amphetamine: supersensitivity to locomotor effects	ND
	Reduced response to novelty Increased rotarod performances Elevated midbrain dopamine levels	ND	ND

[a]cAMP, 3',5'-monophosphate ; DAT, dopamine transporter ; ND, not determined.

way. This is illustrated by a decrease affinity of D1 receptor for DA in the striatum and an abolition of the locomotor effect of D1 receptor agonist in the mutant mice. Furthermore, it is demonstrated that the locomotor stimulant action of cocaine and its induction of the immediate early gene *c-fos* in the STR is absent in $G_{olf\alpha}$ null mice, supporting the view of an action of cocaine through G-protein stimulation.

Thus, these findings reveal an anatomical segregation of D1 signaling that might have further relevance for psychostimulant action.

3.2. DARPP-32 Null Mice

DARPP-32 (dopamine and cyclic adenosine 3',5'-monophosphate-regulated phosphoprotein, 32 kDa) plays a central role in the biology of dopaminoceptive neurons. In the neostriatum, it has been shown that dopamine, principally acting at D1 receptor and stimulating PKA, affects the phosphorylation and/or dephosphorylation of the protein. In its phosphorylated form, DARPP-32 is a potent inhibitor of the protein phosphatase-1 (PP-1), which regulates the phosphorylation state and activity of various downstream physiological effectors *(102)*. Thus, the DARPP-32/PP-1 pathway integrates information in part from the DA neurotransmitter and coordinates responses involving numerous different neurotransmitter receptors or voltage-gated ion channels. This pathway is therefore an important target of psychostimulant drugs that have been found to increase DARPP-32 phosphorylation in the neostriatum.

Using DARPP-32 null mice, *(103)* observed that the ability of a D1 agonist to regulate in the neostriatum the excitability of dopaminoceptive neurons via Na^+/K^+ pumps and/or Ca^{2+} channels is reduced as well as the inhibition via D1 receptor of glutamate-evoked activity of medium spiny neurons. Further evidence for an alteration in the properties of DA neurons in DARPP-32 mutant mice is sustained by the attenuated ability of DA to induce GABA release from nerve terminals of medium spiny neurons.

These authors also reported a reduction in biochemical and behavioral actions of acute administration of psychostimulants. This was demonstrated by a decreased *c-fos* induction by amphetamine and an attenuated locomotor effect of cocaine and amphetamine at low doses, in the absence of differences in baseline locomotion between wild-type and DARPP-32 mutant mice *(103)*. These response deficits to acute psychostimulants can be overcome at higher doses, suggesting that an overstimulation of the DA pathway can induce changes sufficient to bypass the need for the DARPP-32/PP-1 cascade. This hypothesis is supported by the observation that a repeated treatment with a high dose of cocaine induces a higher rate of locomotor sensitization in the mutant mice, while the long-term induction of Δ*fos*B isoforms, which is normally seen in response to chronic cocaine administration, is not observed in the knockout animals *(104)*.

These results emphasize the distinct molecular mechanisms involved in the acute vs chronic effects of psychostimulant drugs. They further indicate that DARPP-32 may act as a positive or negative feedback in accordance, respectively, to the acute or chronic drug regimen used.

3.3. FosB Null Mice

*Fos*B is a member of the Fos and Jun families of transcription factors binding to the activator protein 1 (AP1) DNA-binding complex. In the NA and ventral STR, they are induced by acute or chronic psychostimulant exposures. It has been shown that chronic cocaine administration results in the persistent expression of 35- to 37-kDa Fos-related proteins, which have been termed chronic FRAs (Fos-related antigen) and are products of the *fos*B gene, specifically isoforms of Δ*fos*B *(105,106)*.

Investigation of *fos*B null mice behaviors shows an increase spontaneous locomotion and a higher locomotor response to acute cocaine administration, while there is an attenuate locomotor sensitization induced by the higher dose of the drug compared to wild-type littermates *(107)*. However, wild-type and mutant mice exhibit an equivalent conditioned activity when given a saline challenge after the repeated cocaine treatment. This result supports previous work reporting the absence of abnormalities in sensory, motor, or motivational functions in these mutant mice *(108)* and is consolidated by the observation of no abnormalities in the development of STR in these mice. Furthermore, *fos*B mutant mice also exhibit a heightened sensitivity to cocaine in the conditioned place preference paradigm *(107)*. Thus, these mutant mice appears to be in a "presensitized" state to cocaine, with exaggerated response or sensitivity to acute and low dose of cocaine, while there is an inconsistence to further increase this effect after repeated exposures to the drug or using higher doses. Therefore, the authors argued for a "ceiling" effect of repeated cocaine treatment in the absence of accumulation of FRAs and proposed an important compensatory role exerted by Δ*fos*B isoforms to counteract the effects of cocaine. This proposition was recently sustained by the increase responsiveness to the rewarding and locomotor effects of cocaine observed in transgenic mice overexpressing Δ*fos*B *(109)*. Furthermore, in an attempt to identify the specific targets genes that mediate the effects of this transcription factor on behavioral adaptations to cocaine, this group of investigators demonstrate the implication of the cyclin-dependent kinase 5 (Cdk5) as a downstream target gene of Δ*fos*B in the STR of inducible Δ*fos*B transgenic mice *(110)*. This work reveals a specific Cdk5/DARPP-32 pathway that plays a negative-feedback homeostatic role in response to repeated cocaine treatment.

These studies, dissecting the molecular network in which Δ*fos*B isoforms are implicated, offer valuable input into the resulting changes in D1 signaling that contribute to adaptive changes in the brain related to cocaine addiction.

3.4. Neurotrophic Factor Null Mice

Lately, a lot of interest has been devoted for the implication of different members of the nerve growth factor (NGF)-related family of neurotrophins in the mediation of psychostimulant drug actions. So, BDNF (brain-derived neurotrophic factor) and GDNF (glial cell line-derived neurotrophic factor) have been shown to support the survival and function of DA midbrain neurons in vivo and in vitro *(111,112)* and protect them against neurotoxic injuries induced by 6-OHDA or MPTP *(113,114)*. These factors utilize respectively the TrkB receptor-protein tyrosine kinase and the GFR-1 and its associated protein tyrosine kinase Ret as the primary targets of their signal transduction, which are both highly expressed in DA midbrain neurons *(115,116)*. The involvement of BDNF and GDNF in functions associated with DA, and specifically the long-term adaptations induced by repeated exposures to psychostimulants, is supported by the ability of these factors to block some biochemical and morphological changes elicited by the drugs *(117,118)*. Moreover, BDNF and cocaine have been shown to interact on a common cellular signaling pathway, suggesting an overlap regulation of DA neurons responses to chronic drug exposure *(119)*.

Table 4
Lessons from Knockout Strategies: Summary of Recent Advances in Molecular Mechanisms of Psychostimulant Drugs Using Knockout Mice[a]

DAT-deficient mice	Reveal the primordial role of DAT in mediating the psychostimulant locomotor effect of cocaine and amphetamine
	Reveal an unexpected lack of participation in the rewarding action of psychostimulant drugs
SERT-deficient mice	Reveal the lack of contribution of SERT in the locomotor action of psychostimulants
	Reveal an inhibitory action of SERT on the rewarding properties of psychostimulants
NET-deficient mice	Reveal an inhibitory control of NET on the locomotor and rewarding properties of psychostimulants
D1-deficient mice	Reveal the positive necessary participation of D1 receptor in mediating the locomotor effect of psychostimulants
	Reveal the lack of participation of D1 receptor in the rewarding action of psychostimulants
D3-deficient mice	Reveal the important role of D3 receptor in inhibiting the cooperative action of D1/D2 receptors in mediating the locomotor and rewarding properties of psychostimulants
D4-deficient mice	Reveal the inhibitory cortical action of D4 receptors on psychostimulant-induced locomotion
$G_{olf\alpha}$ protein null mice	Reveal a specific D1 signaling mediating the locomotor effects of cocaine
DARPP-32 null mice	Reveal the positive homeostatic action of the DARPP-32/PP-1 cascade on locomotion during acute exposure to psychostimulants
*Fos*B null mice	Reveal the negative counterbalancing action of FosB isoforms on locomotion and induction of conditioned locomotion during acute exposure to psychostimulants
BDNF heterozygous mice	Reveal the positive homeostatic action of BDNF on the locomotor effects induced by acute or repeated exposures to psychostimulants

[a]BDNF, brain-derived neurotrophic factor; DARPP-32/PP-1, dopamine and cyclic adenosine 3',5'-monophosphate-regulated phosphoprotein, 32 kDa/ Protein Phosphatase-1; DAT, dopamine transporter; GDNF, glial cell line-derived neurotrophic factor; NET, norepinephrine transporter; SERT, serotonin transporter.

As a consequence of the functions of BDNF and GDNF, null mice are not viable and only the behaviors of the heterozygous littermates, BDNF+/− and GDNF+/− mice, can be studied.

BDNF+/− mice are reported to be less sensitive to the locomotor stimulant effect of an acute administration of cocaine and show a delayed development of locomotor sensitization to the drug *(120)*. Thus, BDNF seems to be required for the normal locomotor response to acute cocaine and development of behavioral sensitization, supporting the possibility that the neurotrophic factor and cocaine share common molecular pathway to modulate DA neurons functioning. This is further sustained by the facilitating action of chronic BDNF injected in the nucleus accumbens of rats on the locomotor stimulant effect of cocaine, the development of locomotor sensitization, and long-term persistence of conditioned activity *(120)*. The possibility of a common action of BDNF and cocaine on the DA D3 receptor has recently being suggested *(121)*, given the implication of this receptor in behavioral sensitization and specifically in the maintenance of cues-induced conditioned activity *(57)*. Furthermore, D3 receptor and TrkB colocalize in the shell of the NA, and Guillin et al. *(121)* demonstrated a regulation by BDNF of D3 receptor expression. The apparent synergistic action of BDNF and cocaine may be contradictory but can be explained as the development of a homeostatic system to balance the changes induced by chronic exposure to the drug.

GDNF+/− mice exhibit a normal locomotor response to acute cocaine but an enhanced sensitivity to cocaine when tested in the conditioned place preference paradigm or for the development of locomotor sensitization *(122)*. Furthermore, these mice show dramatically elevated levels of Δ*fos*B in the nucleus accumbens. These results indicate that GDNF blocks the behavioral responses related to long-term exposures to the drug, supporting the neuroprotective property of GDNF against biochemical changes induced by chronic cocaine. This is conversely confirmed by the enhanced responses to cocaine observed in rats infused into the ventral tegmental area by an anti-GDNF antibody and by the fact that chronic cocaine decreases levels of the phospho Ret protein in this brain area, though GDNF, GRF-1, and RET gene expression are not affected *(122)*. Therefore, a feedback loop has been proposed, whereby GDNF pathways in the VTA counteract the action of chronic drug exposures via a mechanism that may interfere with Δ*fos*B isoforms expression in the nucleus accumbens.

Investigation of BDNF and GDNF functions reveal new molecular mechanisms of action for psychostimulant drugs that interfere with the long-term biochemical and morphological functions of DA neurons.

Conclusions

See Table 4.

References

1. Glowinski, J., and Axelrod, J. (1964) Inhibition of uptake of tritiated noradrenaline in the intact rat brain by imipramine and structurally related compounds. *Nature* **204,** 1318–1319.
2. Kuhar, M. J., Ritz, M. C., and Boja, J. W. (1991) The dopamine hypothesis of the reinforcing properties of cocaine. *Trends Neurol. Sci.* **14,** 299–302.
3. Di Chiara, G. and Imperato, A. (1988) Drugs abused by humans preferentially increase synaptic dopamine concentrations in the mesolimbic system of freely moving rats. *Proc. Natl. Acad. Sci. USA* **85,** 5274–5278.
4. Amara, S. G. and Kuhar, M. J. (1993) Neurotransmitter transporters: recent progress. *Annu. Rev. Neurosci.* **16,** 73–93.
5. Giros, B. and Caron, M. G. (1993) Molecular characterization of the dopamine transporter. *Trends Pharmacol. Sci.* **14,** 43–49.
6. Uhl, G. R. and Hartig, P. R. (1992) Transporter explosion: update on uptake. *Trends Pharmacol. Sci.* **13,** 421–425.
7. Giros, B., Jaber, M., Jones, S. R., Wightman, R. M., and Caron, M. G. (1996) Hyperlocomotion and indifference to cocaine and amphetamine in mice lacking the dopamine transporter. *Nature* **379,** 606–612.

8. Bosse, R., Fumagalli, F., Jaber, M., Giros, B., Gainetdinov, R. R., Wetsel, W. C., Missale, C., and Caron, M. G. (1997) Anterior pituitary hypoplasia and dwarfism in mice lacking the dopamine transporter. *Neuron* **19,** 127–138.
9. Spielewoy, C., Gonon, F., Roubert, C., Fauchey, V., Jaber, M., Caron, M. G., Roques, B. P., Hamon, M., Betancur, C., Maldonado, R. and Giros, B. (2000) Increased rewarding properties of morphine in dopamine-transporter knockout mice. *Eur. J. Neurosci.* **12,** 1827–1837.
10. Walker, J. K., Gainetdinov, R. R., Mangel, A. W., Caron, M. G., and Shetzline, M. A. (2000) Mice lacking the dopamine transporter display altered regulation of distal colonic motility. *Am. J. Physiol. Gastrointest. Liver Physiol.* **279,** G311–G318.
11. Ralph, R. J., Paulus, M. P., Fumagalli, F., Caron, M. G., and Geyer, M. A. (2001) Prepulse inhibition deficits and perseverative motor patterns in dopamine transporter knock-out mice: differential effects of D1 and D2 receptor antagonists. *J. Neurosci.* **21,** 305–313.
12. Wisor, J. P., Nishino, S., Sora, I., Uhl, G. H., Mignot, E., and Edgar, D. M. (2001) Dopaminergic role in stimulant-induced wakefulness. *J. Neurosci.* **21,** 1787–1794.
13. Spielewoy, C., Roubert, C., Hamon, M., Nosten-Bertrand, M., Betancur, C., and Giros, B. (2000) Behavioural disturbances associated with hyperdopaminergia in dopamine-transporter knockout mice. *Behav. Pharmacol.* **11,** 279–290.
14. Gainetdinov, R. R., Jones, S. R., and Caron, M. G. (1999) Functional hyperdopaminergia in dopamine transporter knock-out mice. *Biol. Psychiatr.* **46,** 303–311.
15. Jones, S. R., Gainetdinov, R. R., Jaber, M., Giros, B., Wightman, R. M., and Caron, M. G. (1998) Profound neuronal plasticity in response to inactivation of the dopamine transporter. *Proc. Natl. Acad. Sci. USA* **95,** 4029–4034.
16. Jaber, M., Dumartin, B., Sagne, C., Haycock, J. W., Roubert, C., Giros, B., Bloch, B., and Caron, M. G. (1999) Differential regulation of tyrosine hydroxylase in the basal ganglia of mice lacking the dopamine transporter. *Eur. J. Neurosci.* **11,** 3499–3511.
17. Benoit-Marand, M., Jaber, M., and Gonon, F. (2000) Release and elimination of dopamine in vivo in mice lacking the dopamine transporter: functional consequences. *Eur. J. Neurosci.* **12,** 2985–2992.
18. Jones, S. R., Gainetdinov, R. R., Hu, X. T., Cooper, D. C., Wightman, R. M., White, F. J., and Caron, M. G. (1999) Loss of autoreceptor functions in mice lacking the dopamine transporter. *Nat. Neurosci.* **2,** 649–655.
19. Fauchey, V., Jaber, M., Caron, M. G., Bloch, B., and Le Moine, C. (2000) Differential regulation of the dopamine D1, D2 and D3 receptor gene expression and changes in the phenotype of the striatal neurons in mice lacking the dopamine transporter. *Eur. J. Neurosci.* **12,** 19–26.
20. Fauchey, V., Jaber, M., Bloch, B., and Le Moine, C. (2000) Dopamine control of striatal gene expression during development: relevance to knockout mice for the dopamine transporter. *Eur. J. Neurosci.* **12,** 3415–3425.
21. Dumartin, B., Jaber, M., Gonon, F., Caron, M. G., Giros, B., and Bloch, B. (2000) Dopamine tone regulates D1 receptor trafficking and delivery in striatal neurons in dopamine transporter-deficient mice. *Proc. Natl. Acad. Sci. USA* **97,** 1879–1884.
22. Gainetdinov, R. R., Wetsel, W. C., Jones, S. R., Levin, E. D., Jaber, M., and Caron, M. G. (1999) Role of serotonin in the paradoxical calming effect of psychostimulants on hyperactivity. *Science* **283,** 397–401.
23. Spielewoy, C., Biala, G., Roubert, C., Hamon, M., Betancur, C., and Giros, B. (2002) Hypolocomotor effects of acute and chronic D-amphetamine in mice lacking the dopamine transporter. *Psychopharmacology* (Berlin) **159,** 2–9.
24. Sora, I., Wichems, C., Takahashi, N., Li, X. F., Zeng, Z., Revay, R., Lesch, K. P., Murphy, D. L., and Uhl, G. R. (1998) Cocaine reward models: conditioned place preference can be established in dopamine- and in serotonin-transporter knockout mice. *Proc. Natl. Acad. Sci. USA* **95,** 7699–7704.

25. Rocha, B. A., Fumagalli, F., Gainetdinov, R. R., Jones, S. R., Ator, R., Giros, B., Miller, G. W., and Caron M. G. (1998) Cocaine self-administration in dopamine-transporter knockout mice. *Nat. Neurosci.* **1,** 132–137.
26. Sora, I., Hall, F. S., Andrews, A. M., Itokawa, M., Li, X. F., Wei, H. B., Wichems C., Lesch, K. P., Murphy, D. L., and Uhl, G. R. (2001) Molecular mechanisms of cocaine reward: combined dopamine and serotonin transporter knockouts eliminate cocaine place preference. *Proc. Natl. Acad. Sci. USA* **98,** 5300–5305.
27. Carboni, E., Spielewoy, C., Vacca, C., Nosten-Bertrand, M., Giros, B., and Di Chiara, G. (2001) Cocaine and amphetamine increase extracellular dopamine in the nucleus accumbens of mice lacking the dopamine transporter gene. *J. Neurosci.* **21,** RC141: 1–4.
28. Bengel, D., Murphy, D. L., Andrews, A. M., Wichems, C. H., Feltner, D., Heils, A., Mossner, R., Westphal, H. and Lesch, K. P. (1998) Altered brain serotonin homeostasis and locomotor insensitivity to 3,4-methylenedioxymethamphetamine ("Ecstasy") in serotonin transporter-deficient mice. *Mol. Pharmacol.* **53,** 649–655.
29. Xu, F., Gainetdinov, R. R., Wetsel, W. C., Jones, S. R., Bohn, L. M., Miller, G. W., Wang, Y. M. and Caron, M. G. (2000) Mice lacking the norepinephrine transporter are supersensitive to psychostimulants. *Nat. Neurosci.* **3,** 465–471.
30. Henry, J. P. and Scherman, D. (1989) Radioligands of the vesicular monoamine transporter and their use as markers of monoamine storage vesicles. *Biochem. Pharmacol.* **38,** 2395–2404.
31. Henry, J. P., Sagne, C., Bedet, C., and Gasnier, B. (1998) The vesicular monoamine transporter: from chromaffin granule to brain. *Neurochem. Int.* **32,** 227–246.
32. Seedat, Y. K. (2001) The limits of antihypertensive therapy—lessons from Third World to First. *Cardiovasc. J. S. Afr.* **12,** 94–100.
33. Carlsson, A. (1967) Basic actions of psychoactive drugs. *Int. J. Neurol.* **6,** 27–45.
34. Erickson, J. D., Eiden, L. E., and Hoffman, B. J. (1992) Expression cloning of a reserpine-sensitive vesicular monoamine transporter. *Proc. Natl. Acad. Sci. USA* **89,** 10,993–10,997.
35. Liu, Y., Peter, D., Roghani, A., Schuldiner, S., Prive, G. G., Eisenberg, D., Brecha, N., and Edwards, R. H. (1992) A cDNA that suppresses MPP+ toxicity encodes a vesicular amine transporter. *Cell* **70,** 539–551.
36. Takahashi, N., Miner, L. L., Sora, I., Ujike, H., Revay, R. S., Kostic, V., Jackson-Lewis, V., Przedborski, S., and Uhl, G. R. (1997) VMAT2 knockout mice: heterozygotes display reduced amphetamine-conditioned reward, enhanced amphetamine locomotion, and enhanced MPTP toxicity. *Proc. Natl. Acad. Sci. USA* **94,** 9938–9943.
37. Fon, E. A., Pothos, E. N., Sun, B. C., Killeen, N., Sulzer, D., and Edwards, R. H. (1997) Vesicular transport regulates monoamine storage and release but is not essential for amphetamine action. *Neuron* **19,** 1271–1283.
38. Wang, Y. M., Gainetdinov, R. R., Fumagalli, F., Xu, F., Jones, S. R., Bock, C. B., Miller, G. W., Wightman, R. M., and Caron, M. G. (1997) Knockout of the vesicular monoamine transporter 2 gene results in neonatal death and supersensitivity to cocaine and amphetamine. *Neuron* **19,** 1285–1296.
39. Sulzer, D., Chen, T. K., Lau, Y. Y., Kristensen, H., Rayport, S., and Ewing, A. (1995) Amphetamine redistributes dopamine from synaptic vesicles to the cytosol and promotes reverse transport. *J. Neurosci.* **15,** 4102–4108.
40. Jones, S. R., Gainetdinov, R. R., Wightman, R. M., and Caron, M. G. (1998) Mechanisms of amphetamine action revealed in mice lacking the dopamine transporter. *J. Neurosci.* **18,** 1979–1986.
41. Pothos, E. N., Davila, V. and Sulzer, D. (1998) Presynaptic recording of quanta from midbrain dopamine neurons and modulation of the quantal size. *J. Neurosci.* **18,** 4106–4118.
42. Pothos, E. N., Larsen, K. E., Krantz, D. E., Liu, Y., Haycock, J. W., Setlik, W., Gershon, M. D., Edwards, R. H., and Sulzer, D. (2000) Synaptic vesicle transporter expression regulates vesicle phenotype and quantal size. *J. Neurosci.* **20,** 7297–7306.

43. Gingrich, J. A. and Caron, M. G. (1993) Recent advances in the molecular biology of dopamine receptors. *Annu. Rev. Neurosci.* **16,** 299–321.
44. Kebabian, J. W., and Calne, D. B. (1979) Multiple receptors for dopamine. *Nature* **277,** 93–96.
45. Missale, C., Nash, S. R., Robinson, S. W., Jaber, M., and Caron, M. G. (1998) Dopamine receptors: from structure to function. *Physiol. Rev.* **78,** 189–225.
46. Vallone D., Picetti R., and Borrelli, E. (2000) Structure and function of dopamine receptors. *Neurosci. Biobehav. Rev.* **24,** 125–132.
47. Wise R. A. and Bozarth M. A. (1987) A psychomotor stimulant theory of addiction. *Psychol. Rev.* **94,** 469–492.
48. Beninger, R. J., Hoffman, D. C., and Mazurski, E. J. (1989) Receptor subtype-specific dopaminergic agents and conditioned behavior. *Neurosci. Biobehav. Rev.* **13,** 113–122.
49. Caine, S. B. and Koob, G. F. (1994) Effects of dopamine D-1 and D-2 antagonists on cocaine self-administration under different schedules of reinforcement in the rat. *J. Pharmacol. Exp. Ther.* **270,** 209–218.
50. Phillips G. D., Robbins T. W. and Everitt B. J. (1994) Bilateral intra-accumbens self-administration of d-amphetamine: antagonism with intra-accumbens SCH-23390 and sulpiride. *Psychopharmacology* (Berl) **114,** 477–485.
51. Pierce, R. C. and Kalivas, P. W. (1997) A circuitry model of the expression of behavioral sensitization to amphetamine-like psychostimulants. *Brain Res. Rev.* **25,** 192–216.
52. Robertson, G. S., Vincent, S. R., and Fibiger, H. C. (1992) D1 and D2 dopamine receptors differentially regulate c-fos expression in striatonigral and striatopallidal neurons. *Neuroscience* **49,** 285–296.
53. Self, D. W. and Nestler, E. J. (1995) Molecular mechanisms of drug reinforcement and addiction. *Annu. Rev. Neurosci.* **18,** 463–495.
54. Gerfen, C. R., Engber, T. M., Mahan, L. C., Suzel, Z., Chase, T. N., Monsuma, F. J., and Sibley, D. R. (1990) D1 and D2 dopamine receptor-regulated gene expression of striatonigral and striatopallidal neurons. *Science* **250,** 1429–1432.
55. Caine, S. B., Koob, G. F., Parsons, L. H., Everitt, B. J., Schwartz, J. C., and Sokoloff,, P. (1997) D3 receptor test in vitro predicts decreased cocaine self-administration in rats. *Neuroreport* **8,** 2373–2377.
56. Parsons, L. H., Caine, S. B., Sokoloff, P., Schwartz, J. C., Koob, G. F., and Weiss, F. (1996) Neurochemical evidence that postsynaptic nucleus accumbens D3 receptor stimulation enhances cocaine reinforcement. *J. Neurochem.* **67,** 1078–1089.
57. Pilla, M., Perachon, S., Sautel, F., Garrido, F., Mann, A., Wermuth, C. G., Schwartz, J. C., Everitt, B. J., and Sokoloff, P. (1999) Selective inhibition of cocaine-seeking behaviour by a partial dopamine D3 receptor agonist. *Nature* **400,** 371–375.
58. Diaz, J., Levesque, D., Griffon, N., Lammers, C. H., Martres, M. P., Sokoloff, P., and Schwartz, J. C. (1994) Opposing roles for dopamine D2 and D3 receptors on neurotensin mRNA expression in nucleus accumbens. *Eur. J. Neurosci.* **6,** 1384–1387.
59. Xu, M., Moratalla, R., Gold, L. H., Hiroi, N., Koob, G. F., Graybiel, A. M., and Tonegawa, S. (1994) Dopamine D1 receptor mutant mice are deficient in striatal expression of dynorphin and in dopamine-mediated behavioral responses. *Cell* **79,** 729–742.
60. Drago, J., Gerfen, C. R., Lachowicz, J. E., Steiner, H., Hollon, T. R., Love, P. E., et al. (1994) Altered striatal function in a mutant mouse lacking D1A dopamine receptors. *Proc. Natl. Acad. Sci. USA* **91,** 12564–12568.
61. Waddington, J. L., Clifford, J. J., McNamara, F. N., Tomiyama, K., Koshikawa, N., and Croke, D. T. (2001) The psychopharmacology-molecular biology interface: exploring the behavioural roles of dopamine receptor subtypes using targeted gene deletion ('knockout'). *Prog. Neuropsychopharmacol. Biol. Psychiatr.* **25,** 925–964.

62. Xu, M., Hu, X. T., Cooper, D. C., Moratalla, R., Graybiel, A. M., White, F. J., and Tonegawa, S. (1994) Elimination of cocaine-induced hyperactivity and dopamine-mediated neurophysiological effects in dopamine D1 receptor mutant mice. *Cell* **79,** 945–955.
63. Cabib, S., Castellano, C., Cestari, V., Filibeck, U., and Puglisi-Allegra, S. (1991) D1 and D2 receptor antagonists differently affect cocaine-induced locomotor hyperactivity in the mouse. *Psychopharmacology* (Berl.) **105,** 335–339.
64. Tella, S. R. (1994) Differential blockade of chronic versus acute effects of intravenous cocaine by dopamine receptor antagonists. *Pharmacol. Biochem. Behav.* **48,** 151–159.
65. Pradham, S. N., Battacharyya, A. K., and Pradham, S. (1979) Serotonergic manipulation of the behavioral effects of cocaine in rats. *Commun. Psychopharmacol.* **2,** 481–487.
66. Xu, M., Guo, Y., Vorhees, C. V., and Zhang, J. (2000) Behavioral responses to cocaine and amphetamine administration in mice lacking the dopamine D1 receptor. *Brain Res.* **852,** 198–207.
67. Zhang, J., Walsh, R. R., and Xu, M. (2000) Probing the role of the dopamine D1 receptor in psychostimulant addiction. *Ann. N.Y. Acad. Sci.* **914,** 13–21.
68. Crawford, C. A., Drago, J., Watson, J. B., and Levine, M. S. (1997) Effects of repeated amphetamine treatment on the locomotor activity of the dopamine D1A-deficient mouse. *Neuroreport* **8,** 2523–2527.
69. Cador, M., Bjijou, Y., and Stinus, L. (1995) Evidence of a complete independence of the neurobiological substrates for the induction and expression of behavioral sensitization to amphetamine. *Neuroscience* **65,** 385–395.
70. Bjijou, Y., Stinus, L., Le Moal, M., and Cador, M. (1996) Evidence for selective involvement of dopamine D1 receptors of the ventral tegmental area in the behavioral sensitization induced by intra-ventral tegmental area injections of D-amphetamine. *J. Pharmacol. Exp. Ther.* **277,** 1177–1187.
71. Miner, L. L., Drago J., Chamberlain, P. M., Donovan, D., and Uhl, G. R. (1995) Retained cocaine conditioned place preference in D1 receptor deficient mice. *Neuroreport* **6,** 2314–2316.
72. Moratalla, R., Xu, M., Tonegawa, S., and Graybiel, A. M. (1996) Cellular responses to psychomotor stimulant and neuroleptic drugs are abnormal in mice lacking the D1 dopamine receptor. *Proc. Natl. Acad. Sci. USA* **93,** 14928–14933.
73. Drago, J., Gerfen, C. R., Westphal, H., and Steiner, H. (1996) D1 dopamine receptor-deficient mouse: cocaine-induced regulation of immediate-early gene and substance P expression in the striatum. *Neuroscience* **74,** 813–823.
74. Holmes, A., Hollon, T. R., Gleason, T. C., Liu, Z., Dreiling, J., Sibley, D. R., and Crawley, J. N. (2001) Behavioral characterization of dopamine D5 receptor null mutant mice. *Behav. Neurosci.* **115,** 1129–1144.
75. Baik, J. H., Picetti, R., Saiardi, A., Thiriet, G., Dierich, A., Depaulis, A., Le Meur, M., and Borrelli, E. (1995) Parkinsonian-like locomotor impairment in mice lacking dopamine D2 receptors. *Nature* **377,** 424–428.
76. Kelly, M. A., Rubinstein, M., Asa, S. L., Zhang, G., Saez, C., Bunzow, J. R., Allen, R. G., Hnasko, R., Ben-Jonathan, N., Grandy, D. K., and Low, M. J. (1997) Pituitary lactotroph hyperplasia and chronic hyperprolactinemia in dopamine D2 receptor-deficient mice. *Neuron* **19,** 103–113.
77. Mercuri, N. B., Saiardi, A., Bonci, A., Picetti, R., Calabresi, P., Bernardi, G., and Borrelli, E. (1997) Loss of autoreceptor function in dopaminergic neurons from dopamine D2 receptor deficient mice. *Neuroscience* **79,** 323–327.
78. Kelly, M. A., Rubinstein, M., Phillips, T. J., Lessov, C. N., Burkhart-Kasch, S., Zhang, G., Bunzow, J. R., Fang, Y., Gerhardt, G. A., Grandy, D. K., and Low, M. J. (1998) Locomotor activity in D2 dopamine receptor-deficient mice is determined by gene dosage, genetic background, and developmental adaptations. *J. Neurosci.* **18,** 3470–3479.

79. Dickinson, S. D., Sabeti, J., Larson, G. A., Giardina, K., Rubinstein, M., Kelly, M. A., Grandy, D. K., Low, M. J., Gerhardt, G. A., and Zahniser, N. R. (1999) Dopamine D2 receptor-deficient mice exhibit decreased dopamine transporter function but no changes in dopamine release in dorsal striatum. *J. Neurochem.* **72,** 148–156.
80. Maldonado, R., Saiardi, A., Valverde, O., Samad, T. A., Roques, B. P., and Borrelli, E. (1997) Absence of opiate rewarding effects in mice lacking dopamine D2 receptors. *Nature* **388,** 586–589.
81. Phillips, T. J., Brown, K. J., Burkhart-Kasch, S., Wenger, C. D., Kelly, M. A., Rubinstein, M., Grandy, D. K., and Low, M. J. (1998) Alcohol preference and sensitivity are markedly reduced in mice lacking dopamine D2 receptors. *Nat. Neurosci.* **1,** 610–615.
82. Caine, S. B., Negus, S. S., Mello, N. K., Patel, S., Bristow, L., Kulagowski, J., Vallone, D., Saiardi, A., and Borrelli, E. (2000) Role for the dopamine D2 receptor subtype in the regulation of cocaine self-administration in rats and mice. *Society for Neurosci. New Orleans* #681.8, 1833.
83. Accili, D., Fishburn, C. S., Drago, J., Steiner, H., Lachowicz, J. E., Park, B. H., Gauda, E. B., Lee, E. J., Cool, M. H., Sibley, D. R., Gerfen, C. R., Westphal, H., and Fuchs, S. (1996) A targeted mutation of the D3 dopamine receptor gene is associated with hyperactivity in mice. *Proc. Natl. Acad. Sci. USA* **93,** 1945–1949.
84. Xu, M., Koeltzow, T. E., Santiago, G. T., Moratalla, R., Cooper, D. C., Hu, X. T., White, N. M., Graybiel, A. M., White, F. J., and Tonegawa, S. (1997) Dopamine D3 receptor mutant mice exhibit increased behavioral sensitivity to concurrent stimulation of D1 and D2 receptors. *Neuron* **19,** 837–848.
85. Steiner, H., Fuchs, S., and Accili, D. (1997) D3 dopamine receptor-deficient mouse: evidence for reduced anxiety. *Physiol. Behav.* **63,** 137–141.
86. Svensson K., Carlsson, A., and Waters, N. (1994) Locomotor inhibition by the D3 ligand R-(+)-7-OH-DPAT is independent of changes in dopamine release. *J. Neural. Transm. Gen. Sect.* **95,** 71–74.
87. Daly, S. A. and Waddington, J. L. (1993) Behavioural effects of the putative D-3 dopamine receptor agonist 7-OH-DPAT in relation to other "D-2-like" agonists. *Neuropharmacology* **32,** 509,510.
88. Sautel, F., Griffon, N., Sokoloff, P., Schwartz, J. C., Launay, C., Simon, P., et al. (1995) Nafadotride, a potent preferential dopamine D3 receptor antagonist, activates locomotion in rodents. *J. Pharmacol. Exp. Ther.* **275,** 1239–1246.
89. Koeltzow, T. E., Xu, M., Cooper, D. C., Hu, X. T., Tonegawa, S., Wolf, M. E., and White, F. J. (1998) Alterations in dopamine release but not dopamine autoreceptor function in dopamine D3 receptor mutant mice. *J. Neurosci.* **18,** 2231–2238.
90. Carta, A. R., Gerfen, C. R., and Steiner, H. (2000) Cocaine effects on gene regulation in the striatum and behavior: increased sensitivity in D3 dopamine receptor-deficient mice. *Neuroreport* **11,** 2395–2399.
91. Caine, S. B. and Koob, G. F. (1993) Modulation of cocaine self-administration in the rat through D-3 dopamine receptors. *Science* **260,** 1814–1816.
92. Betancur, C., Lepee-Lorgeoux, I., Cazillis, M., Accili, D., Fuchs, S., and Rostene, W. (2001) Neurotensin gene expression and behavioral responses following administration of psychostimulants and antipsychotic drugs in dopamine D(3) receptor deficient mice. *Neuropsychopharmacology* **24,** 170–182.
93. Rubinstein, M., Phillips, T. J., Bunzow, J. R., Falzone, T. L., Dziewczapolski, G., Zhang, G., Fang, Y., Larson, J. L., McDougall, J. A., Chester, J. A., Saez, C., Pugsley, T. A., Gershanik, O., Low, M. J., and Grandy, D. K. (1997) Mice lacking dopamine D4 receptors are supersensitive to ethanol, cocaine, and methamphetamine. *Cell* **90,** 991–1001.
94. Dulawa, S. C., Grandy, D. K., Low, M. J., Paulus, M. P., and Geyer, M. A. (1999) Dopamine D4 receptor-knock-out mice exhibit reduced exploration of novel stimuli. *J. Neurosci.* **19,** 9550–9556.

95. Benjamin, J., Li, L., Patterson, C., Greenberg, B. D., Murphy, D. L., and Hamer, D. H. (1996) Population and familial association between the D4 dopamine receptor gene and measures of novelty seeking. *Nat. Genet.* **12,** 81–84.
96. Ebstein, R. P., Novick, O., Umansky, R., Priel, B., Osher, Y., Blaine, D., Bennett, E. R., Nemanov, L., Katz, M. and Belmaker, R. H. (1996) Dopamine D4 receptor (D4DR) exon III polymorphism associated with the human personality trait of Novelty Seeking. *Nat. Genet.* **12,** 78–80.
97. Nestler, E. J. (1994) Molecular neurobiology of drug addiction. *Neuropsychopharmacology* **11,** 77–87.
98. Nestler, E. J. and Aghajanian, G. K. (1997) Molecular and cellular basis of addiction. *Science* **278,** 58–63.
99. Nestler, E. J. (2001) Molecular basis of long-term plasticity underlying addiction. *Nat. Rev. Neurosci.* **2,** 119–128.
100. Herve, D., Levi-Strauss, M., Marey-Semper, I., Verney, C., Tassin, J. P., Glowinski, J., and Girault, J. A. (1993) G(olf) and Gs in rat basal ganglia: possible involvement of G(olf) in the coupling of dopamine D1 receptor with adenylyl cyclase. *J. Neurosci.* **13,** 2237–2248.
101. Zhuang, X., Belluscio, L., and Hen, R. (2000) GOLFalpha mediates dopamine D1 receptor signaling. *J. Neurosci.* **20,** RC91.
102. Greengard, P., Allen, P. B., and Nairn, A. C. (1999) Beyond the dopamine receptor: the DARPP-32/protein phosphatase-1 cascade. *Neuron* **23,** 435–447.
103. Fienberg, A. A., Hiroi, N., Mermelstein, P. G., Song, W., Snyder, G. L., Nishi, A., Cheramy, A., O'Callaghan, J. P., Miller, D. B., Cole, D. G., Corbett, R., Haile, C. N., Cooper, D. C., Onn, S. P., Grace, A. A., Ouimet, C. C., White, F. J., Hyman, S. E., Surmeier, D. J., Girault, J., Nestler, E. J., and Greengard, P. (1998) DARPP-32: regulator of the efficacy of dopaminergic neurotransmission. *Science* **281,** 838–842.
104. Hiroi, N., Fienberg, A. A., Haile, C. N., Alburges, M., Hanson, G. R., Greengard, P., and Nestler, E. J. (1999) Neuronal and behavioural abnormalities in striatal function in DARPP-32-mutant mice. *Eur. J. Neurosci.* **11,** 1114–1118.
105. Hope, B. T., Nye, H. E., Kelz, M. B., Self, D. W., Iadarola, M. J., Nakabeppu, Y., Duman, R. S., and Nestler, E. J. (1994) Induction of a long-lasting AP-1 complex composed of altered Fos-like proteins in brain by chronic cocaine and other chronic treatments. *Neuron* **13,** 1235–1244.
106. Moratalla, R., Elibol, B., Vallejo, M., and Graybiel, A. M. (1996) Network-level changes in expression of inducible Fos-Jun proteins in the striatum during chronic cocaine treatment and withdrawal. *Neuron* **17,** 147–156.
107. Hiroi, N., Brown, J. R., Haile, C. N., Ye, H., Greenberg, M. E., and Nestler, E. J. (1997) FosB mutant mice: loss of chronic cocaine induction of Fos-related proteins and heightened sensitivity to cocaine's psychomotor and rewarding effects. *Proc. Natl. Acad. Sci. USA* **94,** 10,397–10,402.
108. Brown, J. R., Ye, H., Bronson, R. T., Dikkes, P., and Greenberg, M. E. (1996) A defect in nurturing in mice lacking the immediate early gene fosB. *Cell* **86,** 297–309.
109. Kelz, M. B., Chen, J., Carlezon, W. A., Jr., Whisler, K., Gilden, L., Beckmann, A. M., Steffen, C., Zhang, Y. J., Marotti, L., Self, D. W., Tkatch, T., Baranauskas, G., Surmeier, D. J., Neve, R. L., Duman, R. S., Picciotto, M. R., and Nestler, E. J. (1999) Expression of the transcription factor deltaFosB in the brain controls sensitivity to cocaine. *Nature* **401,** 272–276.
110. Bibb, J. A., Chen, J., Taylor, J. R., Svenningsson, P., Nishi, A., Snyder, G. L., Yan, Z., Sagawa, Z. K., Ouimet, C. C., Nairn, A. C., Nestler, E. J., and Greengard, P. (2001) Effects of chronic exposure to cocaine are regulated by the neuronal protein Cdk5. *Nature* **410,** 376–380.
111. Hyman, C., Hofer, M., Barde, Y. A., Juhasz, M., Yancopoulos, G. D., Squinto, S. P., and Lindsay, R. M. (1991) BDNF is a neurotrophic factor for dopaminergic neurons of the substantia nigra. *Nature* **350,** 230–232.

112. Lin, L. F., Doherty, D. H., Lile, J. D., Bektesh, S., and Collins, F. (1993) GDNF: a glial cell line-derived neurotrophic factor for midbrain dopaminergic neurons. *Science* **260,** 1130–1132.
113. Spina, M. B., Squinto, S. P., Miller, J., Lindsay, R. M., and Hyman, C. (1992) Brain-derived neurotrophic factor protects dopamine neurons against 6-hydroxydopamine and N-methyl-4-phenylpyridinium ion toxicity: involvement of the glutathione system. *J. Neurochem.* **59,** 99–106.
114. Kearns, C. M., and Gash, D. M. (1995) GDNF protects nigral dopamine neurons against 6-hydroxydopamine in vivo. *Brain Res.* **672,** 104–111.
115. Kawamoto, Y., Nakamura, S., Nakano, S., Oka, N., Akiguchi, I., and Kimura, J. (1996) Immunohistochemical localization of brain-derived neurotrophic factor in adult rat brain. *Neuroscience* **74,** 1209–1226.
116. Treanor, J. J., Goodman, L., de Sauvage, F., Stone, D. M., Poulsen, K. T., Beck, C. D., Gray, C., Armanini, M. P., Pollock, R. A., Hefti, F., Phillips, H. S., Goddard, A., Moore, M. W., Buj-Bello, A., Davies, A. M., Asai, N., Takahashi, M., Vandlen, R., Henderson, C. E., and Rosenthal, A. (1996) Characterization of a multicomponent receptor for GDNF. *Nature* **382,** 80–83.
117. Berhow, M. T., Russell, D. S., Terwilliger, R. Z., Beitner-Johnson, D., Self, D. W., Lindsay, R. M., and Nestler, E. J. (1995) Influence of neurotrophic factors on morphine- and cocaine-induced biochemical changes in the mesolimbic dopamine system. *Neuroscience* **68,** 969–979.
118. Robinson, T. E. and Kolb, B. (1997) Persistent structural modifications in nucleus accumbens and prefrontal cortex neurons produced by previous experience with amphetamine. *J. Neurosci.* **17,** 8491–8497.
119. Berhow, M. T., Hiroi, N., and Nestler, E. J. (1996) Regulation of ERK (extracellular signal regulated kinase), part of the neurotrophin signal transduction cascade, in the rat mesolimbic dopamine system by chronic exposure to morphine or cocaine. *J. Neurosci.* **16,** 4707–4715.
120. Horger, B. A., Iyasere, C. A., Berhow, M. T., Messer, C. J., Nestler, E. J., and Taylor, J. R. (1999) Enhancement of locomotor activity and conditioned reward to cocaine by brain-derived neurotrophic factor. *J. Neurosci.* **19,** 4110–4122.
121. Guillin, O., Diaz, J., Carroll, P., Griffon, N., Schwartz, J. C., and Sokoloff, P. (2001) BDNF controls dopamine D3 receptor expression and triggers behavioural sensitization. *Nature* **411,** 86–89.
122. Messer, C. J., Eisch, A. J., Carlezon, W. A., Jr., Whisler, K., Shen, L., Wolf, D. H., Westphal, H., Collins, F., Russell, D. S., and Nestler, E. J. (2000) Role for GDNF in biochemical and behavioral adaptations to drugs of abuse. *Neuron* **26,** 247–257.

7

Opioid Modulation of Psychomotor Stimulant Effects

Toni S. Shippenberg and Vladimir I. Chefer

1. Introduction

The acute administration of psychomotor stimulants, such as cocaine and amphetamine, produce behavioral activation in humans and increased locomotor activity in laboratory animals. These agents are also self-administered by various species by virtue of their reinforcing effects. It is generally accepted that these actions results, at least in part, from an increase in dopaminergic (DAergic) neurotransmission in the nucleus accumbens (NAc), a terminal projection region of dopamine (DA) neurons comprising the mesocorticolimbic system *(1)*. Cocaine increases extracellular DA concentrations by binding to the DA transporter and inhibiting the uptake of DA from the synaptic cleft, whereas amphetamine causes a reversal of the DA transporter and increases DA release *(2,3)*.

Repeated exposure to psychostimulants and other drugs of abuse results in a progressive and enduring enhancement of their locomotor stimulatory effects, a phenomenon referred to as sensitization. A prior history of psychostimulant exposure also results in an augmentation of the reinforcing effects of these agents *(4,5)*, indicating that sensitization also develops to those behavioral effects of drugs more directly related to their abuse liability. Behavioral sensitization can persist for weeks or months after the last drug exposure and is postulated to contribute, in part, to drug craving and relapse to addiction *(6)*. Evidence, although more limited, indicating the development of sensitization to cocaine and amphetamine in human subjects has been obtained *(7,8)*.

Behavioral sensitization is associated with marked and long-lasting alterations in the functional activity of the mesocorticolimbic DA system. These include: a transient subsensitivity of A10 somatodendritic D2 autoreceptors *(9–11)* as well as presynaptic D2 autoreceptors in striatum and nucleus accumbens *(12,13)*; an enhancement of the ability of psychostimulants to increase extracellular DA levels in the Nac *(14–17)*; and a persistent enhancement of D1 receptor sensitivity within the NAc *(18)*. No consistent changes in D1 receptor number or affinity have been observed after chronic cocaine administration, suggesting that supersensitive D1 responses reflect changes in postreceptor signal transduction. Since D1 receptors are positively linked to adenylate cyclase via G_s and G_{olf}, it is noteworthy that chronic cocaine administration increases levels of adenylate cyclase and cAMP-dependent protein kinase in the NAc, while decreasing levels of G_i *(19)*. More recently, in vivo microdialysis studies have shown

From: *Molecular Biology of Drug Addiction*
Edited by: R. Maldonado © Humana Press Inc., Totowa, NJ

that chronic pretreatment with cocaine potentiates the ability of a subsequent amphetamine *(20)* or cocaine challenge (Chefer and Shippenberg, in press) to release DA in the NAc.

The repeated administration of psychostimulants and other drugs that increase DA neurotransmission in the mesocorticolimbic system is also associated with marked alterations in the activity of endogenous opioid peptide systems *(21)*. These opioid systems modulate DA neurotransmission *(22,23)*, and increasing evidence suggests that alterations in the activity of these opioid systems can profoundly affect the response to psychostimulants. This chapter will review data regarding the interaction of opioid peptide systems with mesocorticolimbic neurons and the relevance of this interaction to adaptations in behavior and neurochemistry that occur as a consequence of repeated psychostimulant administration.

2. Anatomy of the Mesocorticolimbic System

DA neurons project to various brain regions, including thc dorsal and ventral striatum, medial prefrontal cortex, amygdala, and hippocampus *(24)*. The mesocorticolimbic DA system, originating in the ventral tegmental area, consists of numerous topographically organized positive and negative feedback circuits (Fig. 1). As our understanding of the organization of these projections has advanced, new insights have emerged as to the neuronal circuitry mediating the acute actions of psychostimulants as well as the sensitized behavioral responses that occur following their repeated administration. The vast majority of the ventral tegmental area (VTA) projections to the NAc are DAergic *(24)*, whereas about 60% of mesoprefrontal projections contain GABA *(25)*. However, prefrontal cortex (PFC) glutamatergic terminals selectively synapse onto GABA cells that project to the NAc and DA cells that project to the PFC *(26)*. These PFC projections exert an excitatory influence on target cells *(27–29)*. There are data indicating the existence of a direct excitatory PFC drive on mesoaccumbens DA neurons *(30,31)*, but recent data indicate that this drive is mediated by indirect glutamatergic projections to the VTA *(26)*. Alterations in the activity of this PFC–VTA projection may be one mechanism leading to the induction of behavioral sensitization. White et al. *(32)* have hypothesized that, during repeated treatment with cocaine, DA-induced inhibition of mPFC excitatory amino acid neurons projecting to the VTA diminishes, leading to greater activation of glutamatergic receptors on VTA DA neurons and the induction of cocaine-induced sensitization. Moreover, it has been shown that abstinence from repeated cocaine administration is associated with a decrease in the mPFC DA response to a subsequent cocaine challenge *(33,34)*. This decrease, by increasing the glutamatergic drive on mesoaccumbens DA neurons, may underlie the increase in cocaine-evoked DA release in the NAc that occurs following repeated cocaine (*20*; Chefer and Shippenberg, in press) or amphetamine *(35)* administration and has been implicated in the long-term expression of sensitization.

The majority (up to 90%) of neurons comprising the NAc are medium spiny neurons, containing GABA as the primary neurotransmitter and a variety of neuropeptides, including neurotensin, enkephalin, dynorphin, substance P, and neurokinin B *(36)*. The local circuit neurons, which make up approximately 10% of all accumbal cells, contain ether acetylcholine or GABA as well as neuropeptides such as cholecystokinin, neurotensin and neuropeptide Y.

The primary excitatory input to the NAc derives from glutamatergic axons of cortical and thalamic origins *(37)* (Fig. 1). Other excitatory inputs arise from the amygdala

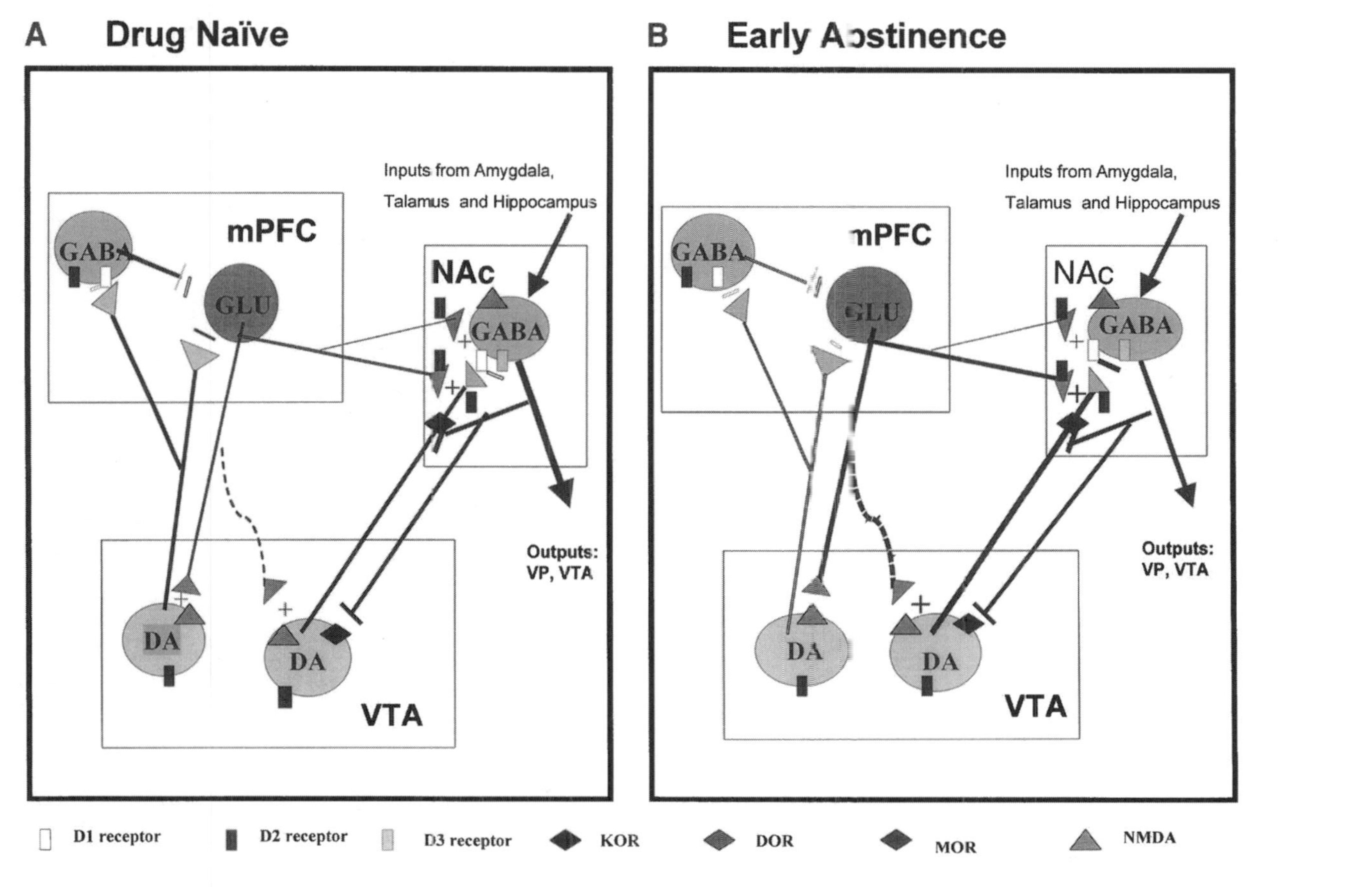

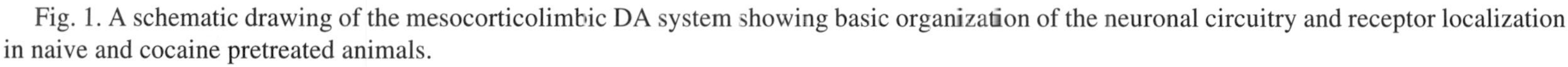
Fig. 1. A schematic drawing of the mesocorticolimbic DA system showing basic organization of the neuronal circuitry and receptor localization in naive and cocaine pretreated animals.

and hippocampus *(36,37)*. DA terminals from the VTA form symmetrical synapses onto dendrites and perikarya of both medium spiny neurons and local circuit neurons. There are distinct differences in innervation of the core and the shell of the NAc *(36,38)* that underlie the functional differences that exist between these two regions. However, this issue is beyond the topic of the present review.

The basic organization of the NAc described above is far from complete, but serves to provide a rudimentary understanding of the functioning of accumbal neuronal network. DA and glutamate from extrinsic sources and acetylcholine and GABA from local interneurons or collaterals of medium spiny neurons converge on the medium spiny neurons and influence their activity. DA terminals are always found in close proximity to those containing glutamate, and therefore can effectively "gate" signals from cortical areas *(36,39–41)*. Moreover, DA can modulate the response of NAc medium spiny neurons to corticoaccumbens fiber stimulation via D2 receptors, located presynaptically on cortical terminals *(42)*. In addition, the corticoacumbens system is also subject to gating influences by other afferents to the NAc as well. The overlapping inputs from the amygdala and hippocampus appear to be capable of gating PFC information via modulation of NAc neuronal activity *(43)*. Thus, taking into consideration the above observations, one can suggest that the complex neural network of topographically organized positive and negative feedback circuits in the NAc can be modified by various alterations in neurotransmission that occur within the limbic system.

3. Pharmacology and Distribution of Opioid Peptides and Receptors

The complexity of the NAc neural network is further complicated by the presence of an array of neuropeptides that are co-localized in medium spiny neurons and also found in various afferent projections. This review will focus on the endogenous opioid peptides: β-endorphin, methionine and leucine enkephalin, the dynorphins, the endomorphins, and orphanin FQ *(44)*. Genes encoding these opioid peptide families and the G-protein-coupled receptors to which they bind (μ-opioid receptor, MOR; δ-opioid receptor, DOR; and κ-opioid receptor, KOR; opioid receptor like-1, ORL1) have been cloned and characterized (*see* refs. *45* and *46* for review). β-Endorphin binds with high affinity to both the MOR and DOR receptors. The enkephalins and the dynorphins bind with highest affinity to the DOR and KOR receptors, respectively, and are considered to be endogenous ligands for these receptor types. However, these peptides bind with varying affinities to other opioid receptor types, and the correspondence between peptide and receptor is not invariant. The recently discovered opioid peptides, endorphin-1 and -2, although structurally distinct from other opioid peptides in which the first four amino acids are Tyr Gly Gly Phe followed by either methionine or leucine, bind with high affinity and selectivity to MOR *(44)*, suggesting that they may be the endogenous ligands for the MOR. In addition, the superfamily of opioid receptors and peptides was recently extended by the discovery of a novel G-protein-coupled receptor, termed opioid receptor like-1 (ORL1) *(47,48)*, which is widely distributed throughout the CNS *(49)*, and its endogenous peptide orphaninFQ or nociceptin (OFQ/N) *(50,51)*.

3.1. Mesoaccumbens System

A dense plexus of β-endorphin processes, arising from the arcuate nucleus of the hypothalamus, is observed in the NAc *(52)*. Fibers rich in endomorphin-1 and -2 nerve fibers are also apparent *(53)*, as are enkephalin-containing cell bodies that exhibit a patchlike distribution. Cell bodies and terminals immunoreactive for dynorphin A and B are also observed *(54)*. OFQ/N and its mRNA are sparsely localized in the cells of the core and shell of the NAc, and more densely localized fibers and terminals are found in these same brain regions, where they are especially dense in the NAc core *(55)*.

An extensive collocalization of opioid peptide precursors and DA receptors is observed in the various subdivisions of the NAc (rostral pole, shell, and core). However, the expression pattern of these proteins is region-dependent *(56)*. In the NAc core, proenkephalin (PENK) and prodynorphin (PDYN) mRNA, the precursors of enkephalin and dynorphin, respectively, are localized in separate cells. PDYN cells in the core express D1 and D3 receptor mRNAs, while PENK cells primarily express D2 mRNA. In contrast, the rostral pole and the shell of the NAc contain specific clusters of cells that express PENK, PDYN, D1, and D3 mRNA. A very small portion of NAc PENK cells that also express PDYN/D1 mRNA has also been observed. Acb PDYN medium spiny neurons primarily send projections to the VTA and substantia nigra (compacta and reticulata). PENK mRNA neurons innervate the ventral pallidum, which can influence mesoaccumbens neurotransmission via projections to the mediodorsal thalamus as well as by shorter feedback loops to the VTA and NAc.

MOR immunoreactivity is observed throughout the NAc. The shell compartment contains vast clusters of cells and fibers, with the highest density seen in fiber clusters along the ventral edge, the lateral border of the NAc, and the dorsal septal pole region. Scattered among the clusters are numerous medium spiny neurons that are MOR-immunoreactive *(57–59)*. Ultrastructural studies *(60–62)* have shown that in the shell, MORs are localized prominently on extrasynaptic somatodendritic plasma membranes (65%) and axon terminals (28%) of GABA-containing neurons *(62)*. By contrast, DORs are localized mainly to non-DAergic terminals and dendritic spines that contain GABA and possibly acetylcholine, as well as one or more neuropeptides *(61)*. The proportion of KOR within dendritic spines is similar to that of MOR and DOR, but the primary sites for KOR are small axons and terminals having the morphological features of DA- or acetylcholine-containing neurons or larger axon terminals that contain either GABA and dynorphin or excitatory amino acids *(60)*. Interestingly, KOR mRNA is expressed in PDYN-, but not in PENK-positive cells, and significantly overlaps with MOR mRNA. A low to moderate density of ORL1 is observed in the core and the shell of the NAc, with slightly greater receptor binding in the shell compared with the core *(49)*.

The findings that distinctive cell clusters in the shell of the NAc contain opioid peptide precursors, KOR, MOR, and D1 and D3 receptors, is consistent with functional data showing extensive interactions between opioid and DA systems in this brain region. Although the distribution of opioid peptide and receptors has been examined in other regions that modulate mesoaccumbal DA neurotransmission, an in-depth knowledge of their ultrastructural localization is lacking. In comparison to the NAc, opioid input to the VTA in the rat is relatively modest. β-Endorphin fibers are present in limited quantities throughout this region. Fibers containing enkephalin are somewhat denser than those for the other opioid peptides, and monosynaptic contacts between axon terminal-containing enkephalin and DAergic neurons have been demonstrated *(63)*. Although largely devoid of dynorphin- and endorphin-containing cell bodies, a modest number of enkephalin-positive perikarya are seen. The VTA contains also numerous neurons moderately stained for orphaninFQ/nociceptin as well as lightly stained fibers *(55)*.

In contrast to other areas in the rat, opioid receptor density is low in the VTA. MOR and MOR RNA are present in both the parabrachial and paranigral regions of the VTA *(57,64)*. Rostral portions have light fiber and cell staining, whereas a high frequency of densely labeled fibers and a few darkly stained perikarya are observed more caudally *(57)*. MOR is preferentially targeted to the plasmalemma of non-DAergic dendrites and axon terminals *(65)*. Although several lines of evidence indicated that these dendrites and axons are GABAergic *(66,67)*, MOR immunoreactivity is also observed in some unmyleinated axons and terminals that form excitatory synapses onto neurons expressing tyrosine hydroxylase or MOR. Scattered DAergic cells expressing MOR are also seen *(65)*. These data indicate that although non-DA cells are principal sites for MOR activation in the VTA, MOR ligands can influence DA output, indirectly, through modulation of presynaptic axon terminals apposing DA terminals and directly in a subpopulation of MOR expressing DA neurons. In contrast to MOR, the density of DOR and ORL1 receptors is low.

3.2. Mesocortical System

PDYN mRNA-expressing neuronal populations are present throughout the neocortex, with a notably higher expression in the medial prefrontal and anterior cingulate cortices, regions that have been implicated in the modulation of motivation and goal-directed behavior. These regions have a strong reciprocal connectivity to the amygdala, NAc, and VTA. PDYN mRNA-expressing cortical neurons are localized primarily in layer V and are presumed to be corticofugal fibers that innervate subscortical regions such as the NAc. The anatomical localization of the KOR mRNA and binding sites is, however, consistent with a greater local intrinsic role of KOR in the deep cortical layers *(68)*, though cortically derived KOR terminals are also evident in the striatum *(69)*. In the rat, the mPFC contains a limited number of fibers and cell bodies positive for dynorphin and enkephalin. Dynorphin staining is found in nonpyramidal cells of layers 2–4 as well as in layer 5, which contains scattered cells and fibers. OFQ/N is found throughout the frontal cortex, with densest staining in area 4 and moderate staining in layers 2 and 6 *(55)*. Although MOR, DOR, and KOR are present in the prefrontal cortex, MOR binding is particularly dense relative to other receptor subtypes *(64)*.

3.3. Amygdaloid Complex

The amygdaloid complex provides substantial inputs to the VTA and NAc and can thus profoundly affect mesocorticolimbic neurotransmission and the response of DA neurons to psychostimulants. This complex is rich in opioid peptides and receptors. In the human brain, expression of PDYN is particularly high in the accessory basal nucleus, amygdalohippocampal area, cortical nucleus, and periamygdaloid cortex *(70)*. The accessory basal and basal nuclei target the NAc, giving rise to some of the KOR binding sites located on the presynaptic terminals of the NAc. These nuclei also have a strong interconnectivity with the medial prefrontal and orbitofrontal cortices and provide the main intra-amygdala input to the central nucleus, which, in turn, receives the prominent amygdala DAergic innervation from the VTA (71-73). In the rat, staining of dynorphin- and enkephalin-containing fibers and cells is also high *(54)* in the central nucleus, as is that for OFQ/N *(55)*. Interestingly, a recent study has shown that OFQ/N decreases neuronal excitability in a large number of spiny projection neurons in the central and lateral amygdala, resulting in a net dampening effect of amygdaloid output *(74)*.

MOR immunoreactivity is high in the central nucleus, intercalated nuclei of amygdala, and the posteriormedial cortical amygdala (57-59) with only a few immunoreactive fibers seen in the basolateral nuclei. In contrast, DOR staining is light in the central nucleus, dense in the basolateral nucleus, and intermediate in the cortical amygdaloid nucleus. Like the MOR, KOR labeling is dense in the central nucleus. However, profuse labeling is also observed in the medial, central, and cortical nuclei. Low to moderate ORL1 binding observed in the anterior part of the medial and basolateral nuclei, and the central nucleus, is devoid of binding at all levels *(49)*.

Taken together, the distribution of opioid systems in the mesocorticolimbic sytem as well as in afferent and efferent projection areas clearly suggests that endogenous opioid peptides systems are strategically situated to modulate DA, glutamate, and GABA neurotransmission in the NAc.

4. Opioid Modulation of Mesocorticolimbic Neurotransmission

Acute systemic administration of KOR agonists decreases DA levels in the NAc and dorsal striatum *(75,76)*. In vitro studies assessing the modulation of electrically evoked [^{3}H] DA by opioid receptor activation revealed that in the NAc, olfactory tubercle, and PFC, DA release can be inhibited by activation of KOR *(77,78)*. Evidence that the acute systemic administration of selective KOR agonists decreases dialysis levels of DA in the NAc and dorsal striatum has also been obtained *(75)*. In vivo microdialysis studies in the rat have also shown that the intra-NAc infusion of the selective KOR agonist U-69593 decreases basal DA overflow in the NAc, whereas the selective blockade of KOR in this region significantly increases basal DA overflow *(22)*. Infusion of KOR ligands into the VTA fails to modify basal DA overflow in the NAc, indicating the existence of a tonically active KOR system in the NAc that regulates basal DA tone in the NAc. In view of the localization of KOR on DA terminals, these effects have been attributed to a direct effect on DA neurons. However, the NAc shell receives significant glutamatergic input *(79)*, and KORs are present on the presynaptic terminals of presumed excitatory synapses as well as on the dendrites of medium spiny neurons *(60,80)*. This anatomical arrangement raises the possibility that the behavioral effects of KOR activation in the NAc are due, in part, to the regulation of glutamatergic excitatory transmission. Indeed, NAc KOR activation produces a dose-dependent inhibition of glutamatergic excitatory postsynaptic currents (EPSCs). KOR activation also causes an increase in the paired-pulse ratio as well as a decrease in the frequency of spontaneous miniature events, consistent with a decrease in presynaptic glutamate release *(81)*. A KOR-mediated inhibition of calcium-dependent glutamate release has also been observed in the dorsal and ventral striatum *(82,83)*.

A recent study has shown that KOR also modulates DA uptake in the NAc. Using the technique of quantitative microdialysis, which permits simultaneous assessment of drug-induced alterations in DA uptake and extracellular DA levels, Thompson et al. *(84)* have shown that acute KOR activation increases DA uptake. This effect is delayed, occurring 1–2 h after drug administration, is dose-dependent, and is reversed by a selective KOR antagonist. These data are particularly interesting in view of ultrastructural studies (A. Svingos, personal communication) showing collocalization of KOR and DA transporter in axon terminals, small axons of NAc neurons, and suggest that KOR agonists regulate mesoaccumbens DA neurotransmission by two distinct mechanisms, inhibition of release and stimulation of uptake. In contrast to acute KOR activation, a decrease in DA uptake is observed 24–72 h following repeated KOR antagonist treatment. This effect is due to a decrease in the maximum capacity of uptake rather than a change in the affinity. As discussed below, these effects on DA uptake and release are functionally opposite to those produced by the acute and repeated administration of psychostimulants and provide an anatomical basis for the modulation of behavioral sensitization by KOR ligands.

In contrast to KOR agonists, systemically administered MOR agonists increase DA overflow in the NAc *(75)*. Infusion of MOR agonists into the VTA produces similar effects indicating an involvement of VTA opioid receptors in producing this effect *(22,85)*, but see *(86)*. Since MOR induces membrane hyperpolarization, this increase is consistent with the location of MOR on VTA GABA interneurons and disinhibition of DA neurotransmission *(67)*. Anatomical and neurophysiological studies indicate that MOR activation may also stimulate the activity of mesoaccumbens DA neurons, indirectly, by enhancing the response of NAc medium spiny neurons to the excitatory effects of NMDA *(87)*. Medium spiny neurons that contain enkephalin and GABA project to the ventral pallidum and onto GABA interneurons in the VTA. An increase in their output would enhance DA release in the NAc. Consistent with this hypothesis, a recent dialysis study showed that the intra- NAc infusion of the MOR agonists DAMGO and fentanyl increases DA overflow in the NAc *(86)*. In contrast to MOR activation, the blockade of MOR receptors in the VTA has been reported to decrease DA overflow in the NAc (22). This finding is consistent with the existence of a tonically active VTA MOR system that functionally opposes the actions of the NAc KOR system.

Like MOR agonists, the intracerebroventricular infusion of the DOR agonist, DPDPE stimulates DA overflow in the NAc *(76)*. A similar effect is also observed in response to intra-VTA infusions, suggesting that the activation of either MOR or DOR receptors in the VTA increases DA release in the NAc *(85)*. Infusion of the enkephalinase inhibitor thiorphan also increases DA overflow, suggesting that the VTA is one site mediating DOR agonist-induced increases in DA neurotransmission *(88)*. Interestingly, however, other studies have shown that the DOR agonists deltorphin and DPDPE are also effective in increasing DA levels when dialyzed into the NAc, suggesting that DOR activation in the VTA or NAc can modulate mesoaccumbens DA neurotransmission *(86,89)*.

Recent studies have shown that OFQ/N can also modulate DA levels in the NAc. Like KOR agonists, this peptide inhibits locomotor activity *(51)*. In accordance with this inhibitory effect, the intracerebroventricular or intra-VTA infusion of OFQ/N reduces DA overflow in the NAc *(90,91)*. This effect is associated with an increase in GABA and glutamate overflow in the VTA. Administration of the $GABA_A$ receptor antagonist bicuculline into the VTA prevents the effect of OFQ/N on NAc DA levels, suggesting that GABAergic interneurons located in the VTA may mediate this effect. Since, however, ORL1 activation, like MOR activation, induces membrane hyperpolarization *(92)*, it appears that OFQ/N reduces the release of another neurotransmitter that inhibits GABA release in the VTA. Likely candidates in this regard are the enkephalin fibers that project to the VTA and synapse with GABA neurons *(93)*.

5. Modulation of Endogenous Opioid Systems by Psychostimulants

In addition to the modulatory effects of opioids on DA neurotransmission, it is also apparent that psychostimulants and other drugs that increase DA neurotransmission, affect the activity of endogenous opioid systems. The acute administration of cocaine, amphetamine, methamphetamine, or the DA uptake inhibitor GBR 12909 increase PDYN expression in the rodent NAc and dorsal striatum *(94,95)*. This increase is delayed relative to the increase in DA levels, produced by these drugs and, in view, of the inhibitory effects of KOR agonists on extracellular DA levels, is consistent with it representing a compensatory mechanism that opposes stimulant-induced increase in DA neurotransmission.

Increases in PDYN gene expression are also observed in the NAc and dorsal striatum immediately following the cessation of repeated cocaine self-administration or a binge pattern of cocaine or amphetamine administration *(96,97)*. Although specific subterritories of the NAc have not been examined, the increase in gene expression in striatum is generally greater than that in the NAc. The increase persists for some days following the cessation of cocaine treatment, and although it is one of the more persistent changes in gene expression that occur in response to repeated psychostimulant administration, it is no longer apparent at later phases of abstinence *(98)*, and, at later time points, a decrease in PDYN gene expression may be seen *(99)*. Interestingly, tissue levels of dynorphin are also increased during the early phase of cocaine abstinence *(100)*. Importantly, manipulations that reduced DA but not serotonin neurotransmission prevented this effect of cocaine, suggesting that the effects on PDYN result from DA uptake inhibition.

KOR density is increased in the NAc, caudate putamen, and ventral tegmental area of rats immediately following repeated cocaine administration *(101)*. These effects persist for at least 48 h following the cessation of cocaine treatment. Importantly, postmortem studies have also revealed an upregulation of KOR in the NAc and its afferent projections (e.g., cingulate cortex, basolateral amygdala) in humans following an overdose of cocaine *(102,103)*.

At present, methodological limitations have precluded the assessment of dynorphin release. Therefore, it is unclear as to whether the observed increases in receptor and endogenous ligand reflect an increase or decrease in the activity of the KOR system. In view, however, of the dysphoric and aversive effects of KOR agonists in humans and experimental animals *(104–106)*, it has been hypothesized that the unopposed actions of an upregulated KOR system may be one mechanism underlying the "crash" that occurs in human subjects following binge cocaine use *(107)*.

To date, the effects of psychostimulants upon changes in POMC or endomorphin-1 and -2 have not been examined. Studies in rats have shown that PENK expression is increased *(97,108,109)* or unchanged *(101,110)* immediately following the cessation of repeated cocaine administration. A very recent study has shown that PENK expression is increased in the NAc for at least 10 d in animals that have been allowed to self-administer cocaine for some weeks, as well as in animals that received passive injections of cocaine *(108)*. Interestingly, and in contrast to the NAc, PENK expression was decreased in the central nucleus of the amygdala.

Several studies have shown that repeated psychostimulant administration affects the density of MOR opioid receptors in the NAc as well as in its afferent and efferent projections. In rats sacrificed immediately after a 2 wk binge pattern of repeated cocaine administration, the density of MOR-binding sites is increased in the rostral nucleus accumbens, basolateral amygdala, cingulate cortex, and caudate putamen *(110)*. In contrast, DOR binding was unchanged *(101)*. An increase in MOR density in the NAc can also be detected when the duration of cocaine exposure is reduced.

Alterations in MOR systems have also been observed in human subjects. Positron emission tomography revealed increased MOR binding in several brain regions, including the frontal cingulate cortex, 1–4 d after the cessation of cocaine use *(111)*. A positive correlation between MOR binding and self-reports of the severity of drug craving was also observed. The upregulation of MOR observed in this study may result from an increase in binding sites or a reduction in endogenous opioid peptide release. In view, however, of the positive reinforcing effects of MOR agonists and the dysphoria that cocaine addicts experience during the early phase of drug abstinence, a reduction in peptide release or in receptor/effector coupling appears likely.

Data regarding the effects of other manipulations that alter opioid gene expression shed some light on the possible mechanisms mediating the above changes. D1 DA receptors are located on medium spiny neurons that contain dynorphin, and their activation increases PDYN gene expression in the caudate putamen *(112)*. In contrast D2 receptors are located predominantly on PENK neurons and their activation inhibits PENK gene expression *(112)*. The repeated administration of cocaine is associated with D1 receptor supersensitivity and a transient downregulation of D2 DA receptors that regulate DA release. If, as has been suggested, the early phase of abstinence from cocaine and other psychostimulants is associated with an increase in DA release, then increased release in the face of supersensitive D1 DA receptors would be expected to increase PDYN gene expression. Although DA release is increased, there appears to be a functional downregulation of D2 receptors. Thus, PENK expression would be unchanged or even increased, depending on the extent of downregulation.

6. Modulation of Psychostimulant-Induced Sensitization by KOR Ligands: *Role of the Mesoprefrontal and Mesolimbic DA Systems*

The systemic administration of KOR agonists with cocaine prevents the sensitization that develops to the locomotor activating and conditioned reinforcing effects of this psychostimulant *(113–116)*. KOR agonists also prevent changes in basal and cocaine-evoked DA levels in the NAc that are associated with the induction and long-term expression of sensitization *(117,118)*. As discussed above, quantitative dialysis studies have shown that repeated administration of KOR agonists produces changes in DA neurotransmission that are opposite to that produced by cocaine. Abstinence from repeated cocaine use is associated with a transient increase in DA uptake and release in the NAc core, whereas abstinence from repeated KOR agonist treatment produces decreases in these parameters *(84)*. In animals that received a 5-d cocaine treatment regimen with the selective KOR agonist, U-69593, DA uptake was not different from controls. Thus, repeated activation of KORs during cocaine administration may prevent cocaine-induced alterations in NAc DA neurotransmission by producing changes in basal DA uptake and release that are opposite to those produced by cocaine.

In the mPFC, the early phase of abstinence from cocaine is associated with an elevation of basal DA uptake and a blunted response of mesocortical DA neurons to a subsequent cocaine challenge *(34)*. This reduction in cocaine-evoked levels of DA in mPFC can be an important element in the mechanism of behavioral sensitization, since disinhibition of EAA neurons in mPFC leads to increased excitatory drive on VTA neurons *(32)*. The co-administration of a selective KOR agonist with cocaine prevents these changes in mPFC DA neurotransmission as well as the sensitized behavioral response *(34)*, possibly by decreasing basal DA uptake and increasing drug-evoked DA release in the mPFC. This, in turn, would oppose and normalize changes in mPFC DA neurotransmission induced by repeated cocaine administration.

Interestingly, although KOR agonist treatment, by itself, does not modify basal DA uptake in the mPFC, the co-administration of a KOR agonist with cocaine is effective in preventing increases in DA uptake, which occur in this brain region during early abstinence. Taken together, these data indicate that repeated activation of KOR profoundly affects the function of the DA transporter in terminal areas of the mesolimbic DA system. At present the mechanism by which KOR activation regulates DA transporter function is unknown. However, KORs are collocalized with DA transporters in the NAc (A. Svingos, personal communication), and their activation modulates protein kinase C and other intracellular cascades that has been implicated in both phosphorylation and internalization of the transporter *(119–121)*. Therefore, activation of these cascades and resulting changes in transporter trafficking may underlie the KOR agonist-induced changes in transporter function that occur in the absence of changes in transporter protein expression.

Anatomical mapping studies provide some insights regarding the localization of KOR that mediate the prevention of sensitization produced by systemic agonists. Infusions of a KOR agonist into the NAc suppressed behavioral sensitization to cocaine and prevented cocaine-induced increases in basal and drug-evoked levels in the NAc (Fig. 2), while activation of KOR in the mPFC potentiated the expression of behavioral sensitization and increased cocaine-evoked DA levels *(122)*. The exact mechanisms mediating these opposing effects of KOR activation in the two projections fields of DA neurons are unclear. However, KOR agonists decrease DA release in vitro *(75)* and in vivo *(123,124)*, and downregulate D2 autoreceptors *(125,126)*. Inhibition of DA release and D2 receptor downregulation in the mPFC would, through D2-mediated changes in GABA release *(127)* or disinhibition pyramidal cell firing, increase glutamatergic drive on VTA DA neurons, and enhance the behavioral and neurochemical effects produced by the repeated administration of cocaine. In contrast, a decrease in DA release and downregulation of D2 autoreceptors within the NAc would oppose the effects of cocaine and attenuate behavioral sensitization.

Recent studies in our laboratory have shown that the selective blockade of KORs in the NAc with the selective KOR antagonist norbinaltorphinine (nor-BNI), does not affect the development of behavioral sensitization to cocaine, but significantly exacerbates the locomotor response to a subsequent cocaine challenge in cocaine naive animals (Fig. 3; Chefer and Shippenberg, unpublished). This fact is consistent with the hypothesis that NAc KOR plays an important role in inhibiting those processes leading to behavioral sensitization. It is also consistent with the notion that changes in DA neurotransmission in axon terminal fields such as the NAc are critical for the expression of sensitization, while the somatodendritic region of VTA is critical for its initiation *(128)*. These findings are particular noteworthy, because selective and irreversible

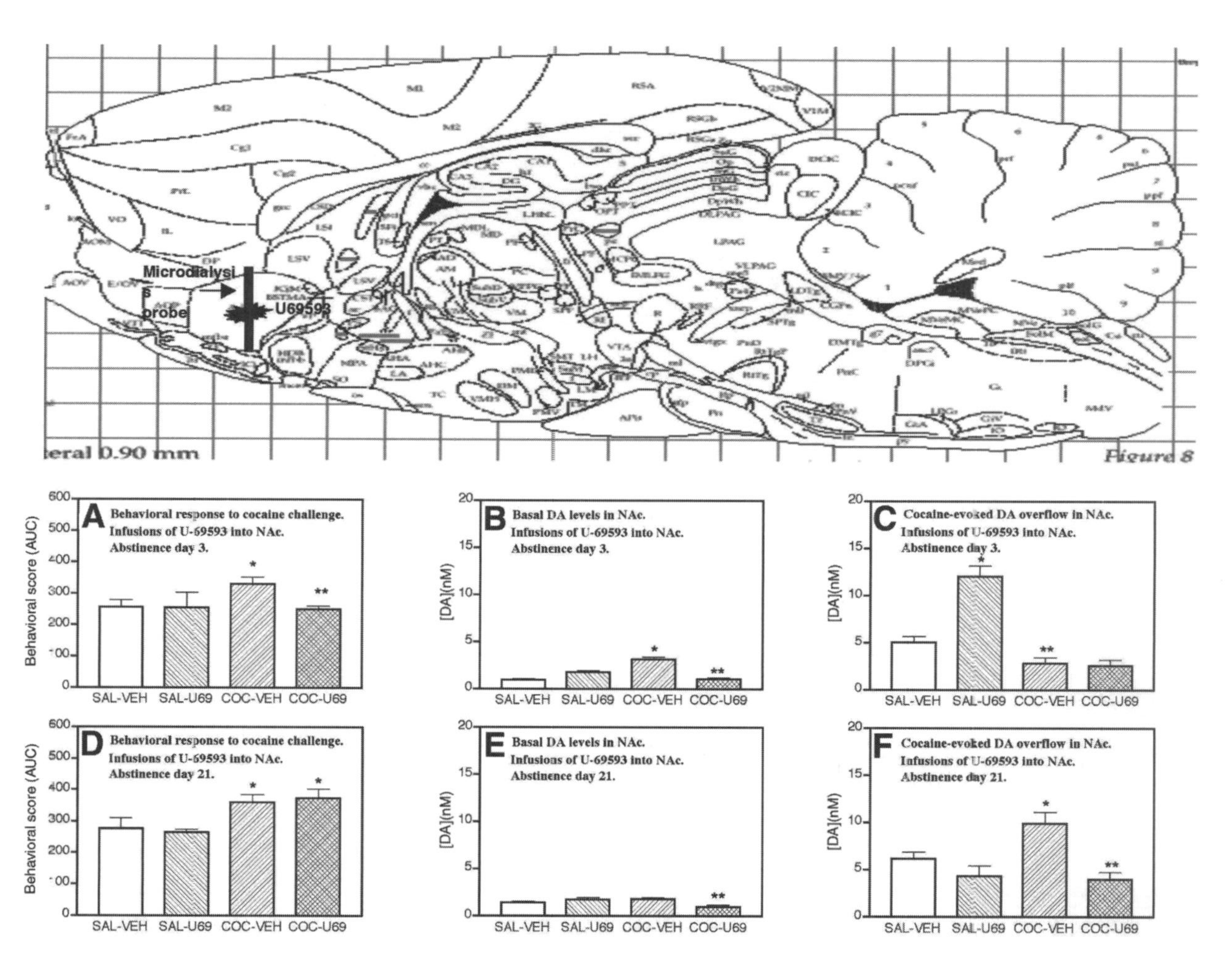

Fig. 2. Influence of intra-NAc infusions of a KOR agonist on alterations in behavior and DA dynamics in the NAc that occur during abstinence from repeated cocaine administration. Rats received once-daily injections of cocaine (20 mg/kg, ip) or saline for 5 d. They received bilateral intra-NAc infusions of the KOR agonist U-69593 (1.0 μg) or vehicle on d 3–5 of the 5-d treatment regimen. Behavioral and microdialysis studies were conducted 3 d after the cessation of the various treatments. *Upper panel:* Sagittal diagram of the rat brain with localization of cannulae and the microdialysis probe in the NAc. *Lower panel:* (**A,D**) The behavioral response to a cocaine challenge (20 mg/kg, ip) in rats previously treated with U-69593 (U69) or its vehicle (VEH) in combination with cocaine (COC) or saline (SAL). (**B,E**) Basal levels of DA in the NAc for each of the pretreatment groups (SAL-VEH, SAL-U69, COC-VEH, COC-U69). (**C,F**) cocaine-evoked DA levels in the NAc for each of the pretreatment groups. *Significant difference between control and cocaine pretreated animals. **Significant difference between vehicle and U-69593-pretreated animals.

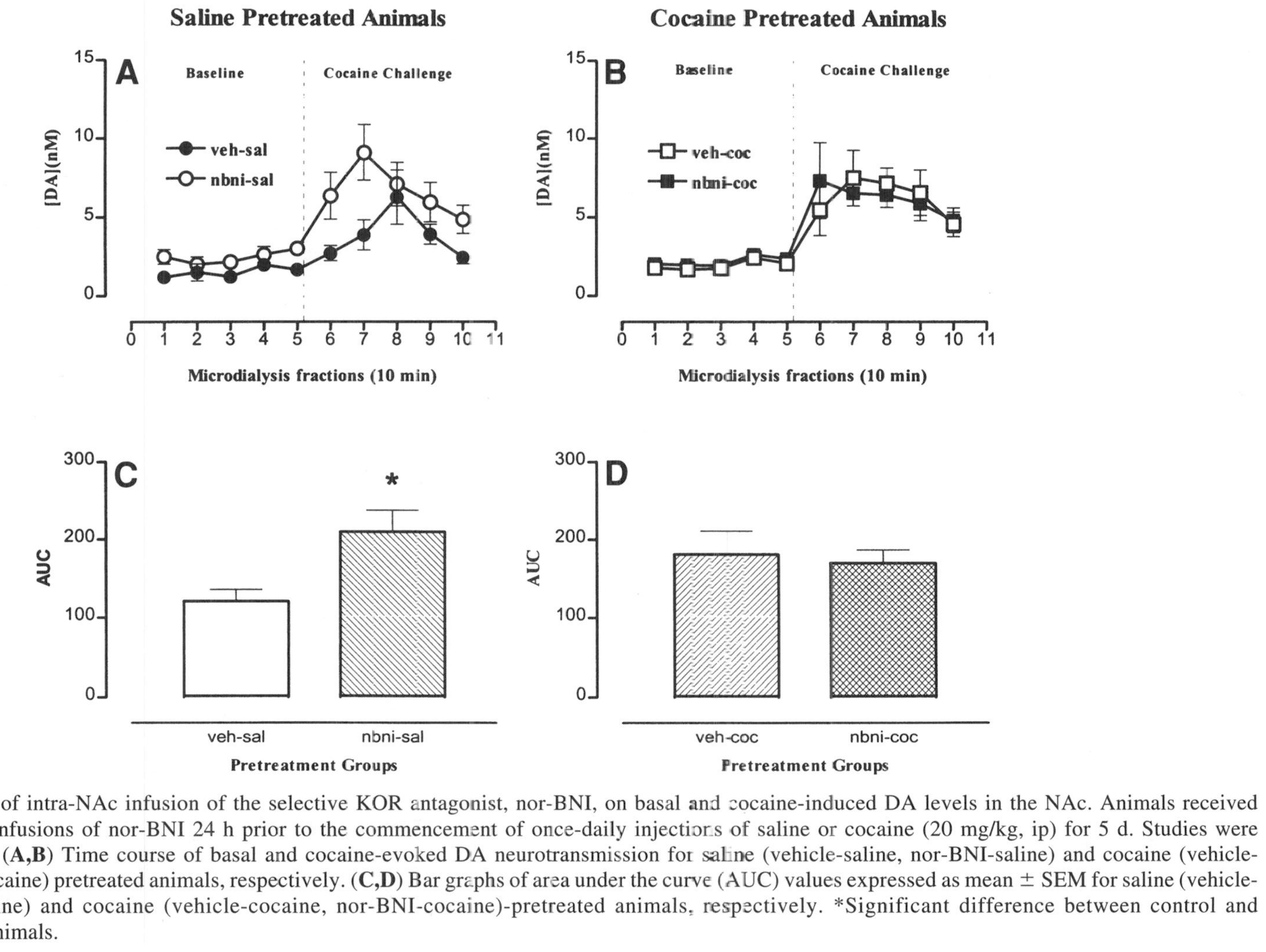

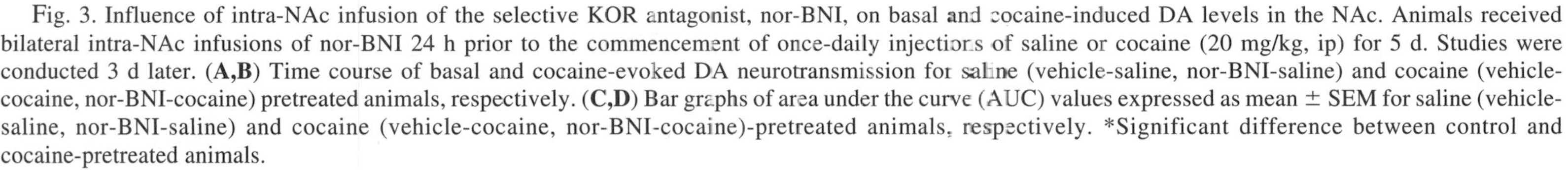
Fig. 3. Influence of intra-NAc infusion of the selective KOR antagonist, nor-BNI, on basal and cocaine-induced DA levels in the NAc. Animals received bilateral intra-NAc infusions of nor-BNI 24 h prior to the commencement of once-daily injections of saline or cocaine (20 mg/kg, ip) for 5 d. Studies were conducted 3 d later. (**A,B**) Time course of basal and cocaine-evoked DA neurotransmission for saline (vehicle-saline, nor-BNI-saline) and cocaine (vehicle-cocaine, nor-BNI-cocaine) pretreated animals, respectively. (**C,D**) Bar graphs of area under the curve (AUC) values expressed as mean ± SEM for saline (vehicle-saline, nor-BNI-saline) and cocaine (vehicle-cocaine, nor-BNI-cocaine)-pretreated animals, respectively. *Significant difference between control and cocaine-pretreated animals.

blockade of KORs by nor-BNI can be seen as a type of functional deletion of KORs and can be compared with emerging data from mice with a constitutive deletion of the KOR1 receptor.

Recent advances in molecular biology have enabled the construction of mice lacking different types of opioid peptides or receptors *(46)*. Since this technique, in contrast to pharmacological approaches, is not limited by the lack of in vivo selectivity and CNS penetration of ligands, it may be particularly useful in dissecting the roles of individual components of the opioid system. On the other hand, one cannot exclude possible compensatory adaptations that occur in response to the absence of particular proteins throughout development in mediating an expressed phenotype. Knockout mice for several genes of the opioid system have been generated: β-endorphin *(129,130)*, enkephalin *(131,132)*, and dynorphin *(133)*, as well as the MOR *(134,135)*, the DOR *(136,137)*, and the KOR *(138–140)*.

To date, only one study has examined the influence of opioid-receptor deletion on the effects of psychostimulants *(140)*. Conventional and no-net-flux microdialysis was used to assess basal and cocaine-evoked levels of extracellular DA as well as the rate of DA uptake in the NAc of KOR knockout, heterozygous, and wild-type mice. Conventional microdialysis revealed that basal as well as cocaine-evoked dialysate levels of DA in the NAc were significantly higher in KOR knockout mice compared with wild-type controls. No-net-flux microdialysis revealed no difference in basal extracellular DA between genotypes. However, the extraction fraction for DA (an indirect measure of the rate of DA uptake) was greater in KOR knockout animals as compared to wild-type controls. Because the extracellular levels of DA are determined by its release and uptake, the fact that the extracellular DA levels were not different, whereas the rate of DA uptake was higher in KOR knockout mice, attest that these animals have greater DA release then wild-type controls. Therefore, one can assume that DA release is increased in the NAc of knockout animals due to the absence of KOR inhibitory control and the apparent increase in DA uptake is a compensatory adaptation to this change in DA neurotransmission.

7. Modulation of Psychostimulant-Induced Sensitization by OFQ/N

The intracerebroventricular application of OFQ/N dose-dependently attenuates both basal and cocaine-induced locomotor activity as well as basal and cocaine-induced DA levels in the NAc *(141)*. In accordance with the notion that changes in mesolimbic DA neurotransmission are involved in mechanisms of behavioral sensitization and the finding that OFQ/N can modulate this neurotransmission *(90,91)*, OFQ/N does not prevent the development of behavioral sensitization when administered into the VTA prior to cocaine injection, but induces behavioral sensitization to a cocaine challenge when administered alone *(142)*. As one of the possible explanations for this effect, the authors hypothesize that OFQ/N, directly or indirectly, increases somatodendritic DA release, thereby reducing DA levels in the NAc *(90,91)*. This increase would, presumably, result in subsensitivity of somatodendritic D2 autoreceptors in the VTA and a sensitized response to a subsequent cocaine challenge, when the peptide is administered alone, but would not be additive to the effect of repeated cocaine. Interestingly, however, preliminary studies in OFQ1 knockout mice revealed no alteration in basal DA overflow in the NAc (N. T. Maidment, personal communication). Whether or not, however, a compensatory increase (e.g., increased KOR activation) occurs in other systems and functionally opposes the primary effect of the gene deletion awaits clarification.

8. Modulation of Psychostimulant Sensitization by MOR and DOR Ligands

Studies examining the influence of repeated administration of MOR and DOR agonists on the locomotor activating effects of amphetamine demonstrate an important role of these opioid receptor types in the modulation of sensitization. The repeated intra-VTA infusion of enkephalin, which results in sensitization to its locomotor-activating effects, also results an enhancement of amphetamine- and cocaine-evoked activity *(143)*. Exposure of animals to an environment previously paired with repeated systemic injections of low, but not high, doses of morphine also results in sensitization to the activating effects of amphetamine *(144)*. Using the conditioned place preference and conditioned reinforcement paradigms, morphine-induced cross-sensitization to the rewarding effects of cocaine and amphetamine has also been demonstrated *(5,145,146)*. Although anatomical mapping studies are limited, the findings that intra-VTA or intra-NAc infusions of MOR and DOR agonists facilitate the development of sensitization *(146)* and increase mesoaccumbens DA neurotransmission *(76)* are consistent with a DA-dependent mechanism of opiate-induced cross-sensitization.

The co-administration of the opioid receptor antagonist, naltrexone, with cocaine attenuates sensitization to cocaine-induced increases in locomotor activity and stereotypy *(147)*. Similarly, in animals that received once-daily injections of cocaine in combination with naltrexone for 14 d, the conditioned reinforcing effects of cocaine were prevented. Although naltrexone produces conditioned aversive effects in rodents *(105,148)*, this action did not underlie the prevention of the conditioned reinforcing effects of cocaine, since lithium chloride, another drug that produces aversive effects, did not modify the conditioned response to cocaine. These findings indicate that naltrexone can prevent locomotor sensitization as well as context-dependent sensitization to the conditioned reinforcing effects of cocaine. Parallel biochemical studies revealed that the ability of DA to stimulate adenylate cylcase is increased in cocaine-treated animals, a finding consistent with previous reports of D1 receptor supersensitivity following repeated psychostimulant administration. This increase was not observed in animals that had received naltrexone in combination with cocaine, indicating that naltrexone also can inhibit a biochemical correlate of sensitization. In view of the relatively large doses of naltrexone employed, it is unclear whether the observed effects reflect an interaction with the MOR and/or the DOR. However, these findings and those described above clearly indicate an involvement of one or both receptor types in the modulation of cocaine-induced behavioral sensitization.

Studies employing selective DOR antagonists suggest the specific involvement of DOR receptors in the development of sensitization to cocaine *(149)*. Co-administration of low doses of naltrindole with cocaine prevented the development of sensitization to the locomotor-activating and conditioned reinforcing effects of cocaine. When, however, naltrindole was administered only once, immediately prior to a challenge dose of cocaine, no alteration in the expression of sensitization was seen *(150)*, suggesting an involvement of endogenous DOR systems in the development, but not the expression of cocaine-induced sensitization.

Evidence that MOR and DOR systems contribute to the sensitization that develops to cocaine-induced seizures and lethality has also been presented *(149)*. Daily injections of a high dose of cocaine (e.g., 50 mg/kg) are associated with a progressive increase in the percentage of rats exhibiting convulsions, indicating the development of sensitization. Intracerebroventricular administration of naltrexone, the MOR antagonist CTOP, or the DOR antagonist naltrindole significantly reduced the incidence of seizures across days as well as the mean seizure score. Both naltrexone and naltrindole, but not CTOP, decreased cocaine-induced lethality.

To date, only one study *(151)* has examined the interaction of endomorphin with psychostimulants. The repeated intra-VTA infusion of either endomorphin-1 or -2 resulted in a sensitized locomotor activity response to an acute challenge dose of amphetamine administered 1–3 d following the cessation of peptide treatments. Thus, in rats that had received repeated infusions of endomorphin alone, amphetamine-evoked activity was significantly greater that that of control (saline-treated) animal and did not differ from that of animals that had received repeated injections of amphetamine. In contrast to the ability of endomorphins to induce sensitization, the intracerebroventricular infusion of these peptides failed to modify its expression when administered following the cessation of repeated amphetamine administration.

9. Opioid Peptide Systems, Sensitization, and Synaptic Plasticity

At present, the molecular mechanisms by which opioid peptide systems modulate the development of psychostimulant sensitization are unknown. It is apparent, however, that psychostimulants alter protein phosphorylation and gene expression in medium spiny neurons of the NAc. Their acute administration is associated with phosphorylation of cAMP response element-binding protein (CREB) and increased expression of both AP1 and non-AP1 family immediate early genes *(19,152)*. As discussed by Berke and Hyman *(153)*, it is likely that these alterations induce a series of downstream molecular events that contribute to the process of sensitization.

D1 DA receptor signaling has been implicated in the initiation of psychostimulant sensitization *(152)*. Medium spiny neurons containing PDYN express D1 DA receptors and psychostimulants induce the expression of PDYN in these cells. PDYN gene expression is also induced in response to other manipulations that elevate extracellular DA *(154)*. In contrast, D1 receptor deletion decreases PDYN gene expression *(155)*. As is observed in the NAc, induction of the PDYN gene by DA in primary cultures is dependent on D1 DA receptor stimulation. It is also correlated with serine-133 phosphorylation of CREB *(156)*. This same serine residue is phosphorylated in response to amphetamine *(152)*. After chronic amphetamine administration, the time course of CREB phosphorylation is prolonged *(156)*, consistent with the documented upregulation of other components of the cAMP system after chronic psychostimulant treatment *(19)*. An upregulated D1 receptor–camp–CREB pathway may serve as an important mechanism of adaptation, contributing not only to PDYN gene induction and an attenuation of sensitization, but also to the regulation of other target genes that contain cAMP response elements and whose induction facilitates or opposes the behavioral and neurochemical consequences of repeated psychostimulant use. The postulated role for CREB and PDYN in mediating neuroadaptations produced by drugs of abuse has a parallel in recent work implicating these proteins in the maintenance (CREB) and attenuation (PDYN) of some forms of long-term synaptic plasticity, including long-term potentiation and long-term depression *(157,158)*. Long-term synaptic plasticity has been demonstrated in the NAc and, more recently, in the ventral tegmental area *(154,159,160)*. Although the role of endogenous opioid peptides in modulating plasticity in the mesocorticolimbic system has not yet been examined, the influence of MOR, DOR, and KOR agonists on long-term potentiation and long-term depression in the hippocampus is not inconsistent with the hypothesis that modulation of synaptic plasticity by opioid peptide systems may occur in regions comprising this system *(161–163)*. Consistent with the opposing effects of MOR—as compared to KOR—agonists on the development of sensitization, MOR agonists facilitate mossy fiber long-term potentiation *(161,162)* and KOR activation blocks its induction *(158,163)*.

10. Summary and Conclusions

As is apparent from this review, interactions of opioid peptide and DA systems have been reported at the cellular, system, and behavioral level. Endogenous opioid peptide systems modulate DA neurotransmission in the NAc and its projections areas, and repeated psychostimulant administration can profoundly affect the activity of these same opioid systems. The repeated administration of cocaine and other drugs that increase DA neurotransmission in the NAc increase PDYN gene expression via D1 DA receptors, and it has been suggested that this increase is a compensatory response that opposes increases in DA that occur in response to these agents. Consistent with this hypothesis, the co-administration of KOR agonists with cocaine attenuates psychostimulant-induced sensitization, whereas KOR blockade results in an enhanced response to cocaine. Data regarding psychostimulant-induced changes in DOR and MOR systems are more limited. However, behavioral studies suggest that an increase in the activity of these opioid systems, in contrast to the PDYN system, promotes psychostimulant-induced behavioral sensitization.

References

1. Le Moal, M. and Simon, H. (1991) Mesocorticolimbic dopaminergic network: functional and regulatory roles. *Physiol. Rev.* **71,** 155–234.
2. White, F. J. and Kalivas, P. W. (1998) Neuroadaptations involved in amphetamine and cocaine addiction. *Drug Alcohol Depend.* **51,** 141–153.
3. Pifl, C., Drobny, H., Reither, H., Hornykiewicz, O., and Singer, E. A. (1995) Mechanism of the dopamine-releasing actions of amphetamine and cocaine: plasmalemmal dopamine transporter versus vesicular monoamine transporter. *Mol. Pharmacol.* **47,** 368–373.
4. Horger, B. A., Shelton, K., and Schenk, S. (1990) Preexposure sensitizes rats to the rewarding effects of cocaine. *Pharmacol. Biochem. Behav.* **37,** 707–711.
5. Shippenberg, T. S. and Heidbreder, C. (1995) Sensitization to the conditioned rewarding effects of cocaine: pharmacological and temporal characteristics. *J. Pharmacol. Exp. Ther.* **273,** 808–815.
6. Robinson, T. E. and Berridge, K. C. (1993) The neural basis of drug craving: an incentive-sensitization theory of addiction. *Brain Res. Brain Res. Rev.* **18,** 247–291.
7. Bartlett, E., Hallin, A., Chapman, B., and Angrist, B. (1997) Selective sensitization to the psychosis-inducing effects of cocaine: a possible marker for addiction relapse vulnerability? *Neuropsychopharmacology* **16,** 77–82.
8. Strakowski, S. M. and Sax, K. W. (1998) Progressive behavioral response to repeated D-amphetamine challenge: further evidence for sensitization in humans. *Biol. Psychiatr.* **44,** 1171–1177.

9. Ackerman, J. M. and White, F. J. (1990) A10 somatodendritic dopamine autoreceptor sensitivity following withdrawal from repeated cocaine treatment. *Neurosci. Lett.* **117,** 181–187.
10. White, F. J. and Wang, R. Y. (1984) Electrophysiological evidence for A10 dopamine autoreceptor subsensitivity following chronic D-amphetamine treatment. *Brain Res.* **309,** 283–292.
11. Lee, T. H., Gao, W. Y., Davidson, C., and Ellinwood, E. H. (1999) Altered activity of midbrain dopamine neurons following 7-day withdrawal from chronic cocaine abuse is normalized by D2 receptor stimulation during the early withdrawal phase. *Neuropsychopharmacology* **21,** 127–136.
12. Yi, S. J. and Johnson, K. M. (1990) Chronic cocaine treatment impairs the regulation of synaptosomal 3H-DA release by D2 autoreceptors. *Pharmacol. Biochem. Behav.* **36,** 457–461.
13. Davidson, C., Ellinwood, E. H., and Lee, T. H. (2000) Altered sensitivity of dopamine autoreceptors in rat accumbens 1 and 7 days after intermittent or continuous cocaine withdrawal. *Brain Res. Bull.* **51,** 89–93.
14. Akimoto, K., Hamamura, T. Kazahaya, Y., Akiyama, K., and Otsuki, S. (1990) Enhanced extracellular dopamine level may be the fundamental neuropharmacological basis of cross-behavioral sensitization between methamphetamine and cocaine—an in vivo dialysis study in freely moving rats. *Brain Res.* **507,** 344–346.
15. Akimoto, K., Hamamura, T., and Otsuki , S. (1989) Subchronic cocaine treatment enhances cocaine-induced dopamine efflux, studied by in vivo intracerebral dialysis [published erratum appears in Brain Res 1989 Aug 21; 495(1): 203]. *Brain Res.* **490,** 339–344.
16. Kalivas, P. W. and Duffy, P. (1990) Effect of acute and daily cocaine treatment on extracellular dopamine in the nucleus accumbens. *Synapse* **5,** 48–58.
17. Kalivas, P. W. and Duffy, P. (1993) Time course of extracellular dopamine and behavioral sensitization to cocaine. I. Dopamine axon terminals. *J. Neurosci.* **13,** 266–275.
18. Henry, D. J. and White, F. J. (1991) Repeated cocaine administration causes persistent enhancement of D1 dopamine receptor sensitivity within the rat nucleus accumbens. *J. Pharmacol. Exp. Ther.* **258,** 882–890.
19. Self, D. W. and Nestler, E. J. (1995) Molecular mechanisms of drug reinforcement and addiction. *Annu. Rev. Neurosci.* **18,** 463–495.
20. Pierce, R. C. and Kalivas, P. W. (1997) Repeated cocaine modifies the mechanism by which amphetamine releases dopamine. *J. Neurosci.* **17,** 3254–3261.
21. Herz, A. (1998) Opioid reward mechanisms: a key role in drug abuse? *Can. J. Physiol. Pharmacol.* **76** 252–258.
22. Spanagel, R., Herz, A., and Shippenberg, T. S. (1992) Opposing tonically active endogenous opioid systems modulate the mesolimbic dopaminergic pathway. *Proc. Natl. Acad. Sci. USA* **89,** 2046–2050.
23. Churchill, L. and Kalivas, P. W. (1996) Dopamine-opioid interactions in the basal forebrain: in *The Modulation of Dopaminergic Neurotransmission by Other Neurotransmitters.* (Ashby, C. R., ed.), CRC Press, New York, pp. 55–86.
24. Swanson, L. W. (1982) The projections of the ventral tegmental area and adjacent regions: a combined fluorescent retrograde tracer and immunofluorescence study in the rat. *Brain Res. Bull.* **9,** 321–353.
25. Carr, D. B. and Sesack, S. R. (2000) GABA-containing neurons in the rat ventral tegmental area project to the prefrontal cortex. *Synapse* **38,** 114–123.
26. Carr, D. B. and Sesack, S. R. (2000) Projections from the rat prefrontal cortex to the ventral tegmental area: target specificity in the synaptic associations with mesoaccumbens and mesocortical neurons. *J. Neurosci.* **20,** 3864–3873.
27. Tong, Z. Y., Overton, P. G., Martinez-Cue, C., and Clark, D. (1998) Do non-dopaminergic neurons in the ventral tegmental area play a role in the responses elicited in A10 dopaminergic neurons by electrical stimulation of the prefrontal cortex? *Exp. Brain Res.* **118,** 466–476.

28. Tong, Z. Y., Overton, P. G., and Clark, D. (1996) Stimulation of the prefrontal cortex in the rat induces patterns of activity in midbrain dopaminergic neurons which resemble natural burst events. *Synapse* **22,** 195–208.
29. Overton, P. G., Tong, Z. Y., and Clark, D. (1996) A pharmacological analysis of the burst events induced in midbrain dopaminergic neurons by electrical stimulation of the prefrontal cortex in the rat. *J. Neural Transm. Gen. Sect.* **103,** 523–540.
30. Murase, S., Grenhoff, J., Chouvet, G., Gonon, F. G., and Svensson, T. H. (1993) Prefrontal cortex regulates burst firing and transmitter release in rat mesolimbic dopamine neurons studied in vivo. *Neurosci. Lett.* **157,** 53–56.
31. Karreman, M. and Moghaddam, B. (1996) The prefrontal cortex regulates the basal release of dopamine in the limbic striatum: an effect mediated by ventral tegmental area. *J. Neurochem.* **66,** 589–598.
32. White, F. J., Hu, X. T., Zhang, X. F., and Wolf, M. E. (1995) Repeated administration of cocaine or amphetamine alters neuronal responses to glutamate in the mesoaccumbens dopamine system. *J. Pharmacol. Exp. Ther.* **273,** 445–454.
33. Sorg, B. A., Davidson, D. L., Kalivas, P. W., and Prasad, B. M. (1997) Repeated daily cocaine alters subsequent cocaine-induced increase of extracellular dopamine in the medial prefrontal cortex. *J. Pharmacol. Exp. Ther.* **281,** 54–61.
34. Chefer, V. I., Moron, J. A., Hope, B., Rea, W., and Shippenberg, T. S. (2000) Kappa opioid receptor activation prevents alterations in mesocortical dopamine neurotransmission that occur during abstinence from cocaine. *Neuroscience* **101,** 619–627.
35. Castaneda, E., Becker, J. B., and Robinson, T. E. (1988) The long-term effects of repeated amphetamine treatment in vivo on amphetamine, KCl and electrical stimulation evoked striatal dopamine release in vitro. *Life Sci.* **42,** 2447–2456.
36. Meredith, G. E. (1999) The synaptic framework for chemical signaling in nucleus accumbens. *Ann. N.Y. Acad. Sci.* **877,** 140–156.
37. Meredith, G. E., Pennartz, C. M., and Groenewegen, H. J. (1993) The cellular framework for chemical signalling in the nucleus accumbens. *Prog. Brain Res.* **99,** 3–24.
38. Meredith, G. E., Agolia, R., Arts, M. P., Groenewegen, H. J., and Zahm, D. S. (1992) Morphological differences between projection neurons of the core and shell in the nucleus accumbens of the rat. *Neuroscience* **50,** 149–162.
39. Hirata, K. and Mogenson, G. J. (1984) Inhibitory response of pallidal neurons to cortical stimulation and the influence of conditioning stimulation of substantia nigra. *Brain Res.* **321,** 9–19.
40. Hirata, K., Yim, C. Y., and Mogenson, G. J. (1984) Excitatory input from sensory motor cortex to neostriatum and its modification by conditioning stimulation of the substantia nigra. *Brain Res.* **321,** 1–8.
41. Sesack, S. R. and Pickel, V. M. (1992) Prefrontal cortical efferents in the rat synapse on unlabeled neuronal targets of catecholamine terminals in the nucleus accumbens septi and on dopamine neurons in the ventral tegmental area. *J. Comp. Neurol.* **320,** 145–160.
42. O'Donnell, P. and Grace, A. A. (1994) Tonic D2-mediated attenuation of cortical excitation in nucleus accumbens neurons recorded in vitro. *Brain Res.* **634,** 105–112.
43. Grace, A. A. (2000) Gating of information flow within the limbic system and the pathophysiology of schizophrenia. *Brain Res. Brain Res. Rev.* **31,** 330–341.
44. Zadina, J. E., Hackler, L., Ge, L. J. and Kastin, A. J. (1997) A potent and selective endogenous agonist for the mu-opiate receptor. *Nature* **386,** 499–502.
45. Akil, H., Meng, F., Mansour, A., Thompson, R., Xie, G. X., and Watson, S. (1996) Cloning and characterization of multiple opioid receptors. *NIDA Res. Monogr.* **161,** 127–140.
46. Kieffer, B. L. (1999) Opioids: first lessons from knockout mice. *Trends Pharmacol. Sci.* **20,** 19–26.
47. Chen, Y., Fan, Y., Liu, J., Mestek, A., Tian, M., Kozak, C. A., and Yu, L. (1994) Molecular cloning, tissue distribution and chromosomal localization of a novel member of the opioid receptor gene family. *FEBS Lett.* **347,** 279–283.

48. Mollereau, C., Parmentier, M., Mailleux, P., Butour, J. L., Moisand, C., Chalon, P., Caput, D., Vassart, G., and Meunier, J. C. (1994) ORL1, a novel member of the opioid receptor family. Cloning, functional expression and localization. *FEBS Lett.* **341,** 33–38.
49. Neal, C. R., Jr., Mansour, A., Reinscheid, R., Nothacker, H. P., Civelli, O., Akil, H., and Watson, S. J., Jr. (1999) Opioid receptor-like (ORL1) receptor distribution in the rat central nervous system: comparison of ORL1 receptor mRNA expression with (125)I-[(14)Tyr]-orphanin FQ binding. *J. Comp. Neurol.* **412,** 563–605.
50. Meunier, J. C., Mollereau, C., Toll, L., Suaudeau, C., Moisand, C., Alvinerie, P. J., Butour, L., Guillemot, J. C., Ferrara, P., Monsarrat, B., et al. (1995) Isolation and structure of the endogenous agonist of opioid receptor-like ORL1 receptor. *Nature* **377,** 532–535.
51. Reinscheid, R. K., Nothacker, H. P., Bourson, A., Ardati, A., Henningsen, R. A., Bunzow, J. R., Grandy, D. K., Langen, H., Monsma, F. J., Jr., and Civelli, O. (1995) Orphanin FQ: a neuropeptide that activates an opioidlike G protein-coupled receptor. *Science* **270,** 792–794.
52. Finley, J. C., Lindstrom, P., and Petrusz, P. (1981) Immunocytochemical localization of beta-endorphin-containing neurons in the rat brain. *Neuroendocrinology* **33,** 28–42.
53. Schreff, M., Schulz, S., Wiborny, D., and Hollt, V. (1998) Immunofluorescent identification of endomorphin-2-containing nerve fibers and terminals in the rat brain and spinal cord. *Neuroreport* **9,** 1031–1034.
54. Fallon, J. H. and Leslie, F. M. (1986) Distribution of dynorphin and enkephalin peptides in the rat brain. *J. Comp. Neurol.* **249,** 293–336.
55. Neal, C. R., Jr., Mansour, A., Reinscheid, R., Nothacker, H. P., Civelli, O., and Watson, S. J., Jr. (1999) Localization of orphanin FQ (nociceptin) peptide and messenger RNA in the central nervous system of the rat. *J. Comp. Neurol.* **406,** 503–547.
56. Curran, E. J. and Watson, S. J., Jr. (1995) Dopamine receptor mRNA expression patterns by opioid peptide cells in the nucleus accumbens of the rat: a double in situ hybridization study. *J. Comp. Neurol.* **361,** 57–76.
57. Mansour, A., Fox, C. A., Burke, S., Akil, H., and Watson, S. J. (1995) Immunohistochemical localization of the cloned mu opioid receptor in the rat CNS. *J. Chem. Neuroanat.* **8,** 283–305.
58. Mansour, A., Fox, C. A., Akil, H., and Watson, S. J. (1995) Opioid-receptor mRNA expression in the rat CNS: anatomical and functional implications. *Trends Neurosci.* **18,** 22–29.
59. Mansour, A., Fox, C. A., Burke, S., Meng, F., Thompson, R. C., Akil, H., and Watson, S. J. (1994) Mu, delta, and kappa opioid receptor mRNA expression in the rat CNS: an in situ hybridization study. *J. Comp. Neurol.* **350,** 412–438.
60. Svingos, A. L., Colago, E. E., and Pickel, V. M. (1999) Cellular sites for dynorphin activation of kappa-opioid receptors in the rat nucleus accumbens shell. *J. Neurosci.* **19,** 1804–1813.
61. Svingos, A. L., Clarke, C. L., and Pickel, V. M. (1999) Localization of the delta-opioid receptor and dopamine transporter in the nucleus accumbens shell: implications for opiate and psychostimulant cross-sensitization. *Synapse* **34,** 1–10.
62. Svingos, A. L., Moriwaki, A., Wang, J. B., Uhl, G. R., and Pickel, V. M. (1997) Mu-opioid receptors are localized to extrasynaptic plasma membranes of GABAergic neurons and their targets in the rat nucleus accumbens. *J. Neurosci.* **17,** 2585–2594.
63. Sesack, S. R. and Pickel, V. M. (1992) Dual ultrastructural localization of enkephalin and tyrosine hydroxylase immunoreactivity in the rat ventral tegmental area: multiple substrates for opiate-dopamine interactions. *J. Neurosci.* **12,** 1335–1350.
64. Mansour, A., Khachaturian, H., Lewis, M. E., Akil, H., and Watson, S. J. (1987) Autoradiographic differentiation of mu, delta, and kappa opioid receptors in the rat forebrain and midbrain. *J. Neurosci.* **7,** 2445–2464.
65. Garzon, M. and Pickel, V. M. (2001) Plasmalemmal mu-opioid receptor distribution mainly in nondopaminergic neurons in the rat ventral tegmental area. *Synapse* **41,** 311–328.

66. Klitenick, M. A., DeWitte, P., and Kalivas, P. W. (1992) Regulation of somatodendritic dopamine release in the ventral tegmental area by opioids and GABA: an in vivo microdialysis study. *J. Neurosci.* **12,** 2623–2632.
67. Johnson, S. W., and North, R. A. (1992) Two types of neurone in the rat ventral tegmental area and their synaptic inputs. *J. Physiol.* (Lond.) **450,** 455–468.
68. Peckys, D. and Landwehrmeyer, G. B. (1999) Expression of mu, kappa, and delta opioid receptor messenger RNA in the human CNS: a 33P in situ hybridization study. *Neuroscience* **88,** 1093–1135.
69. Pickel, V. M., Chan, J., and Sesack, S. R. (1993) Cellular substrates for interactions between dynorphin terminals and dopamine dendrites in rat ventral tegmental area and substantia nigra. *Brain Res.* **602,** 275–289.
70. Hurd, Y. L. (1996) Differential messenger RNA expression of prodynorphin and proenkephalin in the human brain. *Neuroscience* **72,** 767–783.
71. Aggleton, J. P. (1985) A description of intra-amygdaloid connections in old world monkeys. *Exp. Brain Res.* **57,** 390–399.
72. Aggleton, J. P., Friedman, D. P., and Mishkin, M. (1987) A comparison between the connections of the amygdala and hippocampus with the basal forebrain in the macaque. *Exp. Brain Res.* **67,** 556–568.
73. Fudge, J. L. and Haber, S. N. (2000) The central nucleus of the amygdala projection to dopamine subpopulations in primates. *Neuroscience* **97,** 479–494.
74. Meis, S. and Pape, H. C. (2001) Control of glutamate and GABA release by nociceptin/orphanin FQ in the rat lateral amygdala. J. Physiol 532, 701–712.
75. Di Chiara, G. and Imperato, A. (1988) Drugs abused by humans preferentially increase synaptic dopamine concentrations in the mesolimbic system of freely moving rats. *Proc. Natl. Acad. Sci. USA* **85,** 5274–5278.
76. Spanagel, R., Herz, A., and Shippenberg, T. S. (1990) The effects of opioid peptides on dopamine release in the nucleus accumbens: an in vivo microdialysis study. *J. Neurochem.* **55,** 1734–1740.
77. Heijna, M. H., Bakker, J. M., Hogenboom, F., Mulder, A. H., and Schoffelmeer, A. N. (1992) Opioid receptors and inhibition of dopamine-sensitive adenylate cyclase in slices of rat brain regions receiving a dense dopaminergic input. *Eur. J. Pharmacol.* **229,** 197–202.
78. Heijna, M. H., Hogenboom, F., Mulder, A. H., and Schoffelmeer, A. N. (1992) Opioid receptor-mediated inhibition of ^{3}H-dopamine and ^{14}C-acetylcholine release from rat nucleus accumbens slices. A study on the possible involvement of K^{+} channels and adenylate cyclase. *Naunyn Schmiedebergs Arch. Pharmacol.* **345,** 627–632.
79. Pennartz, C. M., Groenewegen, H. J., and Lopes da Silva, F. H. (1994) The nucleus accumbens as a complex of functionally distinct neuronal ensembles: an integration of behavioural, electrophysiological and anatomical data. *Prog. Neurobiol.* **42,** 719–761.
80. Meshul, C. K. and McGinty, J. F. (2000) Kappa opioid receptor immunoreactivity in the nucleus accumbens and caudate-putamen is primarily associated with synaptic vesicles in axons. *Neuroscience* **96,** 91–99.
81. Hjelmstad, G. O. and Fields, H. L. (2001) Kappa opioid receptor inhibition of glutamatergic transmission in the nucleus accumbens shell. *J. Neurophysiol.* **85,** 1153–1158.
82. Rawls, S. M., and McGinty, J. F. (1998) Kappa receptor activation attenuates l-*trans*-pyrrolidine-2,4-dicarboxylic acid-evoked glutamate levels in the striatum. *J. Neurochem.* **70,** 626–634.
83. Gray, A. M., Rawls, S. M., Shippenberg, T. S., and McGinty, J. F. (1999) The kappa-opioid agonist, U-69593, decreases acute amphetamine-evoked behaviors and calcium-dependent dialysate levels of dopamine and glutamate in the ventral striatum. *J. Neurochem.* **73,** 1066–1074.
84. Thompson, A. C., Zapata, A., Justice, J. B., Vaughan, R. A., Sharpe, L. G., and Shippenberg, T. S. (2000) Kappa-opioid receptor activation modifies dopamine uptake in the nucleus accumbens and opposes the effects of cocaine. *J. Neurosci.* **20,** 9333–9340.

85. Devine, D. P., Leone, P., Pocock, D., and Wise, R. A. (1993) Differential involvement of ventral tegmental mu, delta and kappa opioid receptors in modulation of basal mesolimbic dopamine release: in vivo microdialysis studies. *J. Pharmacol. Exp. Ther.* **266,** 1236–1246.
86. Yoshida, Y., Koide, S., Hirose, N., Takada, K., Tomiyama, K., Koshikawa, N., and Cools, A. R. (1999) Fentanyl increases dopamine release in rat nucleus accumbens: involvement of mesolimbic mu- and delta-2-opioid receptors. *Neuroscience* **92,** 1357–1365.
87. Martin, G., Nie, Z., and Siggins, G. R. (1997) mu-Opioid receptors modulate NMDA receptor-mediated responses in nucleus accumbens neurons. *J. Neurosci.* **17,** 11–22.
88. Dauge, V., Kalivas, P. W., Duffy, T., and Roques, B. P. (1992) Effect of inhibiting enkephalin catabolism in the VTA on motor activity and extracellular dopamine. *Brain Res.* **599,** 209–214.
89. Longoni, R., Spina, L., Mulas, A., Carboni, E., Garau, L., Melchiorri, P., and Di Chiara, G. (1991) (D-Ala2)deltorphin II: D1-dependent stereotypies and stimulation of dopamine release in the nucleus accumbens. *J. Neurosci.* **11,** 1565–1576.
90. Murphy, N. P., Ly, H. T., and Maidment, N. T. (1996) Intracerebroventricular orphanin FQ/nociceptin suppresses dopamine release in the nucleus accumbens of anaesthetized rats. *Neuroscience* **75,** 1–4.
91. Murphy, N. P. and Maidment, N. T. (1999) Orphanin FQ/nociceptin modulation of mesolimbic dopamine transmission determined by microdialysis. *J. Neurochem.* **73,** 179–186.
92. Henderson, G. and McKnight, A. T. (1997) The orphan opioid receptor and its endogenous ligand—nociceptin/orphanin FQ. *Trends Pharmacol. Sci.* **18,** 293–300.
93. Sesack, S. R. and Pickel, V. M. (1995) Ultrastructural relationships between terminals immunoreactive for enkephalin, GABA, or both transmitters in the rat ventral tegmental area. *Brain Res.* **672,** 261–275.
94. Hurd, Y. L. and Herkenham, M. (1992) Influence of a single injection of cocaine, amphetamine or GBR 12909 on mRNA expression of striatal neuropeptides. *Brain Res. Mol. Brain Res.* **16,** 97–104.
95. Wang, J. Q. and McGinty, J. F. (1995) Dose-dependent alteration in zif/268 and preprodynorphin mRNA expression induced by amphetamine or methamphetamine in rat forebrain. *J. Pharmacol. Exp. Ther.* **273,** 909–917.
96. Turchan, J., Przewlocka, B., Lason, W., and Przewlocki, R. (1998) Effects of repeated psychostimulant administration on the prodynorphin system activity and kappa opioid receptor density in the rat brain. *Neuroscience* **85,** 1051–1059.
97. Hurd, Y. L., Brown, E. E., Finlay, J. M., Fibiger, H. C., and Gerfen, C. R. (1992) Cocaine self-administration differentially alters mRNA expression of striatal peptides. Brain Res. Mol. Brain Res. 13, 165–170.
98. Breiter, H. C., Gollub, R. L., Weisskoff, R. M., Kennedy, D. N., Makris, N., Berke, J. D., Goodman, J. M., Kantor, H. L., Gastfriend, D. R., Riorden, J. P., Mathew, R. T., Rosen, B. R., and Hyman, S. E. (1997) Acute effects of cocaine on human brain activity and emotion. *Neuron* **19,** 591–611.
99. Svensson, P. and Hurd, Y. L. (1998) Specific reductions of striatal prodynorphin and D1 dopamine receptor messenger RNAs during cocaine abstinence. *Brain Res. Mol. Brain Res.* **56,** 162–168.
100. Smiley, P. L., Johnson, M., Bush, L., Gibb, J. W., and Hanson, G. R. (1990) Effects of cocaine on extrapyramidal and limbic dynorphin systems. *J. Pharmacol. Exp. Ther.* **253,** 938–943.
101. Unterwald, E. M., Rubenfeld, J. M., and Kreek, M. J. (1994) Repeated cocaine administration upregulates kappa and mu, but not delta, opioid receptors. *Neuroreport* **5,** 1613–1616.
102. Staley, J. K., R. B. Rothman, K. C. Rice, J. Partilla, and D. C. Mash (1997) Kappa2 opioid receptors in limbic areas of the human brain are upregulated by cocaine in fatal overdose victims. *J. Neurosci.* **17,** 8225–8233.

103. Mash, D. C. and Staley, J. K. (1999) D3 dopamine and kappa opioid receptor alterations in human brain of cocaine-overdose victims. *Ann. N.Y. Acad. Sci.* **877,** 507–522.
104. Pfeiffer, A., Brantl, V., Herz, A., and Emrich, H. M. (1986) Psychotomimesis mediated by kappa opiate receptors. *Science* **233,** 774–776.
105. Bals-Kubik, R., Herz, A., and . Shippenberg, T. S (1989) Evidence that the aversive effects of opioid antagonists and kappa-agonists are centrally mediated. *Psychopharmacology* (Berl.) **98,** 203–206.
106. Bals-Kubik, R., Ableitner, A., Herz, A., and Shippenberg, T. S. (1993) Neuroanatomical sites mediating the motivational effects of opioids as mapped by the conditioned place preference paradigm in rats. *J. Pharmacol. Exp. Ther.* **264,** 489–495.
107. Gawin, F. H. and Ellinwood, E. H., Jr. (1989) Cocaine dependence. *Annu. Rev. Med.* **40,** 149–161.
108. Crespo, J. A., Manzanares, J., Oliva, J. M., Corchero, J., Palomo, T., and Ambrosio, E. (2001) Extinction of cocainc sclf-administration produces a differential time-related regulation of proenkephalin gene expression in rat brain. *Neuropsychopharmacology* **25,** 185–194.
109. Mathieu-Kia, A. M. and Besson, M. J. (1998) Repeated administration of cocaine, nicotine and ethanol: effects on preprodynorphin, preprotachykinin A and preproenkephalin mRNA expression in the dorsal and the ventral striatum of the rat. *Brain Res. Mol. Brain Res.* **54,** 141–151.
110. Unterwald, E. M., Hornc-King, J., and Kreck, M. J. (1992) Chronic cocaine alters brain mu opioid receptors. *Brain Res.* **584,** 314–318.
111. Zubieta, J. K., Gorelick, D. A., Stauffer, R., Ravert, H. T., Dannals, R. F., and Frost, J. J. (1996) Increased mu opioid receptor binding detected by PET in cocaine-dependent men is associated with cocaine craving. *Nat. Med.* **2,** 1225–1229.
112. Gerfen, C. R., Engber, T. M., Mahan, L. C., Susel, Z., Chase, T. N., Monsma, F. J., Jr., and Sibley, D. R. (1990) D1 and D2 dopamine receptor-regulated gene expression of striatonigral and striatopallidal neurons. *Science* **250,** 1429–1432.
113. Heidbreder, C. A., Goldberg, S. R., and Shippenberg, T. S. (1993) The kappa-opioid receptor agonist U-69593 attenuates cocaine-induced behavioral sensitization in the rat. *Brain Res.* **616,** 335–338.
114. Heidbreder, C. A., Babovic-Vuksanovic, D., Shoaib, M., and Shippenberg, T. S. (1995) Development of behavioral sensitization to cocaine: influence of kappa opioid receptor agonists. *J. Pharmacol. Exp. Ther.* **275,** 150–163.
115. Collins, S. L., Gerdes, R. M., D'Addario, C., and Izenwasser, S. (2001) Kappa opioid agonists alter dopamine markers and cocaine-stimulated locomotor activity. *Behav. Pharmacol.* **12,** 237–245.
116. Shippenberg, T. S. and W. Rea (1997) Sensitization to the behavioral effects of cocaine: modulation by dynorphin and kappa-opioid receptor agonists. *Pharmacol. Biochem. Behav.* **57,** 449–455.
117. Heidbreder, C. A. and Shippenberg, T. S. (1994) U-69593 prevents cocaine sensitization by normalizing basal accumbens dopamine. *Neuroreport.* **5,** 1797–1800.
118. Heidbreder, C. A., Thompson, A. C., and Shippenberg, T. S. (1996) Role of extracellular dopamine in the initiation and long-term expression of behavioral sensitization to cocaine. *J. Pharmacol. Exp. Ther.* **278,** 490–502.
119. Bohn, L. M., Belcheva, M. M., and Coscia, C. J. (2000) Mitogenic signaling via endogenous kappa-opioid receptors in C6 glioma cells: evidence for the involvement of protein kinase C and the mitogen-activated protein kinase signaling cascade. *J. Neurochem.* **74,** 564–573.
120. Saunders, C., Ferrer, J. V., Shi, L., Chen, J., Merrill, G., Lamb, M. E., Leeb-Lundberg, L. M., Carvelli, L., Javitch, J. A., and Galli, A. (2000) Amphetamine-induced loss of human dopamine transporter activity: an internalization-dependent and cocaine-sensitive mechanism. *Proc. Natl. Acad. Sci. USA* **97,** 6850–6855.

121. Vaughan, R. A., Huff, R. A., Uhl, G. R., and Kuhar, M. J. (1997) Protein kinase C-mediated phosphorylation and functional regulation of dopamine transporters in striatal synaptosomes. *J. Biol. Chem.* **272,** 15541–15546.
122. Chefer, V., Thompson, A. C., and Shippenberg, T. S. (1999) Modulation of cocaine-induced sensitization by kappa-opioid receptor agonists. Role of the nucleus accumbens and medial prefrontal cortex. *Ann. N.Y. Acad. Sci.* **877,** 803–806.
123. Heijna, M. H., Padt, M., Hogenboom, F., Schoffelmeer, A. N., and Mulder, A. H. (1991) Opioid-receptor-mediated inhibition of [3H]dopamine but not [^{3}H]noradrenaline release from rat mediobasal hypothalamus slices. *Neuroendocrinology* **54,** 118–126.
124. Heijna, M. H., Padt, M., Hogenboom, F., Portoghese, P. S., Mulder, A. H., and Schoffelmeer , A. N. (1990) Opioid receptor-mediated inhibition of dopamine and acetylcholine release from slices of rat nucleus accumbens, olfactory tubercle and frontal cortex. *Eur. J. Pharmacol.* **181,** 267–278.
125. Izenwasser, S., Acri, J. B., Kunko, P. M., and Shippenberg , T. (1998) Repeated treatment with the selective kappa opioid agonist U-69593 produces a marked depletion of dopamine D2 receptors. *Synapse* **30,** 275–283.
126. Acri, J. B., Thompson, A. C., and Shippenberg, T. (2001) Modulation of pre- and postsynaptic dopamine D2 receptor function by the selective kappa-opioid receptor agonist U69593. *Synapse* **39,** 343–350.
127. Seamans, J. K., Gorelova, N., Durstewitz, D., and Yang, C. R. (2001) Bidirectional dopamine modulation of GABAergic inhibition in prefrontal cortical pyramidal neurons. *J. Neurosci.* **21,** 3628–3638.
128. Kalivas, P. W. and Stewart, J. (1991) Dopamine transmission in the initiation and expression of drug- and stress-induced sensitization of motor activity. *Brain Res. Brain Res. Rev.* **16,** 223–244.
129. Grisel, J. E., Mogil, J. S., Grahame, N. J., Rubinstein, M., Belknap, J. K., Crabbe, J. C., and Low, M. J. (1999) Ethanol oral self-administration is increased in mutant mice with decreased beta-endorphin expression. *Brain Res.* **835,** 62–67.
130. Rubinstein, M., Mogil, J. S., Japon, M., Chan, E. C., Allen, R. G., and Low, M. J. (1996) Absence of opioid stress-induced analgesia in mice lacking beta-endorphin by site-directed mutagenesis. *Proc. Natl. Acad. Sci. USA* **93,** 3995–4000.
131. Brady, L. S., Herkenham, M., Rothman, R. B., Partilla, J. S., Konig, M., Zimmer, A. M., and Zimmer, A. (1999) Region-specific up-regulation of opioid receptor binding in enkephalin knockout mice. *Brain Res. Mol. Brain Res.* **68,** 193–197.
132. Konig, M., Zimmer, A. M., Steiner, H. P., Holmes, V., Crawley, J. N., Brownstein, M. J., and Zimmer, A. (1996) Pain responses, anxiety and aggression in mice deficient in preproenkephalin. *Nature* **383,** 535–538.
133. Sharifi, N., Diehl, N., Yaswen, L., Brennan, M. B., and Hochgeschwender, U. (2001) Generation of dynorphin knockout mice. *Brain Res. Mol. Brain Res.* **86,** 70–75.
134. Matthes, H. W., Maldonado, R., Simonin, F., Valverde, O., Slowe, S., Kitchen, I., Befort, K., Dierich, A., Le Meur, M., Dolle, P., Tzavara, E., Hanoune, J., Roques, B. P., and Kieffer, B. L. (1996) Loss of morphine-induced analgesia, reward effect and withdrawal symptoms in mice lacking the mu-opioid-receptor gene. *Nature* **383,** 819–823.
135. Matthes, H. W., Smadja, C., Valverde, O., Vonesch, J. L., Foutz, A. S., Boudinot, E., Denavit-Saubie, M., Severini, C., Negri, L., Roques, B. P., Maldonado, R., and Kieffer, B. L. (1998) Activity of the delta-opioid receptor is partially reduced, whereas activity of the kappa-receptor is maintained in mice lacking the mu-receptor. *J. Neurosci.* **18,** 7285–7295.
136. Zhu, Y., King, M. A., Schuller, A. G., Nitsche, J. F., Reidl, M., Elde, R. P., Unterwald, E. G., Pasternak, W., and Pintar, J. E. (1999) Retention of supraspinal delta-like analgesia and loss of morphine tolerance in delta opioid receptor knockout mice. *Neuron* **24,** 243–252.
137. Filliol, D., Ghozland, S., Chluba, J., Martin, M., Matthes, H. W., Simonin, F., Befort, K., Gaveriaux-Ruff, C., Dierich, A., LeMeur, M., Valverde, O., Maldonado, R., and Kieffer, B. L. (2000) Mice deficient for delta- and mu-opioid receptors exhibit opposing alterations of emotional responses. *Nat. Genet.* **25,** 195–200.

138. Slowe, S. J., Simonin, F., Kieffer, B., and Kitchen, I. (1999) Quantitative autoradiography of mu-, delta- and kappa1 opioid receptors in kappa-opioid receptor knockout mice. *Brain Res.* **818,** 335–345.
139. Simonin, F., Valverde, O., Smadja, C., Slowe, S., Kitchen, I., Dierich, A., Le Meur, M., Roques, B. P., Maldonado, R., and Kieffer, B. L. (1998) Disruption of the kappa-opioid receptor gene in mice enhances sensitivity to chemical visceral pain, impairs pharmacological actions of the selective kappa-agonist U-50,488H and attenuates morphine withdrawal. *EMBO J.* **17,** 886–897.
140. Chefer, V., Czyzyk, T., Pintar, J. E., and Shippenberg, T. S. (2000) Dopaminergic neurotransmission in the nucleus accumbens of kappa-opioid receptor (KOR) knockout mice: an in vivo microdialysis study. Proceedings International Narcotics Research Conference, Seattle, WA.
141. Lutfy, K., Do, T., and Maidment, N. T. (2001) Orphanin FQ/nociceptin attenuates motor stimulation and changes in nucleus accumbens extracellular dopamine induced by cocaine in rats. *Psychopharmacology* (Berl.) **154,** 1–7.
142. Narayanan, S. and Maidment, N. T. (1999) Orphanin FQ and behavioral sensitization to cocaine. *Pharmacol. Biochem. Behav.* **63,** 271–277.
143. DuMars, L. A., Rodger, L. D., and Kalivas, P. W. (1988) Behavioral cross-sensitization between cocaine and enkephalin in the A10 dopamine region. *Behav. Brain Res.* **27,** 87–91.
144. Vezina, P., Giovino, A. A., Wise, R. A., and Stewart, J. (1989) Environment-specific cross-sensitization between the locomotor activating effects of morphine and amphetamine. *Pharmacol. Biochem. Behav.* **32,** 581–584.
145. Lett, B. T. (1989) Repeated exposures intensify rather than diminish the rewarding effects of amphetamine, morphine, and cocaine. *Psychopharmacology* (Berl.) **98,** 357–362.
146. Cunningham, S. T., Finn, M., and Kelley, A. E. (1997) Sensitization of the locomotor response to psychostimulants after repeated opiate exposure: role of the nucleus accumbens. *Neuropsychopharmacology* **16,** 147–155.
147. Sala, M., Braida, D., Colombo, M., Groppetti, A., Sacco, S., Gori, E., and Parenti, M. (1995) Behavioral and biochemical evidence of opioidergic involvement in cocaine sensitization. *J. Pharmacol. Exp. Ther.* **274,** 450–457.
148. Mucha, R. F. (1987) Is the motivational effect of opiate withdrawal reflected by common somatic indices of precipitated withdrawal? A place conditioning study in the rat. *Brain Res.* **418,** 214–220.
149. Braida, D., Paladini, E., Gori, E., and Sala, M. (1997) Naltrexone, naltrindole, and CTOP block cocaine-induced sensitization to seizures and death. *Peptides* **18,** 1189–1195.
150. Heidbreder, C., Shoaib, M., and Shippenberg, T. S. (1996) Differential role of delta-opioid receptors in the development and expression of behavioral sensitization to cocaine. *Eur. J. Pharmacol.* **298,** 207–216.
151. Chen, J. C., Liang, K. W., and Huang, E. Y. (2001) Differential effects of endomorphin-1 and -2 on amphetamine sensitization: neurochemical and behavioral aspects. *Synapse* **39,** 239–248.
152. Konradi, C., Cole, R. L., Heckers, S., and Hyman, S. E. (1994) Amphetamine regulates gene expression in rat striatum via transcription factor CREB. *J. Neurosci.* **14,** 5623–5634.
153. Berke, J. D., and Hyman, S. E. (2000) Addiction, dopamine, and the molecular mechanisms of memory. *Neuron* **25,** 515–532.
154. Giros, B., Jaber, M., Jones, S. R., Wightman, R. M., and Caron, M. G. (1996) Hyperlocomotion and indifference to cocaine and amphetamine in mice lacking the dopamine transporter. *Nature* **379,** 606–612.
155. Xu, M., Hu, X. T., Cooper, D. C., Moratalla, R., Graybiel, A. M., White, F. J., and Tonegawa, S. (1994) Elimination of cocaine-induced hyperactivity and dopamine-mediated neurophysiological effects in dopamine D1 receptor mutant mice [see comments]. *Cell* **79,** 945–955.
156. Cole, R. L., Konradi, C., Douglass, J., and Hyman, S. E. (1995) Neuronal adaptation to amphetamine and dopamine: molecular mechanisms of prodynorphin gene regulation in rat striatum. *Neuron* **14,** 813–823.

157. Sutton, M. A., Masters, S. E., Bagnall, M. W., and Carew, T. J. (2001) Molecular mechanisms underlying a unique intermediate phase of memory in aplysia. *Neuron* **31,** 143–154.
158. Terman, G. W., Wagner, J. J., and Chavkin, C. (1994) Kappa opioids inhibit induction of long-term potentiation in the dentate gyrus of the guinea pig hippocampus. *J. Neurosci.* **14,** 4740–4747.
159. Bonci, A. and Malenka, R. C. (1999) Properties and plasticity of excitatory synapses on dopaminergic and GABAergic cells in the ventral tegmental area. *J. Neurosci.* **19,** 3723–3730.
160. Thomas, M. J., Malenka, R. C., and Bonci A. (2000) Modulation of long-term depression by dopamine in the mesolimbic system. *J. Neurosci.* **20,** 5581–5586.
161. Derrick, B. E. and Martinez, J. L., Jr. (1994) Opioid receptor activation is one factor underlying the frequency dependence of mossy fiber LTP induction. *J. Neurosci.* **14,** 4359–4367.
162. Jin, W. and Chavkin, C. (1999) Mu opioids enhance mossy fiber synaptic transmission indirectly by reducing GABAB receptor activation. *Brain Res.* **821,** 286–293.
163. Weisskopf, M. G., Zalutsky, R. A., and Nicoll, R. A. (1993) The opioid peptide dynorphin mediates heterosynaptic depression of hippocampal mossy fibre synapses and modulates long-term potentiation. *Nature* **362,** 423–427.

8

Influence of Environmental and Hormonal Factors in Sensitivity to Psychostimulants

Michela Marinelli and Pier Vincenzo Piazza

1. Aims and Scope

This review will analyze the interactions among glucocorticoid hormones, life events, and behavioral effects of pyschostimulant drugs. After an introductory section on the physiology of the hypothalamus–pituitary–adrenal (HPA) axis, we will analyze the influence of glucocorticoid hormones on the behavioral effects of drugs of abuse. This interaction will first be analyzed in basal conditions (third section) and then in stress conditions (fourth section). We will then describe possible mechanisms of glucocorticoid action and in particular the implication of the dopamine system. Finally, we will try to address the question of the possible physiological meaning of the interaction between glucocorticoid hormones and drugs of abuse.

2. The Hypothalamus–Pituitary–Adrenal Axis

Glucocorticoid hormones (cortisol in humans and corticosterone in rodents) are the final step of the activation of the hypothalamic–pituitary–adrenal (HPA) axis, one of the major systems implicated in responding to environmental modifications. Activation of the HPA axis is determined by brain inputs to the hypothalamus, which releases corticotropin-releasing hormone (CRH); CRH reaches the hypophysis via the hyphophyseal portal system and activates the release of ACTH in the bloodstream, which, in turn, triggers the secretion of glucocorticoids by the cortical part of the adrenal gland (for review, *see* ref. *1*). The secretion of glucocorticoids is characterized by a circadian cycle. Concentrations of these hormones are low during the inactive phase (dark phase in humans and light phase in rodents) and rise during the first hours that precede the active phase *(2)*. The secretion of glucocorticoids is also activated by practically all forms of stress, which induces a rapid and large increase in hormone levels (for review, *see* ref. *3*). Stress-induced glucocorticoid secretion has been extensively studied and is considered one of the principal adaptive responses to environmental challenges (for review, *see* ref. *4*).

Glucocorticoids have many peripheral effects that principally involve modulation of energy metabolism and of the immune system. These hormones also readily reach the brain, where they exercise a negative feedback on their own secretion and where they regulate many behavioral and neurobiological activities (for review, *see* ref. *5*). These effects are mediated by binding of glucocorticoids to two receptors: the mineralocorti-

From: *Molecular Biology of Drug Addiction*
Edited by: R. Maldonado © Humana Press Inc., Totowa, NJ

coid receptor (MR) and the glucocorticoid receptor (GR) *(1,5,6)*. These receptors are hormone-activated transcription factors belonging to the family of the nuclear receptors. In the inactive state, GRs and MRs are located in the cytoplasm. Upon activation by glucocorticoids, these receptors migrate to the nucleus, where they bind to specific responsive elements located on the promoters of many genes and activate or repress transcription (for review, *see* ref. *5*). In the brain, MRs are principally located in the septo-hippocampal system, whereas GRs have a more widespread distribution. Because of their different affinity for glucocorticoids, MRs are practically saturated by low basal levels of the hormone; in contrast, GRs are only activated during the circadian elevation of the hormones and after stress *(5)*. Consequently, it is generally believed that the effects of stress-induced corticosterone secretion are mediated by GRs.

3. Influence of Glucocorticoids on the Behavioral Effects of Psychostimulant Drugs

Studies of the role of glucocorticoids on drug responding have been performed by manipulating glucocorticoid levels in animals and by studying the behavioral consequences of these manipulations. In this section we will review the role of glucocorticoids on behavioral responses to drugs such as locomotor activity, self-administration, and relapse.

3.1. Locomotor Response to Psychostimulants

3.1.1. Acute Injection of Psychostimulants

Locomotor response to drugs is an interesting parameter, as it constitutes an unconditioned response to drugs that correlates well with drug self-administration *(7)*. Studies on the role of corticosterone in drug-induced locomotor activity are consistent, showing a facilitatory effect of this hormone. This has been shown over a wide range of drug doses and with different methods of manipulating HPA activity.

Suppression of glucocorticoids by adrenalectomy reduces the psychomotor stimulant effects of cocaine *(8)* and amphetamine *(9–11)* These effects are corticosterone-dependent, as they are reversed by administration of corticosterone. Detailed dose-response studies have shown that adrenalectomy does not modify the locomotor response to low doses of cocaine, but decreases the response to higher doses; that is, adrenalectomy produces a vertical downward shift in the effects of psychostimulants, and decreases the maximal locomotor response to these drugs by approximately 50% (Fig. 1A). The decrease in drug effects caused by adrenalectomy is reversed by corticosterone replacement (a subcutaneous corticosterone pellet delivering constant basal levels of the hormone) in a dose-dependent manner *(10,12)*, and the response to cocaine is fully restored when basal concentrations of corticosterone are reached (Fig. 1B).

The decrease in drug effects following adrenalectomy has been confirmed using pharmacological manipulations that reduce corticosterone levels. Thus, acute or repeated treatment with metyrapone, a corticosterone synthesis inhibitor, has been shown to decrease the locomotor response to cocaine (*13,14*, although *see* ref. *15*).

Further confirmation that decreased drug responding after adrenalectomy or pharmacological blockade of corticosterone synthesis is truly due to suppression of circulating glucocorticoids comes from studies using corticosteroid receptor antagonists. These studies show that blockade of central corticosteroid receptors (both MRs and

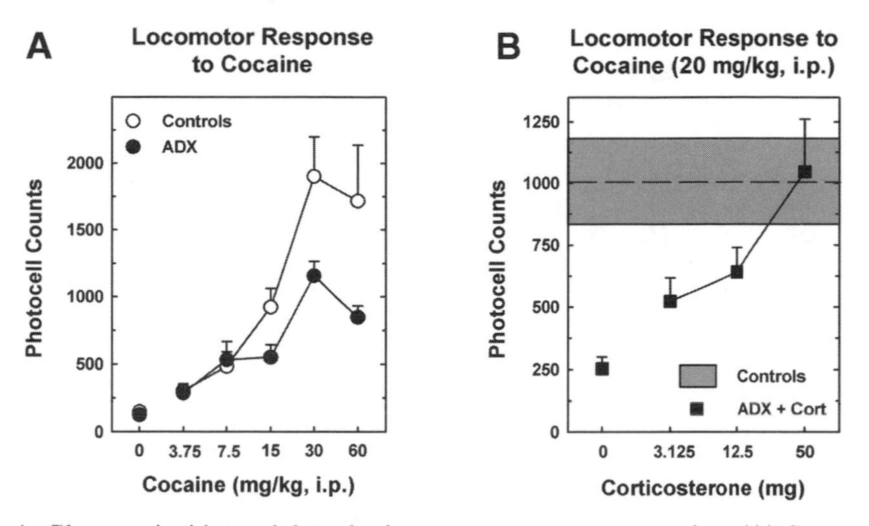

Fig. 1. Glucocorticoids modulate the locomotor response to cocaine. **(A)** Suppression of corticosterone by adrenalectomy (ADX) decreases the maximal locomotor response to cocaine over a wide range of cocaine doses. **(B)** The effects of ADX are reversed dose dependently by corticosterone replacement (subcutaneous implantation of corticosterone pellet; ADX + Cort). The response to cocaine (20 mg/kg, ip) is fully restored when animals receive a replacement treatment reproducing basal levels of corticosterone. (Modified from Marinelli et al. *[12]*).

GRs), by central administration of the MR antagonist spironolactone and of the GR antagonist RU 38486, also decreases the locomotor response to an injection of cocaine *(12)*. This decrease is similar to that observed after adrenalectomy.

3.1.2. Repeated Injections of Psychostimulants

Repeated injections of psychostimulants produce an increase in their psychomotor effects, which is referred to as behavioral sensitization (for review, *see* ref. *16*). Although suppression of glucocorticoids reduces the acute locomotor response to psychostimulants, its effect on repeated drug treatment is more controversial. For example, regarding amphetamine sensitization, Badiani and colleagues *(9)* have shown that although adrenalectomy attenuates the psychomotor effects of an acute injection of amphetamine, it does not prevent the development of sensitization to this drug. Rivet et al. *(17)*, however, concluded that adrenalectomy does inhibit development of amphetamine sensitization. Regarding cocaine sensitization, Prasad et al. *(18)* reported that, similarly to amphetamine sensitization, adrenalectomy does not prevent behavioral sensitization tested at late withdrawal times. At early withdrawal times, however, different groups have shown that adrenalectomy does prevent the development of cocaine-induced sensitization *(18,19)*. In these studies, adrenalectomy decreased sensitization only if it was performed before, but not after, the sensitizing paradigm. However, blockade of GRs by selective antagonists after sensitization has been shown to prevent the expression of amphetamine sensitization *(20)*. These results illustrate the controversial role of glucocorticoids on drug-induced sensitization. One factor that could explain some of the discrepancies among different studies is the parameter used to measure sensitization. For example, the studies by both Badiani et al. and Rivet et al. *(9,17)* show a similar phenomenon: adrenalectomized rats, compared to controls, have a lower locomotor response to amphetamine during the first and last injections of the drug. However, both groups of rats show an increase in the psychomotor effects of the drug between the first and the last injections. Consequently, when a within-group comparison is performed, as in the case of Badiani et al., one concludes that sensitization is still present after adrenalectomy. Instead, if one compares, as in the case of Rivet et al., the response between groups, it could be concluded that sensitization is reduced.

Drugs of abuse, and in particular psychostimulants, induce an important increase in corticosterone secretion (for review, *see* ref. *21*). However, this response does not seem to be involved in mediating the effects of glucocorticoids on the locomotor response to psychostimulants. Thus, animals with basal levels of corticosterone and that cannot secrete corticosterone in response to the drug challenge (i.e., adrenalectomized animals with corticosterone replacement treatments reproducing basal levels of the hormone) show a locomotor response to the drug that is similar to the one seen in control animals *(8,12)*. This finding is consistent with the observation that there is no correlation between drug-induced locomotion and drug-induced corticosterone increase *(22,23)*.

Finally, it is also noteworthy to specify that the reduction in drug effects induced by a decrease in glucocorticoids is not due to possible differences in the bioavailability of cocaine; thus cerebral concentrations of cocaine are not decreased in adrenalectomized animals, or in animals treated with metyrapone *(12,14)*.

3.2. Self-Administration of Psychostimulants

Drug self-administration in animals undoubtedly is one of the best models of drug-taking in humans. Animals self-administer most drugs that are addictive in humans *(24–26)* and, like humans, show large individual differences in drug responding *(7)*.

Several studies in rodents have shown that suppression or reduction of circulating corticosterone decreases the reinforcing effects of psychostimulants as measured by self-administration. For example, suppression of glucocorticoids by adrenalectomy abolishes acquisition of cocaine self-administration over a wide range of cocaine doses *(27,28)*, an effect that is reversed by corticosterone replacement. Other dose–response studies have shown that adrenalectomy induces a vertical downward shift of the dose–response curve to cocaine *(29)* during the maintenance phase (Fig. 2A). This reduction in drug effects is reversed dose-dependently by exogenous administration of corticosterone to adrenalectomized rats and the response to cocaine is fully restored when stress levels of corticosterone are reached (Fig. 2B). Blockade of corticosterone secretion by metyrapone also reduces self-administration of cocaine, in both the acquisition and retention phases *(27,28)*. Another corticosterone synthesis inhibitor, ketoconazole, has similar effects and reduces acquisition of self-administration of low but not high doses of cocaine *(30)*. These effects are not due to nonspecific decreases in motor behavior or motivation, as these treatments do not modify seeking behavior in food-related tasks *(31)*.

These findings that suppression of corticosterone decreases self-administration behavior are in contrast with studies by Suzuki and colleagues showing that adrenalectomy does not modify cocaine-induced place preference *(32)*, a behavior that is often considered to be an index of the rewarding effects of drugs of abuse *(33–35)*. However, this discrepancy can be explained by the fact that place preference induced by intraperitoneal injections of cocaine (as in the studies by Suzuki and colleagues *[32]*) does not depend on the same neurobiological systems as those induced by intravenous injec-

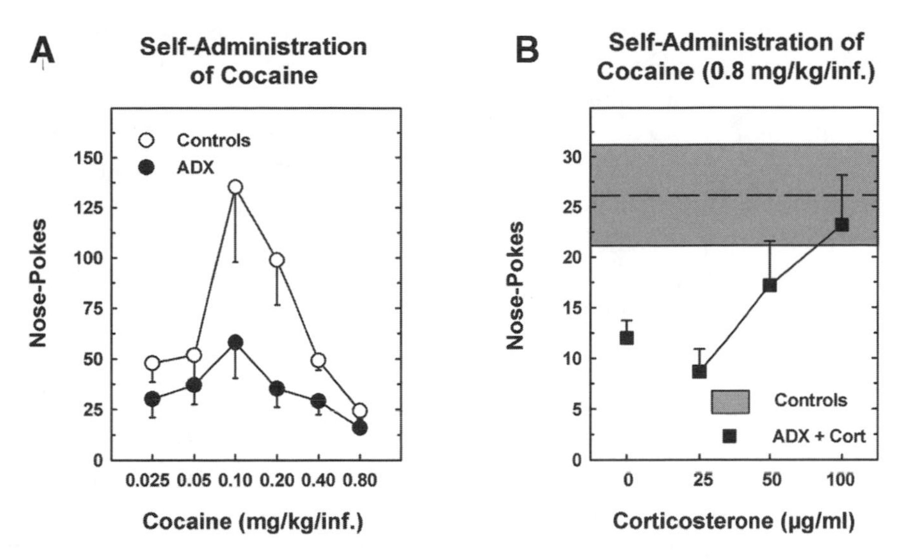

Fig 2. Glucocorticoids modulate the reinforcing effects of cocaine. (**A**) Suppression of corticosterone by adrenalectomy (ADX) produces a vertical downward shift in self-administration behavior (number of nose pokes for cocaine) over a wide range of cocaine doses. (**B**) The effects of ADX are reversed dose-dependently by corticosterone replacement (corticosterone added to the drinking solution of the animals; ADX + Cort). The response to cocaine (0.8 mg/kg/inf) is fully restored when animals receive a replacement treatment reproducing stress levels of corticosterone. (Modified from Deroche et al. *[29]*).

tions of cocaine (as used in self-administration experiments). In particular, the dopamine system does not seem to mediate the place preference induced by intraperitoneal injections of cocaine *(36,37)*, although it mediates psychostimulant self-administration *(38–40)*. Furthermore, place conditioning is a test that is not very well adapted to measure changes in the intensity of reinforcing or rewarding effect of drugs (for review, *see* ref. *41*). Thus, with this test, changes in the threshold dose of psychostimulant required to produce conditioning can be measured, but once the response is induced, the intensity of its effects does not change as a function of drug dose *(42)*. In other words, place conditioning is adapted to measure horizontal shifts in dose–response functions, but not vertical ones. In contrast, in drug self-administration, the intensity of responding is highly dependent on changes in drug dose, and this test can clearly evaluate both horizontal and vertical shifts in dose–response functions. Because suppression of glucocorticoids induces a vertical shift in the dose–response to cocaine self-administration *(29)*, it is not surprising that this manipulation has no effects of cocaine-induced place conditioning.

3.3. Relapse in Psychostimulants Self-Administration

Relapse behavior is generally studied following an extinction period from drug self-administration. During this period, the drug is not available, so the animals extinguish responding to obtain the drug that is no longer available. Different priming factors such as exposure to the drug or the drug cue can induce the reinstatement of drug-seeking behavior. Although corticosterone plays a facilitatory role in many behavioral responses to drugs of abuse, its role in relapse is a little more controversial. Suppression of glucocorticoids does not seem to have important effects on relapse induced by drug priming. Thus, cocaine-induced reinstatement of cocaine self-administration is only minimally decreased by adrenalectomy *(43)*, and is not modified by ketoconazole, a corticosterone synthesis inhibitor that reduces circulating levels of corticosterone *(44)*. Instead, corticosterone plays a significant role in cue-induced reinstatement of seeking. Thus, treatment with ketoconazole prevents conditioned cue-induced reinstatement of cocaine-seeking behavior produced by contingent exposure to a light and tone previously paired with cocaine during self-administration (for review, *see* ref. *45*). Furthermore, as we will see in the next section, corticosterone also seems to play an important role in stress-induced reinstatement.

3.4. Methodological Considerations

Although most studies examining the effects of adrenalectomy on the behavioral responses to drugs are consistent, some studies have not observed decreased drug effects following ablation of the adrenal glands. For example, as mentioned previously, adrenalectomy does not decrease sensitization to the locomotor effects of cocaine when it is performed after a sensitizing paradigm *(18,19)*. In addition, adrenalectomy does not reduce cocaine facilitation of brain stimulation *(46)* or drug-induced reinstatement of seeking behavior (for review, *see* ref. *43*). The nature of this discrepancy is unclear. However, it is important to point out that corticosterone levels at the time of adrenalectomy (i.e., circulating levels of corticosterone at the time when the adrenal glands are removed) could play a fundamental role in determining whether adrenalectomy will or not reduce drug effects. Thus, it has been shown *(12,47)* that adrenalectomy has no effects on the locomotor response to cocaine or on the analgesic effects of morphine if it is performed when corticosterone levels are elevated, such as during the dark phase, following stress, an injection of corticosterone, or when animals are anesthetized with pentobarbital (because of the longer induction of anesthesia with barbiturates, adrenals are removed several minutes after the animals have been removed from the colony room—a time long enough for corticosterone levels to rise). Adrenalectomy seems most efficient in reducing drug effects when levels of the hormone are low, that is, when it is performed rapidly, under inhalant anesthetics *(12,47)*. Although the mechanisms underlying this state-dependent effect of adrenalectomy are not known, it is likely that this effect could explain, at least in part, the discrepancies in the literature. For example, it is possible that high levels of corticosterone after withdrawal from drug self-administration could prevent the effects of adrenalectomy on drug-induced reinstatement of seeking behavior, or that removing the adrenal glands during a drug-sensitizing paradigm (when levels of the hormone are elevated) could also prevent the locomotor-suppressing effects of adrenalectomy.

4. Influence of Glucocorticoids on Stress-Induced Sensitization to the Behavioral Effects of Psychostimulant Drugs

As we will see in this section, stress increases the behavioral effects of psychostimulant drugs. We will review how glucocorticoids mediate these effects.

4.1. Effects of Stress on the Behavioral Response to Psychostimulants

Stressful conditions have been shown to increase the psychomotor stimulant effects of drugs of abuse *(48–51)*, a phenomenon often referred to as stress-induced sensitization. Mild food restriction (10–20% of body-weight loss) increases locomotor activa-

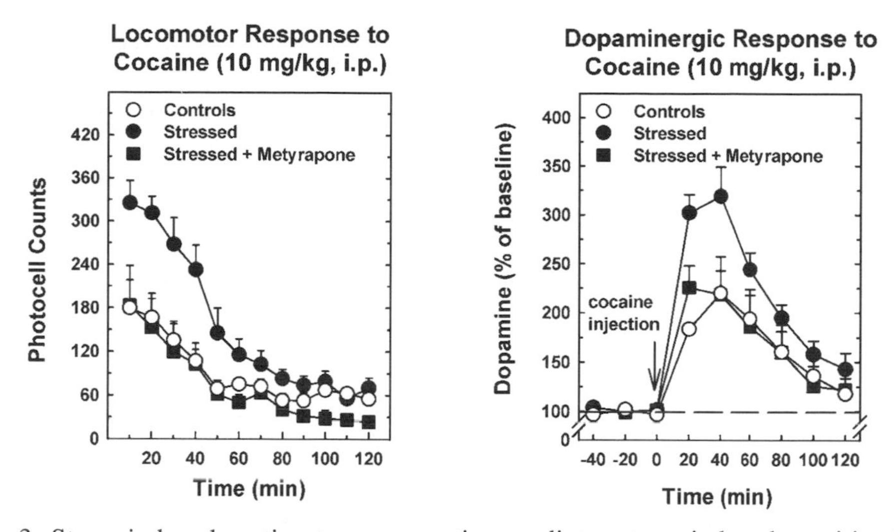

Fig 3. Stress-induced corticosterone secretion mediates stress-induced sensitization to cocaine. **(A)** Stress (food restriction) increases the locomotor response to an injection of cocaine (10 mg/kg). Blockade of stress-induced secretion of corticosterone by treatment with metyrapone (100 mg/kg) prevents this stress-induced effect. **(B)** Food restriction increases dopamine levels in response to cocaine (10 mg/kg). Blockade of stress-induced secretion of corticosterone by treatment with metyrapone (100 mg/kg) prevents this stress-induced effect. (Modified from Marinelli et al. *[13]* and Rougé-Pont et al. *[71]*).

tion induced by cocaine and amphetamine *(52,53)*, and other stressors such as mild tail pressure, foot shock, restraint/immobilization, handling, social isolation, or social defeat also have similar effects *(48,49,51,54–56)*.

The enhancing effects of stress on drug self-administration have been described for decades. Again, different stressors such as tail pinch *(50)*, foot shock *(57)*, social isolation *(58,59)*, social stress *(60–62)*, and food restriction *(63–67)* all increase intravenous self-administration of amphetamine and cocaine. These effects have been observed for different doses of the drugs, during the acquisition phase, the retention one, as well as in progressive ratio schedules.

4.2. Locomotor Response to Psychostimulants

Several studies have shown that stress-induced sensitization to the locomotor effects of psychostimulants depends on the increase in corticosterone levels induced by the stressor. Thus, treatments that block stress-induced corticosterone secretion, but maintain basal levels of the hormone, have been shown to inhibit stress-induced sensitization. For example, adrenalectomy associated with replacement of basal levels of glucocorticoids (via subcutaneous corticosterone pellets) prevents the increase in locomotor response to amphetamine observed after food-restriction stress *(53)*. A similar effect is seen on the locomotor response to amphetamine following social isolation *(68)*, restraint stress *(69)*, or different repeated stressors *(70)*. These findings are confirmed by the observation that pharmacological blockade of stress levels of corticosterone (with acute or repeated metyrapone treatment) also decreases sensitization to the psychomotor effects of cocaine or amphetamine *(15,71)* (Fig. 3A). Overall, using different stressors and different methods to block stress-induced corticosterone secretion, these works show that stress levels of corticosterone secretion are essential to produce stress-induced sensitization.

In addition to these studies, the role of stress levels of corticosterone on drug responding has been confirmed by research examining the effects of repeated exposure to corticosterone on drug effects. Thus, repeated administration of corticosterone at doses producing blood levels of the hormone that are comparable to those observed in stressful conditions also increases (similarly to stress) the psychomotor effects of amphetamine *(72)*. It is noteworthy to point out, however, that these glucocorticoid treatments were performed repeatedly, and not acutely. It is therefore possible that a long-term exposure to high levels of corticosterone is necessary for stress-induced sensitization to develop.

4.3. Self-Administration of Psychostimulants

Similar to locomotor activity, the increase in self-administration induced by stress also depends on stress-induced corticosterone secretion. Thus, blocking the rise in corticosterone during stress prevents the increase in drug responding induced by the stressors. To our knowledge, these effects have only been examined following treatment with ketoconazole, the corticosterone synthesis inhibitor. Repeated ketoconazole treatment reduces both the rate of acquisition of cocaine self-administration and the percentage of rats meeting acquisition criterion following food-restriction stress *(73)*. The effects of corticosterone reduction are not due to nonspecific changes in motivation or motor behavior, because operant responding for food is not modified in groups whose corticosterone levels have been modified *(31,74)*.

A larger number of studies examined the consequences of administering stress levels of corticosterone. These studies show that repeated administration of high (stress-level) doses of glucocorticoids increases drug self-administration, and reproduces the increase in drug responding seen during stressful situations. For example, rats receiving repeated injections of corticosterone intraperitoneally acquire cocaine self-administration at a lower dose compared with vehicle-treated controls *(75)*. An intravenous injection of corticosterone before a self-administration session can also increase drug responding in animals that would not readily acquire cocaine self-administration *(76)*.

The importance of stress levels of corticosterone in drug self-administration is confirmed by work showing that the decrease in cocaine intake induced by adrenalectomy is reversed by administration of high (stresslike) doses of corticosterone, via corticosterone in the drinking solution of the animals *(29)*. It is still uncertain whether these effects of stress levels of corticosterone are mediated by MRs or GRs. Mantsch et al. *(75)* suggested a role for MRs. In their study, treatment with GR agonist dexamethasone prevented acquisition of cocaine self-administration. This effect, attributed to the decrease in corticosterone following dexamethasone, was reversed by the MR agonist aldosterone. It is also possible, however, that GRs are involved in these effects. Adrenalectomized rats receiving replacement of low levels of corticosterone (probably only saturating MRs) do not maintain self-administration. Instead, high levels of the hormone (probably activating GRs) are necessary for cocaine self-administration *(29)*. In addition, blockade of GRs by RU 38486 dose-dependently decreases cocaine self-administration (Aouizerate et al., unpublished observations), further suggesting a role of GRs in drug self-administration.

4.4. Relapse in Psychostimulant Self-Administration

Following extinction training, reinstatement of drug-seeking behavior can be elicited by exposure to different stressors such as foot shock or food restriction (*see* refs. *77–79*). Regarding foot shock stress, it was shown that basal levels of corticosterone are necessary for foot shock to induce cocaine seeking, but that stress-induced increase in corticosterone does not play an important role on this type of reinstatement. Thus, adrenalectomy decreases foot shock-induced reinstatement, but basal levels of corticosterone are sufficient to reverse this effect *(43)*. Similarly, in a different study, treatment with ketoconazole has been shown to decrease foot shock-induced reinstatement while only partially decreasing stress-induced corticosterone secretion *(80)*. Studies on foot shock-induced reinstatement show that this behavior is blocked by administration of CRH antagonists *(43,81)* or α_2-adrenergic receptor agonists *(82)*, suggesting that extra-hypothalamic CRH and the central noradrenergic system are the important players in this type of relapse. Using a different stressor to induce reinstatement of drug-seeking behavior (food restriction), the role of corticosterone was reconsidered. Preliminary data show that adrenalectomy prevents the effects of food restriction-induced reinstatement to cocaine-seeking behavior. These effects are not reversed by restoring basal levels of the hormone, but only by restoring higher levels, suggesting that more elevated levels of corticosterone might be necessary for food restriction to produce reinstatement of drug-seeking behavior *(83)*. Overall, these results suggest that corticosterone plays an important role in relapse to drug-seeking behavior induced by stress. However, it is possible that different stressors require different threshold doses of corticosterone to produce reinstatement.

The implication of stress levels of corticosterone in drug-seeking behavior has also been assessed in studies using other relapse paradigms. This work shows that intravenous injections of corticosterone can precipitate reinstatement of drug-seeking behavior following extinction training. This effect is dose-dependent, and shows an inverted U-shaped curve. Thus, lower doses of corticosterone produce mild reinstatement, higher doses produce greater effects, and even higher doses will decrease reinstatement (Fig. 4). Peak effects are obtained for doses of corticosterone that are similar to those observed during stress *(29)*, whereas the descending limb of the curve is obtained for supra-physiological doses of corticosterone, suggesting that physiological stress levels of corticosterone facilitate drug-seeking behavior. In another study, it has been shown that after acquisition and stabilization of cocaine self-administration, food-restricted animals treated with metyrapone during withdrawal and reexposure to drug self-administration show decreased drug-taking during the reexposure phase *(31)*, suggesting that corticosterone facilitates relapse to drug-taking in animals undergoing food-restriction stress.

5. The Mesolimbic Dopamine System: *A Possible Substrate Mediating the Effects of Glucocorticoids on Drug Responses*

5.1. The Mesolimbic Dopamine System

Research on the substrates of the addictive effects of drugs of abuse has underlined the role of dopamine cells, and in particular of the neurons originating in the ventral tegmental area (VTA) and projecting to the nucleus accumbens (NAc). This dopamine

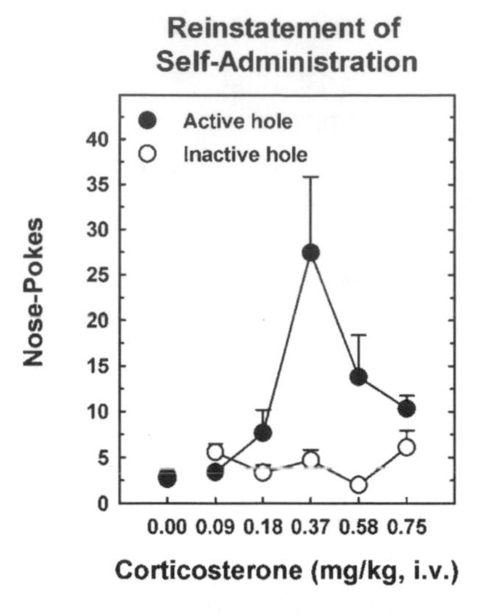

Fig 4. Injection of corticosterone induces reinstatement of drug-seeking behavior. Administration of corticosterone induces an increase in the number of responses (nose pokes) in the device previously associated with cocaine (active hole), without modifying responding in the device without scheduled consequences (inactive hole). Peak effects are obtained for doses of corticosterone that are similar to those observed during stress; the descending limb of the curve is obtained for supra-physiological doses of corticosterone. (Modified from Deroche et al. *[29]*).

pathway plays an important role in natural reward-related behaviors, such as seeking for food, sexual partners, or novel stimuli *(84,85)*. Numerous studies have shown that different addictive drugs, such as psychostimulants, opioids, nicotine, and alcohol, all share the common property of increasing NAc dopamine *(86–89)*. In addition, lesions of the NAc, or of neurons projecting to the NAc, decrease self-administration of these drugs *(38,39,90,91)*, and dopamine receptor agonists or antagonists, can respectively increase or decrease the rewarding properties of psychostimulants *(92–99)*. The role of the meso-accumbens dopamine projection in drug addiction is further underlined by findings showing that the activity of dopamine neurons codes for the rewarding aspects of environmental stimuli *(100)* and that increased activity of these cells is associated with enhanced vulnerability to drug addiction *(101)*. In addition, animals with increased vulnerability to drugs also display increased dopamine levels in the NAc in basal conditions, in response to psychostimulant drugs and to stress *(102–105)*.

Dopaminergic hyperactivity also seems to play an important role in stress-mediated increase in vulnerability to drugs. Following the work by Thierry and collaborators *(106)*, numerous studies have analyzed the relationship between stress and dopaminergic activity (for review, *see* ref. *107*). It has been hypothesized that dopaminergic hyperactivity could be a mechanism by which stress increases drug responding. Thus, stress-induced increase in drug effects is associated with increased dopamine cell activity *(65,108)* as well as with increased levels of dopamine in the NAc in basal conditions, following administration of psychostimulants, or a subsequent stressor *(62,71,109–116)*.

In the past decade, studies on drug addiction have given a particularly important role to the "shell" of the NAc, a functionally and anatomically distinct subset of this nucleus (for review, *see* ref. *117*). Thus, addictive drugs preferentially increase dopamine transmission in the shell of the NAc *(102,118–121)*, and dopamine or dopamine receptor antagonists administered in the shell can modify drug self-administration behavior *(122)*. In addition, animals will self-administer drugs with addictive potential preferentially in the shell of the NAc *(123,124)*. It should be noted that although the shell seems to play a key role in brain reward mechanisms, the "core" of the NAc also has an important role; this substructure seems particularly involved in the processing of reward-related cues (for review, *see* ref. *125*).

5.2. Glucocorticoids and the Dopamine System

Here we will review several observations suggesting that glucocorticoid hormones could facilitate drug-related behaviors by acting on the dopamine system.

5.2.1. Glucocorticoids and Dopamine-Dependent Responses to Drugs

The first evidence of dopamine involvement in the interaction between glucocorticoids and psychostimulants probably comes from studies investigating changes in behavioral effects of centrally injected psychostimulants. Thus, the locomotor response induced by the injection of psychostimulants in the NAc is dopamine-dependent *(126,127)*. It was shown that the locomotor response to intra-NAc cocaine is decreased by suppressing glucocorticoid hormones *(8)* and reestablished by restoring basal levels of the hormone. In addition, this response is also modulated by glucocorticoids in stress conditions. In food-restricted animals, blockade of stress-induced increase in corticosterone (by adrenalectomy and replacement of basal levels of corticosterone) prevents the stress-induced increase in locomotor activity following intra-NAc amphetamine. These effects are reversed by reproducing stress levels of the hormone *(128)*.

5.2.2. Glucocorticoids and Dopamine Transmission in the NAc

As noted above, glucocorticoids facilitate dopamine-dependent behaviors. Several studies tried to determine whether these hormones act by modulating dopamine transmission in the NAc. The effects of both suppression and enhancement of corticosterone have been examined using in vivo microdialysis and expression of Fos-related proteins.

Suppression of glucocorticoids by adrenalectomy reduces extracellular concentrations of dopamine in the NAc, both in basal conditions and in response to psychostimulants *(129,130)*. These effects are corticosterone-dependent, as they are reversed by corticosterone replacement. Interestingly, glucocorticoids have a specificity of action in the NAc. Thus, as Fig. 5 shows, adrenalectomy selectively and dramatically (over 50%) decreases dopamine in the shell of the NAc, without modifying dopamine concentrations in the core *(129)*. The reduction in NAc shell dopamine following adrenalectomy has been observed for basal levels of dopamine, as well as for dopamine increases following cocaine or stress *(129)*. Detailed studies of the role of MRs and GRs in these effects have proposed a role for GRs. Thus, the administration of the MR antagonist spironolactone does not modify extracellular levels of dopamine, whereas the administration of GR receptor antagonist RU 38486 or RU 39305 dose-dependently decreases basal levels of dopamine in the shell of the NAc *(131)*. These effects are very similar to those produced by adrenalectomy, as these antagonists reduce the basal levels of dopamine by more than 50%.

The decrease in dopamine levels in the NAc shell following glucocorticoid suppression is also translated postsynaptically. Thus, in adrenalectomized animals, Fos expression, an index of cellular activation that depends mainly on dopamine D1 receptor

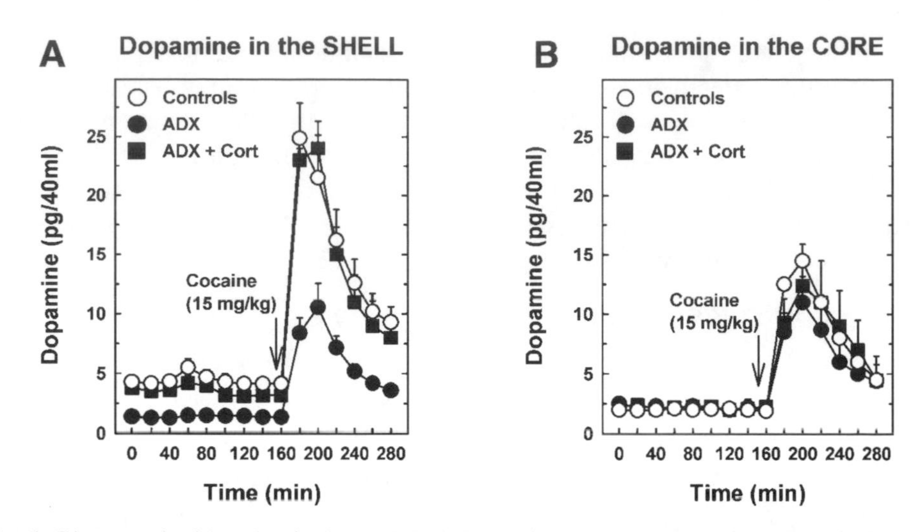

Fig. 5. Glucocorticoids selectively modulate dopamine concentrations in the shell of the NAc. **(A)** Suppression of corticosterone by adrenalectomy (ADX) decreases dopamine concentrations in the shell of the NAc, both in basal conditions and in response to cocaine (15 mg/kg). These effects are reversed by administration of corticosterone (ADX + Cort). **(B)** Adrenalectomy does not modify dopamine concentrations in the core of the NAc. (Modified from Barrot et al. *[129]*).

activation *(132,133)*, is also decreased in the shell of the NAc after administration of cocaine, whereas no changes are observed in the core *(129)*. In other words, decreased levels of dopamine in the NAc shell following adrenalectomy produce decreased postsynaptic activation. On the other hand, adrenalectomized and sham controls show similar Fos activation in the shell in response to the D1 receptor agonist SKF 82958, suggesting that postsynaptic D1 receptors are functionally unaltered in this structure. Changes in NAc D1 receptors have, however, been reported following corticosterone manipulations. Thus, adrenalectomy decreases D1 receptors in the NAc *(134)*, and this effect is reversed by administration of the GR agonist dexamethasone, suggesting, once again, a role for GRs. Furthermore, decrease in circulating corticosterone by administration of metyrapone, the corticosterone synthesis inhibitor, also decreases D1 receptor binding and mRNA in the NAc *(135)*. Overall, these results indicate that glucocorticoids, through GRs, have a facilitatory action on dopamine transmission presynaptically, probably by modulating dopamine release, and that this effect results in reduced postsynaptic activity.

Studies on the effects of stress levels of corticosterone on NAc dopamine are more controversial. Thus, using in vivo microdialysis, Imperato and co-workers *(109,110)* found that adrenalectomy does not prevent stress-induced increase in NAc dopamine, and Reid et al. *(15)* reported that blockade of stress levels of corticosterone by treatment with metyrapone enhances amphetamine-induced dopamine release. Instead, we found that blockade of stress-induced corticosterone secretion by either adrenalectomy or metyrapone treatment prevents the increase in NAc dopamine induced by stress *(71,136)* (Fig. 3B). It is possible that the location of the microdialysis probe (core vs shell), which was not clearly determined in these studies, could explain these discrepancies.

Studies on dopamine levels following administration of corticosterone are also controversial. Using in vivo microdialysis, Imperato and co-workers *(109,110)* reported that corticosterone administration produces a modest increase in NAc dopamine, but these effects are only obtained with concentrations of corticosterone that are well above the physiological range observed during stress. Instead, voltammetry studies by Mittleman et al. *(137)* have found that dopamine release is increased following administration of stresslike levels of corticosterone.

The variability in the effects of glucocorticoids on NAc dopamine may be explained by possible state-dependent effects of these hormones. Thus, it has been shown that corticosterone administration increases NAc dopamine when it is administered during the dark phase, but not during the light phase, and these effects are greater in the dark phase if the hormone is administered just before eating *(138)*. In addition, after administration of corticosterone, there is a greater increase in NAc dopamine in rats that spontaneously show higher dopamine release than in those with lower dopaminergic activity *(136)*. In other words, it appears that corticosterone can only increase NAc dopamine if the hormone is administered in conditions in which the dopamine system is activated, such as during the dark phase *(139)*, during food intake *(140,141)*, or in animals with a spontaneously increased dopaminergic tone *(105)*.

In conclusion, glucocorticoids, via GRs, modulate extracellular concentrations of dopamine in the NAc. These hormones specifically modulate dopamine transmission in the shell of the NAc, without influencing dopamine transmission in the core. In addition, the effects of glucocorticoids are state-dependent, and are greater when the dopamine system is activated. These observations suggest that glucocorticoid hormones could enhance drug responding by selectively facilitating dopamine transmission in the shell of the NAc.

5.2.3. Possible Mechanims Underlying the Effects of Glucocorticoids on NAc Dopamine

The mechanisms by which glucocorticoids facilitate dopamine transmission and increase NAc dopamine are unknown, but various hypotheses can be made on possible effects of glucocorticoids on dopamine metabolism, dopamine reuptake, and dopamine cell activity. It is still unclear whether these effects of glucocorticoids are direct or indirect, but the presence of glucocorticoid receptors in a subset of dopamine neurons *(142)* in the VTA suggests a possible direct action of glucocorticoid hormones on these cells. A strict relationship between glucocorticoids and dopamine neurons is also comforted by the observation that mesencephalic dopamine cell cultures respond to blockade of corticosteroid receptors by decreasing dopamine release in basal and stimulated conditions *(143)*.

One possible action of glucocorticoids is modulation of dopamine synthesis and degradation. Thus, it has been shown that suppression of corticosterone by adrenalectomy decreases the activity of tyrosine hydroxylase (TH), the rate-limiting enzyme in dopamine synthesis *(144)*. Conversely, administration of glucocorticoids increases TH activity in mouse brain *(145)*, in the locus coeruleus *(146)*, and in the VTA *(147)*. However, total TH protein content in the midbrain was not modified by adrenalectomy in other studies *(148)*. Regarding dopamine catabolism, it has been shown that hydrocortisone or the GR receptor agonist dexamethasone decreases the activity of monoamine oxidases (MAO) in vivo and in vitro, but have no effects on catechol-O-methyltransferase (COMT) *(149–152)*. This is in agreement with the effects of glucocorticoids on the degradation products of MAO and COMT. Thus, treatment with dexamethasone decreases

DOPAC (the deamination product of dopamine by MAO), and increases 3MT (the O-methylated product of dopamine by COMT) *(152,153)*.

It is also possible that glucocorticoids increase dopamine levels by blocking dopamine transporter sites, or that these hormones modulate the sensitivity or the number of dopamine transporter sites in dopamine terminal regions. For example, Gilad et al. *(154)* have shown that dopamine reuptake is decreased in striatal synaptosomes incubated with methylprednisolone, a glucocorticoid analog. More recently, Sarnyai et al. *(155)* reported that removal of corticosterone by adrenalectomy decreases the number of dopamine-binding sites in the shell of the NAc, without modulating those in the core. These effects depend on corticosterone, as they are reversed by replacing basal levels of corticosterone. In addition, these findings are consistent with the selective role of glucocorticoids in the shell of the NAc.

Finally, glucocorticoids could modulate dopamine transmission by acting on the impulse activity of dopamine cells. An increase in the impulse activity of dopamine cells, and in particular the bursting mode, is associated with increased dopamine release in dopamine terminal regions *(156,157)*. Although there are only a few electrophysiological studies about the influence of glucocorticoids on the activity of midbrain dopamine cells, in vivo extracellular recordings by Overton and co-workers *(158)* have shown that glucocorticoids facilitate glutamate-induced bursting of midbrain dopamine cells. Thus, adrenalectomy reduces the activity of VTA dopamine cells when these cells are stimulated by iontophoretically administered glutamate, and corticosterone administration reverses this effect. In addition, in vitro extracellular recordings by Cho and Little *(159)* also showed that corticosterone potentiates the response of dopamine cells to excitatory amino acid activation. Thus, corticosterone enhances NMDA-induced increase in dopamine cell impulse activity, but has no effects on spontaneous firing. Once again, these studies confirm the state-dependent effect of glucocorticoids on dopamine transmission. It is probable that the effects of glucocorticoids on the activity of dopamine cells are mediated by GRs, as in the work by Cho and Little *(159)* the effects of corticosterone were reversed by the GR antagonist RU 38486. Overton and co-workers, however, suggested an involvement of MRs *(159)*, but in this study plasma levels of corticosterone were not measured, so it is possible that the dose of corticosterone used to reverse the effects of adrenalectomy was high enough to induce GR activation.

Corticosterone's facilitation of cell activity could also depend on intrinsic components of dopamine cells that modulate bursting activity, such as impulse-regulating somatodendritic receptors *(160–166)*. These autoreceptors, primarily of the D2 subtype *(164,167,168)*, are activated by somatodendritically released dopamine *(169–171)* and reduce neuronal activity by hyperpolarizing the cell by activating K^+ channels, possibly G-protein inwardly rectifying potassium channels (GIRKs) *(172–177)*. Other ion-gated channels, such as L-type Ca^{2+} channels and Ca^{2+}-activated K^+ channels, are also important modulators of dopamine cell impulse activity *(178)* and could also be involved *(179)*. Although no studies have yet examined the mechanisms by which glucocorticoids could modulate the impulse activity of midbrain dopamine cells, we could speculate that glucocorticoids alter cell excitability by modifying these ion-gated channels, or even the intracellular signaling cascades, such as pertussis toxin-sensitive G proteins coupled to potassium channels. For example, in the hippocampus, it has been shown that low levels of glucocorticoids increase cell excitability, possibly by modifying Ca^{2+} conductances, Ca^2-activated K^+ channels, and the binding of the Ca^{2+}–calmodulin complex to cell membranes *(180–189)*. In addition, in the hippocampus, these hormones are able to modulate GIRK protein levels (190) and could thus participate in regulating cell excitability. It is also possible that glucocorticoids modulate dopamine activity by acting on the density or function of impulse-regulating dopamine autoreceptors in the VTA; however, to our knowledge, no data are available on this possible action of glucocorticoids in the midbrain.

Whatever the mechanism by which glucocorticoids influence dopamine transmission and facilitate dopamine release in the NAc, their effect does not seem to be derived from an action on the integrity of the dopamine system. In fact, although suppression of glucocorticoids reduces dopamine transmission, this manipulation does not induce changes in dopamine cell death or gliosis. Thus, the number of cells immunostained with TH or with glial fibrillary acidic protein in the midbrain is similar between adrenalectomized and control rats *(130)*.

The above studies suggest that glucocorticoids could modulate dopamine activity by acting directly on dopamine neurons, but we cannot exclude that these hormones could modify dopamine transmission via indirect mechanisms. Glucocorticoids could influence, for example, glutamatergic, opioid, and GABAergic systems *(5)*, which, in turn, are susceptible of modulating the activity of midbrain dopamine cells (for review, *see* ref. *191*). Although we cannot exclude this effect, the hypothesis is rather improbable, as most of theses excitatory and inhibitory afferences to dopamine cells seem to implicate MRs *(192)*, whereas many effects on dopamine activity have been shown to depend on GRs *(193,194)*.

In summary, the mechanisms by which glucocorticoid hormones facilitate dopamine transmission could be diverse and could vary from direct actions on dopamine neurons to indirect ones involving afferent regulation of dopamine transmission. Although a direct modulation of dopamine cell activity seems plausible, further electrophysiological, molecular, and biochemical investigations are definitely necessary to dissect the mechanisms underlying the interaction between glucocorticoids and dopamine that mediates the addictive properties of drugs of abuse.

6. Why Do Glucocorticoid Hormones and Drugs of Abuse Interact?

The data reviewed above indicate that glucocorticoid hormones have a profound influence on the behavioral effects of psychostimulant drugs. In particular, glucocorticoids seem to modify the motivation of the subject to self-administer drugs. Thus, after suppression or inhibition of endogenous glucocorticoids, animals still self-administer psychostimulants, but the amount of work the subjects are ready to provide is reduced. An action of glucocorticoids on the motivational effects of drugs of abuse is also suggested by the ability of these hormones to induce reinstatement of drug-seeking after a period of extinction. Finally, this idea is also supported by the profound control that glucocorticoid hormones exert on drug-induced dopamine release in the shell of the NAc, a brain region that seems to be involved in the extremely powerful incentive value of these drugs and in their capacity to induce addiction *(195)*.

This striking relationship between glucocorticoids and drugs of abuse prompts one question: why it is so? In other words, which physiological function of these hormones allows them to interact so profoundly with drug-taking?

Glucocorticoids hormone are generally considered as "the" stress hormones, and it is certainly true that their secretion constitutes one of the major responses to external challenges. However, this does not necessarily imply, as is generally believed, that glucocorticoids mediate the primary effects of stress, that is, its aversive and avoidance-inducing properties. In fact, in our opinion, glucocorticoids have the very opposite effects. These hormones are more likely a component of the endogenous reward system whose primary role would be to "energize" goal-directed behaviors. Activation of glucocorticoids during stress would thus be a secondary compensatory response aimed at reducing the aversive effects of stress and increasing, in this way, the copying capabilities of the subject. In fact, glucocorticoids are not only activated by stress, their secretion also precedes many goal-directed behaviors and in particular food seeking (for example, *see* refs. *140* and *141*, and *196–199*). Furthermore, glucocorticoids, in the range of the concentrations induced by stress, have positive reinforcing effects and stimulate dopamine release *(200,201)*.

This theory of the role of the central effects of glucocorticoids allows for the unification of the central and the peripheral function of these hormones. Indeed, it has already been hypothesized that the action of glucocorticoid hormones in the periphery is not part of the primary response to stress but a way of the organisms to protect itself against the primary responses to stress *(3,4,202)*. For example, glucocorticoids increase lipolysis and glyconeogenesis, which increases blood glucose levels and help the organism under stressful conditions by increasing the availability of energy substrates. Similarly, the immunosuppressive effects of glucocorticoids allow avoiding potential adverse effects of overactivating the immune system and the inflammatory responses to external aggressions (for a more extensive review on this issue, *see* refs. *4* and *203*).

In this context, the increase in vulnerability to drugs of abuse observed after repeated stress should be reinterpreted as the activation of a protective compensatory response during environmental challenges. During chronic stress the repeated activation of glucocorticoid hormones and dopamine, aimed to compensate the aversive effects of acute stress, would result in sensitization of the reward system. This sensitized state, which can persist after the end of the stress, would render the subject more responsive to drugs of abuse and consequently more vulnerable to develop addiction.

In conclusion, glucocorticoid hormones are an essential component of our capacity to endure stressful situations, probably by attenuating their aversive impact. The role of these hormones in drug abuse is likely related to the long-lasting sensitization they induce in the reward system when repeatedly activated during stress. Consequently, understanding the molecular mechanisms that mediate such long-lasting effects of glucocorticoids could help us to better understand addiction and to develop new treatments of this condition.

References

1. McEwen, B. S., de Kloet, E. R., and Rostene, W. (1986) Adrenal steroid receptors and actions in the nervous system. *Physiol. Rev.* **66,** 1121–1188.
2. Akana, S. F., Cascio, C. S., Du, J. Z., Levin, N., and Dallman, M. F. (1986) Reset of feedback in the adrenocortical system: an apparent shift in sensitivity of adrenocorticotropin to inhibition by corticosterone between morning and evening. *Endocrinology* **119,** 2325–2332.
3. Dallman, M. F., Darlington, D. N., Suemaru, S., Cascio, C. S., and Levin, N. (1989) Corticosteroids in homeostasis. *Acta Physiol. Scand. Suppl.* **583,** 27–34.

4. Munck, A., Guyre, P. M., and Holbrook, N. J. (1984) Physiological functions of glucocorticoids in stress and their relation to pharmacological actions. *Endocrinol. Rev.* **5,** 25–44.
5. Joels, M. and de Kloet, E. R. (1994) Mineralocorticoid and glucocorticoid receptors in the brain. Implications for ion permeability and transmitter systems. *Prog. Neurobiol.* **43,** 1–36.
6. Reul, J. M. and de Kloet, E. R. (1985) Two receptor systems for corticosterone in rat brain: microdistribution and differential occupation. *Endocrinology* **117,** 2505–2511.
7. Piazza, P. V., Deminiere, J. M., Le Moal, M., and Simon, H. (1989) Factors that predict individual vulnerability to amphetamine self-administration. *Science* **245,** 1511–1513.
8. Marinelli, M., Piazza, P. V., Deroche, V., Maccari, S., Le Moal, M., and Simon, H. (1994) Corticosterone circadian secretion differentially facilitates dopamine-mediated psychomotor effect of cocaine and morphine. *J. Neurosci.* **14,** 2724–2731.
9. Badiani, A., Morano, M. I., Akil, H., and Robinson, T. E. (1995) Circulating adrenal hormones are not necessary for the development of sensitization to the psychomotor activating effects of amphetamine. *Brain Res.* **673,** 13–24.
10. Cador, M., Dulluc, J., and Mormede, P. (1993) Modulation of the locomotor response to amphetamine by corticosterone. *Neuroscience* **56,** 981–988.
11. Mormede, P., Dulluc, J., and Cador, M. (1994) Modulation of the locomotor response to amphetamine by corticosterone. *Ann. N.Y. Acad. Sci.* **746,** 394–397.
12. Marinelli, M., Rouge-Pont, F., Deroche, V., Barrot, M., Jesus-Oliveira, C., Le Moal, M., and Piazza, P. V. (1997) Glucocorticoids and behavioral effects of psychostimulants. I: locomotor response to cocaine depends on basal levels of glucocorticoids. *J. Pharmacol. Exp. Ther.* **281,** 1392–1400.
13. Marinelli, M., Le Moal, M., and Piazza, P. V. (1996) Acute pharmacological blockade of corticosterone secretion reverses food restriction-induced sensitization of the locomotor response to cocaine. *Brain Res.* **724,** 251–255.
14. Marinelli, M., Rouge-Pont, F., Jesus-Oliveira, C., Le Moal, M., and Piazza, P. V. (1997) Acute blockade of corticosterone secretion decreases the psychomotor stimulant effects of cocaine. *Neuropsychopharmacology* **16,** 156–161.
15. Reid, M. S., Ho, L. B., Tolliver, B. K., Wolkowitz, O. M., and Berger, S. P. (1998) Partial reversal of stress-induced behavioral sensitization to amphetamine following metyrapone treatment. *Brain Res.* **783,** 133–142.
16. Robinson, T. E. and Becker, J. B. (1986) Enduring changes in brain and behavior produced by chronic amphetamine administration: a review and evaluation of animal models of amphetamine psychosis. *Brain Res.* **396,** 157–198.
17. Rivet, J. M., Stinus, L., Le Moal, M., and Mormede, P. (1989) Behavioral sensitization to amphetamine is dependent on corticosteroid receptor activation. *Brain Res.* **498,** 149–153.
18. Prasad, B. M., Ulibarri, C., Kalivas, P. W., and Sorg, B. A. (1996) Effect of adrenalectomy on the initiation and expression of cocaine-induced sensitization. *Psychopharmacology* (Berl.) **125,** 265–273.
19. Przegalinski, E., Filip, M., Siwanowicz, J., and Nowak, E. (2000) Effect of adrenalectomy and corticosterone on cocaine-induced sensitization in rats. *J. Physiol Pharmacol.* **51,** 193–204.
20. De Vries, T. J., Schoffelmeer, A. N., Tjon, G. H., Nestby, P., Mulder, A. H., and Vanderschuren, L. J. (1996) Mifepristone prevents the expression of long-term behavioural sensitization to amphetamine. *Eur. J. Pharmacol.* **307,** R3–R4.
21. Mello, N. K. and Mendelson, J. H. (1997) Cocaine's effects on neuroendocrine systems: clinical and preclinical studies. *Pharmacol. Biochem. Behav.* **57,** 571–599.
22. Schmidt, E. D., Tilders, F. J., Binnekade, R., Schoffelmeer, A. N., and De Vries, T. J. (1999) Stressor- or drug-induced sensitization of the corticosterone response is not critically involved in the long-term expression of behavioural sensitization to amphetamine. *Neuroscience* **92,** 343–352.

23. Spangler, R., Zhou, Y., Schlussman, S. D., Ho, A., and Kreek, M. J. (1997) Behavioral stereotypies induced by "binge" cocaine administration are independent of drug-induced increases in corticosterone levels. *Behav. Brain Res.* **86,** 201–204.
24. Pickens, R. and Harris, W. C. (1968) Self-administration of d-amphetamine by rats. *Psychopharmacologia* **12,** 158–163.
25. Schuster, C. R. and Thompson, T. (1969) Self administration of and behavioral dependence on drugs. *Annu. Rev. Pharmacol.* **9,** 483–502.
26. Weeks, J. R. (1962) Experimental morphine addiction: method for automatic intravenous injections in unrestrained rats. *Science* **138,** 143–144.
27. Goeders, N. E. and Guerin, G. F. (1996) Role of corticosterone in intravenous cocaine self-administration in rats. *Neuroendocrinology* **64,** 337–348.
28. Goeders, N. E. and Guerin, G. F. (1996) Effects of surgical and pharmacological adrenalectomy on the initiation and maintenance of intravenous cocaine self-administration in rats. *Brain Res.* **722,** 145–152.
29. Deroche, V., Marinelli, M., Le Moal, M., and Piazza, P. V. (1997) Glucocorticoids and behavioral effects of psychostimulants. II: cocaine intravenous self-administration and reinstatement depend on glucocorticoid levels. J. Pharmacol. Exp. Ther. 281, 1401–1407.
30. Goeders, N. E., Peltier, R. L., and Guerin, G. F. (1998) Ketoconazole reduces low dose cocaine self-administration in rats. Drug Alcohol Depend. 53, 67–77.
31. Piazza, P. V., Marinelli, M., Jodogne, C., Deroche, V., Rouge-Pont, F., Maccari, S., Le Moal, M., and Simon, H. (1994) Inhibition of corticosterone synthesis by Metyrapone decreases cocaine-induced locomotion and relapse of cocaine self-administration. *Brain Res.* **658,** 259–264.
32. Suzuki, T., Sugano, Y., Funada, M., and Misawa, M. (1995) Adrenalectomy potentiates the morphine- but not cocaine-induced place preference in rats. *Life Sci.* **56,** L339–L344.
33. Carr, G. D., Fibiger, H. C., and Phillips, A. G. (1989). Conditioned place preference as a measure of drug reward, in *The Neuropharmacological Basis of Reward* (Liebman, J. M. and Cooper, S. J., eds.), Clarendon Press, Oxford, UK, pp. 264–319.
34. Hoffman, D. C. (1989) The use of place conditioning in studying the neuropharmacology of drug reinforcement. *Brain Res. Bull.* **23,** 373–387.
35. Tzschentke, T. M. (1998) Measuring reward with the conditioned place preference paradigm: a comprehensive review of drug effects, recent progress and new issues. *Prog. Neurobiol.* **56,** 613–672.
36. Spyraki, C., Fibiger, H. C., and Phillips, A. G. (1982) Cocaine-induced place preference conditioning: lack of effects of neuroleptics and 6-hydroxydopamine lesions. *Brain Res.* **253,** 195–203.
37. Spyraki, C., Fibiger, H. C., and Phillips, A. G. (1982) Dopaminergic substrates of amphetamine-induced place preference conditioning. *Brain Res.* **253,** 185–193.
38. Roberts, D. C., Koob, G. F., Klonoff, P., and Fibiger, H. C. (1980) Extinction and recovery of cocaine self-administration following 6-hydroxydopamine lesions of the nucleus accumbens. *Pharmacol. Biochem. Behav.* **12,** 781–787.
39. Roberts, D. C. and Koob, G. F. (1982) Disruption of cocaine self-administration following 6-hydroxydopamine lesions of the ventral tegmental area in rats. *Pharmacol. Biochem. Behav.* **17,** 901–904.
40. Roberts, D. C., Loh, E. A., and Vickers, G. (1989) Self-administration of cocaine on a progressive ratio schedule in rats: dose-response relationship and effect of haloperidol pretreatment. *Psychopharmacology* (Berl.) **97,** 535–538.
41. Bardo, M. T. and Bevins, R. A. (2000) Conditioned place preference: what does it add to our preclinical understanding of drug reward? *Psychopharmacology* (Berl.) **153,** 31–43.
42. Costello, N. L., Carlson, J. N., Glick, S. D., and Bryda, M. (1989) Dose-dependent and baseline-dependent conditioning with d-amphetamine in the place conditioning paradigm. *Psychopharmacology* (Berl.) **99,** 244–247.

43. Erb, S., Shaham, Y., and Stewart, J. (1998) The role of corticotropin-releasing factor and corticosterone in stress- and cocaine-induced relapse to cocaine seeking in rats. *J. Neurosci.* **18,** 5529–5536.
44. Mantsch, J. R. and Goeders, N. E. (1999) Ketoconazole does not block cocaine discrimination or the cocaine-induced reinstatement of cocaine-seeking behavior. *Pharmacol. Biochem. Behav.* **64,** 65–73.
45. Goeders, N. E. (2002) The HPA axis and cocaine reinforcement. *Psychoneuroendocrinology* **27,** 13–33.
46. Carr, K. D. and Abrahamsen, G. C. (1998) Effect of adrenalectomy on cocaine facilitation of medial prefrontal cortex self-stimulation. *Brain Res.* **787,** 321-327.
47. Ratka, A., Sutanto, W., and de Kloet, E. R. (1988) Long-lasting glucocorticoid suppression of opioid-induced antinociception. *Neuroendocrinology* **48,** 439-444.
48. Antelman, S. M., Eichler, A. J., Black, C. A., and Kocan, D. (1980) Interchangeability of stress and amphetamine in sensitization. *Science* **207,** 329–331.
49. Herman, J. P., Stinus, L., and Le Moal, M. (1984) Repeated stress increases locomotor response to amphetamine. *Psychopharmacology* (Berl.) **84,** 431–435.
50. Piazza, P. V., Deminiere, J. M., Le Moal, M., and Simon, H. (1990) Stress- and pharmacologically-induced behavioral sensitization increases vulnerability to acquisition of amphetamine self-administration. *Brain Res.* **514,** 22–26.
51. Robinson, T. E., Angus, A. L., and Becker, J. B. (1985) Sensitization to stress: the enduring effects of prior stress on amphetamine-induced rotational behavior. *Life Sci.* **37,** 1039–1042.
52. Bell, S. M., Stewart, R. B., Thompson, S. C., and Meisch, R. A. (1997) Food-deprivation increases cocaine-induced conditioned place preference and locomotor activity in rats. *Psychopharmacology* (Berl.) **131,** 1–8.
53. Deroche, V., Piazza, P. V., Casolini, P., Le Moal, M., and Simon, H. (1993) Sensitization to the psychomotor effects of amphetamine and morphine induced by food restriction depends on corticosterone secretion. *Brain Res.* **611,** 352–356.
54. Miczek, K. A., Nikulina, E., Kream, R. M., Carter, G., and Espejo, E. F. (1999) Behavioral sensitization to cocaine after a brief social defeat stress: c-fos expression in the PAG. *Psychopharmacology* (Berl.) **141,** 225–234.
55. Nikulina, E. M., Marchand, J. E., Kream, R. M., and Miczek, K. A. (1998) Behavioral sensitization to cocaine after a brief social stress is accompanied by changes in fos expression in the murine brainstem. *Brain Res.* **810,** 200–210.
56. Stohr, T., Almeida, O. F., Landgraf, R., Shippenberg, T. S., Holsboer, F., and Spanagel, R. (1999) Stress- and corticosteroid-induced modulation of the locomotor response to morphine in rats. *Behav. Brain Res.* **103,** 85–93.
57. Goeders, N. E. and Guerin, G. F. (1994) Non-contingent electric footshock facilities the acquisition of intravenous cocaine self-administration in rats. *Psychopharmacology* (Berl.) **114,** 63–70.
58. Schenk, S., Lacelle, G., Gorman, K., and Amit, Z. (1987) Cocaine self-administration in rats influenced by environmental conditions: implications for the etiology of drug abuse. *Neurosci. Lett.* **81,** 227–231.
59. Schenk, S., Hunt, T., Klukowski, G., and Amit, Z. (1987) Isolation housing decreases the effectiveness of morphine in the conditioned taste aversion paradigm. *Psychopharmacology* (Berl.) **92,** 48–51.
60. Haney, M., Maccari, S., Le Moal, M., Simon, H., and Piazza, P. V. (1995) Social stress increases the acquisition of cocaine self-administration in male and female rats. *Brain Res.* **698,** 46–52.
61. Miczek, K. A. and Mutschler, N. H. (1996) Activational effects of social stress on IV cocaine self-administration in rats. *Psychopharmacology* (Berl.) **128,** 256–264.
62. Tidey, J. W. and Miczek, K. A. (1997) Acquisition of cocaine self-administration after social stress: role of accumbens dopamine. *Psychopharmacology* (Berl.) **130,** 203–212.

63. Carroll, M. E., France, C. P., and Meisch, R. A. (1979) Food deprivation increases oral and intravenous drug intake in rats. *Science* **205,** 319–321.
64. Macenski, M. J. and Meisch, R. A. (1999) Cocaine self-administration under conditions of restricted and unrestricted food access. *Exp. Clin. Psychopharmacol.* **7,** 324–337.
65. Marinelli, M., Cooper D. C., and White, F. J. (2002) A brief period of reduced food availability increases dopamine neuronal activity and enhances motivation to self-administer cocaine. *Behav. Pharmacol.* **12**(S1), S62.
66. Papasava, M. and Singer, G. (1985) Self-administration of low-dose cocaine by rats at reduced and recovered body weight. *Psychopharmacology* (Berl.) **85,** 419–425.
67. Papasava, M., Singer, G., and Papasava, C. L. (1986) Intravenous self-administration of phentermine in food-deprived rats: effects of abrupt refeeding and saline substitution. *Pharmacol. Biochem. Behav.* **25,** 623–627.
68. Deroche, V., Piazza, P. V., Le Moal, M., and Simon, H. (1994) Social isolation-induced enhancement of the psychomotor effects of morphine depends on corticosterone secretion. *Brain Res.* **640,** 136–139.
69. Deroche, V., Piazza, P. V., Casolini, P., Maccari, S., Le Moal, M., and Simon, H. (1992) Stress-induced sensitization to amphetamine and morphine psychomotor effects depend on stress-induced corticosterone secretion. *Brain Res.* **598,** 343–348.
70. Prasad, B. M., Ulibarri, C., and Sorg, B. A. (1998) Stress-induced cross-sensitization to cocaine: effect of adrenalectomy and corticosterone after short- and long-term withdrawal. *Psychopharmacology* (Berl.) **136,** 24–33.
71. Rouge-Pont, F., Marinelli, M., Le Moal, M., Simon, H., and Piazza, P. V. (1995) Stress-induced sensitization and glucocorticoids. II. Sensitization of the increase in extracellular dopamine induced by cocaine depends on stress-induced corticosterone secretion. *J. Neurosci.* **15,** 7189–7195.
72. Deroche, V., Piazza, P. V., Maccari, S., Le Moal, M., and Simon, H. (1992) Repeated corticosterone administration sensitizes the locomotor response to amphetamine. *Brain Res.* **584,** 309–313.
73. Campbell, U. C., and Carroll, M. E. (2001) Effects of ketoconazole on the acquisition of intravenous cocaine self-administration under different feeding conditions in rats. *Psychopharmacology* (Berl.) **154,** 311–318.
74. Micco, D. J., Jr., McEwen, B. S., and Shein, W. (1979) Modulation of behavioral inhibition in appetitive extinction following manipulation of adrenal steroids in rats: implications for involvement of the hippocampus. *J. Comp. Physiol. Psychol.* **93,** 323–329.
75. Mantsch, J. R., Saphier, D., and Goeders, N. E. (1998) Corticosterone facilitates the acquisition of cocaine self-administration in rats: opposite effects of the type II glucocorticoid receptor agonist dexamethasone. *J. Pharmacol. Exp. Ther.* **287,** 72–80.
76. Piazza, P. V., Maccari, S., Deminiere, J. M., Le Moal, M., Mormede, P., and Simon, H. (1991) Corticosterone levels determine individual vulnerability to amphetamine self-administration. *Proc. Natl. Acad. Sci. USA* **88,** 2088–2092.
77. Erb, S., Shaham, Y., and Stewart, J. (1996) Stress reinstates cocaine-seeking behavior after prolonged extinction and a drug-free period. *Psychopharmacology* (Berl.) **128,** 408–412.
78. Shaham, Y., Erb, S., and Stewart, J. (2000) Stress-induced relapse to heroin and cocaine seeking in rats: a review. *Brain Res. Brain Res. Rev.* **33,** 13–33.
79. Shalev, U., Highfield, D., Yap, J., and Shaham, Y. (2000) Stress and relapse to drug-seeking in rats: studies on the generality of the effect. *Psychopharmacology* (Berl.) **150,** 337–346.
80. Mantsch, J. R., and Goeders, N. E. (1999) Ketoconazole blocks the stress-induced reinstatement of cocaine-seeking behavior in rats: relationship to the discriminative stimulus effects of cocaine. *Psychopharmacology* (Berl.) **142,** 399–407.
81. Shaham, Y., Erb, S., Leung, S., Buczek, Y., and Stewart, J. (1998) CP-154,526, a selective, non-peptide antagonist of the corticotropin-releasing factor1 receptor attenuates stress-induced

relapse to drug-seeking in cocaine- and heroin-trained rats. *Psychopharmacology* (Berl.) **137,** 184–190.

82. Erb, S., Hitchcott, P. K., Rajabi, H., Mueller, D., Shaham, Y., and Stewart, J. (2000) Alpha-2 adrenergic receptor agonists block stress-induced reinstatement of cocaine seeking. *Neuropsychopharmacology* **23,** 138–150.
83. Shalev, U., Marinelli, M., Piazza, P. V., and Shaham, Y. (2002) The role of corticosterone in food deprivation-induced reinstatement of cocaine seeking. *Psychopharmacology*, in press.
84. Bozarth, M. A. (1991). The mesolimbic dopamine system as a model reward system, in *The Mesolimbic Dopamine System: From Motivation to Action* (Willner, P. and Scheel-Krüger, J., eds.), John Wiley and Sons, Chichester, UK, pp. 301–330.
85. Wise, R. A. and Rompre, P. P. (1989) Brain dopamine and reward. *Annu. Rev. Psychol.* **40,** 191–225.
86. Di Chiara, G. and Imperato, A. (1986) Preferential stimulation of dopamine release in the nucleus accumbens by opiates, alcohol, and barbiturates: studies with transcerebral dialysis in freely moving rats. *Ann. N.Y. Acad. Sci.* **473,** 367–381.
87. Di Chiara, G. and Imperato, A. (1988) Drugs abused by humans preferentially increase synaptic dopamine concentrations in the mesolimbic system of freely moving rats. *Proc. Natl. Acad. Sci. USA* **85,** 5274–5278.
88. Heikkila, R. E., Orlansky, H., and Cohen, G. (1974) Studies on the distinction between uptake inhibition and release of [3H] dopamine in rat brain slices. *Biochem. Pharmacol.* **24,** 847–852.
89. Hurd, Y. L., Weiss, F., Koob, G. F., And N. E., and Ungerstedt, U. (1989) Cocaine reinforcement and extracellular dopamine overflow in rat nucleus accumbens: an in vivo microdialysis study. *Brain Res.* **498,** 199–203.
90. Caine, S. B. and Koob, G. F. (1994) Effects of mesolimbic dopamine depletion on responding maintained by cocaine and food. *J. Exp. Anal. Behav.* **61,** 213–221.
91. Roberts, D. C., Corcoran, M. E., and Fibiger, H. C. (1977) On the role of ascending catecholaminergic systems in intravenous self-administration of cocaine. *Pharmacol. Biochem. Behav.* **6,** 615–620.
92. Caine, S. B. and Koob, G. F. (1994) Effects of dopamine D-1 and D-2 antagonists on cocaine self-administration under different schedules of reinforcement in the rat. *J. Pharmacol. Exp. Ther.* **270,** 209–218.
93. Caine, S. B., Heinrichs, S. C., Coffin, V. L., and Koob, G. F. (1995) Effects of the dopamine D-1 antagonist SCH 23390 microinjected into the accumbens, amygdala or striatum on cocaine self-administration in the rat. *Brain Res.* **692,** 47–56.
94. Davis, W. M. and Smith, S. G. (1977) Catecholaminergic mechanisms of reinforcement: direct assessment by drug-self-administration. *Life Sci.* **20,** 483–492.
95. Koob, G. F., Le, H. T., and Creese, I. (1987) The D1 dopamine receptor antagonist SCH 23390 increases cocaine self-administration in the rat. *Neurosci. Lett.* **79,** 315–320.
96. Maldonado, R., Robledo, P., Chover, A. J., Caine, S. B., and Koob, G. F. (1993) D1 dopamine receptors in the nucleus accumbens modulate cocaine self-administration in the rat. *Pharmacol. Biochem. Behav.* **45,** 239–242.
97. Ranaldi, R. and Wise, R. A. (2001) Blockade of D1 dopamine receptors in the ventral tegmental area decreases cocaine reward: possible role for dendritically released dopamine. *J. Neurosci.* **21,** 5841–5846.
98. Roberts, D. C. and Vickers, G. (1984) Atypical neuroleptics increase self-administration of cocaine: an evaluation of a behavioural screen for antipsychotic activity. *Psychopharmacology* (Berl.) **82,** 135–139.
99. Weissenborn, R., Deroche, V., Koob, G. F., and Weiss, F. (1996) Effects of dopamine agonists and antagonists on cocaine-induced operant responding for a cocaine-associated stimulus. *Psychopharmacology* (Berl.) **126,** 311–322.
100. Schultz, W. (2001) Reward signaling by dopamine neurons. *Neuroscientist* **7,** 293–302.

101. Marinelli, M. and White, F. J. (2000) Enhanced vulnerability to cocaine self-administration is associated with elevated impulse activity of midbrain dopamine neurons. *J. Neurosci.* **20,** 8876–8885.
102. Ferraro, T. N., Golden, G. T., Berrettini, W. H., Gottheil, E., Yang, C. H., Cuppels, G. R., and Vogel, W. H. (2000) Cocaine intake by rats correlates with cocaine-induced dopamine changes in the nucleus accumbens shell. *Pharmacol. Biochem. Behav.* **66,** 397–401.
103. Hooks, M. S., Jones, G. H., Smith, A. D., Neill, D. B., and Justice, J. B., Jr. (1991) Response to novelty predicts the locomotor and nucleus accumbens dopamine response to cocaine. *Synapse* **9,** 121–128.
104. Piazza, P. V., Rouge-Pont, F., Deminiere, J. M., Kharouby, M., Le Moal, M., and Simon, H. (1991) Dopaminergic activity is reduced in the prefrontal cortex and increased in the nucleus accumbens of rats predisposed to develop amphetamine self-administration. *Brain Res.* **567,** 169–174.
105. Rouge-Pont, F., Piazza, P. V., Kharouby, M., Le Moal, M., and Simon, H. (1993) Higher and longer stress-induced increase in dopamine concentrations in the nucleus accumbens of animals predisposed to amphetamine self-administration. A microdialysis study. *Brain Res.* **602,** 169–174.
106. Thierry, A. M., Tassin, J. P., Blanc, G., and Glowinski, J. (1976) Selective activation of mesocortical DA system by stress. *Nature* **263,** 242–244.
107. Cabib, S. and Puglisi-Allegra, S. (1996) Stress, depression and the mesolimbic dopamine system. *Psychopharmacology* (Berl.) **128,** 331–342.
108. Moore, H., Rose, H. J., and Grace, A. A. (2001) Chronic cold stress reduces the spontaneous activity of ventral tegmental dopamine neurons. *Neuropsychopharmacology* **24,** 410–419.
109. Imperato, A., Puglisi-Allegra, S., Casolini, P., Zocchi, A., and Angelucci, L. (1989) Stress-induced enhancement of dopamine and acetylcholine release in limbic structures: role of corticosterone. *Eur. J. Pharmacol.* **165,** 337–338.
110. Imperato, A., Puglisi-Allegra, S., Casolini, P., and Angelucci, L. (1991) Changes in brain dopamine and acetylcholine release during and following stress are independent of the pituitary-adrenocortical axis. *Brain Res.* **538,** 111–117.
111. Kalivas, P. W. and Duffy, P. (1989) Similar effects of daily cocaine and stress on mesocorticolimbic dopamine neurotransmission in the rat. *Biol. Psychiatry* **25,** 913–928.
112. Kalivas, P. W. and Stewart, J. (1991) Dopamine transmission in the initiation and expression of drug- and stress-induced sensitization of motor activity. *Brain Res. Brain Res. Rev.* **16,** 223–244.
113. Kalivas, P. W., and Duffy, P. (1995) Selective activation of dopamine transmission in the shell of the nucleus accumbens by stress. *Brain Res.* **675,** 325–328.
114. Robinson, T. E. (1988) Stimulant drugs and stress: factors influencing individual differences in the susceptibility to sensitization, in *Sensitization of the Nervous System* (Kalivas, P. W. and Barnes, C., eds.), Telford Press, Caldwell, NJ, pp. 145–173.
115. Tidey, J. W. and Miczek, K. A. (1996) Social defeat stress selectively alters mesocorticolimbic dopamine release: an in vivo microdialysis study. *Brain Res.* **721,** 140–149.
116. Wilcox, R. A., Robinson, T. E., and Becker, J. B. (1986) Enduring enhancement in amphetamine-stimulated striatal dopamine release in vitro produced by prior exposure to amphetamine or stress in vivo. *Eur. J. Pharmacol.* **124,** 375–376.
117. McBride, W. J., Murphy, J. M., and Ikemoto, S. (1999) Localization of brain reinforcement mechanisms: intracranial self-administration and intracranial place-conditioning studies. *Behav. Brain Res.* **101,** 129–152.
118. Barrot, M., Marinelli, M., Abrous, D. N., Rouge-Pont, F., Le Moal, M., and Piazza, P. V. (1999) Functional heterogeneity in dopamine release and in the expression of Fos-like proteins within the rat striatal complex. *Eur. J. Neurosci.* **11,** 1155–1166.
119. Hedou, G., Feldon, J., and Heidbreder, C. A. (1999) Effects of cocaine on dopamine in subregions of the rat prefrontal cortex and their efferents to subterritories of the nucleus accumbens. *Eur. J. Pharmacol.* **372,** 143–155.

120. Pontieri, F. E., Tanda, G., and Di Chiara, G. (1995) Intravenous cocaine, morphine, and amphetamine preferentially increase extracellular dopamine in the "shell" as compared with the "core" of the rat nucleus accumbens. *Proc. Natl. Acad. Sci. USA* **92,** 12304–12308.
121. Pontieri, F. E., Tanda, G., Orzi, F., and Di Chiara, G. (1996) Effects of nicotine on the nucleus accumbens and similarity to those of addictive drugs. *Nature* **382,** 255–257.
122. Hurd, Y. L. and Ponten, M. (2000) Cocaine self-administration behavior can be reduced or potentiated by the addition of specific dopamine concentrations in the nucleus accumbens and amygdala using in vivo microdialysis. *Behav. Brain Res.* **116,** 177–186.
123. Carlezon, W. A., Jr. and Wise, R. A. (1996) Rewarding actions of phencyclidine and related drugs in nucleus accumbens shell and frontal cortex. *J. Neurosci.* **16,** 3112–3122.
124. McKinzie, D. L., Rodd-Henricks, Z. A., Dagon, C. T., Murphy, J. M., and McBride, W. J. (1999) Cocaine is self-administered into the shell region of the nucleus accumbens in Wistar rats. *Ann. N.Y. Acad. Sci.* **877,** 788–791.
125. Everitt, B. J., Dickinson, A., and Robbins, T. W. (2001) The neuropsychological basis of addictive behaviour. *Brain Res. Brain Res. Rev.* **36,** 129–138.
126. Delfs, J. M., Schreiber, L., and Kelley, A. E. (1990) Microinjection of cocaine into the nucleus accumbens elicits locomotor activation in the rat. *J. Neurosci.* **10,** 303–310.
127. Vezina, P. and Stewart, J. (1984) Conditioning and place-specific sensitization of increases in activity induced by morphine in the VTA. *Pharmacol. Biochem. Behav.* **20,** 925–934.
128. Deroche, V., Marinelli, M., Maccari, S., Le Moal, M., Simon, H., and Piazza, P. V. (1995) Stress-induced sensitization and glucocorticoids. I. Sensitization of dopamine-dependent locomotor effects of amphetamine and morphine depends on stress-induced corticosterone secretion. *J. Neurosci.* **15,** 7181–7188.
129. Barrot, M., Marinelli, M., Abrous, D. N., Rouge-Pont, F., Le Moal, M., and Piazza, P. V. (2000) The dopaminergic hyper-responsiveness of the shell of the nucleus accumbens is hormone-dependent. *Eur. J. Neurosci.* **12,** 973–979.
130. Piazza, P. V., Barrot, M., Rouge-Pont, F., Marinelli, M., Maccari, S., Abrous, D. N., Simon, H., and Le Moal, M. (1996) Suppression of glucocorticoid secretion and antipsychotic drugs have similar effects on the mesolimbic dopaminergic transmission. *Proc. Natl. Acad. Sci. USA* **93,** 15445–15450.
131. Marinelli, M., Aouizerate, B., Barrot, M., Le Moal, M., and Piazza, P. V. (1998) Dopamine-dependent responses to morphine depend on glucocorticoid receptors. *Proc. Natl. Acad. Sci. USA* **95,** 7742–7747.
132. Graybiel, A. M., Moratalla, R., and Robertson, H. A. (1990) Amphetamine and cocaine induce drug-specific activation of the *c-fos* gene in striosome-matrix compartments and limbic subdivisions of the striatum. *Proc. Natl. Acad. Sci. USA* **87,** 6912–6916.
133. Liu, J., Nickolenko, J., and Sharp, F. R. (1994) Morphine induces c-fos and junB in striatum and nucleus accumbens via D1 and N-methyl-D-aspartate receptors. *Proc. Natl. Acad. Sci. USA* **91,** 8537–8541.
134. Biron, D., Dauphin, C., and Di Paolo, T. (1992) Effects of adrenalectomy and glucocorticoids on rat brain dopamine receptors. *Neuroendocrinology* **55,** 468–476.
135. Czyrak, A., Wedzony, K., Michalska, B., Fijal, K., Dziedzicka-Wasylewska, M., and Mackowiak, M. (1997) The corticosterone synthesis inhibitor metyrapone decreases dopamine D1 receptors in the rat brain. *Neuroscience* **79,** 489–495.
136. Rouge-Pont, F., Deroche, V., Le Moal, M., and Piazza, P. V. (1998) Individual differences in stress-induced dopamine release in the nucleus accumbens are influenced by corticosterone. *Eur. J. Neurosci.* **10,** 3903–3907.
137. Mittleman, G., Blaha, C. D., and Phillips, A. G. (1992) Pituitary-adrenal and dopaminergic modulation of schedule-induced polydipsia: behavioral and neurochemical evidence. *Behav. Neurosci.* **106,** 408–420.

138. Piazza, P. V., Rouge-Pont, F., Deroche, V., Maccari, S., Simon, H., and Le Moal, M. (1996) Glucocorticoids have state-dependent stimulant effects on the mesencephalic dopaminergic transmission. *Proc. Natl. Acad. Sci. USA* **93,** 8716–8720.
139. Paulson, P. E. and Robinson, T. E. (1994) Relationship between circadian changes in spontaneous motor activity and dorsal versus ventral striatal dopamine neurotransmission assessed with on-line microdialysis. *Behav. Neurosci.* **108,** 624–635.
140. Hoebel, B. G., Hernandez, L., Schwartz, D. H., Mark, G. P., and Hunter, G. A. (1989) Microdialysis studies of brain norepinephrine, serotonin, and dopamine release during ingestive behavior. Theoretical and clinical implications. *Ann. N.Y. Acad. Sci.* **575,** 1713–191.
141. Taber, M. T. and Fibiger, H. C. (1997) Feeding-evoked dopamine release in the nucleus, accumbens: regulation by glutamatergic mechanisms. *Neuroscience* **76,** 1105–1112.
142. Harfstrand, A., Fuxe, K., Cintra, A., Agnati, L. F., Zini, I., Wikstrom, A. C., Okret, S., Yu, Z. Y., Goldstein, M., Steinbusch, H., et al. (1986) Glucocorticoid receptor immunoreactivity in monoaminergic neurons of rat brain. *Proc. Natl. Acad. Sci. USA* **83,** 9779–9783.
143. Rouge-Pont, F., Abrous, D. N., Le Moal, M., and Piazza, P. V. (1999) Release of endogenous dopamine in cultured mesencephalic neurons: influence of dopaminergic agonists and glucocorticoid antagonists. *Eur. J. Neurosci.* **11,** 2343–2350.
144. Dunn, A. J., Gildersleeve, N. B., and Gray, H. E. (1978) Mouse brain tyrosine hydroxylase and glutamic acid decarboxylase following treatment with adrenocorticotrophic hormone, vasopressin or corticosterone. *J. Neurochem.* **31,** 977–982.
145. Iuvone, P. M., Morasco, J., and Dunn, A. J. (1977) Effect of corticosterone on the synthesis of [^{3}H]catecholamines in the brains of CD-1 mice. *Brain Res.* **120,** 571–576.
146. Markey, K. A., Towle, A. C., and Sze, P. Y. (1982) Glucocorticoid influence on tyrosine hydroxylase activity in mouse locus coeruleus during postnatal development. *Endocrinology* **111,** 1519–1523.
147. Ortiz, J., DeCaprio, J. L., Kosten, T. A., and Nestler, E. J. (1995) Strain-selective effects of corticosterone on locomotor sensitization to cocaine and on levels of tyrosine hydroxylase and glucocorticoid receptor in the ventral tegmental area. *Neuroscience* **67,** 383–397.
148. Lindley, S. E., Bengoechea, T. G., Schatzberg, A. F., and Wong, D. L. (1999) Glucocorticoid effects on mesotelencephalic dopamine neurotransmission. *Neuropsychopharmacology* **21,** 399–407.
149. Caesar, P. M., Collins, G. G., and Sandler, M. (1970) Catecholamine metabolism and monoamine oxidase activity in adrenalectomized rats. *Biochem. Pharmacol.* **19,** 921–926.
150. Ho-Van-Hap, A., Babineau, L. M., and Berlinguet, L. (1967) Hormonal action on monoamine oxidase activity in rats. *Can. J. Biochem.* **45,** 355–362.
151. Parvez, H. and Parvez, S. (1973) The regulation of monoamine oxidase activity by adrenal cortical steroids. *Acta Endocrinol.* (Copenh.) **73,** 509–517.
152. Veals, J. W., Korduba, C. A., and Symchowicz, S. (1977) Effect of dexamethasone on monoamine oxidase inhibiton by iproniazid in rat brain. *Eur. J. Pharmacol.* **41,** 291–299.
153. Rothschild, A. J., Langlais, P. J., Schatzberg, A. F., Miller, M. M., Saloman, M. S., Lerbinger, J. E., Cole, J. O., and Bird, E. D. (1985) The effects of a single acute dose of dexamethasone on monoamine and metabolite levels in rat brain. *Life Sci.* **36,** 2491–2501.
154. Gilad, G. M., Rabey, J. M., and Gilad, V. H. (1987) Presynaptic effects of glucocorticoids on dopaminergic and cholinergic synaptosomes. Implications for rapid endocrine-neural interactions in stress. *Life Sci.* **40,** 2401–2408.
155. Sarnyai, Z., McKittrick, C. R., McEwen, B. S., and Kreek, M. J. (1998) Selective regulation of dopamine transporter binding in the shell of the nucleus accumbens by adrenalectomy and corticosterone-replacement. *Synapse* **30,** 334–337.
156. Chergui, K., Suaud-Chagny, M. F., and Gonon, F. (1994) Nonlinear relationship between impulse flow, dopamine release and dopamine elimination in the rat brain in vivo. *Neuroscience* **62,** 641–645.

157. Gonon, F. G. (1988) Nonlinear relationship between impulse flow and dopamine released by rat midbrain dopaminergic neurons as studied by in vivo electrochemistry. *Neuroscience* **24,** 19–28.
158. Overton, P. G., Tong, Z. Y., Brain, P. F., and Clark, D. (1996) Preferential occupation of mineralocorticoid receptors by corticosterone enhances glutamate-induced burst firing in rat midbrain dopaminergic neurons. *Brain Res.* **737,** 146–154.
159. Cho, K. and Little, H. J. (1999) Effects of corticosterone on excitatory amino acid responses in dopamine-sensitive neurons in the ventral tegmental area. *Neuroscience* **88,** 837–845.
160. Aghajanian, G. K. and Bunney, B. S. (1977) Dopamine "autoreceptors": pharmacological characterization by microiontophoretic single cell recording studies. *Naunyn Schmiedebergs Arch. Pharmacol.* **297,** 1–7.
161. Aghajanian, G. K. and Bunney, B. S. (1977) Pharmacological characterization of dopamine "autoreceptors" by microiontophoretic single-cell recording studies. *Adv. Biochem. Psychopharmacol.* **16,** 433–438.
162. Bunney, B. S., Sesack S. R., and Silva, N. L. (1987). Midbrain dopaminergic systems: neurophysiology and electrophysiological pharmacology. In *Psychopharmacology: The Third Generation of Progress* (H. Y. Meltzer, ed.), pp. 113–126, Raven Press, New York.
163. Chiodo, L. A. (1988) Dopamine-containing neurons in the mammalian central nervous system: electrophysiology and pharmacology. *Neurosci. Biobehav. Rev.* **12,** 49–91.
164. White, F. J. and Wang, R. Y. (1984) Pharmacological characterization of dopamine autoreceptors in the rat ventral tegmental area: microiontophoretic studies. *J. Pharmacol. Exp. Ther.* **231,** 275–280.
165. White, F. J. and Wang, R. Y. (1984) A10 dopamine neurons: role of autoreceptors in determining firing rate and sensitivity to dopamine agonists. *Life Sci.* **34,** 1161–1170.
166. White, F. J., (1996) Synaptic regulation of mesocorticolimbic dopamine neurons. *Annu. Rev. Neurosci.* **19,** 405–436.
167. Koeltzow, T. E., Xu, M., Cooper, D. C., Hu, X. T., Tonegawa, S., Wolf, M. E., and White, F. J. (1998) Alterations in dopamine release but not dopamine autoreceptor function in dopamine D3 receptor mutant mice. *J. Neurosci.* **18,** 2231–2238.
168. Mercuri, N. B., Saiardi, A., Bonci, A., Picetti, R., Calabresi, P., Bernardi, G., and Borrelli, E. (1997) Loss of autoreceptor function in dopaminergic neurons from dopamine D2 receptor deficient mice. *Neuroscience* **79,** 323–327.
169. Beart, P. M., McDonald, D., and Gundlach, A. L. (1979) Mesolimbic dopaminergic neurones and somatodendritic mechanisms. *Neurosci. Lett.* **15,** 165–170.
170. Cheramy, A., Leviel, V., and Glowinski, J. (1981) Dendritic release of dopamine in the substantia nigra. *Nature* **289,** 537–542.
171. Kalivas, P. W. and Duffy, P. (1991) A comparison of axonal and somatodendritic dopamine release using in vivo dialysis. *J. Neurochem.* **56,** 961–967.
172. Kim, K. M., Nakajima, Y., and Nakajima, S. (1995) G protein-coupled inward rectifier modulated by dopamine agonists in cultured substantia nigra neurons. *Neuroscience* **69,** 1145–1158.
173. Lacey, M. G., Mercuri, N. B., and North, R. A. (1987) Dopamine acts on D2 receptors to increase potassium conductance in neurones of the rat substantia nigra zona compacta. *J. Physiol.* **392,** 397–416.
174. Liu, L., Shen, R. Y., Kapatos, G., and Chiodo, L. A. (1994) Dopamine neuron membrane physiology: characterization of the transient outward current (IA) and demonstration of a common signal transduction pathway for IA and IK. *Synapse* **17,** 230–240.
175. Liu, L. X., Monsma, F. J., Jr., Sibley, D. R., and Chiodo, L. A. (1996) D2L, D2S, and D3 dopamine receptors stably transfected into NG108-15 cells couple to a voltage-dependent potassium current via distinct G protein mechanisms. *Synapse* **24,** 156–164.
176. Mercuri, N. B., Calabresi, P., and Bernardi, G. (1992) The electrophysiological actions of dopamine and dopaminergic drugs on neurons of the substantia nigra pars compacta and ventral tegmental area. *Life Sci.* **51,** 711–718.

177. Silva, N. L. and Bunney, B. S. (1988) Intracellular studies of dopamine neurons in vitro: pacemakers modulated by dopamine. *Eur. J. Pharmacol.* **149,** 307–315.
178. Overton, P. G. and Clark, D. (1997) Burst firing in midbrain dopaminergic neurons. *Brain Res. Brain Res. Rev.* **25,** 312–334.
179. Shepard, P. D. and Bunney, B. S. (1991) Repetitive firing properties of putative dopamine-containing neurons in vitro: regulation by an apamin-sensitive Ca(2+)-activated K+ conductance. *Exp. Brain Res.* **86,** 141–150.
180. Joels, M. and de Kloet, E. R. (1992) Control of neuronal excitability by corticosteroid hormones. *Trends Neurosci.* **15,** 25–30.
181. Joels, M., Hesen, W., Karst, H., and de Kloet, E. R. (1994) Steroids and electrical activity in the brain. *J. Steroid Biochem. Mol. Biol.* **49,** 391–398.
182. Karst, H., Wadman, W. J., and Joels, M. (1993) Long-term control by corticosteroids of the inward rectifier in rat CA1 pyramidal neurons, in vitro. *Brain Res.* **612,** 172–179.
183. Karst, H., Wadman, W. J., and Joels, M. (1994) Corticosteroid receptor-dependent modulation of calcium currents in rat hippocampal CA1 neurons. *Brain Res.* **649,** 234–242.
184. Karst, H., Werkman, T. R., Struik, M., Bosma, A., and Joels, M. (1997) Effects of adrenalectomy on Ca^{2+} currents and Ca^{2+} channel subunit mRNA expression in hippocampal CA1 neurons of young rats. *Synapse* **26,** 155–164.
185. Karst, H. and Joels, M. (2001) Calcium currents in rat dentate granule cells are altered after adrenalectomy. *Eur. J. Neurosci.* **14,** 503–512.
186. Kerr, D. S., Campbell, L. W., Thibault, O., and Landfield, P. W. (1992) Hippocampal glucocorticoid receptor activation enhances voltage-dependent Ca^{2+} conductances: relevance to brain aging. *Proc. Natl. Acad. Sci. USA* **89,** 8527–8531.
187. Sze, P. Y. and Iqbal, Z. (1994) Glucocorticoid actions on synaptic plasma membranes: modulation of [125I]calmodulin binding. *J. Steroid Biochem. Mol. Biol.* **48,** 179–186.
188. Sze, P. Y. and Iqbal, Z. (1994) Regulation of calmodulin content in synaptic plasma membranes by glucocorticoids. *Neurochem. Res.* **19,** 1455–1461.
189. Sze, P. Y. and Iqbal, Z. (1994) Glucocorticoid action on depolarization-dependent calcium influx in brain synaptosomes. *Neuroendocrinology* **59,** 457–465.
190. Muma, N. A. and Beck, S. G. (1999) Corticosteroids alter G protein inwardly rectifying potassium channels protein levels in hippocampal subfields. *Brain Res.* **839,** 331–335.
191. Kalivas, P. W. (1993) Neurotransmitter regulation of dopamine neurons in the ventral tegmental area. *Brain Res. Brain Res. Rev.* **18,** 75–113.
192. Chao, H. M.,and McEwen, B. S. (1991) Glucocorticoid regulation of neuropeptide mRNAs in the rat striatum. *Brain Res. Mol. Brain Res.* **9,** 307–311.
193. Joels, M. and Fernhout, B. (1993) Decreased population spike in CA1 hippocampal area of adrenalectomized rats after repeated synaptic stimulation. *J. Neuroendocrinol.* **5,** 537–543.
194. Lucas, L. R., Pompei, P., Ono, J., and McEwen, B. S. (1998) Effects of adrenal steroids on basal ganglia neuropeptide mRNA and tyrosine hydroxylase radioimmunoreactive levels in the adrenalectomized rat. *J. Neurochem.* **71,** 833–843.
195. Di Chiara, G., Tanda, G., Bassareo, V., Pontieri, F., Acquas, E., Fenu, S., Cadoni, C., and Carboni, E. (1999) Drug addiction as a disorder of associative learning. Role of nucleus accumbens shell/extended amygdala dopamine. *Ann. N.Y. Acad. Sci.* **877,** 461–485.
196. Bassareo, V. and Di Chiara, G. (1999) Differential responsiveness of dopamine transmission to food-stimuli in nucleus accumbens shell/core compartments. *Neuroscience* **89,** 637–641.
197. Salamone, J. D., Cousins, M. S., McCullough, L. D., Carriero, D. L., and Berkowitz, R. J. (1994) Nucleus accumbens dopamine release increases during instrumental lever pressing for food but not free food consumption. *Pharmacol. Biochem. Behav.* **49,** 25–31.
198. Westerink, B. H., Teisman, A., and de Vries, J. B. (1994) Increase in dopamine release from the nucleus accumbens in response to feeding: a model to study interactions between drugs

and naturally activated dopaminergic neurons in the rat brain. *Naunyn Schmiedebergs Arch. Pharmacol.* **349,** 230–235.
199. Wilson, C., Nomikos, G. G., Collu, M., and Fibiger, H. C. (1995) Dopaminergic correlates of motivated behavior: importance of drive. *J. Neurosci.* **15,** 5169–5178.
200. Deroche, V., Piazza, P. V., Deminiere, J. M., Le Moal, M., and Simon, H. (1993) Rats orally self-administer corticosterone. *Brain Res.* **622,** 315–320.
201. Piazza, P. V., Deroche, V., Deminiere, J. M., Maccari, S., Le Moal, M., and Simon, H. (1993) Corticosterone in the range of stress-induced levels possesses reinforcing properties: implications for sensation-seeking behaviors. *Proc. Natl. Acad. Sci. USA* 90, 11,738–11,742.
202. Reul, J. M., Sutanto, W., van Eekelen, J. A., Rothuizen, J., and de Kloet, E. R. (1990) Central action of adrenal steroids during stress and adaptation. *Adv. Exp. Med. Biol.* **274,** 243–256.
203. Piazza, P. V. and Le Moal, M. (1996) Pathophysiological basis of vulnerability to drug abuse: role of an interaction between stress, glucocorticoids, and dopaminergic neurons. *Annu. Rev. Pharmacol. Toxicol.* **36,** 359–378.

9

Development and Expression of Behavioral Sensitization

Temporal Profile of Changes in Gene Expression

Peter W. Kalivas

Sensitization can be defined as a behavioral, physiological, or cellular response to a stimulus that is augmented as a result of previous exposure to that stimulus *(1)*. This definition is derived initially from behavioral models of learning, and in the field of addiction, most work with sensitization refers to the progressive increase in the motor stimulant effect of amphetamine-like psychostimulants that is elicited by repeated drug administration *(2,3)*. Notably, behavioral sensitization persists for weeks or months after a moderate treatment regimen of repeated psychostimulant administration *(2,4)*. The relative permanence of behavioral sensitization has resulted in frequent use of behavioral sensitization as an animal model of addiction, specifically the enduring changes in behavior that are associated with a diagnosis of psychostimulant addiction, such as the development of craving-mediated drug seeking and paranoia *(5,6)*.

The veracity of behavioral sensitization as an animal model of addiction has been disputed on both theoretical and empirical grounds *(7)*. Similarly, both theory and empirical observation have been marshaled to support a role for sensitization processes in addiction. These arguments have been reviewed elsewhere and are not the topic of the present chapter *(7,8)*. Rather, this chapter will focus on the certitude that sensitization to psychostimulants is enduring and must therefore be accompanied by long-term changes in the brain. Do these changes precisely mirror the situation in the human addict? Probably not, since sensitized motor activity in experimental animals is not a precise homolog of behaviors characteristic of psychostimulant addiction, such as craving and paranoia. Nonetheless, there is emerging literature indicating an overlap in the neuronal substrates at both circuitry and cellular levels of analysis between sensitization protocols and drug self-administration *(9,10)*. Thus, sensitization appears to be an appropriate and efficient model for studying a behavioral correlate of the neuronal alterations produced by repeated psychostimulant exposure. Moreover, as an efficient behavioral screening protocol for identifying mechanisms of psychostimulant-induced neuronal plasticity, sensitization has contributed greatly toward focusing neurobiological efforts using models of addiction having greater face validity than motor sensitization, but that are substantially more cumbersome to establish.

From: *Molecular Biology of Drug Addiction*
Edited by: R. Maldonado © Humana Press Inc., Totowa, NJ

1. Development and Expression Are Temporally Distinct Mechanisms of Sensitization

Early on, sensitization was described as a process that became progressively more robust following repeated administration, and following an extended withdrawal period *(11–14)*. This led to early conceptualizations that sensitization may studied in a least two temporally distinct domains *(4)*. (A) The development of sensitization results from a sequelae of neuroadaptations that are transient and occur during the repeated injection process, but ultimately establish an enduring change in the response of the brain to subsequent drug injections. (B) The enduring neuroadaptations become a relatively permanent element of the addicted nervous system and change subsequent responding to the drug or stimuli associated with drug administration. This latter component of sensitization is typically termed expression, and arises from changes in cellular function that emerge during withdrawal.

2. The Development of Sensitization

Given that a primary molecular mechanism whereby psychostimulants produce motor activation is by inhibiting the elimination of dopamine from the synaptic cleft and/or promoting the vesicular or nonvesicular release of dopamine *(15,16)*, it is appropriate that the last 15 years of research into the development of sensitization has focused on the dopaminergic projections from the ventral mesencephalon (ventral tegmental area, VTA) to the forebrain, notably the nucleus accumbens and prefrontal cortex (PFC). As a result of these studies it is now clear that the VTA is one site of action where psychostimulant drugs can act to produce behavioral sensitization *(9,17)*. While dopamine transporters constitute the primary binding site for psychostimulant-induced motor activity, it is well established that enhanced dopamine transmission must work in concert with glutamatergic afferents to the VTA in order to develop behavioral sensitization *(17)*. One critical source of glutamatergic afferents is the PFC, and electrical stimulation of the PFC can enhance, while lesions prevent the development of behavioral sensitization *(18–20)*. Based on these observations, two general scenarios of how the VTA may be involved in the development of sensitization have emerged.

The first scenario is that by blocking dopamine transporters directly in the VTA and increasing the extracellular concentration of dopamine in the VTA, it is possible to initiate the sequence of cellular events required to develop sensitization. The primary evidence for this action is that psychostimulant administration directly into the VTA produces behavioral sensitization *(21,22)*. Likewise, the administration of dopamine D1 receptor antagonists or *N*-methyl-D-aspartic acid NMDA glutamate receptor antagonists into the VTA prevents the development of sensitization to a systemic injection of amphetamine *(17,23,24)*. Moreover, it has been argued that D1 receptor administration may facilitate the presynaptic release or postsynaptic effects of glutamate in the VTA *(25,26)*. The second scenario is that by blocking dopamine transporters in certain dopamine terminal fields, notably the PFC or nucleus accumbens, this produces a change in projections from these nuclei back to the VTA, which then initiates sensitization *(27)*. As described above, this scenario is particularly viable for the PFC.

In the broad view, both scenarios are likely correct since manipulations in either the VTA or PFC have been shown to elicit or inhibit the development of behavioral sensitization to psychostimulants. The extent to which one or the other mechanism is critical probably depends on a number of factors. Notably, there is emerging evidence that different psychostimulants may preferentially involve one or the other scenario *(28,29)*. Thus, there is evidence for an action by amphetamine directly in the VTA, with less involvement from the PFC, while cocaine-induced behavioral sensitization appears to rely more on effects in the PFC (although an action directly in the VTA has also been shown). A distinction between amphetamine and cocaine is perhaps more dramatically emphasized by the fact that dopamine receptor antagonists can prevent the development of sensitization to amphetamine, but are generally ineffective at preventing cocaine sensitization *(30,31)*. Given the role of dopamine transporter blockade in the acute motor stimulant effect of cocaine, this is a remarkable observation and presumably points to a critical role for cocaine blockade of the serotonin or norepinephrine transporter(s) in the development of sensitization. Another important factor is the involvement of environmental stimuli that are associated with the drug administration. It is well established that associated environmental cues can augment the development of sensitization *(32)*. The PFC as well as the amygdala have been postulated as cortical and allocortical afferents to the VTA that may be especially important in the development of sensitization that has been specifically paired with environmental stimuli *(33–35)*.

The cellular underpinnings of the development of sensitization have been difficult to establish. This predicament arises from the fact that development is mediated by a sequence of temporary events. Thus, which cellular neuroadaptations are identified and deemed critical will differ between studies, depending on the dosage regimen and the point in the development that the measurements are made. This conundrum is exemplified in the hunt for changes in gene expression that may mediate the development of sensitization. This search has proceeded aggressively for the last 15 years, and three distinction temporal patterns of gene expression have been described that may play different roles in the development of sensitization.

1. A number of immediate-early gene (IEG) responses have been shown in response to an acute injection of cocaine or amphetamine, including changes in c-fos, narp, Homer1a and ARC *(36–39)*. However, following repeated administration, the expression induced by each drug administration diminishes until near-complete tolerance to the IEG response is produced. Notably, the IEG response to acute drug administration returns after a period of withdrawal. Both nuclear and cytosolic IEGs have shown this pattern of expression in response to repeated drug administration. The nuclear IEGs generally function as transcription factors to rapidly regulate the expression of other genes, while the cytosolic IEGs, such as *Homer1a*, act in part to rapidly adjust the trafficking of proteins within the cytoplasm and plasma membrane. These IEG changes tend to be widely distributed in the nervous system, often corresponding to dopamine terminal fields.
2. The next temporal pattern includes alterations in gene expression that gradually shift during the course of repeated drug exposure, and the change dissipates over the course of a few hours to a few days following the last drug injection *(40–42)*. There are many examples of this type of change in gene expression that include nuclear, cytosolic and membrane-bound proteins. Notably, the majority of adaptations having this pattern of expression are located in the VTA (although in many cases other brain areas have not been extensively examined). Well-established changes in the VTA include a reduction in $G_{i/o}$ and increases in tyrosine hydroxylase and GluR1. In addition, cellular processes involving interactions between proteins have been shown to undergo transient alterations in the VTA, notably a desensitization of D2 autoreceptors and enhanced releasability of somatodendritic dopamine *(40,43)*.

3. The final temporary change in gene expression that may be critical for the development of sensitization includes changes that progressively develop during repeated drug administration and do not dissipate for many days or weeks after discontinuing repeated drug exposure *(42–46)*. Notable in this class of neuroadaptation is the increase in delta-fosB, p-CREB, and preprodynorphin in the nucleus accumbens and striatum. Although temporary, these changes appear to involve a restructuring of expression patterns in the patch and matrix of the striatum. Moreover, the moderately enduring neuroadaptations have been postulated to be critical in the transition from development of sensitization or addiction to the more permanent steady-state neuroadaptations that actually mediate sensitization or other behaviors associated with addiction *(42)*.

A general pattern that overlies the temporal subcategories of changes in gene expression is that the earliest IEG changes are widespread in the brain, the short-lived changes that accumulate with repeated injection tend to be focused in the VTA, while the moderately enduring changes are not generally in the VTA but are most abundant in dopamine axon terminal fields. As outlined below, this temporal shift in the location of changes in gene expression in the brain terminates with alterations that are predominately localized to dopamine terminal fields such as the nucleus accumbens and PFC (although other dopamine terminal fields have not been so intensely studied).

3. The Expression of Sensitization

Behavioral sensitization is an enduring change in the brain that can be manifested weeks and months after discontinuing repeated drug administration. In contrast to the development of sensitization, for which the focus has been on the VTA and cortical afferents to the VTA, the study of the expression of sensitization has predominately involved dopamine terminal fields, in particular the nucleus accumbens and to a lesser extent the prefrontal cortex *(9)*. Thus, the expression of sensitization can be revealed by psychostimulant injection into the nucleus accumbens *(47)*. Conversely, sensitization can be blocked by intra-accumbens administration of dopamine or glutamate antagonists *(48)*. The dependence of the expression of sensitization on dopamine and glutamate transmission in the nucleus accumbens is thought to involve both pre- and postsynaptic neuroadaptations.

One of the hallmark observations in the field of sensitization is that the capacity of a psychostimulant to increase extracellular dopamine in the nucleus accumbens is augmented, and that this augmentation increases progressively during withdrawal *(13,14,49)*. This increase has been shown to arise in early withdrawal, in part from desensitization of autoreceptors. However, as in the VTA, this desensitization of D2 receptors is transient. At later withdrawal times an enduring elevation in vesicular release of dopamine predominates and occurs via an increase in calcium signaling, probably involving CaMKII *(50,51)*. Some studies have also reported an increase in the releasibility of glutamate in the nucleus accumbens *(52,53)*. However, this enhanced release is associated with a reduced basal level of extracellular glutamate, and when the reduced basal levels are normalized, an acute drug challenge no longer elevates glutamate *(54)*. These data point to the likelihood that the presynaptic changes in glutamate transmission may involve nonvesicular release, perhaps through the cystine/glutamate exchanger. Indeed, cocaine-induced neuroadaptations in the regulation of the cystine/glutamate exchanger were recently reported *(54)*.

Another classic observation in the field of sensitization is the augmentation in the postsynaptic response to D1 receptor stimulation *(55)*. This is most clearly shown in electrophysiological studies in which the inhibitory effect produced by the iontophoretic administration of D1 dopamine agonist is augmented. This effect endures in parallel with behavioral sensitization. In contrast, electrophysiological studies show a reduction in the postsynaptic responses to AMPA glutamate receptor stimulation *(56)*. This is paralleled by neurochemical studies showing a reduction in the responsiveness of group I metabotropic glutamate receptors (mGluRs) *(57)*.

Whereas enduring neuroadaptations in both glutamate and dopamine transmission in the nucleus accumbens have been revealed in animal models of behavioral sensitization, it has been argued that the manifestation of changes in glutamate transmission in particular depend on the presence of environmental stimuli that the subject associates with the drug. This dependence on learned associations was most clearly revealed in a study showing that the augmented glutamate release occurred only in an environment that had been specifically paired with daily drug administration *(58)*. Similarly, blockade of AMPA receptors in the nucleus accumbens prevented the expression of sensitization only when the drugs were given in a drug-paired environment *(48)*. One source of glutamate involved in the expression of sensitization and changes in glutamate transmission in the nucleus accumbens is the PFC. Thus, lesions of the PFC were found simultaneously to inhibit the expression of sensitization and to reduce the increased release of glutamate in response to a drug challenge *(59)*.

Given the enduring changes in neurotransmission in the nucleus accumbens, and potentially in afferents to the nucleus accumbens from the PFC, it is not surprising that the search for changes in gene expression that may underlie the neuroadaptations in transmission has focused on the nucleus accumbens. The hallmark of a change in gene expression that may be critical for the expression of enduring sensitization is that the alteration be present for weeks after discontinuing repeated drug administration. In addition, a particularly compelling temporal characteristic would be a change in gene expression that is not manifest during the repeated injection period or during the first few days of withdrawal. A number of genes fall into this temporal category and many are related to pre- and postsynaptic glutamate transmission, including mGluR5, Homer1bc, and mGluR2/3 *(57,60)*. In addition, other proteins with this temporal characteristic include AGS-3 and the adenosine 1 and TrkB receptor (unpublished observation). Finally, NAC-1 is a unique example of a change in gene expression that includes an IEG profile similar to c-fos and other IEG transcription factors, as well as an enduring elevation that is not present at 24 h of withdrawal but is present at 3 wk after discontinuing repeated drug administration *(61)*.

4. Conclusions

The data outlined above provide a strategy for studying the importance of alterations in gene expression relative to the temporal pattern of the change. Thus, four general patterns were identified that form a sequence of events from very rapid changes in the expression of IEGs to acute drug administration to relatively permanent alterations that generally manifest beginning only after a few days of drug discontinuation. Although it is possible to speculate on how the various changes in gene expression may be playing a role in the development and expression of sensitization, these syntheses must be regarded as formative. An organized analysis of the patterns of gene expression corresponding to the various temporal patterns of expression is only just beginning. However, the process will be facilitated by the fact that earlier studies have demonstrated the relative importance of the VTA and its afferents in the development of sensitization, and the nucleus accumbens and its cortical afferents in the expression of sensitization. While likely an oversimplified circuit for either process, the temporal patterns of gene expression identified to date would seem to support this anatomical organization. Finally, it is important to remember that once the genetic adaptations for behavioral sensitization have been catalogued, motor sensitization will almost assuredly include or lack some of the changes that are important in other behavioral neuroadaptations characteristic of human psychostimulant addiction, such as drug craving and paranoia. Nonetheless, sensitization offers a relatively simple behavioral model that is certainly the result of enduring drug-induced adaptations in gene expression and can be used to provide a template for focusing studies employing more complicated, albeit more precise, models of drug addiction.

Acknowledgments

Aspects of the research described in this contribution were funded in part by USPHS grants DA-03960 and DA-06074.

References

1. Kalivas, P. W. and Barnes, C. D. (1988) *Sensitization in the Nervous System.* Telford Press, Caldwell, NJ.
2. Robinson, T. E. and Becker, J. B. (1986) Enduring changes in brain and behavior produced by chronic amphetamine administration: a review and evaluation of animal models of amphetamine psychosis. *Brain Res. Rev.* **11,** 157–198.
3. Post, R. M. and Rose, H. (1976) Increasing effects of repetitive cocaine administration in the rat. *Nature* (Lond.) **260,** 731–732.
4. Kalivas, P. W. and Stewart, J. (1991). Dopamine transmission in the initiation and expression of drug- and stress-induced sensitization of motor activity. *Brain Res. Rev.* **16,** 223–244.
5. Robinson, T. E. and Berridge, K. C. (1993) The neural basis of drug craving: an incentive-sensitization theory of addiction. *Brain Res. Rev.* **18,** 247–291.
6. Kalivas, P. W., Pierce, R. C., Cornish, J., and Sorg, B. A. (1998) A role for sensitization in craving and relapse in cocaine addiction. *J. Psychopharmacol.* **12,** 49–53.
7. Koob, G. and LeMoal, M. (2001) Drug addiction, dysregulation of reward and allostasis. *Neuropsychopharmacology* **24,** 97–129.
8. Robinson, T. and Berridge, K. (2000) The psychology and neurobiology of addiction: an incentive-sensitization view. *Addiction* **95** (Suppl. 2), S91–S117.
9. Pierce, R. C., and Kalivas, P. W. (1997) A circuitry model of the expression of behavioral sensitization to amphetamine-like psychostimulants. *Brain Res. Rev.* **25,** 192–216.
10. McFarland, K., and Kalivas, P. W. (2001) The circuitry mediating cocaine-induced reinstatement of drug-seeking behavior. *J. Neurosci.* **21**(21), 8655–8663.
11. Antelman, S. M., Eichler, A. J., Black, C. A., and Kocan, D. (1980) Interchangeability of stress and amphetamine in sensitization. *Science* **207,** 329–331.
12. Paulson, P. E. and Robinson, T. E. (1995) Amphetamine-induced time-dependent sensitization of dopamine neurotransmission in the dorsal and ventral striatum: a microdialysis study in behaving rats. *Synapse* **19,** 56–65.

13. Kalivas, P. W. and Duffy, P. (1993) Time course of extracellular dopamine and behavioral sensitization to cocaine. I. Dopamine axon terminals. *J. Neurosci.* **13,** 266–275.
14. Heidbreder, C. A., Thompson, A. C., and Shippenberg, T. S. (1996) Role of extracellular dopamine in the initiation and longterm expression of behavioral sensitization to cocaine. *J. Pharmacol. Exp. Ther.* **278,** 490–502.
15. Seiden, L. S., Sabol, K. E., and Ricuarte, G. A. (1993) Amphetamine: effects on catecholamine systems and behavior. *Annu. Rev. Pharmacol. Toxicol.* **33,** 639–677.
16. Reith, M. E. A., Meisler, B. E., Sershen, H., and Lajtha, A. (1986) Structural requirements for cocaine congeners to interact with dopamine and serotonin uptake sites in mouse brain and to induce stereotyped behavior. *Biochem. Pharmacol.* **35,** 1123–1129.
17. Vezina, P. and Queen, A. (2000) Pharmacological reversal of behavioral and cellular indices of cocaine sensitization in the rat. *Psychopharmacology* **151,** 184–191.
18. Wolf, M. E., Dahlin, S. L., Hu, X.-T., Xue, C.-J., and White, K. (1995) Effects of lesions of prefrontal cortex, amygdala, or fornix on behavioral sensitization to amphetamine; comparison with N-methyl-D-asparatate antagonists. *Neuroscience* **69,** 417–439.
19. Carr, D. and Sesack, S. (2000) Projections from the rat prefrontal cortex to the ventral tegmental area: target specificity in the synaptic associations with mesoaccumbens and mesocortical neurons. *J. Neurosci.* **20,** 3864–3873.
20. Li, Y., Hu, X.-T., Berney, T. G., Vartanian, A. J., Stine, C. D., Wolf, M. E., and White, F. J. (1999) Both glutamate receptor antagonists and prefrontal cortex lesions prevent induction of cocaine sensitization and associated neuroadaptations. *Synapse* **34,** 169–80.
21. Perugini, M. and Vezina, P. (1994) Amphetamine administered to the ventral tegmental area sensitizes rats to the locomotor effects of nucleus accumbens amphetamine. *J. Pharmacol. Exp. Ther.* **270,** 690–696.
22. Kalivas, P. W. and Weber, B. (1988) Amphetamine injection into the A10 dopamine region sensitizes rats to peripheral amphetamine, and cocaine. *J. Pharmacol. Exp. Ther.* **245,** 1095–1102.
23. Vezina, P. (1996) D1 dopamine receptor activation is necessary for the induction of sensitization by amphetamine in the ventral tegmental area. *J. Neurosci.* **16,** 2411–2420.
24. Kalivas, P. W. and Alesdatter, J. E. (1993) Involvement of NMDA receptor stimulation in the VTA and amygdala in behavioral sensitization to cocaine. *J. Pharmacol. Exp. Ther.* **267,** 486–495.
25. Kalivas, P. W. and Duffy, P. (1995) D1 receptors modulate glutamate transmission in the ventral tegmental area. *J. Neurosci.* **15,** 5379–5388.
26. Kim, J. and Vezina, P. (1999) Rats pre-exposed to amphetamine show enhanced locomotion to a D1 dopamine receptor agonist in the nucleus accumbens when glutamate reuptake is inhibited. *Soc. Neurosci. Abstr.* **25,** 221.1.
27. Volkow, N. D., Fowler, J. S., Wang, G. J., Hitzemann, R., Logan, J., Schyler, D. J., Dewey, S. L., and Wolf, A. P. (1993) Decreased dopamine D2 receptor availability is associated with reduced frontal metabolism in cocaine abusers. *Synapse* **14,** 167–177.
28. Wolf, M. E. (1998) The role of excitatory amino acids in behavioral sensitization to psychomotor stimulants. *Prog. Neurobiol.* **54,** 679–720.
29. White, F. J. and Kalivas, P. W. (1998) Neuroadaptations involved in amphetamine and cocaine addiction. *Drug Alcohol Depend.* **51,** 141–154.
30. White, F. J., Joshi, A., Koeltzow, T. E., and Hu, X.-T. (1998) Dopamine receptor antagonists fail to prevent induction of cocaine sensitization. *Neuropsychopharmacology* **18,** 26–40.
31. Mattingly, B. A., Hart, T. C., Lim, K., and Perkins, C. (1994) Selective antagonism of dopamine D1 and D2 receptors does not block the development of behavioral sensitization to cocaine. *Psychopharmacology* **114,** 239–242.

32. Stewart, J. (1992) Neurobiology of conditioning to drugs of abuse. *Ann. N.Y. Acad. Sci.* **654,** 335–346.
33. Everitt, B. J. and Robbins, T. W. (1992) Amygdala-ventral striatal interactions and reward-related processes, in *The Amygdala: Neurobiological Aspects of Emotion, Memory and Mental Dysfunction* (Aggleton, J. P., ed.), Wiley-Liss, New York, pp. 401–429.
34. Berke, J. and Hyman, S. (2000) Addiction, dopamine, and the molecular mechanisms of memory. *Neuron* **25,** 515–532.
35. Childress, A. R., Mozley, P. D., McElgin, W., Fitzgerald, J., Reivich, M., and O'Brien, C. P. (1999) Limbic activation during cue-induced cocaine craving. *Am. J. Psychiatry* **156,** 11–18.
36. Dunais, J. B. and McGinty, J. F. (1994) Acute and chronic cocaine administration differentially alters striatal opioid and nuclear transcription factor mRNAs. *Synapse* **18,** 35–45.
37. Brakeman, P. R., Lanahan, A. A., O'Brien, R., Roche, K., Barnes, C. A., Huganir, R. L., and Worley, P. F. (1997) Homer: a protein that selectively binds metabotropic glutamate receptors.*Nature* **386,** 221–223.
38. Tsui, C., Copeland, N., Gilbert, D., Jenkins, N., and Barnes, C. (1996) Narp, a novel member of the pentraxin family, promotes neurite outgrowth and is dynamically regulated by neuronal activity. *J. Neurosci.* **16,** 2463–2478.
39. Moratalla, R., Vickers, E. A., Robertson, H. A., Cochran, B. H., and Graybiel, A. M. (1993) Coordinate expression of c-fos and jun B is induced in the rat striatum by cocaine. *J. Neurosci.* **13,** 423–433.
40. Wolf, M. E., White, F. J., Nassar, R., Brooderson, R. J., and Khansa, M. R. (1993) Differential development of autoreceptor subsensitivity and enhanced dopamine release during amphetamine sensitization. *J. Pharmacol. Exp. Ther.* **264,** 249–255.
41. Striplin, C. and Kalivas, P. W. (1993) Robustness of G protein changes in cocaine sensitization shown with immunoblotting. *Synapse* **14,** 10–15.
42. Nestler, E. (2001) Molecular basis of long-term plasticity underlying addiction. *Nature Rev.* **2,** 119–128.
43. Kalivas, P. W. and Duffy, P. (1993) Time course of extracellular dopamine and behavioral sensitization to cocaine. II. Dopamine perikarya. *J. Neurosci.* **13,** 276–284.
44. Hope, B. T., Nye, H. E., Kelz, M. B., Self, D. W., Iadorola, M. J., Nakabeppu, Y., Duman, R. S., and Nestler, E. J. (1994) Induction of a long-lasting AP-1 complex composed of altered Fos-like proteins in brain by chronic cocaine and other chronic treatments. *Neuron* **13,** 1235–1244.
45. Ketz, M., Chen, J., Carlezon, W., Whisler, K., Gilden, L., Beckmann, A., Steffen, C., Zhang, Y., Marotti, L., Self, D., Tkatch, T., Baranauskas, B., Surmeier, D., Nene, R., Duman, R., Picciotto, M., and Nestler, E. (1999) Expression of the transcription factor dFosB in the brain controls sensitivity to cocaine. *Nature* **401,** 272–276.
46. Moratalla, R., Elibol, B., Vallejo, M., and Graybiel, A. M. (1996) Network-level changes in expression of inducible fos-jun proteins in the striatum during chronic cocaine treatment and withdrawal. *Neuron* **17,** 147–156.
47. Pierce, R. C. and Kalivas, P. W. (1996) Amphetamine produces sensitized locomotion and dopamine release preferentially in the nucleus accumbens shell of rats administered repeated cocaine. *J. Pharmacol. Exp. Ther.* **275,** 1019–1029.
48. Bell, K. and Kalivas, P. W. (1996) Context-specific cross sensitization between systemic cocaine and intra-accumbens AMPA infusion in rats. *Psychopharmacology* **127,** 377–383.
49. Robinson, T., Jurson, P., Bennett, J., and Bentgen, K. (1988) Persistent sensitization of dopamine neurotransmission in ventral straitum(nucleus accumbens) produced by prior experience with (+)-amphetamine: a microdialysis study in freely moving rats. *Brain Res.* **462,** 211–222.
50. Pierce, R. C. and Kalivas, P. W. (1997) Repeated cocaine modifies the mechanism by which ampehtamine releases dopamine *J. Neurosci.* **17,** 3254–3261.

51. Kantor, L. and Gnegy, M. (1998) Ca^{2+}, K^+ and calmodulim kinase II affect amphetamine-mediated dopamine release in sensitized rats. *FASEB J.* **12,** A159.
52. Pierce, R. C., Bell, K., Duffy, P., and Kalivas, P. W. (1996) Repeated cocaine augments excitatory amino acid transmission in the nucleus accumbens only in rats having developed behavioral sensitization. *J. Neurosci.* **16,** 1550–1560.
53. Reid, M. S. and Berger, S. P. (1996) Evidence for sensitization of cocaine-induced nucleus accumbens glutamate release. *Neuroreport* **7,** 1325–1329.
54. Baker, D., Xi, Z.-X., Shen, H., Swanson, C., and Kalivas, P. (2002) The primary source and neuronal function of in vivo extracellular glutamate. Submitted.
55. Henry, D. J. and White, F. J. (1995) The persistence of behavioral sensitization to cocaine parallels enhanced inhibition of nucleus accumbens neurons. *J. Neurosci.* **15,** 6287–6299.
56. White, F., Hu, X., Zhang, X., and Wolf, M. (1995) Repeated administration of cocaine or amphetamine alters neuronal responses to glutamate in the mesoaccumbens dopamine system. *J. Pharmacol. Exp. Ther.* **273,** 445–454.
57. Swanson, C., Baker, D., Carson, D., Worley, P., and Kalivas, P. (2001) Repeated cocaine administration attenuates group I metabotropic glutamate receptor-mediated glutamate release and behavioral activation: a potential role for Homer 1b/c. *J. Neurosci.* **21,** 9043-9052.
58. Bell, K., Duffy, P., and Kalivas, P. W. (2000) Context-specific enhancement of glutamate transmission by cocaine. *Neuropsychopharmacology* **23,** 335–344.
59. Pierce, R. C., Reeder, D. C., Hicks, J., Morgan, Z. R., and Kalivas, P. W. (1998) Ibotenic acid lesions of the dorsal prefrontal cortex disrupt the expression of behavioral sensitization to cocaine. *Neuroscience* **82,** 1103–1114.
60. Xi, Z.-X., Baker, D. A., Shen, H., Carson, D. S., and Kalivas, P. W. (2002) Group II metabotropic glutamate receptors modulate extracellular glutamate in the nucleus accumbens. *J. Pharmacol. Exp. Ther.* **300,** 162–171.
61. Cha, X.-Y., Pierce, R. C., Kalivas, P. W., and Mackler, S. A. (1997) NAC-1, a rat brain mRNA, is increased in the nucleus accumbens three weeks after chronic cocaine self-administration *J. Neurosci.* **17,** 6864–6871.

Part III

Cannabinoid Addiction

10

New Advances in the Identification and Physiological Roles of the Components of the Endogenous Cannabinoid System

Ester Fride and Raphael Mechoulam

1. Summary

Many aspects of the physiology and pharmacology of anandamide (arachidonoyl ethanolamide), the first endogenous cannabinoid ligand ("endocannabinoid") isolated from pig brain, have been studied since its discovery in 1992. Ethanolamides from other fatty acids have also been identified as endocannabinoids with similar in vivo and in vitro pharmacological properties. 2-Arachidonoyl glycerol and noladin ether (2-arachidonyl glyceryl ether), isolated in 1995 and 2001, respectively, thus far display pharmacological properties in the central nervous system similar to those of anandamide. The endocanabinoids are widely distributed in brain, they are synthesized and released upon neuronal stimulation, and undergo reuptake and are hydrolyzed intracellularly by fatty acid amide hydrolase (FAAH). Pharmacological effects of the endocannabinoids are very similar, yet not identical, to those of the plant-derived and synthetic cannabinoid receptor ligands. In addition to pharmacokinetic explanations, direct or indirect interactions with other receptors have been considered to explain some of these differences, including activities at serotonin and γ-aminobutyric acid (GABA) receptors.

Binding affinities for additional receptors such as the vanilloid receptor have to be taken into account in order to fully understand endocannabinoid physiology. Moreover, possible interactions with receptors for the lysophosphatidic acids deserve attention in future studies.

Endocannabinoids have been implicated in a variety of physiological functions. These areas of central activity include pain reduction, motor regulation, learning/memory, and reward. Neuroprotective effects of anandamide and 2-arachidonoyl glycerol have also been reported. Finally, the role of the endocannabinoid system in appetite stimulation in the adult organism, and perhaps more important, its critical involvement in milk ingestion and survival of the newborn, may further our understanding of the physiology of food intake and growth.

2. Introduction

After the identification *(1)* and cloning *(2)* of the first cannabinoid (CB1) receptor and in view of the existence of an endogenous opioid–opiate receptor signaling system,

From: *Molecular Biology of Drug Addiction*
Edited by: R. Maldonado © Humana Press Inc., Totowa, NJ

it was only natural to start the search for an endogenous ligand for the cannabinoid receptor. However, it took the insight that the putative ligand might be lipophilic *(3)* in order to realize the discovery of the first endogenous cannabinoid ligand ("endocannabinoid"), which turned out to be the ethanol amide of arachidonic acid (20:4, *n*-6) and was denoted "anandamide" *(4)*. A second type of endocannabinoid was discovered in 1995, also a derivative of arachidonic acid, but which is an ester (2-arachidonoyl glycerol or 2-AG *[5,6]*). Very recently, a third type has been reported, this time an ether of arachidonic acid (2-arachidonyl glyceryl ether, denoted noladin ether *[7]*).

The endocannabinoid receptor signaling system, consisting of endocannabinoids, their receptor(s), uptake mechanism, and hydrolyzing enzyme, is phylogenetically old, occurring across vertebrates *(8–10)* and invertebrate species *(11)*. Endocannabinoid molecules per se have been detected in mammals, including humans *(12)*, dogs *(5)*, rats *(13,14)* and pigs *(4)*, and in fish *(11,15)*. Invertebrate species in which the presence of endocannabinoids have been observed include molluscs *(16)*, *Hydra vulgaris (17)*, and sea urchins *(18)*. However, no components of the endocannabinoid system have been detected in insects *(8,19)*.

Brain tissue concentrations of 2-AG are approximately 200-fold higher than those of anandamide *(20)*. The rank order for the distribution of both endocannabinoids in different areas is similar: highest in brainstem, striatum, and hippocampus, and lower in cortex, diencephalon, and cerebellum. No correlation was found between endocannabinoid concentrations and CB1 receptor distribution. Since receptor concentration and receptor activation were not correlated either *(21)*, the lack of association between endocannabinoid concentration and CB1 receptor distribution is not surprising. However, additional explanations for disparities between CB1 receptor distribution and activity have been offered and include the existence of non-CB1 receptor molecular targets for the endocannabiniods (*see* below).

3. Endocannabinoids as Signaling Molecules in the Central Nervous System

Anandamide is synthesized "on demand" (upon stimulation) *(22–24)*, and released from neurons immediately afterward *(23–25)*. Anandamide is inactivated by reuptake via a membranal transport molecule, the "anandamide membrane transporter" (AMT) and subsequent intracellular enzymatic degradation *(23,26,27)* by fatty acid amide hydroxylase (FAAH)-mediated hydrolysis *(9,28,29)*. 2-Arachidonoylglycerol (2-AG) undergoes similar FAAH-mediated hydrolysis *(9,30)* and carrier-mediated transmembranal transport *(31)*, probably through the same anandamide membrane transporter *(32)*.

FAAH and AMT are distributed in brain areas in a patterns corresponding to that of the CB1 receptor, that is, high concentrations in hippocampus, cerebellum, and cerebral cortex *(29,30,33,34)*, thus further supporting the position that the endocannabinoids are true neurotransmitters *(35)*.

It has been argued, based on structure-activity relationships, that 2-AG is the natural ligand at the CB1 receptor *(14,36)*. On the other hand, the observation that anandamide, but not 2-AG, was released upon depolarization in the rat striatum *(25)*, suggests that anandamide rather than 2-AG is the primary ligand for the CB1 receptors, at least in the rat striatum controlling motor activity.

The nature of endocannabinoid (anandamide and 2-AG) neurotransmission has greatly been clarified in a recent series of papers *(36–38)*. These new sets of data have been summarized by Christie and Vaughan *(39)*: endocannabinoids are released from a postsynaptic neuron upon stimulation, diffuse back to presynaptic neurons, where they act on CB1 receptors resulting in a reduced probability of neurotransmitters (such as glutamate and GABA) to be released. Removal of the endocannabinoids is accomplished by uptake into neuronal or glial cells which they enter via endocannabinoid transporters. Once inside the presynaptic cells, the endocannnabinoids are broken down by FAAH. It is too early to generalize these principles to noladin ether.

Endocannabinoids in the central nervous system bind $G_{i/o}$-coupled CB1 receptors that modulate adenylyl cyclase, ion channels, and extracellular signal-regulated kinases *(21,24,35)*. Recently it was determined that the CB1 receptor is coupled to ceramide, a lipid second messenger, which in turn mediates cannabinoid induced apoptosis. Such a mechanism opens up new avenues of investigation for the ways by which the endocannabinoids control cell function *(40)*.

4. Differences Between the Pharmacology of Exo- and Endocannabinoids in the Central Nervous System

In vivo activity of cannabinoids is assessed in mice based on a battery of four assays designed by Martin and colleagues (motor activity, "catalepsy," body temperature, and analgesia) *(41)*. Overall, anandamide displayed similar pharmacologocal effects compared to tetrahydrocannabinol (THC) *(42)*. Anandamide's shorter duration of action in vivo compared to that of the plant-derived and synthetic cannabinoids (by 1 h, compared to several hours at least) has been attributed to anandamide's facile degradation *(23,43)* by FAAH *(28)*.

Further differences between anandamide and THC, albeit subtle, became apparent very soon too. Thus, anandamide has partial agonist activity in vitro for inhibition of forskolin-stimulated release of adenylate cyclase *(44)* and for inhibition of calcium currents in N18 neuroblastoma cells *(45)*, and in vivo for some aspects of the "tetrad" (body temperature and analgesia) *(3,42)*. When different routes of administering anandamide were compared, a complex pattern of full and partial activities was observed *(46)*. Further, Δ^9-THC but not anandamide produced conditioned place avoidance *(47)*.

Two more ethanol amides from fatty acids have been isolated from brain tissue. Like the "original" anandamide (ethanol amide of arachidonic acid), these molecules bind to CB1 receptors and have THC-like activities in the "tetrad." Hence, it was decided to denote all three ethanolamides "anandamides," each one derived from a different fatty acid: 20:4, *n*-6, archidonoyl ethanol amide; 22:4, n-6, docosatetraenyl ethanol amide; and 20:3, *n*-6, homo-linolenyl ethanol amide *(48)*. The latter two were shown to have even lower efficacies compared to anandamide (20:4, *n*-6) *(49)*. Further, in vivo tolerance and cross-tolerance to central effects of THC were detected upon repeated doses of anandamide *(50,51)*, but unlike THC-induced tolerance, no cross-tolerance to the dynorphic system was observed *(51)*.

4.1. Low Doses of Anandamide

Further differences between the endocannabinoids and the prototypical THC include antagonistic activity of the anandamides at the CB1 receptor at very low doses *(52)*. Thus preincubation of N18TG2 neuroblastoma cells with 1 n*M* anandamide antagonized forskolin-stimulated inhibition of adenylate cyclase activity, while pretreatment of mice with 0.0001–0.1 mg/kg of anandamide antagonized THC-induced cannabimimetic effects. Interestingly, the reverse (pretreatment with low-dose THC and testing for anandamide-induced effects) did not show inhibitory activity *(52)*. A possibly related phenomenon is the stimulatory activity of low doses (0.01 mg/kg) of anandamide in the tetrad of cannabimimetic effects as well as in a chemiluminescence assay for phagocytic activity *(42,53)*. It was suggested that this low-dose anandamide-induced stimulation may be ascribed to activation of G_s proteins *(52)*, which are known to stimulate adenylate cyclase activity opposite to the neurobehavioral depression which is mediated by $G_{i/o}$ protein-induced inhibition of adenylate cyclase *(54)*. Indeed, subsequent studies have provided support for this hypothesis *(55)*.

Non-CB1 receptor-mediated stimulation of NMDA receptors by low concentrations of anandamide has been observed *(56)*. It is not known whether this mechanism underlies the behavioral observations.

Much less is known about the in vivo pharmacology of 2-AG compared to anandamide. However, being deactivated by similar mechanisms as anandamide (*see* previous section), it is not surprising that 2-AG also has a short duration of action compared to plant-derived or synthetic cannabinoids, while partial agonist properties are also apparent for 2-AG *(5,56)*.

4.2. Entourage Effects

Additional natural ethanol amides and glycerols, analogs of anandamide and 2-AG, respectively, but that do not bind CB1 receptors, have been detected in several biological tissues including brains *(57,58)*. Oleoylethanol amide and linoleoyl ethanolamide are prominent among the anandamide analogs, found in neural tissue *(23,57,58)* and by themselves, have weak but significant cannabimimetic effects, presumably by enhancing the extracellular levels and half-life of anandamide *(59,60,61,62,63)*.

Oleamide or "sleep factor" induces sleep, or at least sleeplike behaviors *(64)* and cannabimimetic effects *(65,66)*. Since oleamide does not bind to the CB1 receptor, these effects have been ascribed, in part at least, to oleamide's ability to inhibit anandamide hydrolysis and to enhance anandamide's affinity for the CB1 receptor *(65)*.

The two 2-AG analogs palmitoyl glycerol and linoleyl glycerol coexist and are coreleased with 2-AG, but do not bind CB1 receptors. They potentiate in vivo and in vitro effects of 2-AG, thus enabling low concentrations of the endocannabinoid, which by themselves have no overt activity, to have potent effects. They do this by potentiating 2-AG binding to the CB1 receptor while 2-linoleoyl-glycerol also inhibits the inactivation of 2-AG in neuronal cells *(57,67)*. These effects were called "entourage" effects *(24,67)*. Whether the concept of entourage effect should be extended to effects of the (ethanol) amides discussed above will have to be determined.

The accumulating knowledge about the endocannabinoid deactivating mechanisms—facilitated transport by the amide transporter (AMT) and hydrolysis by the enzyme FAAH—has prompted research into the possibilities of developing AMT- and FAAH-inhibiting drugs as a means to enhance endocannabinoid availability, that is, as indirect agonists *(29,68,69)*. It has been suggested that such inhibitors may have more selective therapeutic effects, since their action would only be evident at sites where endocannabinoid production and release are taking place *(70)*.

5. Receptors Other than CB1

In general, cannabinoid-induced pharmacological effects are effectively inhibited by the CB1 receptor antagonist SR141716A *(71)*. Thus, significant blockade of THC-induced effects were observed in the mouse tetrad *(72,73)* and in the mouse tail-flick test for pain perception when supraspinal CB1 receptors were exposed to the antagonist *(74)*. Also, THC- or anandamide-induced memory impairment was attenuated by SR141716A *(75)*. In rats, SR141716A effectively antagonized tetrad-like central effects induced by anandamide *(76)*. Moreover, anandamide-increased appetite was inhibited by SR141716A *(77)*. More recently, 2-AG-induced effects have been included in studies on the CB1 antagonist. Thus SR141716A blocked the antiepileptiform effect of 2-AG, similarly to that of anandamide, in rat hippocampal slices *(78,79)*. Hence we have evidence now that both exogenous and endogenous cannabinoid-induced effects can be blocked by the CB1 antagonist.

However, contrary to expectation, a number of anandamide-induced effects, although they are similar to THC-induced effects, could not be inhibited by the CB1 receptor antagonist. For example, in mice, anandamide-induced effects in the tetrad were antagonized by neither SR141716A *(73,80)* nor by LY320135 (Fride et al., unpublished observations). Further, SR141716A blocked the antinociceptive effects of THC in mice much more efficiently than those of anandamide *(74)*. This phenomenon has been explained as a pharmacokinetic effect, since inhibition of anandamide-induced effects in the tetrad was accomplished when either a (nonspecific) FAAH inhibitor, phenylmethylsulphonyl fluoride (PMSF), was co-administered with anandamide in order to enhance its half-life, or when a stable analog was administered instead of anandamide *(81)* (*see* also review by Nakamura-Palacios and colleagues *[82]*). It is still not clear, however, why the CB1 antagonist should only reverse the effects induced by anandamide when its half-life is sufficiently prolonged (e.g., by phenylmethylsulphonyl fluoride, PMSF). Does anandamide act via (an)other receptor in addition to the CB1 receptor?

Observations on CB1 receptor knockout CB / mice support such possibility. Thus Δ^9-THC-induced hypoalgesia in the tail-flick test was present, despite the gene deletion, in two knockout strains that were developed in different laboratories and from different parent strains *(83,84)*. Since the tail-flick test is presumably measuring spinal pain perception, while the hot-plate test assays supraspinal mechanisms of pain *(85)*, this observation suggests that mainly higher-level pain mechanisms are affected by CB1 receptor deletion. This is compatible with the observation that SR141716A, when injected intraperitoneally or intracerebrally, fully antagonized cannabinoid-induced analgesia, but only partially when injected at the spinal level *(74)*. Since cannabinoid receptor-mediated pain has a spinal component *(86)*, these data suggest a non-cannabinoid receptor mechanism at the spinal level in addition to the CB1 receptor-mediated transduction.

More recent experiments with the CB1 receptor knockout mice showed, surprisingly, that the CB1−/− mice display anandamide-induced CB1 receptor-mediated response including analgesia, catalepsy, and motor inhibition, despite the absence of CB1 receptors *(87)*. This suggests that anandamide exerts some pharmacological effects that are similar to those induced by exogenous cannabinoids but that are not CB1 receptor-mediated.

In the next sections, evidence for endocannabinoid mechanisms of action other than via CB1 receptor activation will be outlined.

5.1. Nonreceptor-Mediated Mechanisms for Endocannabinoid Activities in the Central Nervous System

Before the first CB1 receptor was identified, it was generally assumed that—in accordance with their lipid characteristics—the cannabinoids exert their pharmacological activities by non-receptor mechanisms *(3)*. Since then many cannabinoid-induced effects have been ascribed to receptor activation, yet, non-receptor-mediated effects of anandamide have been demonstrated too *(88)*. It has previously been suggested that the "sleep factor" oleamide may exert at least some of its actions by its ability to inhibit anandamide's degradation and reuptake *(65)*. However, nonspecific membrane perturbation has been posited as a possible mechanism for oleamide's pharmacological effects *(89)*. Is it possible that anandamide is also such endogenous fluidity transmitter? Another potential mode of action, previously studied, is the inhibition of gap junctions *(90,91)*.

5.2. Receptor-Mediated Mechanisms for Endocannabinoid Activities in the Central Nervous System

5.2.1. Known Receptors as Potential Targets for Endocannabinoids

5.2.1.1. 5-HT_3 Receptors

Fan *(92)* has shown that cannabinoids including anandamide, inhibit 5-HT_3 receptor-mediated currents. These data indicated that the 5-HT_3 receptor ion channel is a site of action of cannabinoids and endocannabinoids. The direction of the effect is compatible with the antiemetic potential of cannabinoid agonists *(93)*, since 5-HT_3 receptor antagonist are well-established antiemetic medicinal drugs. Fan's observation went largely unchallenged but was also not supported further by experimental evidence. However, we have observed that the 5-HT_3 antagonists MDL72222 and granisetron show cannabinoid-like profiles in the mouse tetrad *(73)*. In assays for synaptosomal receptor binding, MDL72222 did not bind CB1 receptors, and vice versa, neither anandamide nor HU210 bound to 5-HT_3 receptors (Fride et al., in preparation). Thus, the nature of the interaction between CB1 and 5-HT_3 receptors needs to be clarified further.

5.2.1.2. 5-HT_2 Receptors

The current state of knowledge is complex. Mice injected with the 5-HT_2 receptor antagonist ketanserin displayed cannabimimetic effects on the tetrad, with potencies at least as high as those of anandamide *(73)*. Anandamide*(94)* and oleamide *(95)* have been found to bind to 5-HT_2 receptors, thus raising the possibility that endocannabinoids (and oleamide) may act by 5-HT_2 receptor blockade. In another study, however, no ^{3}H-ketanserin displacement at the 5-HT_2 receptor by oleamide was found *(96)*. Thus the nature and physiological significance of endocannabinoid–5-HT_2 receptor interaction needs to be further clarified.

5.2.1.3. NMDA Receptors

Anandamide (but not THC) was found to have dual effects on NMDA receptor activity *(56)*. First, like THC, anandamide reduced calcium flux via CB1 receptors, this effect being reversed by SR141716A. Second, at low concentrations, anandamide but not THC stimulated calcium influx by directly modulating the NMDA receptor. More recently, inhibition of glutamatergic neurotransmission by the synthetic cannabinoid agonist WIN55,212-2 was reported in CB1 knockout mice *(97)*. It remains to be determined whether this finding has relevance to the observations at the NMDA receptor.

5.2.1.4. Lysophosphatidic Acid (LPA)

LPA bears structural similarities to 2-AG. Contos et al. *(98)* have demonstrated that targeted deletion of the receptor gene for LPA resulted in a defective suckling response in the knockout mice. This phenomenon is strikingly similar to the mortality of SR141716A-treated pups, which also die within days after birth due to a lack of milk ingestion from birth *(99)*. Therefore it is possible that LPA and cannabinoids crossreact with their respective receptors. The sparse data available thus far do not support such a hypothesis. Thus SR141716A-induced inhibition of cannabinoid-stimulated p38 mitogen-activated protein kinases did not alter the effects of LPA on p38-MAPK phosphorylation *(100)*. Moreover, whereas THC completely reversed the effects of neonatally applied SR141716A, LPA did not *(99)*. In both these reports, oleoyl-sn-glycero-3-phosphate were used. Thus it remains to be seen whether other LPA species, notably LPA from arachidonic acid, will display cross-reactivity with cannabinoids.

5.2.1.5. Vanilloid Receptors

Vanilloid type 1 (VR1) receptors are found not only on sensory neurons where they are partly coexpressed with CB1 receptors *(101)*, but also in several central nuclei including hypothalamus and basal ganglia, hippocampus, and cerebellum *(102,103)*. In all these brain areas, CB1 receptors are found as well *(104)*. Anandamide is a full agonist at VR1 receptors *(101,105,106)*. Although still somewhat controversial *(107,108,109)*, it appears now that sufficient amounts of anandamide are available in vivo to stimulate VR1 receptors under physiological conditions *(103,108,109)*. Summing up the evidence available at present, Di Marzo et al. *(103)* have suggested that anandamide interacts with both receptors at binding sites that are situated extra-or intracellularly for CB1 and VR1, respectively *(110)*. The specific dominant interaction depends on a number of factors such as ATP acting to enhance anandamide's effects at the VR1 receptor, levels of anandamide, tissue receptor distribution, and accessibility to the receptor *(103,110)*. These observations suggest that anandamide may be not only an endocannnabinoid but also an "endovanilloid" *(103)*.

5.2.2. Putative Novel Receptors for Endocannabinoids in the Central Nervous System

Several recent reports present evidence suggesting the existence of a new, unknown CB receptor in the brain. One report describes a reduction in amplitudes of excitatory postsynaptic currents by cannabinoids responsible for glutamatergic neurotransmisson in the hippocampus of wild type mice as well as CB1−/− knockout mice *(97)*. Further, Di Marzo and colleagues *(87)* showed that anandamide effectively produced major aspects of the tetrad and stimulated GTP-S binding in CB−/− mice; these effects were not inhibited by SR141716A. These findings were elaborated by Breivogel et al. *(111)*, who observed that the putative receptor is not distributed in the brain in a fashion similar to that of CB1 receptors. Thus, anandamide and WIN55212-1 bound to some brain regions of CB1−/− knockout mice such as cortex, hippocampus, and brainstem, but not in the basal ganglia and cerebella of these mice. It possible that, due to unknown compensatory mechanisms and/or other changes in the knockout mice, receptor types that are physiologically irrelevant in the normal organism may become overexpressed in CB−/− mice. Hence evidence for new CB receptors would be strengthened greatly by experiments using normal tissue or animals. It is of benefit, however, that the two CB1 receptor knockout models are of different genetic backgrounds *(83,84)*, thereby allowing for some degree of generalizaton. Indeed, very recently, WIN55,212-2 was also shown to stimulate GTP S binding in Ledent et al.'s knockout mice. The regional distribution where this was observed, however (cerebellum, not hippocampus) *(112)*, was different from Breivogel et al.'s findings, where, for example, WIN55,212-2 stimulated GTP S binding in hippocampal, but not cerebellar tissue of CB−/− mice *(111)*. Future studies will have to determine whether the putative CB receptors in the different models are the same or different entities.

6. Physiological Functions of Endocannabinoids in the Central Nervous System

6.1. Pain

Evidence for the use of cannabis as an analgesic medicine during childbirth was described as early as 1500 years ago *(113)*. All three types of endocannabinoids (anandamide, 2-AG, and noladin ether) have been shown to inhibit central pain perception (on a "hot plate"), albeit not as efficaciously as THC *(5,7,42)*.

As has been elegantly shown by Walker and colleagues *(114)* using in vivo microdialysis, anandamide is released in the periaqueductal gray (PAG), a midbrain area playing a pivotal role in pain perception *(115,116)*, in response to pain stimuli and to electrical stimulation of PAG. Moreover, hyperalgesia was observed after adminstration of the CB1 receptor antagonist SR141716A *(117)*. These findings strongly suggest that endocannabinioids maintain a tonic inhibition of pain. It should be noted, however, that in CB1 receptor knockout mice, pain response to exposure to a hot plate was either unaffected *(83)* or reduced *(84)*. As noted above, compensatory or other changes in these knockout mice may explain this discrepancy.

Opiate receptors are also richly distributed in PAG, as well as in other areas where CB1 receptors are found and thought to mediate pain *(118)*. Interactions between anandamide and opiates in the pain response have been demonstrated *(51)*.

6.2. Motor Functions and Schizophrenia

CB1 receptors are richly distributed in the basal ganglia and cerebral cortex, regions that play a pivotal role in motor control *(119,120)*. Cannabinoids and endocannabinoids affect motor behavior in a bi- or even triphasic fashion *(52,121)*. An intimate interactivity between the dopamine system and the endocannabinoids has been uncovered. Thus, earlier studies include cannabinoid-induced increases in dopamine release in the frontal brain regions *(122)*, while chronic treatment with dopamine D2 receptor antagonists resulted in increased CB1 receptor expression in the striatum *(123)*. Furthermore, localized application of cannabinoids into the nigrostriatal system in the rat counteracted the motor response to dopamine D2 receptor agonists *(124,125)*. Direct studies on the endocannabinoid system indicated that stimulation of D2 receptors enhances anandamide in the rat striatum *(126)*, while the anandamide transport blocker

AM404 counteracted D2 receptor-mediated responses such as apomorphine-induced yawning *(127)*.

Based on the above, it is not surprising that investigations are conducted into the possibility of using cannabinoid-based medicines for the treatment of impaired motor functions, many of which are thought to involve the dopamine system. Such conditions include Parkinson *(128)* and Huntington's diseases *(129)*, Tourette syndrome *(130)*, multiple sclerosis *(131)*, and schizophrenia *(132)*. However, especially in the case of schizophrenia, the complexity and chronicity of the condition and its treatments *(133,134)* warrant further experimental work until widespread clinical applications may be endorsed *(135)*.

6.3. Cognitive Functions

6.3.1. Hippocampus

The hippocampus, a brain area with a high CB1 receptor density *(136)* fulfils a central role in learning and memory formation *(137)*. Both anandamide *(138)* and 2-AG *139)* interfere with long-term potentiation (LTP) in hippocampal slices, a physiological model for learning and memory.

In vivo, anandamide impaired performance in a non-match-to-position task testing for working memory *(140)*. This was reversed by SR141716A *(75)*. If anandamide acts in similar ways as exocannabinoids in this mnemonic task, it would seem that the memory impairment occurs by interfering with the encoding of information that takes place in the hippocampus. As a result, short-term memory cannot be formed *(141)*. There is also evidence, however, that anandamide impairs memory consolidation *(142)*. Comparisons between mouse strains for an avoidance memory task indicated that dramatic strain-specific differences exist for the effects of anandamide on memory consolidation, causing inhibition *or* enhancement, depending on the strain *(143)*. Recent studies may clarify this complex situation. High doses disrupt the development of LTP, as has been thought for a long time; however, endocannabinoid release may enhance memory by triggering depolarization-induced depression of inhibition (DSI) *(144)*. Thus, different responses to cannabinoids in memory tasks may be explained by strain-dependent differences in concentrations of components of the endocannabinoid system. Opposing effects of high and low doses of anandamide have been described *(52,53)* (*see* above).

6.3.2. Prefrontal Cortex

Another important brain structure in cognitive function is the prefrontal cortex (PFC). Density of CB1 receptors in the PFC is high compared to other G-protein-coupled receptors *(145)*. This region is thought to integrate cognitive and emotional functions and may be the primary dysfunctional area in schizophrenia and the site of action for antischizophrenic drugs *(146)*. Interestingly, Δ^9-THC produces schizophrenia-like symptoms in humans *(132)*, while anandamide levels are higher in schizophrenics than in controls *(134)*. More specifically, THC increased presynaptic dopamine efflux and utilization in the PFC and impaired spatial memory *(122,147)*. Recently, it has been demonstrated in PFC slices that cannabinoids influence glutamatergic synaptic transmission and plasticity *(148)*. Preliminary observations indicated that the PFC of acutely stressed mice (30 min of noise stress) contained four times as much anandamide as those of unstressed mice. Such increase was not seen in the hippocampi of these mice *(149)*. An interesting set of observations on children of marijuana-smoking mothers indicated that these children develop impaired "executive functioning," which is thought to be a cognitive deficit of the PFC *(150)*.

Taken together, these observations suggest that the endocannabinoid and dopamine systems are closely cooperating in the regulatory role of the PFC in stress, cognition, and schizophrenia.

6.4. Sleep

Anandamide has been shown to increase slow-wave and REM sleep in rats at the expense of wakefulness *(142)*, while conversely, the CB1 receptor antagonist SR141716A increased wakefulness at the expense of slow-wave and REM sleep *(151)*. These findings support a role for the endocannabinoids in sleep regulation. As to the pharmacological basis of such action, anandamide has been found to bind to 5-HT_2 receptors *(94)*. However, behaviorally, clear differences can be detected between anandamide- and 5-HT_2-receptor antagonist-injected mice. It has been suggested, therefore, that the unique profile of anandamide (different from both that of Δ^9-THC, a "pure" CB1 agonist, and ketanserin, a 5-HT_2 receptor antagonist), results from a combination of its interactions at CB1 *and* 5-HT_2 receptors *(66)*.

6.5. Feeding and Appetite

Cannabinoids enhance appetite *(93,152)*. Indeed Δ^9-THC is used clinically for this purpose, particularly in acquired immunodeficiency syndrome (AIDS) and cancer patients *(93)*. Anandamide increased food intake in rats *(77)*, while SR141716A has been reported to inhibit the intake of palatable food *(153–155)*. Evidence for a function of the endogenous cannabinoid system in the feeding response has been obtained for the primitive invertebrate *Hydra vulgaris (17)*. These data point to a very ancient history of the endocannabinoid system in the regulation of feeding.

Interestingly, preliminary clinical data suggest that SR141716A is a promising weight reducing agent for the treatment of obesity *(156)*.

Leptin is considered to be a key signal through which the hypothalamus senses the nutritional state of the body *(157)*. In a recent article, experiments were reported which indicate that leptin and endocannabinoids counterbalance each other's control of food intake. For example, it was demonstrated that leptin reduced endocannabinoid levels in the hypothalamus but not the cerebellum of rats, and that endocannabinoid levels increase in animals with defective leptin signaling *(158)*. The possible role of CB1 receptors in maintaining food intake after fasting was suggested by experiments carried out using CB1 knockout mice *(158)*.

In addition to these central mechanisms, non-CB1-receptor fatty acid ethanol amides have been shown recently to be involved in food intake. Thus oleoylethanolamide, which does not bind to CB1 receptors *(60)*, induced sever hypophagia when injected peripherally in high doses (20 mg/kg) in rats *(159)*. Previously it had been shown in mice that such doses produce central effects on the "tetrad," presumably by inhibiting the enzymatic breakdown of endocannabinoids in the CNS *(160)*. Thus it appears that effects of oleoylethanolamide on feeding differ from those of anandamide in its localization (peripheral vs central), the receptor mechanism (non-CB1 receptor vs CB1 receptor mediation) and in the direction (hyperhagic vs hypohagic).

Endocannabinoids are present in milk, with 2-AG found in human milk in higher concentrations than anandamide *(57)*. 2-AG when administered orally, albeit in high doses, is active in the mouse "tetrad" *(57)*. These findings indicate that 2-AG in milk may, in part at least, reach the central nervous system. Moreover, the observation that the levels of 2-AG, but not of anandamide, in rodent pup brain peak immediately after birth *(161)* may indicate a role for 2-AG in suckling in the newborn. We have recently reported that administration of SR141716A to mouse pups, within 24 h after birth, completely inhibited milk intake from the dam, thereby arresting growth and resulting in death within the first week of life. Injecting SR141716A on d 2 after birth resulted in only 50% mortality *(99)*. These data strongly suggest that endocannabinoids (probably 2-AG) play a critical role in survival of the newborn mouse by displaying an absolute control over milk ingestion. The generalizability to other species and the precise mechanism by which milk intake is blocked awaits further clarification. In conclusion, clinical application for endocannabinoids or their direct or indirect agonists for infant "failure to thrive" conditions deserves investigation.

6.6. Neuroprotection

A considerable amount of in vitro, and recently, in vivo work indicates that the endocannabinoids are neuroprotective and that possibly neuroprotection is a major physiological role of this class of compounds (for a review, *see* ref. *162*).

Cannabinoid receptor agonists have been shown to protect cultured rat hippocampal neurons from excitotoxicity and cerebral cortical neurons from in vitro ischemia in rats *(163,164)*. Hampson et al. *(56)* have found that NMDA-induced Ca^{2+} flux could be reduced by anandamide and that SR141716A, a CB1 receptor antagonist, counteracts the activity of the endocannabinoid. Excitatory neurotransmission is associated with activation of the NMDA receptor, which is a glutamate-controlled ion channel. Abood et al. *(165)* have recently reported that activation of the CB1 cannabinoid receptor modulates kainate toxicity in primary neuronal cultures prepared from mouse spinal cord. This effect was blocked by SR141716A. In vivo results support the in vitro data.

Nagayama et al. have found that the synthetic cannabinoid WIN 55212 reduces ischemic damage in rat brain *(166)*, and Van der Stelt and colleagues have reported that THC reduces neuronal injury in neonatal rats injected intracerebrally with the Na/K-ATPase inhibitor ouabain *(167)*.

Panikashvili et al. *(168)* have observed that the levels of 2-AG sharply increase after closed head injury in mice, and that synthetic 2-AG, when administered after closed head injury in mice, caused significant reduction of brain edema, better clinical recovery, reduced infarct volume, and reduced hippocampal cell death compared with controls found. The neuroprotective effect of 2-AG was attenuated by the CB1 receptor antagonist SR141716A, indicating that the mechanisms of the processis apparently cannabinoid receptor-mediated. 2-Acyl glycerols, such as 2-palmitoyl glycerol and lineoyl glycerol, which are present in brain but do not bind to the cannabinoid receptors, enhance the activity of 2-AG as a neuroprotective agent (for a discussion of the entourage effect, *see* above). 2-AG is not the only endocannabinoid involved in neuroprotection. Van der Stelt et al. *(167)* have reported that anandamide, like THC, reduces neuronal injury in a dose-dependent manner in a rat model of ouabain-induced excitoxicity. Hansen et al. *(168,170)* have found that anandamide as well as anandamide precursors, but not 2-AG, accumulated in rat brain after mild to moderate brain injury. The relationship between endocannabinoids and neuroprotection is apparently a complicated one. In view of the lack of efficient neuroprotective drugs, further research in this field may lead to new therapeutic leads. A synthetic cannabinoid, HU-211, is currently already in Phase III clinical trials against brain trauma *(171,172)*.

7. Developmental Aspects of the Endocannabinoid System in the Central Nervous System

Initial reports studying development of the cannabinoid receptor system during the first weeks of postnatal life in the rat described a gradual increase in brain CB1 receptor mRNA *(173)* and in the density of CB1 receptors *(174,175)*. In later studies, investigating the gestational period, CB1 receptor mRNA was detected from gestational d 11 in the rat *(176)*. Additional studies have uncovered more complex developmental patterns. Thus, whereas the highest levels of mRNA expression of the CB1 receptor are seen at adulthood in regions such as the caudate-putamen and the cerebellum, other areas such as the cerebral cortex, the hippocampus, and the ventromedial hypothalamus display the highest mRNA CB1 receptor levels on the first postnatal day *(161,177)*. Endocannaboinoids were also detected from the gestational period in rodents, 2-AG at 1000-fold higher concentrations than anandamide. Interestingly, while anandamide displayed a gradual increase, 2-AG displayed constant levels throughout development, with a single peak on the first postnatal day *(161)*. Is it possible that the high levels of CB1 receptor mRNA and 2-AG that have been observed on the first day of life in structures including the hypothalamic ventromedial nucleus *(161)* (which is associated with feeding behavior) comprise a major stimulus for the newborn to initiate milk intake? (*See* also Section 5.5.)

Atypical patterns (i.e., different from those in adult) of CB1 receptor densities were also observed: a transient presence of CB1 receptors was detected in white matter regions including the corpus callosum and anterior commisure (connecting neuronal pathways between the left and right hemispheres) between gestational d 21 and postnatal d 5, suggesting a role for endocannabinoids in brain development *(178)*.

AMT and FAAH levels were higher in the brains of 6-mo-old CB1-receptor knockout mice compared to wild-type animals *(179)*. More data are required before the biological significance of these findings can be fully understood.

8. Implications for Addiction to Cannabinoids

8.1. Tolerance

Tolerance to cannabinoids developed in all species studied, with varying duration and onset, depending, for example, on the parameter studied *(72,180–184)*. In humans, development of tolerance to the psychoactive effects of marijuana is clearly seen with "heavy" (daily) use, but usually not with casual or moderate use *(72,185)*. Tolerance to ANA has been shown in animal studies *(50,51)*.

The behavioral tolerance is accompanied, analogous to other classes of drugs, by a decrease in CB1 receptors in all brain areas that are relevant for the CB1-tolerant behaviors *(186–187)*.

8.2. Craving and Reinforcement

Addictive potential of marijuana was long thought to be very weak or absent *(72)*. However, although addictive behaviors such as compulsive drug seeking (due to "craving") is rarely induced by marijuana use, preparations containing higher Δ^9-THC concentrations, such as hashish, have been shown to induce addictive behaviors, especially in populations at risk *(188–189)*. Hence one may speculate that marijuana, as obtained at the turn of the millenium, may be addictive as well, since it often contains much higher concentrations of Δ^9-THC than in the 1960s and 1970s *(185,190)*.

From animal studies it has gradually become clear that cannabinoids interact with the same neural substrates that are thought to be responsible for the euphoriant and rewarding effects of other drugs of abuse such as cocaine, opiates, and alcohol *(189,191)*. These neural substrates of addiction include the medial forebrain bundle, containing the dopamine pathways, leading from the mesencephalic ventral tegmentum to the nucleus accumbens and the prefrontal cortex. It appears that cannabinoids, like other drugs of abuse, increase dopamine activity in these neural circuits *(122,147,191–195)*. In behavioral tests of addiction, Δ^9-THC significantly lowered brain reward thresholds in the median forebrain bundle *(189,191,196)*.

Furthermore, Δ^9-THC was shown to be appetitive in the "conditioned place preference" test, but only after the appropriate timing and dosing *(197)*. Thus aversive effects of cannabinoids have been repeatedly shown as well *(47,189,198–201)*. These biphasic effects are well known from human experience; low doses of Δ^9-THC produce a "high" feeling, while high doses may be aversive *(189,190)*. Similarly, self-administration of cannabinoids has been hard to show in animal studies *(72,189)*, possibly due to masking anxiogenic effects of cannabinoids *(202–204)*. Confirming this suspicion in a recent study using the synthetic CB1 antagonist WIN 55,212-2, a robust but biphasic effect on self-administration in mice was demonstrated, suggesting rewarding effects at lower and aversive effect at high doses of WIN 55,212-2 *(205)*.

Thus overall, despite earlier doubts, recent studies have produced convincing evidence for the mesolimbic–mesocortical dopamine system as a substrate for cannabinoid abuse potential. Moreover, a common opioid receptor mechanism appears to mediate both cannabinoid- and heroin-induced activation of the mesolimbic dopamine activation *(195)*. It has also been shown in an alcohol craving paradigm that SR141716A can block the "craving" for alcohol in rats *(206)*, again suggesting a common abuse mechanism for various types of substances. Recently, a role for endocannabinoids in relapse to cocaine seeking was demonstrated in rats *(207)*.

In summary, it has become clear that cannabis has addictive properties similar to other drugs of abuse. This realization lends biological support for the controversial "gateway" theory, which states that cannabis often introduces new users to more destructive and addictive drugs. One should not overlook, however, possible genetic variation in cannabis abuse. Thus studies indicating genetic variation in the reward system *(189)* and in the emotional effects of cannabinoids *(202,203,208)* support a genetic predisposition to cannabis abuse. Whether a certain individual will eventually succumb to the addictive potential of cannabis will obviously be the outcome of a combination of various factors.

9. Conclusions

The newly discovered endocannabinoid system has been found to play a role in many physiological processes. Although anandamide and 2-AG have been found, and their actions have been investigated, in many biological systems, it seems that we have only scratched the surface of the effects of these fascinating biological modulators. And as has happened previously with essentially all neurotransmitter systems, the knowledge accumulated has led to new therapeutics. Will the endocannabinoid system yield a similar crop?

Acknowledgments

This work was supported, in part, by the Israeli Ministry of Health and by the Israel Science Foundation.

References

1. Devane, W. A., Dysarz, F. A., 3d., Johnson, M. R., Melvin, L. S., and Howlett, A. C. (1988) Determination and characterization of a cannabinoid receptor in rat brain. *Mol. Pharmacol.* **34,** 605–613.
2. Matsuda, L. A., Lolait, S. J., Brownstein, M. J., Young, A. C., and Bonner, T. I. (1990) Structure of a cannabinoid receptor and functional expression of the cloned cDNA. *Nature* **346,** 561–564.
3. Mechoulam, R. and Fride, E. (1995) The unpaved road to the endogenous brain cannabinoid ligands, the anandamides. In *Cannabinoid Receptors* (R. Pertwee, ed.), pp. 233–258, Academic Press, London, UK.
4. Devane, W. A., Hanus, L., Breuer, A., Pertwee, R. G., Stevenson, L. A., Griffin, G., Gibson, D., Mandelbaum, A., Etinger, A., and Mechoulam, R. (1992) Isolation and structure of a brain constituent that binds to the cannabinoid receptor. *Science* **258,** 1946–1949.
5. Mechoulam, R., Ben-Shabat, S., Hanus, L., Ligumsky, M., Kaminski, N. E., Schatz, A. R., Gopher, A., Almog, S., Martin, B. R., and Compton, D. R. (1995) Identification of an endogenous 2-monoglyceride, present in canine gut, that binds to cannabinoid receptors. *Biochem. Pharmacol.* **50,** 83–90.
6. Sugiura, T., Kondo, S., Sukagawa, A., Nakane, S., Shinoda, A., Itoh, K., Yamashita, A., and Waku, K. (1995) 2-Arachidonoylglycerol: a possible endogenous cannabinoid receptor ligand in brain. *Biochem. Biophys. Res. Commun.* **215,** 89–97.
7. Hanus, L., Abu-Lafi, S., Fride, E., Breuer, A., Vogel, Z., Shalev, D. E., Kustanovich, I., and Mechoulam, R. (2001) 2-arachidonyl glyceryl ether, an endogenous agonist of the cannabinoid CB1 receptor. *Proc. Natl. Acad. Sci. USA* **98,** 3662–3665.
8. Elphick, M. R. and Egertova M. (2001) The neurobiology and evolution of cannabinoid signalling. *Phil. Trans. R. Soc. Lond. B Biol. Sci.* **356,** 381–408.
9. Fowler, C. J., Nilsson, O., Andersson, M., Disney, G., Jacobsson, S. O., and Tiger, G. (2001) Pharmacological properties of cannabinoid receptors in the avian brain: similarity of rat and chicken cannabinoid1 receptor recognition sites and expression of cannabinoid2 receptor-like immunoreactivity in the embryonic chick brain. *Pharmacol. Toxicol.* **88,** 213–222.
10. Soderstrom, K., Leid, M., Moore, F. L., and Murray, T. F. (2000) Behavioral, pharmacological, and molecular characterization of an amphibian cannabinoid receptor. *J. Neurochem.* **75,** 413–423.
11. Salzet, M., Breton, C., Bisogno, T., and Di Marzo, V. (2000) Comparative biology of the endocannabinoid system. *Eur. J. Biochem.* **267,** 4917–4927.
12. Felder, C. C., Nielsen, A., Briley, E. M., Palkovits, M., Priller, J., Axelrod, J., Nguyen, D. N., Richardson, J. M., Riggin, R. M., Koppel, G. A., Paul, S. M., and Becker, G. W. (1996) Isola-

tion and measurement of the endogenous cannabinoid receptor agonist, anandamide, in brain and peripheral tissues of human and rat. *FEBS Lett.* **393,** 231–235.
13. Hansen, H.-H., Ikonomidou, C., Bittigau, P., Hansen, S. H., and Hansen, H. S. (2001) Accumulation of the anandamide precursor and other N-acylethanolamine phospholipids in infant rat models of in vivo necrotic and apoptotic neuronal death. *J. Neurochem.* **76,** 39–46.
14. Sugiura, T. and Waku, K. (2000) 2-Arachidonoylglycerol and the cannabinoid receptors. *Chem. Phys. Lipids* **108,** 89–106.
15. Schmid, H. H., Schmid, P. C., and Natarajan, V. (1990) N-acylated glycerophospholipids and their derivatives. *Prog. Lipid Res.* **29,** 1–43.
16. Sepe, N., De Petrocellis, L., Montanaro, F., Cimino, G., and Di Marzo, V. (1998) Bioactive long chain N-acylethanolamines in five species of edible bivalve molluscs. Possible implications for mollusc physiology and sea food industry. *Biochim. Biophys. Acta* **1389,** 101–111.
17. De Petrocellis, L., Melck, D., Bisogno, T., Milone, A., and Di Marzo, V. (1999) Finding of the endocannabinoid signalling system in *Hydra*, a very primitive organism: possible role in the feeding response. *Neuroscience* **92,** 377–387.
18. Bisogno, T., Ventriglia, M., Milone, A., Mosca, M., Cimino, G., and Di Marzo, V. (1997) Occurrence and metabolism of anandamide and related acyl-ethanolamides in ovaries of the sea urchin Paracentrotus lividus. *Biochim. Biophys. Acta* **1345,** 338–348.
19. McPartland, J., Di Marzo, V., De Petrocellis, L., Mercer, A., and Glass, M. (2001) Cannabinoid receptors are absent in insects. *J. Comp. Neurol.* **436,** 423–429.
20. Bisogno, T., Berrendero, F., Ambrosino, G., Cebeira, M., Ramos, J.-A., Fernandez-Ruiz, J.-J., and Di Marzo, V. (1999) Brain regional distribution of endocannabinoids: implications for their biosynthesis and biological function. *Biochem. Biophys. Res. Commun.* **256,** 377–380.
21. Breivogel, C.-S., Sim, L.-J., and Childers, S.-R. (1997) Regional differences in cannabinoid receptor/G-protein coupling in rat brain. *J. Pharmacol. Exp. Ther.* **282,** 1632–1642.
22. Cadas, H., diTomaso, E., and Piomelli, D. (1997) Occurrence and biosynthesis of endogenous cannabinoid precursor, N-arachidonoyl phosphatidylethanolamine, in rat brain. *J. Neurosci.* **17,** 1226–1242.
23. Di Marzo, V., Fontana, A., Cadas, H., Schinelli, S., Cimino, G., Schwartz, J. C., and Piomelli, D. (1994) Formation and inactivation of endogenous cannabinoid anandamide in central neurons. *Nature* **372,** 686–691.
24. Mechoulam R., Fride, E., and Di Marzo, V. (1998) Endocannabinoids. *Eur. J. Pharmacol.* **359,** 1–18.
25. Giuffrida,A., Parsons, L. H., Kerr, T. M., Rodriguez-de-Fonseca, F., Navarro, M., and Piomelli, D. (1999) Dopamine activation of endogenous cannabinoid signaling in dorsal striatum. *Nat. Neurosci.* **2,** 358–363.
26. Day, T. A., Rakhshan, F., Deutsch, D. G., and Barker, E.-L. (2001) Role of fatty acid amide hydrolase in the transport of the endogenous cannabinoid anandamide. *Mol. Pharmacol.* **59,** 1369–1375.
27. Deutsch D. G., Glaser S. T., Howell J. M., Kunz J. S., Puffenbarger R. A., Hillard C. J., and Abumrad N. (2001) The cellular uptake of anandamide is coupled to its breakdown by fatty-acid amide hydroxylase. *J. Biol. Chem.* **276,** 6967–6973.
28. Cravatt B. F., Giang D. K., Mayfield, S. P., Boger, D. L., Lerner, R. A., and Gilula, N. B. (1996) Molecular characterization of an enzyme that degrades neuromodulatory fatty-acid amides. *Nature* **384,** 83–87.
29. Giuffrida, A., Beltramo, M., and Piomelli, D. (2001) Mechanisms of endocannabinoid inactivation: biochemistry and pharmacology. *J. Pharmacol. Exp. Ther.* **298,** 7–14.
30. Ueda, N. and Yamamoto, S. (2000) Anandamide amidohydrolase (fatty acid amide hydrolase). *Prostaglandins Other Lipid Mediat.* **61,** 19–28.

31. Beltramo, M., and Piomelli, D. (2000) Carrier-mediated transport and enzymatic hydrolysis of the endogenous cannabinoid 2-arachidonylglycerol. *Neuroreport* **11,** 1231–1235.
32. Bisogno T., MacCarrone M., De Petrocellis L., Jarrahian A., Finazzi-Agro A., Hillard C., and Di Marzo, V. (2001) The uptake by cells of 2-arachidonoylglycerol, an endogenous agonist of cannabinoid receptors. *Eur. J. Biochem.* **268,** 1982–1989.
33. Egertova, M., Giang, D. K., Cravatt, B. F., and Elphick, M. R. (1998) A new perspective on cannabinoid signalling: complementary localization of fatty acid amide hydrolase and the CB1 receptor in rat brain. *Proc. R. Soc. Lond. B Biol. Sci.* **265,** 2081–2085.
34. Tsou, K., Nogueron, M. I., Muthian, S., Sanudo-Pena, M. C., Hillard, C. J., Deutsch, D. G., and Walker, J. M. (2001) Fatty acid amide hydrolase is located preferentially in large neurons in the rat central nervous system as revealed by immunohistochemistry. *Neurosci. Lett.* **254,** 137–140.
35. Self, D. W. (1999) Anandamide: a candidate neurotransmitter heads for the big leagues. *Nature Neurosci.* **2,** 303–304.
36. Kreitzer, A. C. and Regehr, W. G. (1998) Retrograde inhibition of presynaptic calcium influx by endogenous cannabinoids at excitatory synapses onto Purkinje cells. *Neuron* **29,** 717–727.
37. Ohno-Shosaku, T., Maejima, T., and Kano, M. (2001) Endogenous cannabinoids mediate retrograde signals from depolarized postsynaptic neurons to presynaptic terminals. *Neuron* **29,** 729–738.
38. Wilson, R. I. and Nicoll, R. A. (2001) Endogenous cannabinoids mediate retrograde signalling ar hippocampal synapses. *Nature* **410,** 588–592.
39. Christie, M. J. and Vaughan, C. W. (2001) Cannabinoids act backwards. *Nature* **410,** 527–530.
40. Guzman M., Galve-Roperh I., and Sanchez C. (2001) Ceramide: a new second messenger of cannabinoid action. *Trends Pharmacol. Sci.* **22,** 19–22.
41. Martin, B. R., Compton, D. R., Thomas, B. F., Prescott, W. R., Little, P. J., Razdan, R. K., Johnson, M. R., Melvin, L. S., Mechoulam, R., and Ward, S. J. (1991) Behavioral, biochemical, and molecular modeling evaluations of cannabinoid analogs. *Pharmacol. Biochem. Behav.* **40,** 471–478.
42. Fride, E. and Mechoulam, R. (1993) Pharmacological activity of the cannabinoid receptor agonist, anandamide, a brain constituent. *Eur. J. Pharmacol.* **231,** 313–314.
43. Deutsch, D. G. and Chin, S. A. (1993) Enzymatic synthesis and degradation of anandamide, a cannabinoid receptor agonist. *Biochem. Pharmacol.* **46,** 791–796.
44. Vogel, Z., Barg, J., Levy, R., Saya, D., Heldman, E., and Mechoulam, R. (1993) Anandamide, a brain endogenous compound, interacts specifically with cannabinoid receptors and inhibits adenylate cyclase. *J. Neurochem.* **61,** 352–355.
45. Mackie, K., Devane, W. A., and Hille, B. (1993) Anandamide, an endogenous cannabinoid, inhibits calcium currents as a partial agonist in N18 neuroblastoma cells. *Mol. Pharmacol.* **44,** 498–503.
46. Smith, P. B., Compton, D. R., Welch, S. P., Razdan, R. K., Mechoulam, R., and Martin, B. R. (1994) The pharmacological activity of anandamide, a putative endogenous cannabinoid, in mice. *J. Pharmacol. Exp. Ther.* **270,** 219–227.
47. Mallet, P. E. and Beninger, R. J. (1998) Delta^9-tetrahydrocannabinol, but not the endogenous cannabinoid receptor ligand anandamide, produces conditioned place avoidance. *Life Sci.* **62,** 2431–2439.
48. Hanus, L., Gopher, A., Almog, S., and Mechoulam, R. (1993) Two new unsaturated fatty acid ethanolamides in brain that bind to the cannabinoid receptor. *J. Med. Chem.* **36,** 3032–3034.
49. Barg, J., Fride, E., Hanus, L., Levy, R., Matus-Leibovitch, N., Heldman, E., Bayewitch, M., Mechoulam, R., and Vogel, Z. (1995) Cannabinomimetic behavioral effects of and adenylate cyclase inhibition by two new endogenous anandamides. *Eur. J. Pharmacol.* **287,** 145–152.
50. Fride, E. (1995) Anandamides: tolerance and cross-tolerance to delta 9-tetrahydrocannabinol. *Brain Res.* **697,** 83–90.

51. Welch, S. P. (1997) Characterization of anandamide-induced tolerance: comparison to delta 9-THC-induced interactions with dynorphinergic systems. *Drug Alcohol. Depend.* **45,** 39–45.
52. Fride, E., Barg, J., Levy, R., Saya, D., Heldman, E., Mechoulam, R., and Vogel, Z. (1995) Low doses of anandamides inhibit pharmacological effects of delta 9-tetrahydrocannabinol. *J. Pharmacol .Exp. Ther.* **272,** 699–707.
53. Sulcova, E., Mechoulam, R., and Fride, E. (1998) Biphasic effects of anandamide. *Pharmacol. Biochem. Behav.* **59,** 347–352.
54. Howlett, A. C., Qualy, J. M., and Khachatrian, L. L. (1986) Involvement of Gi in the inhibition of adenylate cyclase by cannabimimetic drugs. *Mol. Pharmacol.* **29,** 307–313.
55. Glass, M., and Felder, C.-C. (1997) Concurrent stimulation of cannabinoid CB1 and dopamine D2 receptors augments cAMP accumulation in striatal neurons: evidence for a Gs linkage to the CB1 receptor. *J. Neurosci.* **17,** 5327–5333.
56. Hampson, A. J., Bornheim, L. M., Scanziani, M., Yost, C. S., Gray, A. T., Hansen, B. M., Leonoudakis, D. J., and Bickler, P. E. (1998) Dual effects of anandamide on NMDA receptor-mediated responses and neurotransmission. *J. Neurochem.* **70,** 671–676.
57. Di Marzo, V., Sepe, N., De Petrocellis, L., Berger, A., Crozier, G., Fride, E., and Mechoulam, R. (1998) Trick or treat from food endocannabinoids? *Nature* **396,** 636–637.
58. Schmid, P. C., Krebsbach, R. J., Perry, S. R., Dettmer, T. M., Maasson, J. L., and Schmid, H. H. (1995) Occurrence and postmorten generation of anandamide and other long-chain N-acylethanolamines in mammalian brain. *FEBS Lett.* **375,** 117–120.
59. Maccarrone, M., van der Stelt, M., Rossi, A., Veldink, G. A., Vliegenthart, J. F., and Agro, A. F. (1998) Anandamide hydrolysis by human cells in culture and brain. *J. Biol. Chem.* **273,** 32,332–32,339.
60. di Tomaso, E., Beltramo, M., and Piomelli, D. (1996) Brain cannabinoids in chocolate. *Nature* **382,** 677–678.
61. Fride, E., Bisogno, T., Di Marzo, V., Vogel, Z., and Mechoulam, R. (1997) Anandamide: mediator of the effects of oleamide (a sleep factor) and of chocolate? *Soc. Neurosci. Abstr.* **23,** 1230.
62. Hillard, C. J., Edgemond, W. S., Jarrahian, A., and Campbell, W. B. (1997) Accumulation of N-arachidonoylethanolamine (anandamide) into cerebellar granule cells occurs via facilitated diffusion. *J. Neurochem.* **69,** 631–638.
63. Maurelli, S., Bisogno, T., De Petrocellis, L., Di Luccia, A., Marino, G., and Di Marzo, V. (1995) Two novel classes of neuroactive fatty acid amides are substrates for mouse neuroblastoma "anandamide amidohydrolase." *FEBS Lett.* **377,** 82–86.
64. Cravatt, B. F., Prospero-Garcia, O., Siuzdak, G., Gilula, N. B., Henriksen, S. J., Boger, D. L., and Lerner, R. A. (1995) Chemical characterization of a family of brain lipids that induce sleep. *Science* **268,** 1506–1509.
65. Mechoulam, R., Fride, E., Hanus, L., Sheskin, T., Bisogno, T., Di Marzo, V., Bayewitch, M., and Vogel, Z. (1997) Anandamide may mediate sleep induction. *Nature* **389,** 25,26.
66. Fride, E. (1999) Anandamide and oleamide: no pot, no sleep. *1999 Symposium on the Cannabinoids, Burlington, Vermont, International Cannabinoid Research Society,* p. 23.
67. Ben-Shabat, S., Fride, E., Sheskin, T., Tamiri, T., Rhee, M. H., Vogel, Z., Bisogno, T., De Petrocellis, L., Di Marzo, V., and Mechoulam, R. (1998) An entourage effect: inactive endogenous fatty acid glycerol esters enhance 2-arachidonoyl-glycerol cannabinoid activity. *Eur. J. Pharmacol.* **353,** 23–31.
68. Boger, D. L., Sato, H., Lerner, A. E., Hedrick, M. P., Fecik, R.-A., Miyauchi, H., Wilkie, G. D., Austin, B.-J., Patricelli, M.-P., and Cravatt, B.-F. (2000) Exceptionally potent inhibitors of fatty acid amide hydrolase: the enzyme responsible for degradation of endogenous oleamide and anandamide. *Proc. Natl. Acad. Sci. USA* **97,** 5044–5049.
69. Fowler, C. J., Jonsson, K. O., and Tiger, G. (2001) Fatty acid amide hydrolase: biochemistry, pharmacology, and therapeutic possibilities for an enzyme hydrolyzing anandamide, 2-arachidonoylglycerol, palmitoylethanolamide, and oleamide. *Biochem. Pharmacol.* **62,** 517–526.

70. Pertwee, R. G. (2001) Cannabinoid receptor ligands. *Tocris Rev.* **16.**
71. Rinaldi-Carmona, M., Barth, F., Heaulme, M., Shire, D., Calandra, B., Congy, C., et al. (1994) SR141716A, a potent and selective antagonist of the brain cannabinoid receptor. *FEBS Lett.* **350,** 240–244.
72. Compton, D. R., Aceto, M. D., Lowe, J., and Martin, B. R. (1996) In vivo characterization of a specific cannabinoid receptor antagonist (SR141716A): inhibition of delta 9-tetrahydrocannabinol-induced responses and apparent agonist activity. *J. Pharmacol. Exp. Ther.* **277,** 586–594.
73. Fride, E., Ben-Shabat, S., and Mechoulam, R. (1998) Pharmacology of anandamide: Interaction with serotonin systems? *1998 Symposium on the Cannabinoids, Burlington, Vermont, International Cannabinoid Research Society* p. 76.
74. Welch, S. P., Huffman, J. W., and Lowe, J. (1998) Differential blockade of the antinociceptive effects of centrally administered cannabinoids by SR141716A. *J. Pharmacol. Exp. Ther.* **286,** 1301–1308.
75. Mallet, P. E. and Beninger, R. J. (1998) The cannabinoid CB1 receptor antagonist SR141716A attenuates the memory impairment produced by Δ^9-tetrahydrocannabinol or anandamide. *Psychopharmacol. Berl.* **140,** 11–19.
76. Costa, B., Vailati, S., and Colleoni, M. (1999) SR 141716A, a cannabinoid receptor antagonist, reverses the behavioural effects of anandamide-treated rats. *Behav. Pharmacol.* **10,** 327–331.
77. Williams, C. M. and Kirkham, T. C. (1999) Anandamide induces overeating: mediation by central cannabinoid (CB1) receptors. *Psychopharmacol. Berl.* **143,** 315–317.
78. Ameri, A., and Simmet, T. (2000) Effects of 2-arachidonylglycerol, an endogenous cannabinoid, on neuronal activity in rat hippocampal slices. *Naunyn Schmiedebergs Arch. Pharmacol.* **361,** 265–272.
79. Ameri, A., Wilhelm, A., and Simmet, T. (1999) Effects of the endogeneous cannabinoid, anandamide, on neuronal activity in rat hippocampal slices. *Br. J. Pharmacol.* **126,** 1831–1839.
80. Adams, I. B., Compton, D. R., and Martin, B. R. Assessment of anandamide interaction with the cannabinoid brain receptor: SR 141716A antagonism studies in mice and autoradiographic analysis of receptor binding in rat brain. *J. Pharmacol. Exp. Ther.* **284,** 1209–1217.
81. Wiley, J. L., Mahadevan, A., Razdan, R. K., and Martin, B. R. (2001) Anandamide antagonist: do pharmacokinetics play a role? *2001 Symposium on the Cannabinoids, Burlington, Vermont, International Cannabinoid Research Society,* p. 28.
82. Nakamura-Palacios, E.-M., Moerschbaecher, J. M., and Barker, L. A. (1999) The pharmacology of SR141716A; a review. *CNS Drug Reviews* **5,** 43–58.
83. Ledent, C., Valverde, O., Cossu, G., Petitet, F., Aubert, J. F., Beslot, F., Bohme, G. A., Imperato, A., Pedrazzini, T., Roques, B. P., Vassart, G., Fratta, W., and Parmentier, M. (1999) Unresponsiveness to cannabinoids and reduced addictive effects of opiates in CB1 receptor knockout mice. *Science* **283,** 401–404.
84. Zimmer, A., Zimmer, A. M., Hohmann, A. G., Herkenham, M., and Bonner, T. I. (1999) Increased mortality, hypoactivity, and hypoalgesia in cannabinoid CB1 receptor knockout mice. *Proc. Natl. Acad. Sci. USA* **96,** 5780–5785.
85. Tjolsen, A. and Hole, K. (1997) Animal models of analgesia, iIn *The Pharmacology of Pain (Handbook of Experimental Pharmacology),* Vol. 130 (Dickenson, L. A. and Besson, J.-M., eds.), Springer-Verlag, Heidelberg, pp. 1–19.
86. Martin, B. R. and Lichtman, A. H. (1998) Cannabinoid transmission and pain perception. *Neurobiol. Dis.* **5,** 447–461.
87. Di Marzo, V., Breivogel, C. S., Tao, Q., Bridgen, D.-T., Razdan, R. K., Zimmer, A. M., Zimmer, A., and Martin, B.-R. (2000) Levels, metabolism, and pharmacological activity of anandamide in CB(1) cannabinoid receptor knockout mice: evidence for non-CB(1), non-CB(2) receptor-mediated actions of anandamide in mouse brain. *J. Neurochem.* **75,** 2434–2444.

88. Felder, C. C., Briley, E. M., Axelrod, J., Simpson, J. T., Mackie, K., and Devane, W. (1993) Anandamide, an endogenous cannabimimetic eicosanoid, binds to the cloned human cannabinoid receptor and stimulates receptor-mediated signal transduction. *Proc. Natl. Acad. Sci. USA* **90,** 7656–7660.
89. Lerner, R. A. (1997) A hypothesis about the endogenous analogue of general anesthesia. *Proc Natl. Acad. Sci. USA* **94,** 13,375–13,377.
90. Venance, L., Piomelli, D., Glowinski, J., and Giaume, C. (1995) Inhibition by anandamide of gap junctions and intercellular calcium signalling in striatal astrocytes. *Nature* **376,** 590–594.
91. Boger, D.-L., Sato, H., Lerner, A.-E., Guan, X., and Gilula, N.-B. (1999) Arachidonic acid amide inhibitors of gap junction cell-cell communication. *Bioorg. Med. Chem. Lett.* **9,** 1151–1154.
92. Fan, P. (1995) Cannabinoid agonists inhibit the activation of 5-HT3 receptors in rat nodose ganglion neurons. *J. Neurophysiol.* **73,** 907–910.
93. Mechoulam, R., Hanus, L., and Fride, E. (1998) Towards cannabinoid drugs—revisited. *Prog. Med. Chem.* **35,** 199–243.
94. Kimura T., Ohta T., Watanabe K., Yoshimura H., and Yamamoto I. (1998) Anandamide, an endogenous cannabinoid receptor ligand, also interacts with 5-hydroxytryptamine (5-HT) receptors. *Biol. Pharm. Bull.* **21,** 224–226.
95. Cheer, J. F., Cadogan, A. K., Marsden, C. A., Fone, K. C., and Kendall, D. A. (1999) Modification of 5-HT2 receptor mediated behaviour in the rat by oleamide and the role of cannabinoid receptors. *Neuropharmacology* **38,** 533–541.
96. Basile, A. S., Hanus, L., and Mendelson, W. B. (1999) Characterization of the hypnotic properties of oleamide. *Neuroreport* **10,** 947–951.
97. Hajos, N., Ledent, C. C., and Freund, T. F.(2001) Novel cannabinoid-sensitive receptor mediates inhibition of glutamatergic synaptic transmisson in the hippocampus. *Neuroscience* **106,** 1–4.
98. Contos, J. J., Fukushima, N., Weiner, J. A., Kaushal, D., and Chun, J. (2000) Requirement for the lpA1 lysophosphatidic acid receptor gene in normal suckling behavior. *Proc Natl Acad Sci USA* **97,** 13,384–13,389.
99. Fride, E., Ginzburg, Y., Breuer, A., Bisogno, T., Di Marzo, V., and Mechoulam, R. (2001) Critical role of the endogenous cannabinoid system in mouse pup suckling and growth. *Eur. J. Pharmacol.* **419,** 207–214.
100. Derkinderen, P., Ledent, C., Parmentier, M., and Girault, J.-A. (2001) Cannabinoids activate p38 mitogen-activated protein kinases through CB1 receptors in hippocampus. *J. Neurochem.* **77,** 957–960.
101. Ahluwalia J., Urban L., Capogna M., Bevan S., and Nagy I. (2000) Cannabinoid 1 receptors are expressed in nociceptive primary sensory neurons. *Neuroscience* **100,** 685–688.
102. Mezey, E., Toth, Z. E., Cortright, D. N., Arzubi, M. K., Krause, J.-E., Elde, R., Guo, A., Blumberg, P.-M., and Szallasi, A. (2000) Distribution of mRNA for vanilloid receptor subtype 1 (VR1), and VR1-like immunoreactivity, in the central nervous system of the rat and human. *Proc. Natl. Acad. Sci. USA* **97,** 3655–3660.
103. Di Marzo, V., Bisogno, T., and De Petrocellis, L. (2001) Anandamide: some like it hot. *Trends Pharmacol. Sci.* **22,** 346–349.
104. Ameri , A. (1999) The effects of cannabinoids on the brain. *Prog. Neurobiol.*.**58,** 315–348.
105. Smart, D. and Jerman, J.-C. (2000) Anandamide: an endogenous activator of the vanilloid receptor. *Trends Pharmacol. Sci.* **21,** 134.
106. Zygmunt, P. M., Petersson, J., Andersson, D. A., Chuang, H., Sorgard, M., Di Marzo, V., Julius, D., and Hogestatt, E.-D. (1999) Vanilloid receptors on sensory nerves mediate the vasodilator action of anandamide. *Nature* **400,** 452–457.
107. Szolcsanyi, J. (2000) Anandamide and the question of its functional role for activation of capsaicin receptors. *Trends Pharmacol. Sci.* **21,** 203,204.
108. Szolcsanyi, J. (2000) Are cannabinoids endogenous ligands for the VR1 capsaicin receptor? *Trends Pharmacol. Sci.* **21,** 41,42.

109. Zygmunt, P. M., Julius, I., Di Marzo, V., and Hogestatt, E. D. (2000) Anandamide—the other side of the coin. *Trends Pharmacol. Sci.* **21,** 43–44.
110. De Petrocellis, L., Bisogno, T., Maccarrone, M., Davis, J.-B., Finazzi-Agro, A., and Di Marzo, V. (2001) The activity of anandamide at vanilloid VR1 receptors requires facilitated transport across the cell membrane and is limited by intracellular metabolism. *J. Biol. Chem.* **276,** 12856–12863.
111. Breivogel, C.-S., Griffin, G., Di Marzo, V., and Martin, B.-R. (2001) Evidence for a new G protein-coupled cannabinoid receptor in mouse brain. *Mol. Pharmacol.* **60,** 155–163.
112. Monory, K., Tzavara, E. Th., Ledent, C., Parmentier, M., and Hanoune, J. (2001) Non-conventional cannabinoid stimulated [^{135}S]GTP S binding in CB_1 knockout mice. *2001 Symposium on the Cannabinoids, Burlington, Vermont, International Cannabinoid Research Society,* p. 63.
113. Zias, J., Stark, H., Sellgman, J., Levy, R., Werker, E., Breuer, A., and Mechoulam, R. (1993) Early medical use of cannabis. *Nature* **363,** 215.
114. Walker, J. M., Huang, S. M., Strangman, N. M., Tsou, K., and Sanudo-Pena, M. C. (1999) Pain modulation by release of the endogenous cannabinoid anandamide. *Proc. Natl. Acad. Sci. USA* **96,** 12198–12203.
115. Barbaro, N. M. (1988) Studies of PAG/PVG stimulation for pain relief in humans. *Prog. Brain Res.* **77,** 165–173.
116. Reichling, D.-B., Kwiat, G. C., and Basbaum, A. I. (1988) Anatomy, physiology and pharmacology of the periaqueductal gray contribution to antinociceptive controls. *Prog. Brain. Res.* **77,** 31–46.
117. Richardson, J. D., Aanonsen, L., and Hargreaves, K. M. (1997) SR 141716A, a cannabinoid receptor antagonist, produces hyperalgesia in untreated mice. *Eur. J. Pharmacol.* **319,** R3–R4.
118. Herkenham, M. (1995) Localization of cannabinoid receptors in brain and periphery. In *Cannabinoid Receptors* (R. Pertwee, ed.), pp. 145–166, Academic Press, London, UK.
119. Pertwee, R. G. (1997) Pharmacology of cannabinoid CB1 and CB2 rcccptors. *Pharmacol. Ther.* **74,** 129–180.
120. Piomelli, D., Giuffrida, A., Calignano, A., and Rodriguez-de-Fonseca, F. (2000) The endocannabinoid system as a target for therapeutic drugs. *Trends Pharmacol. Sci.* **21,** 218–224.
121. Sanudo-Pena, M.-C., Romero, J., Seale, G.-E., Fernandez-Ruiz, J.-J., and Walker, J.-M. (2000) Activational role of cannabinoids on movement. *Eur. J. Pharmacol.* **391,** 269–274.
122. Chen J., Paredes, W., Lowinson, J. H., and Gardner, E. L. (1990) Delta 9-tetrahydrocannabinol enhances presynaptic dopamine efflux in medial prefrontal cortex. *Eur. J. Pharmacol.* **190,** 259–262.
123. Mailleux, P., and Vanderhaeghen, J. J. (1993) Dopaminergic regulation of cannabinoid receptor mRNA levels in the rat caudate-putamen: an in situ hybridization study.*J. Neurochem.* **61,** 1705–1712.
124. Sanudo-Pena, M. C., Patrick, S. L., Patrick, R. L., and Walker, J. M. (1996) Effects of intranigral cannabinoids on rotational behavior in rats: interactions with the dopaminergic system. *Neurosci. Lett.* **206,** 21–24.
125. Sanudo-Pena, M. C., Force, M., Tsou, K., Miller, A. S., and Walker, J. M. (1998) Effects of intrastriatal cannabinoids on rotational behavior in rats: interactions with the dopaminergic system. *Synapse* **30,** 221–226.
126. Giuffrida, A., Parsons, L.H., Kerr, T.M., Rodriguez de Fonseca, F.R., Navarro, M., and Piomelli, D. (1999) Dopamine activation of endogenous cannabinoid signaling in dorsal striatum. *Nature Neurosci.* **2,** 358–363.
127. Beltramo, M., Rodriguez-de-Fonseca, F., Navarro, M., Calignano, A., Gorriti, M. A., Grammatikopoulos, G., Sadile, A.-G., Giuffrida, A., and Piomelli, D. (2000) Reversal of dopamine D(2) receptor responses by an anandamide transport inhibitor. *J. Neurosci.* **20,** 3401–3407.

128. Maneuf, Y. P., Crossman, A. R., and Brotchie, J. M. (1997) The cannabinoid receptor agonist WIN 55,212-2 reduces D2, but not D1, dopamine receptor-mediated alleviation of akinesia in the reserpine-treated rat model of Parkinson's disease. *Exp. Neurol.* **148,** 265–270.
129. Richfield, E. K. and Herkenham, M. (1994) Selective vulnerability in Huntington's disease: preferential loss of cannabinoid receptors in lateral globus pallidus. *Ann. Neurol.* **36,** 577.
130. Muller-Vahl, K.-R., Schneider, U., Kolbe, H., and Emrich, H.-M. (1999) Treatment of Tourette's syndrome with delta-9-tetrahydrocannabinol. *Am. J. Psychiatry* **156,** 495.
131. Baker, D., Pryce, G., Croxford, J. L., Brown, P., Pertwee, R. G., Huffman, J. W., and Layward, L. (2000) Cannabinoids control spasticity and tremor in a multiple sclerosis model. *Nature* **404,** 84–87.
132. Emrich, H. M., Leweke, F. M., and Schneider, U. (1997) Towards a cannabinoid hypothesis of schizophrenia: cognitive impairments due to dysregulation of the endogenous cannabinoid system. *Pharmacol. Biochem. Behav.* **56,** 803–807.
133. Gorriti, M. A., Rodriguez-de-Fonseca, F., Navarro, M., and Palomo, T. (1999) Chronic (−)-delta9-tetrahydrocannabinol treatment induces sensitization to the psychomotor effects of amphetamine in rats. *Eur. J. Pharmacol.* **365,** 133–142.
134. Leweke, F. M., Giuffrida, A., Wurster, U., Emrich, H. M., and Piomelli, D. (1999) Elevated endogenous cannabinoids in schizophrenia. *Neuroreport* **10,** 1665–1669.
135. Giuffrida, A. and Piomelli, D. (2000) The endocannabinoid system: a physiological perspective on its role in psychomotor control. *Chem. Phys. Lipids* **108,** 151–158.
136. Herkenham, M., Lynn, A. B., Johnson, M. R., Melvin, L. S., de-Costa, B. R., and Rice, K. C. (1991) Characterization and localization of cannabinoid receptors in rat brain: a quantitative in vitro autoradiographic study. *J. Neurosci.* **11,** 563–583.
137. Deadwyler, S. A., Hampson, R. E., Mu J., Whyte, A., and Childers, S. (1995) Cannabinoids modulate voltage sensitive potassium A-current in hippocampal neurons via a cAMP-dependent process. *J. Pharmacol. Exp. Ther.* **273,** 734–743.
138. Terranova, J. P., Michaud, J. C., Le-Fur, G., and Soubrie, P. (1995) Inhibition of long-term potentiation in rat hippocampal slices by anandamide and WIN55212-2: reversal by SR141716 A, a selective antagonist of CB1 cannabinoid receptors. *Naunyn Schmiedebergs Arch. Pharmacol.* **352,** 576–579.
139. Stella, N., Schweitzer, P., and Piomelli, D. (1997) A second endogenous cannabinoid that modulates long-term potentiation. *Nature* **388,** 773–778.
140. Mallet, P. E. and Beninger, R. J. (1996) The endogenous cannabinoid receptor agonist anandamide impairs memory in rats. *Behav. Pharmacol.* **7,** 276–284.
141. Hampson, R.-E. and Deadwyler, S.-A. (2000) Cannabinoids reveal the necessity of hippocampal neural encoding for short-term memory in rats. *J. Neurosci.* **20,** 8932–8942.
142. Murillo-Rodriguez, E., Sanchez-Alavez, M., Navarro, L., Martinez-Gonzalez, D., Drucker-Colin, R., and Prospero-Garcia, O. (1998) Anandamide modulates sleep and memory in rats. *Brain Res.* **812,** 270–274.
143. Castellano, C., Ventura, R., Cabib, S., and Puglisi-Allegra, S. (1999) Strain-dependent effects of anandamide on memory consolidation in mice are antagonized by naltrexone. *Behav. Pharmacol.* **10,** 453–457.
144. Barinaga, M. (2001) Neurobiology. How cannabinoids work in the brain. *Science* **291,** 2530–2531.
145. Herkenham, M., Lynn, A. B., Little, M. D., Johnson, M. R., Melvin, L. S., de Costa, B. R., and Rice, K. C. (1990) Cannabinoid receptor localization in brain. *Proc. Natl. Acad .Sci. USA* **87,** 1932–1936.
146. Thierry, A.-M., Tassin, J.-P., Blanc, G., and Glowinski, J. (1978) Studies on mesocortical dopamine systems. *Adv. Biochem. Psychopharmacol.* **19,** 205–216.
147. Jentsch, J. D., Andrusiak, E., Tran, A., Bowers, M. B. Jr., and Roth, R. H. (1997) Delta 9-tetrahydrocannabinol increases prefrontal cortical catecholaminergic utilization and impairs

spatial working memory in the rat: blockade of dopaminergic effects with HA966. *Neuropsychopharmacology* **16,** 426–432.
148. Auclair, N., Otani, S., Soubrie P., and Crepel, F. (2000) Cannabinoids modulate synaptic strength and plasticity at glutamatergic synapses of rat prefrontal cortex pyramidal neurons. *J. Neurophysiol.* **83,** 3287–3293.
149. Fride, E. and Sanudo-Pena, M. C. (2002) Cannabinoids and endocannabinoids: behavioural and developmental aspects, in *Biology of Marijuana* (Onaivi, E., ed.), Harwood, London, UK, in press.
150. Fried, P. A., Watkinson, B., and Gray, R. (1998) Differential effects on cognitive functioning in 9- to 12-year olds prenatally exposed to cigarettes and marihuana. *Neurotoxicol. Teratol.* **20,** 293–306.
151. Santucci, V., Storme, J. J., Soubrie, P., and Le-Fur, G. (1996) Arousal-enhancing properties of the CB1 cannabinoid receptor antagonist SR 141716A in rats as assessed by electroencephalographic spectral and sleep-waking cycle analysis. *Life Sci.* **58,** PL103–PL110.
152. Williams, C. M., Rogers, P. J., and Kirkham, T. C. (1998) Hyperphagia in pre-fed rats following oral Δ^9-THC. *Physiol. Behav.* **65,** 343–346.
153. Arnone, M., Maruani, J., Chaperon, F., Thiebot, M. H., Poncelet, M., Soubrie, P., and Le-Fur, G. (1997) Selective inhibition of sucrose and ethanol intake by SR 141716, an antagonist of central cannabinoid (CB1) receptors. *Psychopharmacol. Berl.* **132,** 104–106.
154. Colombo, G., Agabio, R., Diaz, G., Lobina, C., Reali, R., and Gessa, G. L. (1998) Appetite suppression and weight loss after the cannabinoid antagonist SR 141716. *Life Sci.* **63,** PL113–PL117.
155. Simiand, J., Keane, M., Keane, P.-E., and Soubrie, P. (1998) SR 141716, a CB1 cannabinoid receptor antagonist, selectively reduces sweet food intake in marmoset. *Behav. Pharmacol.* **9,** 179–181.
156. Le Fur, G., Arnone, M., Rinaldi-Carmona, M., Barth, F., and Heshmati, H. (2001) SR141716A, a selective antagonist of CB1 receptors and obesity. *2001 Symposium on the Cannabinoids, Burlington, Vermont, International Cannabis Research Society,* p. 101.
157. Mechoulam, R. and Fride, E. (2001) A hunger for cannabinoids. *Nature* **410,** 763–765.
158. Di Marzo, V., Goparaju, S.-K., Wang, L., Liu, J., Batkai, S., Jarai, Z., Fezza, F., Miura, G.-I., Palmiter, R.-D., Sugiura, T., and Kunos, G. (2001) Leptin-regulated endocannabinoids are involved in maintaining food intake. *Nature* **410,** 822–825.
159. Rodriguez-de-Fonseca, F. R., Navarro, M., Gomez, R., Escuredo, L., Nava, F., Fu, J., Murillo-Rodriguez, E., Giuffrida, A., LoVerme, J., Gaetani, S., Kathuria, S., Gall, C., and Piomelli, D. (2001) An anorexic lipid mediator regulated by feeding. *Nature* **414,** 209–212.
160. Fride, E., Bisogno, T., Di Marzo, V., Bayewitch, M., Vogel, Z. and Mechoulam, R. (1997) Anandamide: mediator of the effects of oleamide (a sleep factor) and of chocolate? *Soc. Neurosci. Abstr.* **23,** 1230.
161. Berrendero, F., Sepe, N., Ramos, J. A., Di Marzo, V., and Fernandez-Ruiz, J. J. (1999) Analysis of cannabinoid receptor binding and mRNA expression and endogenous cannabinoid contents in the developing rat brain during late gestation and early postnatal period. *Synapse* **33,** 181–191.
162. Hansen, H.-S., Moesgaard, B., Hansen, H.-H., and Petersen, G. N. (2000) Acylethanolamines and precursor phospholipids—relation to cell injury. *Chem. Phys. Lipids* **108,** 135–150.
163. Shen, M. and Thayer, S.A. (1998) Cannabinoid receptor agonists protect cultured rat hippocampal neurons from excitotoxicity. *Mol. Pharmacol.* **54,** 459–462.
164. Sinor, A.-D., Irvin, S.-M., and Greenberg, D.-A. (2000) Endocannabinoids protect cerebral cortical neurons from in vitro ischemia in rats. *Neurosci. Lett.* **278,** 157–160.
165. Abood, M. E., Rizvi, G., Sallupudi, N., and McAllister, S. D. (2001) Activation of the CB_1 cannabinoid receptor protects cultured mouse spinal neurons against excitotoxicity. *Neurosci. Lett.* **309,** 197–200.

166. Nagayama, T., Sinor, A. D., Simon, R. P., Chen, J., Graham, S. H., Jin, K., and Greenberg, D. A. (1999) Cannabinoids and neuroprotection in global and focal cerebral ischemia and in neuronal cultures. *J. Neurosci.* **19,** 2987–2995.
167. van-der-Stelt, M., Veldhuis, W.-B., Bar, P.-R., Veldink, G.-A., Vliegenthart, J.-F., and Nicolay, K. (2001) Neuroprotection by Delta9-tetrahydrocannabinol, the main active compound in marijuana, against ouabain-induced in vivo excitotoxicity. *J. Neurosci.* **21,** 6475–6479.
168. Panikashvili, D., Simeonidou, C., Ben-Shabat, S., Hanus, L., Breuer, A., Mechoulam, R., and Shohami, E. (2001) An endogenous cannabinoid (2-AG) is neuroprotective after brain injury. *Nature* **413,** 527–531.
169. Hansen, H.-H., Ikonomidou, C., Bittigau, P., Lastres-Becker, I., Berrendero, F., Manzanares, J., Ikonomidou, C., Schmid, H. H. Fernandez-Ruiz, J. J., and Hansen, H.-S. (2001) Accumulation of the anandamide precursor and other N-acylethanol amine phospholipin infant rat models of in vivo necrotic and apoptotic neuronal death *J. Neurochem.* **76,** 39–46.
170. Hansen, H.-H., Schmidt, P. C., Bittigau, P., Hansen, S.-H., and Hansen, H.-S. (2001) Anandamide, but not 2-arachidonoyl glycerol, accumulates during in vivo neurodegenearion. *J. Neurochem.* **78,** 1415–1427.
171. Shohami, E. and Mechoulam, R. (2000) Dexanabinol (HU-211): a nonpsychotic cannaboid with neuroprotective properties. *Drug Dev. Res.*
172. Knoller, N., Levi, L., Shoshan, I., Ereichenthal, E., Razon, N., Rappaport, Z. H., et al. (2002) Deaxanabinol (HU-211) in the treatment of severe closed head injury: a randomized, placebo-controlled, phase II clinical trial. *Crit. Care Med.* **30,** 548–553.
173. McLaughlin, C. R. and Abood, M. E. (1993) Developmental expression of cannabinoid receptor mRNA. *Brain Res. Dev. Brain Res.* **76,** 75–78.
174. Belue, R. C., Howlett, A. C., Westlake, T. M., and Hutchings, D. E. (1995) The ontogeny of cannabinoid receptors in the brain of postnatal and aging rats. *Neurotoxicol. Teratol.* **17,** 25–30.
175. Rodriguez-de-Fonseca, F., Ramos, J. A., Bonnin, A., and Fernandez-Ruiz, J. J. (1993) Presence of cannabinoid binding sites in the brain from early postnatal ages. *Neuroreport* **4,** 135–138.
176. Buckley, N. E., Hansson, G., Harta, G., and Mezey, E. (1998) Expression of the CB1 and CB2 receptor receptor messenger RNAs during embryonic development in the rat. *Neuroscience* **82,** 1131–1149.
177. Fernandez-Ruiz, J., Berrendero, F., Hernandez, M. L., and Ramos, J. A. (2000) The endogenous cannabinoid system and brain development. *Trends Neurosci.* **23,** 14–20.
178. Romero, J., Garcia-Palomero, E., Berrendero, F., Garcia-Gil, L., Hernandez, M. L., Ramos, J. A., and Fernandez-Ruiz, J. J. (1997) Atypical location of cannabinoid receptors in white matter areas during rat brain development. *Synapse* **26,** 317–323.
179. Maccarrone, M., Attina, M., Bari, M., Cartoni, A., Ledentt, C., and Finazzi-Agro, A. (2001) Anandamide degradation and N-acylethanolamines level in wild-type and CB_1 cannabinoid receptor knockout mice of different ages. *J. Neurochem.* **78,** 339–348.
180. Jones, R. T., Benowitz, N. L., and Herning, R. I. (1981) Clinical relevance of cannabis tolerance and dependence. *J. Clin. Pharmacol.* **21,** 143S–152S.
181. Adams, M. D., Chait, L. D., and Earnhardt, J. T. (1976) Tolerance to the cardiovascular effects of delta9-tetrahydrocannabinol in the rat. *Br. J. Pharmacol.* **56,** 43–48.
182. Fitton, A. G. and Pertwee, R. G. (1982) Changes in body temperature and oxygen consumption rate of conscious mice produced by intrahypothalamic and intracerebroventricular injections of delta 9-tetrahydrocannabinol. *Br. J. Pharmacol.* **75,** 409–414.
183. Jarbe, T. U. (1978) delta9-Tetrahydrocannabinol: tolerance after noncontingent exposure in rats. *Arch. Int. Pharmacodyn. Ther.* **231,** 49–56.
184. Webster, C. D., LeBlanc, A. E., Marshman, J. A., and Beaton, J. M. (1973) Acquisitions and loss of tolerance to 1-D9-trans-tetrahydrocannabinol in rats on an avoidance schedule. *Psychopharmacol. Bull.* **30,** 217–226.
185. Iversen, L. L. (2000) *The Science of Marijuana.* Oxford University Press, New York.

186. Breivogel, C. S., Childers, S. R., Deadwyler, S. A., Hampson, R. E., Vogt, L. J., and Sim-Selley, L. J. (1999) Chronic delta9-tetrahydrocannabinol treatment produces a time-dependent loss of cannabinoid receptors and cannabinoid receptor-activated G proteins in rat brain. *J. Neurochem.* **73,** 2447–2459.
187. Romero, J., Garcia-Palomero, E., Castro, J. G., Garcia-Gil, L., Ramos, J. A., and Fernandez-Ruiz, J. J. (1997) Effects of chronic exposure to delta9-tetrahydrocannabinol on cannabinoid receptor binding and mRNA levels in several rat brain regions. *Brain Res. Mol. Brain Res.* **46,** 100–108.
188. Crowley, T. J., Macdonald, M. J., Whitmore, E. A., and Mikulich, S. K. (1998) Cannabis dependence, withdrawal, and reinforcing effects among adolescents with conduct symptoms and substance use disorders. *Drug Alcohol Depend.* **50,** 27–37.
189. Gardner, E. L. and Vorel, S. R. (1998) Cannabinoid transmission and reward-related events. *Neurobiol. Dis.* **5,** 502–533.
190. Ashton, C. H. (1999) Adverse effects of cannabis and cannabinoids. *Br. J. Anaesth.* **83,** 637–649.
191. Gardner, E. L. and Lowinson, J. H. (1991) Marijuana's interaction with brain reward systems: update 1991. *Pharmacol. Biochem. Behav.* **40,** 571–580.
192. Diana, M., Melis, M., Muntoni, A. L., and Gessa, G. L. (1998) Mesolimbic dopaminergic decline after cannabinoid withdrawal. *Proc. Natl. Acad. Sci. USA* **95,** 10,269–10,273.
193. French, E. D. (1997) delta9-Tetrahydrocannabinol excites rat VTA dopamine neurons through activation of cannabinoid CB1 but not opioid receptors. *Neurosci. Lett.* **226,** 159–162.
194. Gessa, G. L., Melis, M., Muntoni, A. L., and Diana, M. (1998) Cannabinoids activate mesolimbic dopamine neurons by an action on cannabinoid CB1 receptors. *Eur. J. Pharmacol.* **341,** 39–44.
195. Tanda, G., Pontieri, F. E., and Di-Chiara, G. (1997) Cannabinoid and heroin activation of mesolimbic dopamine transmission by a common mu1 opioid receptor mechanism. *Science* **276,** 2048–2050.
196. Lepore, M., Liu, X., Savage, V., Matalon, D., and Gardner, E. L. (1996) Genetic differences in delta 9-tetrahydrocannabinol-induced facilitation of brain stimulation reward as measured by a rate-frequency curve-shift electrical brain stimulation paradigm in three different rat strains. *Life Sci.* **58,** PL365–PL372.
197. Lepore, M., Vorel, S. R., Lowinson, J., and Gardner, E. L. (1995) Conditioned place preference induced by delta 9-tetrahydrocannabinol: comparison with cocaine, morphine, and food reward. *Life. Sci.* **56,** 2073–2080.
198. Chaperon, F., Soubrie, P., Puech, A. J., and Thiebot, M. H. (1998) Involvement of central cannabinoid (CB1) receptors in the establishment of place conditioning in rats. *Psychopharmacol. Berl.* **135,** 324–332.
199. Parker, L. A. and Gillies, T. (1995) THC-induced place and taste aversion in Lewis and Sprague-Dawley rats. *Behav. Neurosci.* **109,** 71–78.
200. McGregor, I. S., Issakidis, C. N., and Prior, G. (1996) Aversive effects of the synthetic cannabinoid CP 55,940 in rats. *Pharmacol. Biochem. Behav.* **53,** 657–664.
201. Sañudo-Peña, M. C., Tsou, K., Delay, E. R., Hohman, A. G., Force, M., and Walker, J. M. (1997) Endogenous cannabinoids as an aversive or counter-rewarding system in the rat. *Neurosci. Lett.* **223,** 125–128.
202. Chakrabarti, A., Ekuta, J. E., and Onaivi, E. S. (1998) Neurobehavioral effects of anandamide and cannabinoid receptor gene expression in mice. *Brain Res. Bull.* **45,** 67–74.
203. Onaivi, E. S., Chakrabarti, A., Gwebu, E. T., and Chaudhuri, G. (1995) Neurobehavioral effects of delta 9-THC and cannabinoid (CB1) receptor gene expression in mice. *Behav. Brain Res.* **72,** 115–125.

204. Rodriguez-de-Fonseca, F., Rubio, P., Menzaghi, F., Merlo-Pich, E., Rivier, J., Koob, G. F., and Navarro, M. (1996) Corticotropin-releasing factor (CRF) antagonist [D-Phe12,Nle21,38,C alpha MeLeu37]CRF attenuates the acute actions of the highly potent cannabinoid receptor agonist HU-210 on defensive-withdrawal behavior in rats. *J. Pharmacol. Exp. Ther.* **276,** 56–64.
205. Martellotta, M. C., Cossu, G., Fattore, L., Gessa, G. L., and Fratta, W. (1998) Self-administration of the cannabinoid receptor agonist WIN 55,212-2 in drug-naive mice. *NeuroScience* **85,** 327–330.
206. Gallate, J. E., Saharov, T., Mallet, P. E., and McGregor, I. S. (1999) Increased motivation for beer in rats following administration of a cannabinoid CB1 receptor agonist. *Eur. J. Pharmacol.* **370,** 233–240.
207. De Vries, T. J., Shaham,Y., Homberg, J. R., Crombag, H., Schuurman ,K., Dieben, J., Vanderschuren, L. J. and Scoffelmeer, A. N. (2001) A cannabinoid mechanism in relapse to cocaine seeking. *Nature Med.* **7,** 1151–1154.
208. Onaivi, E. S., Green, M. R., and Martin, B. R. (1990) Pharmacological characterization of cannabinoids in the elevated plus maze. *J. Pharmacol. Exp. Ther.* **253,** 1002–1009.

11

Integration of Molecular and Behavioral Approaches to Evaluate Cannabinoid Dependence

Dana E. Selley, Aron H. Lichtman, and Billy R. Martin

1. Introduction

1.1. The Controversy Over Cannabinoid Dependence

The acute behavioral and pharmacological effects of cannabis (marijuana) have been well characterized and are in relatively little dispute. On the other hand, there is less agreement regarding the consequences of repetitive exposure to cannabis, particularly when it comes to development of dependence. Two events that typically occur concomitantly following chronic exposure to many psychoactive drugs are tolerance and dependence. Although these two events are likely to be mediated through distinct mechanisms, they both demonstrate that some form of adaptation has occurred. As for development of tolerance to cannabis and its main psychoactive constituent, Δ^9-tetrahydrocannabinol (THC), there is ample evidence for its occurrence in both laboratory animals and humans. Actually, animal studies are in good agreement on the development of tolerance to the effects of THC following repeated exposure *(1)*. There is also evidence that chronic heavy cannabis smokers develop tolerance to its subjective and cardiovascular effects *(2)*. Tolerance develops to a variety of THC's effects, following oral administration, including cannabinoid-induced decreases in cardiovascular and autonomic functions, increases in intraoccular pressure, sleep disturbances, and mood changes *(3)*. It should be pointed out that high doses of THC were required for a sustained period of time in order to achieve tolerance to the behavioral effects.

As for cannabis dependence, there is both clinical and epidemiological evidence in heavy, chronic users. Some of these individuals report problems in controlling cannabis use, and they continue to use despite experiencing adverse personal consequences *(4–6)*. It has been reported that the cannabis dependence syndrome is analogous to the alcohol dependence syndrome *(7–9)*. Cannabis dependence as defined in DSM-III was the most common form of illicit drug dependence in one large epidemiological study *(9)*. The risk of becoming dependent on cannabis is probably more like that for alcohol than for nicotine or the opioids *(10)*, with around 10% of those who ever use cannabis meeting criteria for dependence *(10,11)*. Persons who use cannabis on a daily basis over periods of weeks to months are at greatest risk of becoming dependent. It has been estimated that the risk of dependence among near-daily cannabis users (according to approximated DSM-III criteria) is one in three *(12)*.

From: *Molecular Biology of Drug Addiction*
Edited by: R. Maldonado © Humana Press Inc., Totowa, NJ

Although there is little doubt that cannabis dependence involves continued drug taking despite the presence of adverse consequences as discussed above, the extent to which a physical withdrawal syndrome is associated with cannabis dependence is more controversial. It is safe to conclude that cessation of chronic cannabis exposure does not result in severe withdrawal symptoms *(13)*. The development of tolerance and dependence has been studied under rigorous and controlled conditions *(13,14)*. In one study, cessation of treatment with high doses of marijuana extract or THC led to increased irritability, restlessness, insomnia, anorexia, increased sweating, and mild nausea. Objective symptoms were increased body temperature, weight loss, and hand tremor. Readministration of a marijuana cigarette or oral THC alleviated the objective and subjective effects, suggesting the establishment of a withdrawal symptom. In another study, THC administration for 4 d produced ratings of "high," increased food intake, and decreased verbal interaction among participants. Tolerance developed to the subjective effects of THC but not to its effects on food intake or social behavior. Abstinence from THC produced anxious, depressed, and irritable symptoms, decreased the quantity and quality of sleep, and decreased food intake *(15)*. A similar study conducted with marijuana cigarettes resulted in similar effects and led to the conclusion that abstinence symptoms may play a role in maintaining daily marijuana use, even at levels of use that do not produce tolerance *(15)*.

There are epidemiological data to support marijuana dependence, as reviewed by Hall et al. *(11)*. There are numerous cases in which individuals seek treatment for dependence in which marijuana is the primary cause. These patients typically complained of being unable to stop or decrease their use despite experiencing sleepiness, depression, inability to concentrate, and memorization difficulties. In one large epidemiological study *(16)*, approx 4.4% of the population was diagnosed for marijuana abuse and/or dependence and three-fifths of these met the criteria for dependence. Another group of investigators reported that the majority of marijuana users seeking treatment for marijuana dependence experience symptoms consistent with either moderate or severe dependence *(17)*.

One of the difficulties in establishing the presence of cannabinoid dependence was the lack of reliable animal models. One approach has been to elicit drug-seeking behavior in laboratory animals reminiscent of that in humans. The other strategy is to elicit a physical withdrawal syndrome in animals treated chronically with the test drug. Early attempts either to train animals to self-administer cannabinoids or to produce a withdrawal syndrome after chronic administration were unsuccessful. However, appropriate models have now been developed that demonstrate both cannabinoid self-administration and antagonist-precipitated withdrawal. The purpose of this chapter is to provide an overview of advances in the development of these models, and in our understanding of the biological basis of cannabinoid dependence through the implementation of integrative molecular and behavioral investigations.

1.2. Overview of Cannabinoids and Cannabinoid Receptors

Cannabinoids are a group of approx 60 terpinophenolic 21-carbon-containing compounds present in plants of the *Cannabis* genus, particularly *C. sativa* and *C. indica*. Although cannabis preparations have been used for centuries for medicinal, recreational, or religious purposes, it was not until the mid-1960s that the primary psychoactive ingredient was determined to be (−)-*trans*-Δ^9-tetrahydrocannabinol (THC) *(18,19)*. Subsequently, highly potent and selective synthetic cannabinoid agonists, including HU-210, CP-55,940, and the aminoalkylindole WIN 55,212-2, were developed. The study of the biological effects of these and other cannabimimetic compounds led to the hypothesis that the pharmacological actions of cannabinoids are mediated by the activation of specific receptors. Direct evidence for this hypothesis was obtained in the mid-1980s, when cannabinoid agonists were shown to inhibit the enzymatic synthesis of cyclic adenosine monophosphate (cAMP) in neuroblastoma cells in a manner consistent with a hormonal G-protein-coupled receptor-like mechanism that was blocked by pretreatment of the cells with pertussis toxin *(20)*. Identification of a G-protein-coupled cannabinoid receptor in the brain was obtained through the demonstration of guanine nucleotide-sensitive binding of the radiolabeled THC analog [^{3}H]CP 55,940 to rat brain homogenates *(21)*, and has been confirmed by molecular cloning of the CB1 cannabinoid receptor from a rat cerebral cortex cDNA library *(22)*. Thc subsequent cloning of the CB2 receptor *(23)* revealed the presence of this pharmacologically similar cannabinoid receptor in the immune system. For the purpose of this chapter, discussion will focus on the CB1 receptor because it is the major cannabinoid receptor in the central nervous system (CNS).

The CB1 receptor is a heptahelical G-protein-coupled receptor (GPCR) that is coupled primarily through pertussis toxin-sensitive G_i/G_o-type G proteins to signal transduction processes including inhibition of adenylyl cyclase, activation of K^+ channels, inhibition of Ca^{2+} channels, and stimulation of mitogenic kinases *(20,24–27)*. CB1 receptors are the most abundant GPCR in the brain, with levels approximately 10 times higher than most other GPCRs *(21,28,29)*. Anatomical studies have revealed dense localization of CB1 receptors in the basal ganglia, hippocampus, cerebral cortex, and cerebellum, with lower levels present in many other brain regions including thalamus, hypothalamus, amygdala, and periaqueductal gray *(29–33)*. This distribution is consistent with the behavioral effects of cannabinoids, which include memory impairment, antinociception, catalepsy, hypomotility, and hypothermia, as well as alterations of mood and perception in humans *(1,34)*.

Shortly after the discovery of a cannabinoid receptor, the search began for an endogenous agonist. The first reported endogenous cannabinoid, isolated from organic solvent extracts of porcine brain, was the unsaturated lipid amide, arachidonylethanolamide (also termed anandamide) *(35)*. In addition to activation of cannabinoid receptor-mediated cellular signaling pathways *(36–38)*, anandamide also produces THC-like pharmacological effects in isolated organ preparations, such as the mouse vas deferens *(35)*, and in animals *(39)*. Specific depolarization-induced, Ca^{2+}-dependent release of anandamide, as well as its synthesis via a phospholipase D-catalyzed hydrolysis of a phospholipid precursor, N-arachidonyl-phosphatidylethanolamine, has been demonstrated in neurons *(40,41)*. There is also evidence for transporter-mediated uptake of anandamide by neurons and glia *(40,42,43)*, although a specific anandamide transporter has not yet been identified. Anandamide inactivation occurs via hydrolysis by an intracellular membrane-bound enzyme, fatty acid amide hydrolase (FAAH), which has been cloned *(44)*. FAAH is also responsible for inactivation of endogenous fatty acid-derived putative neurotransmitters, such as the sleep-inducing fatty acid amide, *cis*-9-octadecenoamide (oleamide) *(44,45)*. Another endogenous cannabinoid, 2-arachidonylglycerol, has been identified *(46–48)*, and this compound is also a substrate for FAAH *(49)*. Thus, the endogenous agonists identified thus far for the cannabinoid receptors are arachidonic acid-based compounds that satisfy all or most of the accepted criteria for neurotransmitters. The terpinophenolic derivatives of the cannabis plant, especially THC, may produce their psychoactive effects primarily by mimicking the actions of these "endocannabinoid" neurotransmitters at CB1 cannabinoid receptors in the CNS.

2. Animal Models for Measuring Cannabinoid Dependence

Two general procedures that elicit withdrawal responses are abrupt cessation of chronic drug administration and antagonist challenge in animals during chronic drug administration. Adaptation that occurs during chronic drug administration rebounds either to a mild or severe extent and is manifested by a withdrawal syndrome. Most dependence-inducing drugs produce numerous somatic signs that can be quantified. Interestingly, the same somatic signs can be used to characterize the withdrawal syndrome for several different drugs. Typically, a withdrawal syndrome is characterized by hyperactivity, increased grooming, tremors, shaking activity, piloerection, diarrhea, and so on. The withdrawal signs reported for several cannabinoids are presented in Table 1. Given the appearance of some common withdrawal signs for several different drug classes and the apparent ease of quantifying these signs, it is surprising that there can be considerable discrepancy between similar studies carried out in different laboratories. Factors that contribute to these discrepancies include subtle differences in treatment protocols (dose, route of administration, particular agonist used, treatment duration, etc.), as well as differences in the procedures used to quantify withdrawal (criteria for a given withdrawal sign, etc.). The onset, intensity, and duration of a withdrawal syndrome are subject to the pharmacokinetic and pharmacodynamic characteristics of the drug. Actually, most dependence-inducing drugs require an aggressive treatment regimen to produce a robust withdrawal syndrome because of extensive metabolism in laboratory animals. Both opioids and cannabinoids are examples of drugs that are readily metabolized during chronic treatment regimens. Moreover, highly lipophilic drugs such as the cannabinoids are readily stored in fat depots, which greatly influences the onset and duration of action. For lipophilic drugs such as the cannabinoids, they may be chronically infused *(50)* or injected repeatedly *(51)* in order to produce dependence.

2.1. Abrupt Cannabinoid Withdrawal

The sudden termination of chronic treatment with cannabinoids in several different laboratory animal models has not produced uniform results. Kaymakcalan treated rhesus monkeys for 36 d with THC and observed aggressiveness, hyperirritability, tremors, yawning, photophobia, hallucinatory behavior, and anorexia upon abrupt treatment termination *(52)*. This syndrome appeared 12 h after THC was discontinued and lasted for 5 d. Another approach is to determine whether withdrawal can be measured by using a conditioned behavioral paradigm. Beardsley and co-workers were able to demonstrate marked response-time disruption of food-maintained operant behavior in rhesus monkeys. THC was given continuously for 10 d *(53)*. Overt behavioral signs included aggressiveness, bruxism, and hyperactivity.

An abrupt withdrawal syndrome was also described in rats *(54)*; however, numerous investigators were unable to obtain similar findings upon termination of chronic can-

Table 1
Antagonist-Precipitated Withdrawal Syndrome in Cannabinoid-Dependent Animals

Species	Agonist	Quantified withdrawal signs	References
Mouse	Δ^9-THC	Wet-dog shakes, facial rubbing, ataxia, hunched position, tremor, ptosis, piloerection, mastication, global withdrawal score	*(62)*
		Sniffing, wet-dog shakes, paw tremors piloerection, body tremor	*(61)*
		Paw tremors, head shakes	*(51,60)*
		Paw tremors	*(60)*
		Tremor, wet-dog shakes, ptosis, front paw tremor, ataxia, mastication, hunched posture, sniffing, piloerection	*(64)*
		Tremor, ataxia, mastication, front paw tremor, piloerection, wet-dog shakes	*(63)*
Rat	Δ^9-THC	Facial rubs and wet-dog shakes	*(50,59)*
		Forepaw fluttering, wet-dog shakes, grooming, horizontal activity, vertical activity	*(58)*
		Scratching, facial rubbing, licking, wet-dog shakes	*(71)*
		Suppression of food-reinforced operant responding	*(72)*
	CP 55,940	Turning, chewing, digging	*(65)*
	WIN 55,212-2	Wet-dog shakes, facial rubs	*(56)*
	HU-210	Global withdrawal score	*(69)*
	Anandamide	No withdrawal signs observed	*(66)*
		Ptosis, arched back, wet-dog shakes, piloerection, forepaw fluttering, global abstinence score	*(67)*
Dog	Δ^9-THC	Avoidance of human contact, trembling, shaking, shivering, exaggerated reaction to auditory and visual stimuli, excessive salivation, vomiting, diarrhea, global withdrawal score	*(73)*

nabinoid exposure. McMillan and co-workers demonstrated that daily dosing of dogs with THC did not result in overt behavioral effects upon cessation of treatment *(55)*. These same researchers also treated pigeons with daily intramuscular injections of very high doses of THC (up 180 mg/kg) and failed to detect withdrawal signs when the drug exposure was terminated *(55)*. There have also been numerous failures in characterizing cannabinoid dependence in rodents. It is probably the long half-life of THC that makes it particularly challenging to observe abrupt withdrawal. Indeed, cessation of chronic infusion of WIN 55,212-2 in rats led to spontaneous withdrawal 24 h later, as shown in Table 1 *(56)*. Although the pharmacokinetics of WIN 55,212-2 and THC have not been compared in this regimen, it could explain the differences in spontaneous withdrawal between these two drugs. Regardless of the reason for these differences, the results obtained with WIN 55,212-2 suggests that spontaneous withdrawal can occur with cannabinoids.

2.2. Precipitated Cannabinoid Withdrawal

The development of the selective CB1 cannabinoid receptor antagonist, SR 141716A, represented a major breakthrough for cannabinoid research *(57)*. This antagonist has been found to block most pharmacological effects of cannabinoids in different animal species. SR 141716A has been particularly useful for characterizing cannabinoid withdrawal in laboratory animals. In contrast to studies confined to employing abrupt withdrawal techniques, SR 141716A has been demonstrated to elicit reproducible and quantifiable withdrawal reactions following repeated cannabinoid administration in a variety of laboratory animals including mice, rats, and dogs (Table 1). A challenge in quantifying withdrawal is that the syndrome appears in rapidly alternating sequences of different behaviors *(58)*. Moreover, specific withdrawal responses are also affected by methodological procedures employed by different investigators relating to differences in species, strain, agonist, dosing regimen, and aspects of the testing environment. A global abstinence score in which different signs are scored and given a weight depending on the frequency or magnitude of the response leads to significant differences between withdrawal behavior in rats, mice, and dogs, as illustrated in Table 1.

Rats exhibit a variety of somatic withdrawal signs that range in intensity, including wet-dog shakes, facial rubs, horizontal and vertical activity, forepaw fluttering, chewing, tongue rolling, paw shakes, and head shakes, retropulsion, myoclonic spasms, front paw treading, and eyelid ptosis *(58,59)*. In mice, paw tremors and head shakes were found to be the most reliable cannabinoid withdrawal signs in some studies *(51,60)*, whereas others found these signs in addition to hunched posture, mastication, sniffing, and piloerection *(61–64)*. On the other hand, precipitated scratching has been observed in THC-dependent Swiss Webster mice, but not in other strains *(60)*. Actually, SR 141716A alone elicits scratching, an effect that is generally decreased in cannabinoid-dependent mice *(51,65)*. In contrast, writhing and ptosis occurred only sporadically, and diarrhea and jumping are not part of the precipitated withdrawal syndrome. In THC-dependent dogs, SR 141716A-precipitated yet another unique pattern of withdrawal signs that included excessive salivation, vomiting, diarrhea, restless behavior, trembling, and decreases in social behavior (Table 1). These observations taken together indicate that the behavioral signs that occur during cannabinoid withdrawal are species-specific.

Thus far, SR 141716A challenge has been used to precipitate withdrawal responses in animals receiving either THC, WIN 55,212-2, CP 55,940, HU-210, or anandamide. Although it is clear that SR 141716A-precipitated measurable withdrawal responses following THC and the other potent cannabinoid analogs, the results with anandamide are less definitive. SR 141716A failed to precipitate withdrawal in rats that were infused constantly with anandamide (25–100 mg/kg/d) for 4 d *(66)*. It is not too surprising that anandamide lacked dependence liability, given its short half-life. On the other hand, another report showed that a regimen of 15 d of daily intraperitoneal injections of anandamide (20 mg/kg) produced both spontaneous withdrawal and SR 141716A-precipitated withdrawal *(67)*. Future studies should be carried out in order to assess whether readministration of anandamide will reverse these withdrawal signs. Mice lacking FAAH, the primary enzyme responsible for anandamide metabolism, represent another important model for investigating the role of anandamide in cannabinoid dependence *(68)*.

2.3. Intrinsic Effects of SR 141716A

It is important to point out that SR 141716A has been demonstrated to elicit behavioral effects when given alone. In addition to precipitating withdrawal symptoms in cannabinoid-dependent animals and blocking the acute effects of cannabinoid agonists, it has also been reported to produce mild withdrawal-like effects in naïve *(69)* and vehicle-treated rats *(50,59)*. These effects include scratching of the face and body, head shakes, and forepaw fluttering. It should be noted that SR 141716A-induced head shakes and paw tremors are generally significantly less than that found in cannabinoid-dependent animals. Also, SR 141716A-induced scratching has been shown to undergo tolerance following three daily injections of the antagonist *(65)*. The fact that SR 141716A possesses intrinsic activity on its own underscores the importance of including appropriate control groups to ensure that the behavioral effects are indeed a withdrawal response.

2.4. Cannabinoid Self-Administration

The other important characteristic of a dependence-producing substance is that typically animals can be trained to self-administer such compounds. Contrary to the majority of drugs abused by humans, it has been quite difficult to train animals to self-administer cannabinoids. Although the physical characteristics of cannabinoids probably contributed to this difficulty, the general opinion persisted that cannabinoids lack rewarding effects and therefore are devoid of dependence liability. However, a recent study *(70)* demonstrated that the synthetic cannabinoid agonist WIN 55,212-2 was intravenously self-administered by mice in a concentration-dependent manner according to an inverted U-shaped curve. Therefore, it is possible that WIN 55,212-2 may elicit both rewarding and aversive effects depending on the concentration used. It may well be that these dual properties have hindered the development of a THC model of self-administration. Alternatively, decreased response rates at higher doses may be due to the animal reaching a point of satisfaction with total drug intake at lower response rates when higher concentrations are self-administered. In any case, these studies clearly demonstrate that cannabinoid self-administration is not confined to humans.

3. Integrative Investigation of Cannabinoid Dependence

The establishment of experimental animal models to study cannabinoid dependence has made possible the introduction of molecular and biochemical approaches to elucidate the biological basis of this condition. Essentially, two broad applications of the integrative approach have been used. The first involves chronic drug treatment of animals according to procedures shown to cause dependence as determined by the elicitation of antagonist-precipitated withdrawal, followed by investigation of changes in some molecular marker (e.g., gene expression at the level of mRNA or protein, enzymatic activity, or second messenger levels). The second approach involves alteration in the molecular status of the animal by either chemical manipulation (e.g., administration of modulators of molecular signaling pathways) or by genetic alteration (e.g., targeted disruption of specific genes by homologous recombination, also known as gene "knockout"), followed by investigation of the consequences of these manipulations on physiological and behavioral manifestations of cannabinoid dependence. Ultimately, the combined implementation of these approaches may elucidate the cause-and-effect relationship between drug-induced influences on the molecular status of the central nervous system and the consequent alterations in physiology and behavior that result in the drug-dependent state.

3.1. Effects of Chronic Cannabinoid Administration on Molecular Signaling Pathways

Most of the studies employing chronic cannabinoid agonist treatment followed by subsequent examination of the molecular effects of this treatment have not concurrently measured cannabinoid dependence via antagonist-precipitated withdrawal. Indeed, many of these studies report correlation either with cannabinoid tolerance (i.e., the decrease in agonist potency or effectiveness with repeated administration) or have not directly measured any physiological or behavioral endpoint. Nonetheless, the value of this approach is apparent in that most of the chronic drug treatment procedures used in these studies have been shown to produce a drug-dependent state, as determined by antagonist-precipitated withdrawal, in independent studies with the same animal species/strain/sex. The disadvantage of this approach is that interpretation of the results in terms of cannabinoid dependence is limited. Thus, the combination of behavioral, biochemical, and molecular determinations within the same experimental animals or between identically treated animals within the same study is the best approach for determining the applicability of the results to cannabinoid dependence specifically. However, the number of studies in the literature in which both molecular/biochemical and behavioral indications of cannabinoid dependence were reported are comparatively rare, perhaps because animal models of cannabinoid dependence are a relatively recent development.

3.1.1. Effects of Chronic Cannabinoids on Brain CB1 Receptor Expression

The cannabinoid receptor field is somewhat unique in that only 2 yr passed between the first direct demonstration of a specific cannabinoid receptor-binding site *(21)* and cloning of cDNA encoding the CB1 receptor *(22)*. Thus, the tools to study the regulation of CB1 receptor expression using both classical radioligand binding and mRNA hybridization became available almost simultaneously. Early studies of the effects of chronic cannabinoid administration on brain CB1 receptor expression correlated tolerance to the locomotor effects of THC or CP 55,940 with CB1 receptor levels measured by radioligand binding or by *in situ* hybridization or Northern blotting of CB1 receptor mRNA. Results depended on whether whole-brain or a region-specific approach were used to determine receptor levels. Studies in whole mouse brain homogenate found no significant alteration in CB1 receptor levels as determined by [^{3}H]CP55,940 saturation binding or mRNA by Northern blotting *(74)*, despite profound tolerance to the locomotor-inhibiting effects of THC after 1 wk of twice-daily administration of 10 mg/kg of this drug. In contrast, studies using autoradiographic approaches to localize CB1 receptor binding correlated locomotor tolerance to THC or CP 55,940 with reductions in CB1 receptor expression in rat brain sections at the level of the striatum. Reduced receptor levels were detected by [^{3}H]CP55,940 autoradiography in the caudate-putamen, nucleus accumbens, olfactory tubercle, and septum following induction of tolerance to THC or CP 55,940 by 2-wk treatment in rats, and this reduction was due to decreased receptor B_{max} values *(75)*. Similar results were subsequently reported in homogenates prepared from dissected rat brain regions: [^{3}H]CP 55,940 B_{max} values decreased in striatal membranes after several days of injection with THC but not with anandamide *(76,77)*. Similarly, reduced expression of CB1 receptor mRNA in rat caudate-putamen, but not other regions, was obtained using *in situ* hybridization methodology after 11 d of treatment with THC or CP 55,940 *(78)*. Alternatively, a shorter treatment (1 wk) with a higher dose of CP 55,940 in mice produced tolerance in tests of hypoactivity, hypothermia, and catalepsy, as well as a significant reduction in the [^{3}H]CP 55,940 B_{max} value in cerebellar homogenates, but mRNA was observed to increase by Northern blotting *(79)*. These studies demonstrated the utility of combining behavioral measures of chronic cannabinoid effects, in this case tolerance, with molecular indices of adaptation at the level of CB1 receptor expression. Moreover, the advantage of regional approaches to examine molecular changes with anatomical specificity has clearly been demonstrated. However, the need for standardization of experimental animal species, as well as drug, dose, and duration, was highlighted by variation in the findings of these studies. Another technical issue of importance with the lipophilic cannabinoids was controlling for residual drug in the tissue, which produced apparent increases in radioligand K_D measured in tissue sections at 30 min after the final injection *(75)*, but not in membrane homogenates at 1–24 h post-injection *(74,77,79)*.

The apparently disparate results of earlier studies of CB1 receptor mRNA expression *(78,79)* were somewhat resolved in more recent studies examining a time course, ranging from 6 h to 21 d, of daily administration of 10 mg/kg THC in rats. CB1 receptor mRNA was found to increase in cerebellum and hippocampus over the 7–14-d time period *(80)*. It returned to control levels at 21 d, a time point at which tolerance to the memory-impairing effect of THC had been previously reported *(81)*. In contrast, CB1 receptor mRNA in striatum was decreased over approximately 1–14 d of chronic THC administration, when measured by either semiquantitative reverse transcriptase-polymerase chain reaction (RT-PCR) *(80)* or by *in situ* hybridization *(82)*, and returned to baseline at d 21 *(80)*. Similar decreases in striatal CB1 receptor mRNA were found after 7- or 11-d of treatment with CP 55,940 *(65,78)*. Taken together, these results suggest that brain region and drug treatment duration may be more important than species or the specific agonist used in the chronic administration paradigm, at least at the level of CB1 mRNA regulation.

In general, most recent studies of chronic cannabinoid agonist administration have found decreases in CB1 receptor density (downregulation) as measured by radioligand binding using a brain regional approach, although individual results depend somewhat on the specific agonist and administration paradigm employed *(83–88)*. CB1 receptor downregulation is reliably observed in cerebellum and hippocampus, with more variable results reported in cortical regions and basal ganglia. In the former two regions, downregulation in response to daily injections of THC reached a maximum after 7 d and remained at these levels with continued administration of the drug up to 21 d *(83)*. In the basal ganglia, downregulation was observed in the caudate-putamen where cell bodies of CB1 receptor-containing neurons are located *(83–87)*. However, the projection areas of these CB1 receptor-expressing neurons, where the receptors are located presynaptically and at very high density, exhibit less robust responses. Only slight downregulation was observed in the entopeduncular nucleus *(84,85)* and even less downregulation was found in the globus pallidus or substantia nigra *(83,84,86,89)*, unless relatively high doses of THC or methanandamide were used *(85,87)*.

Although decreases in CB1 receptor expression are thought to play a role in tolerance to various behavioral effects of cannabinoids, it is unclear whether CB1 receptor downregulation is involved in cannabinoid dependence. In recent studies comparing the time course of CB1 expression with SR141716A-precipitated withdrawal behaviors after chronic administration of CP 55,940 in rats, it was observed that downregulated CB1 mRNA levels in caudate-putamen returned to baseline levels within 1 h of antagonist injection *(65)*. However, downregulation of receptor-binding sites remained significant up to 24 h after injection of SR141716A, a time point at which precipitated abstinence symptoms were no longer detected *(65,86)*.

3.1.2. Effects of Chronic Cannabinoids on CB1 Receptor-Mediated G-Protein Activation and G-Protein Expression in Brain

The development of an *in vitro* autoradiographic method for examining receptor-mediated $G_{i/o}$-protein activity in tissue sections *(90)* has greatly enhanced the ability to detect region-dependent changes in the functional activity of CB1 receptors in response to chronic drug treatment. The assay is performed in the presence of millimolar concentrations of GDP and utilizes trace (p*M*) concentrations of the hydrolysis-resistant GTP analog [^{35}S]GTPγS. The addition of agonist specifically activates the receptor which stimulates exchange of the radiolabeled GTP analog in place of GDP on only those G proteins that are activated by the receptor. Because GTPγS is poorly hydrolyzed and binds with high affinity to G-protein α subunits, it remains bound, and the ^{35}S radiolabel can be detected by exposure of the sections to film. Using this approach, Sim et al. *(91)* have shown that once-daily injections of 10 mg/kg THC for 21 d in rats produced a significant loss (~30–80%) of CB1 receptor-stimulated G-protein activity (desensitization) in various regions of rat brain, including hippocampus, cerebellum, caudate-putamen, globus pallidus, substantia nigra, septum, and various regions of cortex. This desensitization appeared to be homologous (i.e., affecting only CB1 and not other $G_{i/o}$-coupled receptors) because G-protein activation by $GABA_B$ receptors in regions of co-localization, such as the cerebellum, was unaffected. Subsequent time course experiments revealed that CB1 receptor desensitization reached a maximum in hippocampus and cerebellum within 7–14 d of THC administration, but that in caudate-putamen the desensitization was slow to develop and did not yet appear to reach a maximum at 21 d *(83)*. This study clearly demonstrated that cannabinoid receptor desensitization is both time- and region-dependent. In addition to the utility of a regional approach, another important technical point of this study was that differences in desensitization measurements were obtained between intact brain sections and membrane homogenates prepared from dissected brain regions. For example, although profound (~65%) loss in WIN 55,212-2-stimulated [^{35}S]GTPγS binding was observed in cerebellar sections, no desensitization was observed in cerebellar membranes taken from hemisections of the same brains used in the autoradiographic experiments. These results suggest the importance of a preserved cytoachitecture in determinations of desensitization at the cellular level, and may explain why desensitization at the effector level has not yet been observed in brain membranes from chronic cannabinoid-treated animals (*see* Section 3.1.3.).

The apparent desensitization of CB1 receptors could have been due to either uncoupling of the receptors from G proteins or to downregulation of receptors, based on the similar time course of these two adaptive responses in most regions. However, the relative lack of CB1 receptor downregulation in terminal projection fields of the caudate-putamen (*see* Section 3.1.1.) suggests an uncoupling response in these regions. Alternatively, loss of CB1 receptor-stimulated [^{35}S]GTPγS binding may also result from decreased G-protein expression. Chronic administration of CP 55,940 has been reported to decrease levels of mRNA for the α subunits of both G_i and G_o, as well as G_s, in a region-dependent manner *(65,92)*. Nonetheless, since these changes in G-protein transcription were not associated with significant alterations in translated protein levels, it remains likely that reduced G-protein activation in chronic cannabinoid-treated animals is due to an adaptive response at the CB1 receptor level. The most likely explanation is feedback regulation by G-protein-coupled receptor kinases (GRK), which are known to mediated homologous desensitization of GPCRs. These kinases recognize and phosphorylate GPCRs in their active (agonist-bound) state, resulting in a cascade of events that uncouple the receptors from G-proteins, with subsequent receptor internalization resulting in either resensitization or degradation of the receptors *(93–95)*. Indeed, regulation of CB1 receptors by this mechanism has been demonstrated in an oocyte model *(96)*.

It is unclear whether CB1 receptor desensitization is associated with cannabinoid dependence. By analogy with opioid receptors, which, like CB1 receptors, are $G_{i/o}$-coupled, it might be expected that desensitization is more associated with cannabinoid tolerance than dependence. Recent studies of mice with targeted disruption of the β-arrestin 2 gene, a key protein in the GRK negative regulatory pathway, have demonstrated impairment of μ-opioid receptor desensitization and of the development of analgesic tolerance in response to morphine *(97)*. However, antagonist-precipitated withdrawal symptoms indicative of morphine dependence were unaffected by β-arrestin 2 knockout. Adaptive responses at the G-protein level may also not be directly involved in cannabinoid dependence, as suggested by recent studies comparing the time course of SR141716A-precipitated abstinence symptoms in rats chronically administered CP 55,940 with alterations in G-protein activity *(65,86)*. These rats were rendered tolerant to the antinociceptive effects of CP 55,940, and administration of SR141716A at 1 h after the final CP 55,940 injection elicited withdrawal behavior. At 24 h after cessation of CP 55,940 administration, desensitization of CP 55,940-stimulated [^{35}S]GTPγS binding was observed autoradiographically in several brain regions. At this time point, however, a second SR141716A injection failed to elicit most behavioral symptoms of precipitated withdrawal, indicating that CB1 receptor desensitization was present in the absence of dependence. However, in the CP 55,940-tolerant rats that received SR141716A injection, CB1 receptor-stimulated [^{35}S]GTPγS binding had also recovered to vehicle-treated control levels 24 h later. Because SR141716A-precipitated withdrawal was not tested at the 24-h time point in rats that did not previously receive an SR141716A injection at the 1-h time point, it is not known whether these animals would have exhibited withdrawal signs, which would suggest a correlation in time between cannabinoid dependence and CB1 receptor desensitization. Nonetheless, most data in the literature thus far suggest more of a correlation between GPCR desensitization and drug tolerance than dependence.

3.1.3. Effects of Chronic Cannabinoids on cAMP Signaling in Brain

Despite the fact that inhibition of adenylyl cyclase activity was one of the first effector responses described for cannabinoid receptors in brain *(98–100)*, there have been surprisingly few reports on the effects of chronic cannabinoids on adenylyl cyclase/cAMP signaling. In one study *(79)*, rats highly tolerant to the hypomotilic, hypothermic, and cataleptic effects of CP 55,940 exhibited a 50% downregulation of cannabinoid receptors in cerebellar membranes, but no change in basal or CP 55,940-inhibited adenylyl cyclase activity was detected. More recent studies have focused on changes in basal and stimulated adenylyl cyclase activity in rats or mice chronically administered cannabinoids, and a clear association between elevated adenylyl cyclase activity and the manifestation of SR141716A-precipitated withdrawal symptoms is emerging. In mice rendered tolerant to the hypothermic and antinociceptive effects of THC by twice-daily injection for 1 wk, administration of SR141716A produced both a definite behavioral withdrawal syndrome and significant increases in basal, forskolin-, and Ca^{2+}/calmodulin-stimulated adenylyl cyclase activity in cerebellum *(62)*. In this study, SR141716A had no significant effect on adenylyl cyclase activity in vehicle-pretreated mice, but increased basal and stimulated activity approximately twofold in THC-treated mice. Moreover, this effect appeared to be limited to the cerebellum, as the antagonist produced no significant increases in basal or forskolin-stimulated adenylyl cyclase activity in THC- compared to vehicle-treated mice in any other region examined, including cortex, hippocampus, striatum, and periaqueductal gray. Recent evidence suggests a causal relationship between elevated cAMP signaling and behavioral symptoms of SR141716A-precipitated withdrawal *(63)*. The time course of precipitated withdrawal symptoms correlated with the time course of elevations in both adenylyl cyclase and protein kinase A (PKA) activity in cerebellum, and microinfusion of a PKA inhibitor (Rp-8Br-cAMP) into this region blocked both the elevation in PKA and the antagonist-precipitated withdrawal symptoms. Furthermore, the withdrawal-like behaviors were mimicked by infusion of the cAMP analog, Sp-8Br-cAMP, in vehicle-treated mice. A similar cAMP-dependent mechanism may mediate SR141716A-precipitated withdrawal in CP 55,940-dependent rats, which also exhibit elevations in cerebellar cAMP and PKA activity in response to injection of the antagonist *(86)*. In fact, enhanced cAMP signaling may be a more widespread response in rats than mice, as chronic but not acute THC administration significantly elevated cAMP levels in cortex, striatum, and cerebellum in the absence of SR141716A administration *(87)*. This increase in cAMP was correlated with increased PKA activity in cortex and cerebellum. Furthermore, studies in cell lines co-transfected with CB1 receptors and various adenylyl cyclase isoforms have shown that isoforms acutely inhibited by CB1 receptors, including the Ca^{2+}/calmodulin-stimulated types I and VIII, exhibit this antagonist-precipitated "superactivation" following a period of chronic agonist exposure. Thus, although further studies are needed, it appears that focusing on adaptations in downstream signaling may be a more productive approach to determine the molecular mechanisms of cannabinoid dependence than focusing on the CB1 receptor itself. Moreover, the application of regional approaches to these investigations is crucial because the molecular adaptations underlying cannabinoid dependence, as with tolerance, are region-specific.

3.1.4. Other Molecular Indices of Cannabinoid Dependence

There is molecular evidence for the involvement of the stress-activated corticotropin-releasing factor (CRF) system in cannabinoid withdrawal *(69)*. An SR141716A-precipitated withdrawal syndrome was noted after 2-wk treatment of rats with the potent synthetic cannabinoid agonist, HU-210. This withdrawal behavior was associated with increased CRF concentration in microdialysate from the central nucleus of the amygdala, as well as increased expression of the immediate early gene product, Fos, in this nucleus and several other brain regions. This withdrawal/stress activation pattern, especially in the amygdala, is known to be a common withdrawal phenomenon among several other dependence-inducing drugs, including cocaine, ethanol, and opiates *(101)*. Other recent studies have shown that chronic administration of THC or CP-55,940 increased the expression of CRF mRNA in the hypothalamus and anterior pituitary *(102,103)*. These studies, as well as recent electrophysiological evidence of deficits in dopaminergic neuronal activity upon either abrupt cessation of chronic THC administration or SR141716A-precipitated cannabinoid withdrawal *(71)*, suggest that cannabinoid dependence may involve a variety of neurochemical systems.

The implementation of genomic approaches may prove highly useful in elucidation of the molecular mechanisms underlying cannabinoid dependence. The use of high-density DNA microarrays to perform large-scale analysis of alterations in gene expression in response to cannabinoid administration has recently been reported. In that study, Kittler et al. *(104)* performed microarray analysis of hippocampal preparations from rats subjected to 1, 7, and 21 d of once-daily 10 mg/kg THC administration. Although changes in gene expression were not specifically correlated with any behavioral measurements in this study, this same drug treatment paradigm had been well characterized in previous studies, which showed tolerance to the amnesic effects of THC at d 21 *(81)*, as well as CB1 receptor desensitization and downregulation at d 7–21 *(83,91)* and biphasic changes in CB1 mRNA expression over the entire time course *(80)*. Microarray analysis revealed differences in the expression of several genes between vehicle, acute (d 1) and chronic (d 7 and/or d 21) THC administration. Confirmation that specific transcripts were either up- or downregulated was obtained in independent experiments using RNA dot blots. Many of the affected genes fell into functionally related groups involved in metabolism, cell adhesion and structure, myelination, protein folding and degradation, and signal transduction (including calmodulin). Thus, the simultaneous observance of altered expression in functional clusters of genes is an advantage of the use of large-scale DNA microarrays that may prove enormously useful in elucidating the molecular mechanisms of cannabinoid dependence.

3.2. Effects of Targeted Gene Disruption on the Expression of Cannabinoid Dependence

The application of transgenic approaches to the study of neuropharmacology is providing valuable information on the involvement of various neurotransmitter systems and intracellular signaling pathways in cannabinoid dependence. The recent availability of cannabinoid receptor knockouts *(61,105,106)* has been of immense benefit in defining the roles of both CB1 and CB2 receptors, both in mediating the effects of exogenous cannabinoid drugs and in the functioning of the endogenous cannabinoid system. The potential role of CB1 receptors as important components of the neural substrate, not only for cannabinoid dependence, but also for dependence on opioids and possibly other addictive drugs, is an emerging development in the field. The next two chapters in this volume will describe in greater detail the genetic models that implicate the cannabinoid system in drug dependence and specifically the interaction of cannabinoid and opioid receptor systems in drug dependence. For the purpose of this chapter, discussion of targeted gene disruption will be limited to specific systems implicated in the mediation of cannabinoid dependence.

Although the CB1 receptor has been presumed to mediate the dependence-inducing effects of cannabinoids based on the fact that CB1 is the major cannabinoid receptor in the CNS and that withdrawal symptoms are precipitated by the CB1 antagonist SR141716A, the use of CB1 knockout mice has provided confirmation of this hypothesis. Ledent et al. *(61)* reported that cannabinoid dependence and self-administration were absent in a CB1 receptor knockout mouse line on a CD1 background. Five days of twice-daily intraperitoneal injections of THC produced significant dependence as gauged by SR141716A-precipitated withdrawal signs in the CB1+/+ mice. Most withdrawal signs were absent or significantly reduced in CB1−/− compared to CB1+/+ mice, including rearing, sniffing, wet-dog shakes, paw tremor, piloerection, penile licking, mastication, hunched posture, and body tremor. Although an acute vehicle-injection group was not included to control for the effects of SR141716A in nondependent mice, most behaviors induced by the antagonist (with the exception of piloerection and penile licking) were absent or markedly lower in frequency in the chronic-vehicle group relative to those treated chronically with THC. Moreover, intravenous self-administration of WIN 55,212-2 (0.1 mg/kg/injection), measured using a nose-poke procedure, was obtained in CB1+/+ but not CB1−/− mice, suggesting that CB1 receptors mediate the reinforcing properties of this drug *(61)*. The authors concluded that most, if not all, of the "addictive properties" of cannabinoids are mediated by CB1 receptors. A similar conclusion regarding cannabinoid dependence was reached by Lichtman et al. *(60)*. In this study, 10 mg/kg of THC was injected subcutaneously once daily for 5 d, and SR141716A-precipitated withdrawal behaviors, including paw tremors and head shakes, were found in CB1+/+, but not CB1−/− mice. Importantly, this study was performed using CB1−/− mice on a different background strain, C57Bl/6 *(106)*, further indicating that the deletion of the CB1 receptor rather than other potential genetic differences were responsible for the lack of antagonist-precipitated withdrawal in the CB1 knockout mice used in these studies.

Another contribution of CB1 receptor knockout mice to the study of cannabinoid dependence is in determining whether the intrinsic effects of SR141716A in naïve and vehicle-treated mice represent actions at CB1 receptors. Interestingly, CB1−/− mice chronically treated with vehicle and then injected with SR141716A exhibited less rearing, wet-dog shakes, penile licking, mastication, and hunched posture, but more piloerection, than CB1+/+ mice *(61)*. These results suggest the involvement of CB1 receptors in some, but not all, of the behavioral effects of SR141716A. Unfortunately, no acute vehicle injection controls were included in these experiments (i.e., all mice were injected with SR141716A), so it is difficult to know whether these results represent differences in the intrinsic effects of SR141716A or in baseline behavior between

CB1+/+ and CB1−/− mice. Thus, future studies will be required to determine whether the intrinsic effects of SR141716A are truly CB1-mediated. If so, it will be important to determine whether they result from blockade of endogenous cannabinoid (e.g., anandamide or 2-AG) neurotransmission. The recent development of mice with deletion of the endogenous cannabinoid-degrading enzyme, FAAH *(68)*, will undoubtedly be useful for this purpose.

Two recent studies have provided evidence for the involvement of endogenous opioid systems in cannabinoid dependence. Although these studies will be described in greater detail in Chapter 16, a brief discussion is appropriate here to illustrate how the gene-knockout approach can be used to define the involvement of other neurotransmitter systems in cannabinoid dependence. In the first study *(64)*, mice with disruption of the proenkephalin gene, which encodes the precursor of the endogenous opioid peptides Met- and Leu-enkephalin, exhibited significantly lower frequency of several SR141716A-precipitated withdrawal signs compared to proenkephalin+/+ mice after repeated THC administration. This effect could not be attributed to a reduced functioning of cannabinoid receptors in these mice because no decrement was found in cannabinoid receptor-binding sites or in the potency of THC in several tests of cannabinoid-mediated behaviors. Interestingly, the proenkephalin−/− mice also exhibited an attenuated rate of the development of tolerance to the antinociceptive effects of THC compared to proenkephalin−/− mice. Another recent study has found evidence for decreased cannabinoid withdrawal severity in μ opioid receptor (MOR) knockout mice *(60)*. The frequency of SR141716A-precipitated paw tremors and head shakes were found to be reduced in MOR−/− mice relative to wild-type mice when challenged with SR141716A after repeated treatment with THC. These results suggest a modulatory role of the endogenous opioid peptides and their receptors in the chronic effects of THC, and especially in the manifestation of precipitated withdrawal.

4. Summary and Conclusions

Although marijuana has been used for centuries and the occurrence of psychological dependence in humans has been widely accepted, until recently there was little direct evidence of physiological dependence and no animal models of cannabinoid withdrawal. The establishment of reliable measures of cannabinoid withdrawal was for years hampered by the lack of a specific high-affinity cannabinoid antagonist. The discovery that cannabinoids act through G-protein-coupled receptors and the subsequent cloning of the CB1 and CB2 receptors aided in the development of a selective antagonist of CB1 cannabinoid receptors. Once the antagonist became available, it was possible to use established in vivo pharmacological approaches to define antagonist-precipitated withdrawal behaviors indicative of cannabinoid dependence. Perhaps not surprisingly, the cannabinoid withdrawal syndrome includes many of the same somatic withdrawal signs that are observed with other dependence-inducing psychoactive drugs. The establishment of animal models of psychological dependence on cannabinoids has also been challenging. Laboratory animals do not readily self-administer THC, possibly due to specific pharmacokinetic or pharmacodynamic properties of this drug. However, the recent development of cannabinoid self-administration procedures using the highly potent, efficacious, and relatively short-acting cannabinoid, WIN 55,212-2, has demonstrated that cannabinoid self-administration is not unique to humans. These procedures will undoubtedly be useful in examining the neurobiological substrates underlying the reinforcing properties of cannabinoids.

Integrative research using molecular, biochemical, anatomical, and behavioral approaches is necessary to elucidate the biological basis of cannabinoid dependence. Initial studies in this regard demonstrated the importance of using a brain regional approach to investigate cellular adaptations in response to chronic cannabinoid exposure. Exciting developments in this area have demonstrated that prolonged administration of THC and other cannabinoid agonists produces both desensitization and downregulation of CB1 receptors in brain, but in a highly region-dependent manner. However, some evidence suggests that these adaptive responses may contribute to the development of cannabinoid tolerance rather than dependence. The discovery of elevated cAMP synthesis and consequent increases in PKA activity in some brain regions during SR141716A-precipitated withdrawal has led to the hypothesis that adaptive cellular changes associated with cannabinoid dependence occur downstream of the CB1 receptor–G-protein interaction. Indeed, the ability of chemical modulators of cAMP signaling to block or mimic the expression of cannabinoid withdrawal symptoms lends support to this hypothesis. Additional evidence points to upregulation of the stress-activated CRF system in cannabinoid withdrawal, along with deceased mesolimbic dopaminergic neurotransmission, suggesting commonality between cannabinoid dependence and that of other pyschoactive drugs. The implementation of genomic approaches, such as DNA arrays that screen thousands of genes simultaneously, is beginning to make an impact on our understanding of how cannabinoids alter gene expression in the brain. This approach will undoubtedly lead to new directions in cannabinoid dependence research. Finally, the use of transgenic animal models, such as targeted gene knockouts, is providing a powerful approach to determine the contribution of individual proteins to the process of cannabinoid dependence. Thus far, genetic knockout approaches have demonstrated that cannabinoid dependence requires the CB1 receptor, and that endogenous opioid peptide and receptor systems are probably involved in the development and/or expression of cannabinoid dependence.

Most of these recent discoveries would not have been possible without the development of reliable animal models of cannabinoid dependence and the application of new molecular approaches to begin defining the neurobiological substrates underlying this behavior. As new discoveries are made at the molecular and cellular levels, existing behavioral models may have to be modified to take advantage of these advancements. This integrative approach to cannabinoid research is undoubtedly the key to elucidating the biological basis of cannabinoid dependence.

Acknowledgment

This work was supported by National Institute on Drug Abuse grants DA-03672, DA-05274, DA-02396, and DA-09789.

References

1. Adams, I. B. and Martin, B. R. (1996) Cannabis: pharmacology and toxicology in animals and humans. *Addiction* **91,** 1585–1614.
2. Compton, D. R., Dewey, W. L., and Martin, B. R. (1990) Cannabis dependence and tolerance production. *Adv. Alcohol Subst. Abuse* **9,** 129–147.

3. Jones, R. T., Benowitz, N., and Bachman, J. (1976) Clinical studies of cannabis tolerance and dependence. *Ann. N.Y. Acad. Sci.* **282,** 221–239.
4. Jones, R. T. (1987) Drug of abuse profile: Cannabis. *Clin. Chem.* **33(11(B)),** 72B–81B.
5. Roffman, R. A., Stephens, R. S., Simpson, E. E., and Whitaker, D. L. (1988) Treatment of marijuana dependence: preliminary results. *J. Psychoactive Drugs* **20,** 129–137.
6. Weller, R. A., Halikas, J., and Morse, C. (1984) Alcohol and marijuana: comparison of use and abuse in regular marijuana users. Journal of Clinical Psychiatry **45,** 377–379.
7. Kosten, T. R., Roundsville, B. J., Babor, T. F., Spitzer, R. L., and Williams, J. B. W. (1987) Substance-use disorders in DSM-III-R. *Br. J. Psychiatr.* **151,** 834–843.
8. Newcombe, M. D. (1992) Understanding the mulidimensional nature of drug use and abuse: the role of consumption, risk factors and protective factors, in *Vulnerability to Drug Abuse* (Glantz, M. and Pickens, R., eds.), American Psychological Association, Washington, DC.
9. Rounsaville, B. J., Bryant, K., Babor, T., Kranzler, H., and Kadden, R. (1993) Cross-system agreement for substance use disorders: DSM-111-R, DSM-IV and ICD-10. *Addiction* **88,** 337–348.
10. Anthony, J. C., Warner, L. A. and Kessler, R. C. (1994) Comparative epidemiology of dependence on tobacco, alcohol, controlled substances and inhalants: basic findings from the National Comorbidity Study. *Clin. Exp. Psychopharmacol.* **2,** 244–268.
11. Hall, W., Solowij, N., and Lemon, J. (1994) *The Health and Psychological Consequences of Cannabis Use.* National Drug Strategy Monograph Series No. 25, Australian Government Publication Service, Canberra, Australia.
12. Kandel, D. B. and Davies, M. (1992) Progression to regular marijuana involvement: phenomenology and risk factors for near-daily use. In *Vulnerability to Drug Abuse* (M. Glantz and R. Pickens, eds.), American Psychological Association, Washington, DC.
13. Jones, R. T. (1983) Cannabis tolerance and dependence. In *Cannabis and Health Hazards* (Fehr, K. O. and Kalant, H., eds.), Addiction Research Foundation, Toronto, Canada, pp. 617–689.
14. Georgotas, A. and Zeidenberg, P. (1979) Observations on the effects of four weeks of heavy marihuana smoking on group interaction and individual behavior. *Comp. Psychiatr.* **20,** 427–432.
15. Haney, M., Ward, A. S., Comer, S. D., Foltin, R. W., and Fischman, M. W. (1999) Abstinence symptoms following smoked marijuana in humans. *Psychopharmacology* (Berl.) **141,** 395–404.
16. Robins, L. N. and Regier, D. A. (1991) *Psychiatric Disorders in America.* The Free Press, MacMillan, New York.
17. Budney, A. J., Novy, P. L., and Hughes, J. R. (1999) Marijuana withdrawal among adults seeking treatment for marijuana dependence. *Addiction* **94,** 1311–1322.
18. Gaoni, Y. and Mechoulam, R. (1964) Isolation, structure, and partial synthesis of an active constituent of hashish. *J. Am. Chem. Soc.* **86,** 1646–1647.
19. Mechoulam, R. and Gaoni, Y. (1967) The absolute configuration of delta-1-tetrahydrocannabinol, the major active constituent of hashish. *Tetrahedron Lett.* **12,** 1109–1111.
20. Howlett, A. C., Qualy, J. M., and Khachatrian, L. L. (1986) Involvement of Gi in the inhibition of adenylate cyclase by cannabimimetic drugs. *Mol. Pharmacol.* **29,** 307–313.
21. Devane, W. A., Dysarz, F. A. I., Johnson, M. R., Melvin, L. S., and Howlett, A. C. (1988) Determination and characterization of a cannabinoid receptor in rat brain. *Mol. Pharmacol.* **34,** 605–613.
22. Matsuda, L. A., Lolait, S. J., Brownstein, M. J., Young, A. L., and Bonner, T. I. (1990) Structure of a cannabinoid receptor and functional expression of the cloned cDNA. *Nature* **346,** 561–564.
23. Munro, S., Thomas, K. L., and Abu-Shaar, M. (1993) Molecular characterization of a peripheral receptor for cannabinoids. *Nature* **365,** 61–65.
24. Childers, S. R., Fleming, L., Konkoy, C., Marckel, D., Pacheco, M., Sexton, T., and Ward, S. (1992) Opioid and cannabinoid receptor inhibition of adenylyl cyclase in brain. *Ann. N.Y. Acad. Sci.* **654,** 33–51.

25. Mackie, K. and Hille, B. (1992) Cannabinoids inhibit N-type calcium channels in neuroblastoma-glioma cells. *Proc. Natl. Acad. Sci. USA* **89,** 3825–3829.
26. Mackie, K., Lai, Y., Westenbroek, R., and Mitchell, R. (1995) Cannabinoids activate an inwardly rectifying potassium conductance and inhibit Q-type calcium currents in AtT20 cells transfected with rat brain cannabinoid receptor. *J. Neurosci.* **15,** 6552–6561.
27. Felder, C. C., Joyce, K. E., Briley, E. M., Mansouri, J., Mackie, K., Blond, O., Lai, Y., Ma, A. L., and Mitchell, R. L. (1995) Comparison of the pharmacology and signal transduction of the human cannabinoid CB_1 and CB_2 receptors. *Mol. Pharmacol.* **48,** 443–450.
28. Kuster, J., Stevenson, J., Ward, S., D'Ambra, T., and Haycock, D. (1993) Aminoalkylindole binding in rat cerebellum: selective displacement by natural and synthetic cannabinoids. *J. Pharmacol. Exp. Ther.* **264,** 1352–1363.
29. Herkenham, M., Lynn, A. B., Johnson, M. R., Melvin, L. S., de Costa, B. R., and Rice, K. C. (1991) Characterization and localization of cannabinoid receptors in rat brain: a quantitative *in vitro* autoradiographic study. *J. Neurosci.* **11,** 563–583.
30. Tsou, K., Brown, S., Sanudo-Pena, M. C., Mackie, K., and Walker, J. M. (1998) Immunohistochemical distribution of cannabinoid CB1 receptors in the rat central nervous system. *Neuroscience* **83,** 393–411.
31. Jansen, E. M., Haycock, D. A., Ward, S. J., and Seybold, V. S. (1992) Distribution of cannabinoid receptors in rat brain determined with aminoalkylindoles. *Brain Res.* **575,** 93–102.
32. Glass, M., Dragunow, M., and Faull, R. L. M. (1997) Cannabinoid receptors in the human brain: a detailed anatomical and quantitative autoradiographic study in the fetal, neonatal and adult human brain. *Neuroscience* **77,** 299–318.
33. Herkenham, M., Lynn, A. B., de Costa, B. R., and Richfield, E. K. (1991) Neuronal localization of cannabinoid receptors in the basal ganglia of the rat. *Brain Res.* **547,** 267–274.
34. Dewey, W. L. (1986) Cannabinoid pharmacology. *Pharmacol. Rev.* **38,** 151–178.
35. Devane, W. A., Hanus, L., Breuer, A., Pertwee, R. G., Stevenson, L. A., Griffin, G., Gibson, D., Mandelbaum, A., Etinger, A., and Mechoulam, R. (1992) Isolation and structure of a brain constituent that binds to the cannabinoid receptor. *Science* **258,** 1946–1949.
36. Mackie, K., Devane, W. A., and Hille, B. (1993) Anandamide, an endogenous cannabinoid, inhibits calcium currents as a partial agonist in N18 neuroblastoma cells. *Mol. Pharmacol.* **44,** 498–503.
37. Felder, C. C., Briley, E. M., Axelrod, J., Simpson, J. T., Mackie, K., and Devane, W. A. (1993) Anandamide, an endogenous cannabimimetic eicosanoid, binds to the cloned human cannabinoid receptor and stimulates receptor-mediated signal transduction. *Proc. Nat. Acad. Sci.* **90,** 7656–7660.
38. Childers, S. R., Sexton, T., and Roy, M. B. (1994) Effects of anandamide on cannabinoid receptors in rat brain membranes. *Biochem. Pharmacol.* **47,** 711–715.
39. Smith, P. B., Compton, D. R., Welch, S. P., Razdan, R. K., Mechoulam, R., and Martin, B. R. (1994) The pharmacological activity of anandamide, a putative endogenous cannabinoid, in mice. *J. Pharmacol. Exp. Ther.* **270,** 219–227.
40. Di Marzo, V., Fontana, A., Cadas, H., Schinelli, S., Cimino, G., Schwartz, J. C., and Piomelli, D. (1994) Formation and inactivation of endogenous cannabinoid anandamide in central neurons. *Nature* **372,** 686–691.
41. Cadas, H., di Tomaso, E., and Piomelli, D. (1997) Occurrence and biosynthesis of endogenous cannabinoid precursor, N-arachidonoyl phosphatidylethanolamine, in rat brain. *J. Neurosci.* **17,** 1226–1242.
42. Beltramo, M., Stella, N., Calignano, A., Lin, S. Y., Makriyannis, A., and Piomelli, D. (1997) Functional role of high-affinity anandamide transport, as revealed by selective inhibition. *Science* **277,** 1094–1097.

43. Hillard, C. J., Edgemond, W. S., Jarrahian, A., and Campbell, W. B. (1997) Accumulation of N-arachidonoylethanolamine (anandamide) into cerebellar granule cells occurs via facilitated diffusion. *J. Neurochem.* **69,** 631–638.
44. Cravatt, B. F., Giang, D. K., Mayfield, S. P., Boger, D. L., Lerner, R. A., and Gilula, N. B. (1996) Molecular characterization of an enzyme that degrades neuromodulatory fatty-acid amides. *Nature* **384,** 83–87.
45. Maurelli, S., Bisogno, T., De Petrocellis, L., Di Luccia, A., Marino, G., and Di Marzo, V. (1995) Two novel classes of neuroactive fatty acid amides are substrates for mouse neuroblastoma "anandamide amidohydrolase." *Fed. Eur. Biochem. Soc. Lett.* **377,** 82–86.
46. Stella, N., Schweitzer, P., and Piomelli, D. (1997) A second endogenous cannabinoid that modulates long-term potentiation. *Nature* **388,** 773–778.
47. Di Marzo, V., De Petrocellis, L., Sugiura, T., and Waku, K. (1996) Potential biosynthetic connections between the two cannabimimetic eicosanoids, anandamide and 2-arachidonoylglycerol, in mouse neuroblastoma cells. *Biochem. Biophys. Res. Commun.* **227,** 281–288.
48. Bisogno, T., Sepe, N., Melck, D., Maurelli, S., De Petrocellis, L., and Di Marzo, V. (1997) Biosynthesis, release and degradation of the novel endogenous cannabimimetic metabolite 2-arachidonoylglycerol in mouse neuroblastoma cells. *Biochem. J.* **322,** 671–677.
49. Ueda, N., Goparaju, S. K., Katayama, K., Kurahashi, Y., Suzuki, H., and Yamamoto, S. (1998) A hydrolase enzyme inactivating endogenous ligands for cannabinoid receptors. *J. Med. Invest.* **45,** 27–36.
50. Aceto, M. D., Scates, S. M., Lowe, J. A., and Martin, B. R. (1996) Dependence on delta 9-tetrahydrocannabinol: studies on precipitated and abrupt withdrawal. *J. Pharmacol. Exp. Ther.* **278,** 1290–1295.
51. Cook, S. A., Lowe, J. A., and Martin, B. R. (1998) CB1 receptor antagonist precipitates withdrawal in mice exposed to delta9-tetrahydrocannabinol. *J. Pharmacol. Exp. Ther.* **285,** 1150–1156.
52. Kaymakcalan, S. (1972) Physiological and psychological dependence on THC in rhesus monkeys, in *Cannabis and Its Derivatives* (Paton, W. D. M. and Crown, J., eds.), Oxford University Press, London, UK, pp. 142–149.
53. Beardsley, P. M., Balster, R. L., and Harris, L. S. (1986) Dependence on tetrahydrocannabinol in rhesus monkeys. *J. Pharmacol. Exp. Ther.* **239,** 311–319.
54. Kaymakcalan, K., Ayhan, I. H., and Tulunay, F. C. (1977) Naloxone-induced or postwithdrawal abstinence signs in Δ^9-tetrahydrocannabinol-tolerant rats. Psychopharmacology **55,** 243–249.
55. McMillan, D. E., Dewey, W. L., and Harris, L. S. (1971) Characteristics of tetrahydrocannabinol tolerance. *Ann. N.Y. Acad. Sci.* **191,** 83–99.
56. Aceto, M. D., Scates, S. M., and Martin, B. B. (2001) Spontaneous and precipitated withdrawal with a synthetic cannabinoid, WIN 55212-2. *Eur. J. Pharmacol.* **416,** 75–81.
57. Rinaldi-Carmona, M., Barth, F., Héaulme, M., Shire, D., Calandra, B., Congy, C., Martinez, S., Maruani, J., Néliat, G., Caput, D., Ferrara, P., Soubrié, P., Brelière, J. C. and Le Fur, G. (1994) SR141716A, a potent and selective antagonist of the brain cannabinoid receptor. *Fed. Eur. Biochem. Soc. Lett.* **350,** 240–244.
58. Tsou, K., Patrick, S., and Walker, J. M. (1995) Physical withdrawal in rats tolerant to Δ^9-tetrahydrocannabinol precipated by a cannabinoid receptor antagonist. *Eur. J. Pharmacol.* **280,** R13–R15.
59. Aceto, M. D., Scates, S. M., Lowe, J. A., and Martin, B. R. (1995) Cannabinoid precipitated withdrawal by the selective cannabinoid receptor antagonist, SR 141716A. *Eur. J. Pharmacol.* **282,** R1–2.
60. Lichtman, A. H., Sheikh, S. M., Loh, H. H., and Martin, B. R. (2001) Opioid and cannabinoid modulation of precipitated withdrawal in delta(9)-tetrahydrocannabinol and morphine-dependent mice. *J. Pharmacol. Exp. Ther.* **298,** 1007–1014.

61. Ledent, C., Valverde, O., Cossu, G., Petitet, F., Aubert, J. F., Beslot, F., Bohme, G. A., Imperato, A., Pedrazzini, T., Roques, B. P., Vassart, G., Fratta, W., and Parmentier, M. (1999) Unresponsiveness to cannabinoids and reduced addictive effects of opiates in CB1 receptor knockout mice. *Science* **283,** 401–404.
62. Hutcheson, D. M., Tzavara, E. T., Smadja, C., Valjent, E., Roques, B. P., Hanoune, J., and Maldonado, R. (1998) Behavioral and biochemical evidence for signs of abstinence in mice chemically treated with Δ-9-terahydrocannabinol. *Br. J. Pharmacol.* **125,** 1567–1577.
63. Tzavara, E. T., Valjent, E., Firmo, C., Mas, M., Beslot, F., Defer, N., Roques, B. P., Hanoune, J., and Maldonado, R. (2000) Cannabinoid withdrawal is dependent upon PKA activation in the cerebellum. *Eur. J. Neurosci.* **12,** 1038–1046.
64. Valverde, O., Maldonado, R., Valjent, E., Zimmer, A. M., and Zimmer, A. (2000) Cannabinoid withdrawal syndrome is reduced in preproenkephalin knock-out mice. *J. Neurosci.* **20,** 9284–9289.
65. Rubino, T., Patrini, G., Massi, P., Fuzio, D., Vigano, D., Giagnoni, G., and Parolaro, D. (1998) Cannabinoid-precipitated withdrawal: a time course study of the behavioral aspect and its correlation with cannabinoid receptors and G protein expression. *J. Pharmacol. Exp. Ther.* **285,** 813–819.
66. Aceto, M. D., Scates, S. M., Razdan, R. K., and Martin, B. R. (1998) Anandamide, an endogenous cannabinoid, has a very low physical dependence potential. *J. Pharmacol. Exp. Ther.* **287,** 598–605.
67. Costa, B., Giagnoni, G., and Colleoni, M. (2000) Precipitated and spontaneous withdrawal in rats tolerant to anandamide. *Psychopharmacology* (Berl.) **149,** 121–128.
68. Cravatt, B. F., Demarest, K., Patricelli, M. P., Bracey, M. H., Giang, D. K., Martin, B. R., and Lichtman, A. H. (2001) Supersensitivity to anandamide and enhanced endogenous cannabinoid signaling in mice lacking fatty acid amide hydrolase. *Proc. Natl. Acad. Sci. USA* **98,** 9371–9376.
69. Rodriguez de Fonseca, F., Carrera, M., Navarro, M. R. A., Koob, G. F., and Weiss, F. (1997) Activitation of corticotropin-releasing factor in the limbic system during cannabinoid withdrawal. *Science* **276,** 2050–2054.
70. Martellotta, M. C., Cossu, G., Fattore, L., Gessa, G. L., and Fratta, W. (1998) Self-administration of the cannabinoid receptor agonist WIN 55,212-2 in drug-naive mice. *Neuroscience* **85,** 327–330.
71. Diana, M., Melis, M., Muntoni, A. L., and Gessa, G. L. (1998) Mesolimbic dopaminergic decline after cannabinoid withdrawal. *Proc. Natl. Acad. Sci. USA* **95,** 10269–10273.
72. Beardsley, P. M. and Martin, B. R. (2000) Effects of the cannabinoid CB(1) receptor antagonist, SR141716A, after delta(9)-tetrahydrocannabinol withdrawal. *Eur. J. Pharmacol.* **387,** 47–53.
73. Lichtman, A. H., Wiley, J. L., LaVecchia, K. L., Neviaser, S. T., Arthrur, D. B., Wilson, D. M., and Martin, B. R. (1998) Acute and chronic cannabinoid effects: characterization of precipitated withdrawal in dogs. *Eur. J. Pharmacol.* **357,** 139–148.
74. Abood, M. E., Sauss, C., Fan, F., Tilton, C. L., and Martin, B. R. (1993) Development of behavioral tolerance to Δ^9-THC without alteration of cannabinoid receptor binding or mRNA levels in whole brain. *Pharmacol. Biochem. Behav.* **46,** 575–579.
75. Oviedo, A., Glowa, J., and Herkenham, M. (1993) Chronic cannabinoid administration alters cannabinoid receptor binding in rat brain: a quantitative autoradiographic study. *Brain Res.* **616,** 293–302.
76. Romero, J., Garciá, L., Fernández-Ruiz, J. J., Cebeira, M., and Ramos, J. A. (1995) Changes in rat brain cannabinoid binding sites after acute or chronic exposure to their endogenous agonist, anandamide, or to Δ^9-tetrahydrocannabinol. *Pharmacol. Biochem. Behav.* **51,** 731–737.
77. Rodríguez de Fonseca, F., Gorriti, M. A., Fernandez-Ruiz, J. J., Palomo, T., and Ramos, J. A. (1994) Downregulation of rat brain cannabinoid binding sites after chronic Δ^9-tetrahydrocannabinol treatment. *Pharmacol. Biochem. Behav.* **47,** 33–40.

78. Rubino, T., Massi, P., Patrini, G., Venier, I., Giagnoni, G., and Parolaro, D. (1994) Chronic CP-55,940 alters cannabinoid receptor mRNA in the rat brain: an *in situ* hybridization study. *Neuroreport* **5,** 2493–2496.
79. Fan, F., Tao, Q., Abood, M. E., and Martin, B. R. (1996) Cannabinoid receptor down-regulation without alteration of the inhibitory effect of CP 55,940 on adenylyl cyclase in the cerebellum of CP 55,940-tolerant mice. *Brain Res.* **706,** 13–20.
80. Zhuang, S., Kittler, J., Grigorenko, E. V., Kirby, M. T., Sim, L. J., Hampson, R. E., Childers, S. R., and Deadwyler, S. A. (1998) Effects of long-term exposure to delta9-THC on expression of cannabinoid receptor (CB1) mRNA in different rat brain regions. *Brain Res. Mol. Brain Res.* **62,** 141–149.
81. Deadwyler, S. A., Heyser, C. J., and Hampson, R. E. (1995) Complete adaptation to the memory disruptive effects of delta-9-THC following 35 days of exposure. *Neurosci. Res. Commun.* **17,** 9–18.
82. Corchero, J., Romero, J., Berrendero, F., Fernandez-Ruiz, J., Ramos, J. A., Fuentes, J. A., and Manzanares, J. (1999) Time dependent differences of repeated administration with Delta9-tetrahydrocannabinol in proenkephalin and cannabinoid receptor gene expression and G-protein activation by mu-opioid and CB1-cannabinoid receptors in the caudate-putamen. *Brain Res. Mol. Brain Res.* **67,** 148–157.
83. Breivogel, C. S., Childers, S. R., Deadwyler, S. A., Hampson, R. E., Vogt, L. J., and Sim-Selley, L. J. (1999) Chronic delta9-tetrahydrocannabinol produces a time-dependent loss of cannabinoid receptors and cannabinoid receptor-activated G-proteins in rat brain. *J. Neurochem.* **73,** 2447–2459.
84. Romero, J., Berrendero, F., Garcia-Gil, L., de la Cruz, P., Ramos, J. A., and Fernandez-Ruiz, J. J. (1998) Loss of cannabinoid receptor binding and messenger RNA levels and cannabinoid agonist-stimulated [^{35}S]guanylyl-5'O-(thio)-triphosphate binding in the basal ganglia of aged rats. *Neuroscience* **84,** 1075–1083.
85. Romero, J., Berrendero, F., Garcia-Gil, L., Lin, S. Y., Makriyannis, A., Ramos, J. A., and Fernandez-Ruiz, J. J. (1999) Cannabinoid receptor and WIN-55,212-2-stimulated [35S]GTP-gammaS binding and cannabinoid receptor mRNA levels in several brain structures of adult male rats chronically exposed to R-methanandamide. *Neurochem. Int.* **34,** 473–482.
86. Rubino, T., Vigano, D., Massi, P., and Parolaro, D. (2000) Changes in the cannabinoid receptor binding, G protein coupling, and cyclic AMP cascade in the CNS of rats tolerant to and dependent on the synthetic cannabinoid compound CP55,940. *J. Neurochem.* **75,** 2080–2086.
87. Rubino, T., Vigano, D., Massi, P., Spinello, M., Zagato, E., Giagnoni, G., and Parolaro, D. (2000) Chronic delta-9-tetrahydrocannabinol treatment increases cAMP levels and cAMP-dependent protein kinase activity in some rat brain regions. *Neuropharmacology* **39,** 1331–1336.
88. Rubino, T., Vigano, D., Costa, B., Colleoni, M., and Parolaro, D. (2000) Loss of cannabinoid-stimulated guanosine 5'-O-(3-[(35)S]Thiotriphosphate) binding without receptor down-regulation in brain regions of anandamide-tolerant rats. *J. Neurochem.* **75,** 2478–2484.
89. Romero, J., Garcia-Palomero, E., Castro, J. G., Garcia-Gil, L., Ramos, J. A., and Fernandez-Ruiz, J. J. (1997) Effects of chronic exposure to delta9-tetrahydrocannabinol on cannabinoid receptor binding and mRNA levels in several rat brain regions. *Brain Res. Mol. Brain Res.* **46,** 100–108.
90. Sim, L. J., Selley, D. E., and Childers, S. R. (1995) *in vitro* autoradiography of receptor-activated G-proteins in rat brain by agonist-stimulated guanylyl 5'-[γ-[^{35}S]thio]-triphosphate binding. *Proc. Nat. Acad. Sci.* **92,** 7242–7246.
91. Sim, L. J., Hampson, R. E., Deadwyler, S. A., and Childers, S. R. (1996) Effects of chronic treatment with Δ^9-tetrahydrocannabinol on cannabinoid-stimulated [^{35}S]GTPγS autoradiography in rat brain. *J. Neurosci.* **16,** 8057–8066.
92. Rubino, T., Patrini, G., Parenti, M., Massi, P., and Parolaro, D. (1997) Chronic treatment with a synthetic cannabinoid CP-55,940 alters G-protein expression in the rat central nervous system. Molecular *Brain Res.* **44,** 191–197.

93. Krupnick, J. G. and Benovic, J. L. (1998) The role of receptor kinases and arrestins in G protein-coupled receptor regulation. *Annu. Rev. Pharmacol. Toxicol.* **38,** 289–319.
94. Ferguson, S. S. (2001) Evolving concepts in G protein-coupled receptor endocytosis: the role in receptor desensitization and signaling. *Pharmacol. Rev.* **53,** 1–24.
95. Pitcher, J., Freedman, N., and Lefkowitz, R. (1998) G-protein-coupled receptor kinases. *Annu. Rev. Biochem.* **67,** 653–692.
96. Jin, W., Brown, S., Roche, J. P., Hsieh, C., Celver, J. P., Kovoor, A., Chavkin, C., and Mackie, K. (1999) Distinct domains of the CB1 cannabinoid receptor mediate desensitization and internalization. *J. Neurosci.* **19,** 3773–3780.
97. Bohn, L. M., Gainetdinov, R. R., Lin, F. T., Lefkowitz, R. J., and Caron, M. G. (2000) Mu-opioid receptor desensitization by beta-arrestin-2 determines morphine tolerance but not dependence. *Nature* **408,** 720–723.
98. Pacheco, M., Childers, S. R., Arnold, R., Casiano, F., and Ward, S. J. (1991) Aminoalkylindoles: actions on specific G-protein-linked receptors. *J. Pharmacol. Exp. Ther.* **257,** 170–183.
99. Little, P. J. and Martin, B. R. (1991) The effects of Δ^9-tetrahydrocannabinol and other cannabinoids on cAMP accumulation in synaptosomes. *Life Sci.* **48,** 1133–1141.
100. Bidaut-Russell, M., Devane, W. A., and Howlett, A. C. (1990) Cannabinoid receptors and modulation of cyclic AMP accumulation in the rat brain. *J. Neurochem.* **55,** 21–26.
101. Koob, G. F. (1996) Drug addiction: the yin and yang of hedonic homeostasis. Neuron **16,** 893-896.
102. Corchero, J., Fuentes, J. A., and Manzanares, J. (1999) Chronic treatment with CP-55,940 regulates corticotropin releasing factor and proopiomelanocortin gene expression in the hypothalamus and pituitary gland of the rat. *Life Sci.* **64,** 905–911.
103. Corchero, J., Manzanares, J., and Fuentes, J. A. (1999) Repeated administration of delta9-tetrahydrocannabinol produces a differential time related responsiveness on proenkephalin, proopiomelanocortin and corticotropin releasing factor gene expression in the hypothalamus and pituitary gland of the rat. *Neuropharmacology* **38,** 433–439.
104. Kittler, J. T., Grigorenko, E. V., Clayton, C., Zhuang, S. Y., Bundey, S. C., Trower, M. M., Wallace, D., Hampson, R., and Deadwyler, S. (2000) Large-scale analysis of gene expression changes during acute and chronic exposure to [Delta]9-THC in rats. *Physiol. Genomics* **3,** 175–185.
105. Buckley, N. E., McCoy, K. L., Mezey, E., Bonner, T., Zimmer, A., Felder, C. C., and Glass, M. (2000) Immunomodulation by cannabinoids is absent in mice deficient for the cannabinoid CB(2) receptor. *Eur. J. Pharmacol.* **396,** 141–149.
106. Zimmer, A., Zimmer, A. M., Hohmann, A. G., Herkenham, M., and Bonner, T. I. (1999) Increased mortality, hypoactivity, and hypoalgesia in cannabinoid CB1 receptor knockout mice. *Proc. Natl. Acad. Sci. USA* **96,** 5780–5785.

12

Opioid System Involvement in Cannabinoid Tolerance and Dependence

Rafael Maldonado

1. Introduction

Derivatives of *Cannabis sativa*, such as marijuana and hashish, are the most widely consumed illicit drugs in humans. However, the potential ability of cannabis derivatives to produce dependence in humans is still a controversial issue. Several authors have reported that cannabis derivatives do not induce physical dependence in humans, whereas others describe some symptoms of abstinence in heavy users of strong cannabis preparations *(1–3)*. The rarity of reports of these withdrawal reactions may reflect the fact that they are mild and seldom observed in cannabis users. However, cannabis derivatives produce clear subjective motivational responses in humans, leading to drug-seeking behavior and abuse.

Many studies have used different animal models to clarify the consequences of chronic exposure to cannabinoid agonists and the abuse liability of these compounds. Tolerance and withdrawal syndrome are adaptive responses to the prolonged exposure of neurons to different drugs, but provide only a partial correlate of their addictive properties. The main factor common to all drugs of abuse is their ability to induce drug-seeking behavior, which is due to the positive-reinforcing effects of the drugs. Indirect indices of reinforcement can be evaluated through the ability of a drug to module the reinforcing properties of other rewards (e.g., intracranial self-stimulation techniques) or to impart reinforcing properties on previously neutral stimuli or environments (e.g., place-conditioning paradigm). Drug reinforcement can also be evaluated directly by using operant self-administration paradigms *(4)*. Biochemical and electrophysiological studies can also be helpful to clarify the potential addictive properties of drugs of abuse. Indeed, these kinds of studies have identified the mesolimbic dopaminergic system as the common neuronal substrate for the motivational and rewarding properties of drugs of abuse *(5)*.

The endogenous opioid system has been reported to be a common neurobiological substrate involved in the development of dependence to several drugs of abuse, including cannabinoids. Neurons containing endogenous opioid peptides are largely distributed within the central nervous system, and three different subtypes of opioid receptors, μ-, δ-, and κ-opioid receptors, have been identified and cloned *(6)*. Cannabinoid compounds induce their pharmacological effects by activating two different receptors that have been identified and cloned: the CB1 cannabinoid receptor, which is highly

From: *Molecular Biology of Drug Addiction*
Edited by: R. Maldonado © Humana Press Inc., Totowa, NJ

expressed in the central nervous system *(7)*; and the CB2 cannabinoid receptor, which is localized in the peripheral tissues, mainly at the level of the immune system *(8)*. Anatomical studies have found that CB1 cannabinoid receptor and μ-opioid receptor mRNA are collocalized in brain limbic areas associated with dependence *(9)*. CB1 cannabinoid and μ-opioid receptors are also found in a similar population of striatal GABAergic neurons *(10–12)*. Furthermore, a recent study using electron microscopy has revealed that both CB1 cannabinoid and μ-opioid receptors are collocalized in somata and dendrites of the same striatal neurons, suggesting potential coupling to similar second messenger systems *(13)*.

Cannabinoid and opioid receptors are seven transmembrane domain receptors coupled to G proteins, and through these proteins, the activation of cannabinoid *(14)* and opioid receptors *(15)* produces a similar inhibition of adenylyl cyclase activity. The stimulation of cannabinoid *(16)* and opioid receptors *(17)* also modifies the activity of other second messenger systems, and both produce an increase in the activity of the mitogen-activated protein (MAP) kinase pathway. Activation of cannabinoid and opioid receptors also has similar consequences on the permeability of several ion channels. Thus, both opioid *(18,19)* and cannabinoid *(20)* agonists increase the permeability of potassium channels and inhibit calcium influx. The pharmacological responses induced in vivo by opioid and cannabinoid agonists in different animal species also have some common features. Thus, both groups of compounds share several pharmacological properties, including antinociception, hypothermia, hypolocomotion, hypotension, sedation, and decrease of gastrointestinal motility *(21)*.

Based on these anatomical, biochemical, and pharmacological findings, several pharmacological and molecular studies have investigated the involvement of the endogenous opioid system in the different responses induced by acute and chronic administration of cannabinoid agonists. The involvement of the endogenous opioid system in acute cannabinoid responses has been investigated mainly in the case of antinociceptive effects. Biochemical studies have also revealed that cannabinoids are able to modify the activity of the endogenous opioid system. Furthermore, the opioid system has been reported to be involved in several adaptive and motivational responses induced by repeated cannabinoid administration that are related to the development of addictive processes. This chapter focuses on the role played by the opioid system in these acute and chronic effects induced by cannabinoids. A first section briefly summarizes the extensive literature that has been devoted to the involvement of the opioid system in cannabinoid antinociception. Other sections are devoted to the participation of the opioid system in the various pharmacological responses of cannabinoids related to their addictive properties.

2. Participation of Opioid Mechanisms in Cannabinoid Antinociception

CB1 and CB2 cannabinoid agonists induce antinociceptive responses that have been revealed in several animal species, including mouse, rat, rabbit, cat, dog, and monkeys *(22)*. These antinociceptive properties have been shown in several acute behavioral models of nociception: thermal models such as the radiant heat tail flick *(23)*, tail immersion *(24)*, and the hot plate *(23–25)*, mechanical models evaluating motor *(26)* or reflex responses *(27)*, chemical tests such as the abdominal constrictions induced by phenylbezoquinone *(28,29)*, acetic acid or formic acid *(30)* and electrical stimulation of the paw *(31)*, sciatic nerve *(30)*, or tooth *(32)*. Electrophysiological studies have largely confirmed these antinociceptive properties of cannabinoids *(33)*. Cannabinoid agonists also induce antinociception in inflammatory models of pain, such as hyperalgesia induced by carrageenan *(34)*, capsaicin *(35)*, formalin *(36–38)*, and Freund's adjuvant *(39)*. Doses of cannabinoid agonists required to induce antinociception in inflammatory processes are usually lower than those required in other nociceptive models *(35)*, in agreement with the anti-inflammatory properties of cannabinoid agonists *(22)*. Cannabinoid agonists are also effective in visceral models of pain, such as the bladder wall inflammation induced by turpentine administration *(38)*, and neuropathic models, such as the painful mononeuropathy induced by loose ligation of sciatic nerve *(40–42)*. In contrast with opioid responses, antinociceptive effects of cannabinoids in neuropathic pain are not decreased after repeated administration and seem to be independent of *N*-methyl-D-aspartic acid (NMDA) activity, which is closely involved in the development of tolerance *(42)*.

CB1 cannabinoid receptors are involved in the systemic antinociceptive responses of cannabinoid agonists, as revealed by the use of selective CB1 antagonists *(22)*, oligodeoxyantinucleotide directed against the CB1 receptor *(43)*, and knockout mice lacking CB1 receptors *(44,45)*. However, other cannabinoid receptors different from the CB1 receptor could also be involved in the central antinociceptive effects of cannabinoids *(29,46–50)*. At the peripheral level, both CB1 and CB2 cannabinoid receptors seem to participate in cannabinoid antinociception *(37,51–53)*.

Pharmacological studies have suggested that the endogenous opioid system could be involved in cannabinoid antinociception. Thus, opioid antagonists have been shown to attenuate antinociceptive responses induced by cannabinoids in some experimental conditions *(22)*. Several studies have reported contradictory results concerning the ability of naloxone, a nonselective but preferential μ-opioid antagonist, to antagonize cannabinoid antinociception. Thus, naloxone (1 mg/kg) has been reported to block the antinociceptive effects induced by Δ^9-tetrahydrocannabinol (THC) on the mouse hot-plate test, but very high doses (10 and 20 mg/kg) were required to attenuate the responses of THC in this animal species in the tail flick and the abdominal constrictions induced by phenylbezoquinone *(25)*. Other studies have also reported that THC-induced antinociception in the mouse tail flick was unaffected by naloxone administration *(47,54)*. Similarly, the effects induced by the cannabinoid agonist WIN-55,212 on the electrophysiological responses of rat rostral ventromedial neurons to thermal pain were not modified by naloxone *(55)*. Naloxone was also unable to block the antinociceptive effects induced by intrathecal *(56)* and intracerebroventricular *(25)* administration of THC in the mouse tail flick test. Quadazocine, another preferential μ-opioid antagonist, has been reported not to modify the antinociceptive effects of THC in rhesus monkeys subjected to the tail-immersion test *(53,57)*. However, other studies have reported that naloxone (1 mg/kg) attenuates the antinociceptive responses induced by THC in the mouse tail flick and hot plate tests *(58)* and by 11-OH-Δ^8-THC in the mouse hot-plate test *(59)*.

Selective opioid antagonists have been used to study the specific involvement of the various opioid receptors. The selective δ-opioid antagonist ICI-174,864 failed to block the antinociceptive responses induced by THC in the mouse tail flick *(29,47)*. Naltrindole, another δ-selective antagonist, did not modify THC-induced antinociception in the mouse tail-flick and hot-plate tests *(60)*. The administration of the κ-selective antagonist norbinaltorphimine *(29,46,48,58,61,62)*, antisense oilgodeoxynucleotides directed against the κ-opioid receptor *(63,64)*, or antiserum against endogenous dynorphins *(61,48,58)* has been reported to attenuate cannabinoid-induced antinociception. κ-Opioid receptors seem to be involved in cannabinoid antinociception at the spinal level but do not seem to participate in cannabinoid supraspinal antinociception *(58,65)*.

Recent studies using knockout mice deficient in the various opioid receptors or opioid peptide precursors have provided new highlights to clarify the involvement of the endogenous opioid system in cannabinoid antinociception. Thus, antinociceptive responses induced by a high dose of THC (20 mg/kg) in the tail-immersion and hot-plate tests were not modified in knockout animals deficient in μ-, δ-, or κ-opioid receptors *(66)*. Therefore, the suppression of a single opioid receptor was not enough to alter this acute cannabinoid response. However, an attenuation of THC-induced antinociception in the tail-immersion test was observed in knockout mice deficient in the preproenkephalin gene *(67)*, but derivatives from proenkephalin are not selective agonists of any opioid receptor. THC-induced antinociception in the tail-immersion test was also reduced in pro-dynorphin knockout mice, whereas the effects of THC in the hot-plate test remained unaffected in these animals *(68)*. Most of the peptides derived from prodynorphin are preferential agonists of the κ-opioid receptor but are not selective compounds and can therefore activate other opioid receptors. The different results obtained in studies using selective opioid antagonists and knockout mice could be due to the doses of opioid antagonists used. Indeed, most of these previous pharmacological studies used rather high doses, which could produce a crossreactivity with other opioid receptors.

Synergistic effects of opioid and cannabinoid agonists have been reported on antinociception *(26,29,48,61,69,70)*. This synergism has been shown after the administration of these compounds by intrathecal *(29,48,61,69,)*, intracerebroventricular *(29)*, and systemic routes *(26)*, suggesting that cannabinoid–opioid synergism occurs at both spinal and supraspinal levels *(22)*. μ-Opioid receptors are selectively involved in the antinociceptive responses induced by morphine *(71–73)*. However, the facilitatory effects of THC on morphine-induced antinociception seem to be mediated by different opioid receptors, that is, δ and κ receptors *(61)*, which could have important therapeutic implications. Another study has revealed that δ-opioid receptors are involved in THC-facilitated opioid antinociception, but it suggests that κ-opioid receptors would not be involved in this interaction *(74)*. Finally, it has been also suggested that μ-opioid receptors could be involved in THC-potentiated morphine antinociception at the supraspinal level, whereas κ-opioid receptors would be involved at the spinal level *(58,65)*. Further studies must be performed in order to clarify the exact mechanism involved in this cannabinoid–opioid synergism.

3. Similar Neuroanatomical Sites are Involved in Cannabinoid and Opioid Antinociception

Supraspinal, spinal, and peripheral mechanisms seem to be involved in the antinociceptive properties of cannabinoid agonists. Thus, intracerebreventricular *(29,49,50)*, intrathecal *(29,39,75)*, and local peripheral administration of cannabinoids *(37,75)* are able to induce potent antinociceptive responses. Studies using local administration of cannabinoid agonists in various brain structures have identified the central areas involved specifically in cannabinoid antinociception. Thus, antinociceptive responses have been induced by local microinjection of cannabinoid agonists in the periaqueductal gray matter *(39,76)*, rostral ventromedial medulla *(77)*, submedius and lateral posterior nuclei of the thalamus *(78)*, superior colliculus, central and basolateral nuclei of the amygdala, and A5 noradrenergic group in the brainstorm *(39,77)*. Autoradiographic *(79,80)* and immunocytochemistry studies *(10)* have revealed that these brain areas contain CB1 cannabinoid receptors. These neuroanatomical structures are involved in pain transmission and perception as well as in the descending pain control system *(81)*, and are also sensitive to local opioid microinjection-induced antinociception *(82)*. Indeed, both cannabinoid and opioid compounds modulate the activity of the descending pathways controlling spinal nociceptive neurons *(55,83)*. These modulatory effects are responsible, at least in part, for the antinociceptive responses of cannabinoids and opioids. The rostral ventromedial medulla is a brain structure involved in these modulatory effects. Thus, endogenous cannabinoids and opioids seem to be tonically released and to control basal nociceptive thresholds through the modulation of neuronal activity in this brain structure. However, cannabinoids and opioids act in the rostral ventromedial medulla by independent mechanisms *(55)*. The periaqueductal gray matter has also been reported to participate in the modulatory effects induced by both cannabinoids and opioids in the descending control of spinal nociceptive neurons. Noxious stimuli increase the release of endogenous anandamide in the dorso-lateral portion of the periaqueductal gray matter, which would inhibit pain transmission *(84)*. The site of opioid action in the periaqueductal gray matter is different. Indeed, electrical stimulation of the ventral part of the periaqueductal gray matter, but not the dorso-lateral portion, produces antinociception mediated by the release of endogenous opioid peptides *(75,85)*. Cannabinoids as well as opioids activate this pain-control descending pathway in the rostral ventromedial medulla *(86)* and the periaqueductal gray matter *(87)* by blocking the inhibitory GABA inputs.

Spinal cord is another important neuroanatomical site for cannabinoid and opioid antinociception *(83,88)*. Intrathecal administration of cannabinoids blocks *c-fos* expression induced in the dorsal horn of the spinal cord by noxious stimuli *(83)*. Similarly, opioid agonists are also able to inhibit nociception-induced *c-fos* expression in the dorsal horn of the spinal cord *(89)*. CB1 cannabinoid *(90)* and opioid receptors *(91)* are abundant in the dorsal horn of the spinal cord responsible for pain transmission. CB1 cannabinoid receptors are also present in dorsal root *(83,92,93)*. In contrast with opioid receptors, only a minority of cannabinoid receptors within the spinal cord seem to be localized on small-diameter primary afferent fibers responsible for pain transmission, and most primary afferent neurons that express CB1 receptor mRNA are those with larger-diameter fibers involved in the transmission of non-nociceptive sensitive inputs *(90)*. In agreement with this hypothesis, CB1 mRNA cannabinoid receptor is not collocalized extensively in rat dorsal root ganglia with neuropeptides involved in pain transmission, such as substance P, calcitonin gene-related peptide, and somatostatin *(83)*. This localization is different from opioid receptors that are mainly collocalized in neurons containing substance P *(94)*. In spite of the spinal localization of CB1 cannabinoid receptors, cannabinoids are more effective in inhibiting the activity of nociceptive C fibers than the activity of non-nociceptive Aβ- or Aδ-fibers *(51)*, and seem to inhibit the release of neurotransmitters responsible for pain transmission such as substance P or calcitonin gene-related peptide *(22)*.

CB1 cannabinoid receptors are also located on the peripheral terminals of the primary afferent neurons, and an anterograde axonal transport of cannabinoid receptors has been revealed from dorsal root ganglia toward the peripheral terminals of sensory nerves *(83)*. On these peripheral terminals, CB1 and CB2 cannabinoid receptors have been reported to inhibit nociceptive transmission, and both receptors seem to be activated by an endogenous cannabinoid tone *(37,51–53)*. Similarly, the endogenous opioid system also participates in the control of pain at the peripheral level. Thus, the presence of peripheral opioid receptors has been reported in peripheral tissues *(95–97)*, and endogenous opioid peptides seem to participate in the control of pain at this level *(98–100)*.

Therefore, cannabinoid and opioid agonists are able to inhibit pain transmission at supraspinal, spinal, and peripheral levels, and both groups of compounds modify the activity of the descending pathways controlling spinal nociceptive neurons, which also participates in their antinociceptive effects.

4. Effects of Cannabinoids on Endogenous Opioid Peptides

Biochemical studies have revealed that cannabinoid administration can increase the release of several endogenous peptides, which can contribute to the acute and chronic pharmacological responses induced by these compounds. Thus, acute intrathecal administration of THC and other cannabinoid agonists has been reported to enhance in vivo the extracellular levels of endogenous dynorphins in the spinal cord *(48,61,62,70,101,)*. These dynorphins seem to play an important role in the antinociception induced by cannabinoids at the spinal level *(48,62)*. Different dynorphins seem to be released by the administration of several cannabinoid agonists. Thus, an increase of the extracellular levels of dynorphin A has been observed after THC administration, whereas CP-55,940, another cannabinoid agonist, has been reported to enhance the release of dynorphin B *(48)*. However, the administration of the endogenous cannabinoid ligand anandamide did not modify the extracellular concentration of dynorphins *(102)*.

The antinociceptive responses induced by THC are potentiated by the central administration of inhibitors of enkephalin catabolism, suggesting that cannabinoids are able to enhance the release of endogenous enkephalins *(60)*. A recent in vivo microdialysis study has demonstrated this hypothesis. Thus, acute systemic administration of THC has been reported to increase the extracellular levels of endogenous enkephalins in the nucleus accumbens *(103)*. However, acute intrathecal THC administration did not modify spinal concentrations of Leu-enkephalin *(104)*.

Cannabinoid administration has also been reported to enhance mRNA levels of endogenous opioid peptide precursors. Indeed, repeated THC administration increases preproenkephalin and prodynorphin mRNA levels in the spinal cord *(105)*. Preproenkephalin mRNA levels were also enhanced in several brain structures, including the hypothalamus, after chronic THC treatment *(106)*. An increase in the expression of preproenkephalin mRNA has also been reported in the striatum after acute THC administration *(80)*. Levels of pro-opiomelanocortin mRNA were also enhanced in the hypothalamus after repeated THC treatment *(107)*.

5. Participation of Opioid Mechanisms in the Development of Cannabinoid Tolerance

Chronic administration of cannabinoid agonists develops tolerance to most of their pharmacological responses. Indeed, several studies have shown tolerance to cannabinoid effects on antinociception *(24,25)* locomotion *(24,108–110)*, hypothermia *(24,111)*, catalepsy *(112)*, suppression of operant behavior *(113,114)*, gastrointestinal transit *(115)*, body weight *(24)*, cardiovascular system *(116,117)*, anticonvulsivant activity *(118)*, ataxia *(119)*, and corticosterone levels *(120)*. This tolerance has been reported in rodents but also in pigeons, dogs, and monkeys *(121)*. The development of cannabinoid tolerance is particularly rapid, and an important decrease of the acute response has been already observed after the second administration of a cannabinoid agonist *(24,121,122)*. Tolerance reaches a maximal degree rather soon during chronic cannabinoid treatment *(123)*.

Different pharmacokinetic mechanisms have been suggested to be involved on cannabinoid tolerance, such as changes in drug absorption, distribution, biotransformation, and excretion. However, the role of such pharmacokinetic mechanisms, if any, seems minor *(108,119,124)*. In contrast, pharmacodynamic events play a crucial role in cannabinoid tolerance. Indeed, a significant decrease in the total number of CB1-binding sites *(125–130)* and the levels of mRNA coding for the CB1 cannabinoid receptor have been reported in several brain areas during chronic administration of cannabinoid agonists *(131,132)*. A widespread decrease in mRNA levels of $G_{\alpha i}$ and $G_{\alpha s}$ proteins has also been reported in rats treated chronically with cannabinoids *(133)*. Changes in G-protein expression have been postulated to be related to desensitization of CB1 cannabinoid receptors. Accordingly, a strong reduction of cannabinoid agonist-stimulated [^{35}S]GTPγS binding has been obtained in most brain regions of rats treated chronically with cannabinoids *(129,134)*.

Several studies have revealed the presence of a crosstolerance among different exogenous CB1 cannabinoid agonists for their main pharmacological responses, such as antinociception, hypolocomotion, catalepsy, and hypothermia *(135,136)*. The presence of crosstolerance between opioid and cannabinoid compounds has also been revealed. Thus, systemic THC administration can induce tolerance to the antinociceptive *(137–139)* and cardiac rhythm effects *(138)* produced by systemically administered morphine. However, tolerance to the antinociception induced by systemic morphine was not always observed in THC-tolerant animals *(140)*. Morphine pretreatment can also induce tolerance to the antinociceptive effects of systemic THC *(138,140,141)*. However, tolerance to antinociception induced by systemic and intrathecal THC administration was not always observed in morphine-tolerant rats *(25,42)*, and even a potentiation of the antinociceptive effects of cannabinoids has been reported in animals pretreated with morphine *(142)*. Crosstolerance between CB1 cannabinoid agonists and κ-opioid agonists on antinociceptive responses has also been reported. Thus, systemic repeated THC administration induces tolerance to antinociception produced by dynorphins (preferential κ agonists) and by the selective κ agonists U-50,488H and CI-977 *(46,143)*. Systemic repeated administration of U-50,488H and CI-977 also renders mice tolerant to the antinociceptive effect of intrathecal THC administration *(46)*. Furthermore, administration of antisense oligodeoxynucleotides that selectively blocked the expression of κ-opioid receptors increased the development of tolerance induced by chronic THC *(64)*. These results are in agreement with the release of the endogenous κ-opioid agonist dynorphin induced by THC administration and reported in the previous section *(70,102)*. However, no correlation between THC-induced dynorphin A release and development of tolerance to THC antinociception has been reported during chronic intrathecal THC administration *(101)*. Development of tolerance to the antinociceptive effects of the endogenous cannabinoid anandamide seems to involve different mechanisms to those implicated in tolerance to other cannabinoids. Indeed, anandamide-tolerant animals did not show crosstolerance to the antinociceptive responses induced by μ-, δ-, or κ-opioid agonists, while THC-tolerant animals showed crosstolerance to κ-opioid agonists under similar experimental conditions *(143)*. Furthermore, controversial results have been reported on the possible crosstolerance between anandamide and other cannabinoid agonists *(22)*.

Recent studies using knockout mice deficient in the different components of the endogenous opioid system have provided new data about the mechanisms involved in cannabinoid tolerance. The development of tolerance to the hypothermic effects of THC was not modified in knockout mice lacking the preproenkephalin gene. However, these knockout mice showed a decrease in the development of tolerance to THC antinociceptive effects and a slight attenuation of tolerance to THC-induced hypolocomotion *(67)*. The development of tolerance to the different pharmacological responses of THC was not significantly modified in knockout mice deficient in the prodynorphin gene *(68)*. THC tolerance has been also investigated in knockout mice deficient in the various opioid receptors *(144)*. Thus, the development of tolerance to THC-induced hypothermia, hypolocomotion and antinociception was not modified in knockout mice lacking μ- or δ-opioid receptor. κ-Opioid receptor knockout mice showed a slight decreased tolerance to THC hypolocomotor effects. However, the development of tolerance to THC-induced antinociception, and hypothermia was not significantly modified in κ knockout mice *(144)*. These results indicate that the suppression of a single opioid receptor does not have significant consequences on the development of cannabinoid tolerance. Opioid peptide derivatives from the prodynorphin gene do not seem to participate in cannabinoid tolerance. However, peptide derivatives from the preproenkephalin gene participate in the development of tolerance to antinociceptive effects.

6. Participation of Opioid Mechanisms in Cannabinoid-Induced Dependence

Several studies have reported the absence of somatic signs of spontaneous withdrawal syndrome after chronic THC treatment in rodents, pigeons, dogs, and monkeys, even after the administration of extremely high doses of this compound *(145–151)*. However, a recent study has reported somatic signs of spontaneous abstinence after the abrupt interruption of chronic treatment with the cannabinoid agonist WIN 55,212-2 *(151)*. Differences in the pharmacokinetic properties of THC and WIN 55,212-2 could explain these results. In contrast, the administration of the selective CB1 cannabinoid receptor antagonist SR 141716A in animals (mouse, rat, and dog) treated chronically with THC has been shown to precipitate somatic manifestations of cannabinoid withdrawal. In rodents, this cannabinoid withdrawal syndrome is characterized by the presence of a large number of somatic signs and the absence of vegetative manifestations. The most characteristic somatic manifestations of cannabinoid withdrawal in rodents are wet-dog shakes, head shakes, facial rubbing, front paw tremor, ataxia, hunched posture, body tremor, ptosis, piloerection, hypolocomotion, mastication, licking, rubbing, and scratching *(24,44,151–155)*. It is important to point out the dramatic motor impairment occurring during the cannabinoid withdrawal *(24,155)*. Doses of THC required to induce dependence in rodents are very high, and SR 141716A-precipitated withdrawal is currently reported after chronic administration of doses from 10 to 100 mg/kg daily of THC *(24,44,152–155)*. CB1 cannabinoid receptors are responsible for the somatic manifestations of cannabinoid withdrawal. Thus, SR 141716A administration in CB1 knockout mice receiving chronic THC treatment did not precipitate any manifestation of cannabinoid abstinence *(44)*.

Bidirectional interactions between cannabinoid and opioid dependence have been reported. Thus, administration of the CB1 cannabinoid antagonist SR 141716A is able to precipitate several behavioral and biochemical manifestations of opioid withdrawal in morphine-dependent rats *(9)*. Additionally, the opioid antagonist naloxone precipitated behavioral signs of withdrawal in rats treated chronically with cannabinoid agonists *(9,156)*, but the severity of the abstinence was less intense than the withdrawal syndrome precipitated by the homologous antagonist of each system: SR 141716A in cannabinoid-dependent rats and naloxone in opioid-dependent ones *(9)*. Accordingly, SR 141716A administration in morphine-dependent mice did not precipitate jumping, the most important behavioral manifestation of opioid withdrawal in this animal species *(157)*. In contrast with the previous results obtained in rats, a study has recently reported that SR 141716A administration did not precipitate any behavioral signs of withdrawal in morphine-dependent mice, and naloxone challenge did not produce any behavioral changes in cannabinoid-dependent mice *(158)*.

Molecular studies have reported a significant decrease in the severity of morphine withdrawal syndrome in knockout mice deficient in CB1 cannabinoid receptors *(44)*. In addition, acute administration of several cannabinoid agonists *(158–162)*, including anandamide *(163)*, has been reported to attenuate the severity of naloxone-precipitated morphine withdrawal. A similar effect was observed when THC was administered for a long period of time before starting opioid dependence. Thus, chronic pretreatment with THC (10 mg/kg once daily during 3 wk) before starting chronic opioid administration decreased the somatic manifestations of naloxone-precipitated morphine withdrawal and did not modify morphine rewarding effects *(67)*. Therefore, long term preexposure to cannabinoids does not seem to modify motivational responses of opioids related to their addictive properties.

Recent studies have investigated the involvement of the endogenous opioid system on the somatic expression of cannabinoid abstinence by using knockout mice. Thus, the severity of SR 141716A-precipitated cannabinoid withdrawal was decreased in THC-dependent knockout mice lacking the preproenkephalin gene *(67)*, but was not modified in knockout mice deficient in the prodynorphin gene *(68)*. The behavioral expression of SR 141716A-precipitated THC withdrawal was also evaluated in knockout mice lacking the various opioid receptors. Cannabinoid withdrawal syndrome precipitated by SR 141716A administration was not modified in μ-, δ-, or κ-opioid receptor knockout mice treated chronically with 20 mg/kg (twice daily) of THC *(144)*. Another recent study has reported a decrease in the severity of cannabinoid withdrawal syndrome in μ-knockout mice treated chronically once daily with 30 and 100 mg/kg of THC, but not those receiving 10 mg/kg of THC once daily *(158)*. Therefore, endogenous opioid peptides derivative from preproenkephalin are important for the somatic expression of cannabinoid abstinence by acting on μ- and probably other opioid receptors. The use of combinatorial opioid-receptor knockout mice lacking two or three opioid receptors will clarify these findings in the near future.

Common features of the withdrawal syndrome to several drugs of abuse, such as opioids, psychostimulants, and ethanol, include an important elevation in extracellular levels of corticotropin-releasing factor in the mesolimbic system *(164)*, and a marked inhibition of mesolimbic dopamine activity *(165)*. Similar behavioral changes have been reported during cannabinoid withdrawal. Thus, an increased release of corticotropin-releasing factor and an enhancement of Fos immunoreactivity have been found in the central amygdala during SR 141716A-precipitated cannabinoid withdrawal *(166)*. This alteration of corticotropin-releasing factor function in the limbic system may have a motivational role in mediating the stresslike symptoms and negative affect that accompany cannabinoid withdrawal. In agreement with this hypothesis, the spontaneous firing rate of ventral tegmental area dopamine neurons has been reported to be attenuated during cannabinoid abstinence *(150)*. This decreased dopaminergic activity also seems to be related to the aversive/dysphoric consequences of cannabinoid withdrawal. However, changes in the activity of the dopaminergic system are not involved in the expression of the somatic signs of cannabinoid withdrawal *(92)*.

Cannabinoid and opioid withdrawal syndromes are associated with compensatory changes in the cyclic AMP pathway. Initially, acute activation of cannabinoid *(14)* and opioid receptors *(167)* leads to an inhibition of adenylyl cyclase activity. In contrast, SR 141716A-precipitated THC withdrawal *(24)* and naloxone-precipitated morphine withdrawal *(168)* produce an increase in adenylyl cyclase activity in vivo in the central nervous system in rodents. In spite of these common biochemical mechanisms, different brain structures have been reported to be involved in the physical manifestations of opioid and cannabinoid withdrawal. Thus, locus coeruleus *(169)* and other brainstem structures, such as the periaqueductal gray matter *(170)*, are responsible for the somatic signs of opioid withdrawal. In the case of cannabinoid dependence, the cerebellum seems to play a crucial role for the somatic expression of THC withdrawal *(24,144,155)*. Indeed, basal, forskolin- and calcium/calmodulin-stimulated adenylyl cyclase activities were selectively increased in the cerebellum, but not in other brain structures (frontal cortex, hippocampus, striatum, and periaqueductal gray matter) *(24)*. Furthermore, the behavioral manifestations of cannabinoid abstinence are markedly reduced by preventing the activation of protein kinase A (PKA) by microinfusion of the selective cyclic AMP inhibitor Rp-8Br-cAMPS onto the surface of the cerebellum *(155)*. Similar increases in cyclic AMP levels and PKA activity have been reported during chronic THC treatment in the cerebellum, striatum, and cortex *(130)*.

7. Participation of Opioid Mechanisms in Cannabinoid-Induced Rewarding Effects

7.1. Intracranial Self-Stimulation

Intracranial self-stimulation is an experimental procedure that is widely used to explore brain circuits involved in reward as well as the rewarding properties of drugs of abuse. When an animal is equipped with an electrode placed in a brain rewarding circuit and given the opportunity to perform a behavioral response, such as pressing a lever, that is followed by a short-pulse train of electrical current via the electrode, the animal will initiate and maintain responding *(171)*. A common property of most of drugs of abuse is to acutely facilitate electrical stimulation of brain reward loci, presumably due to their euphorigenic properties *(172,173)*. Acute administration of THC has been reported to lower intracranial self-stimulation thresholds in rats, suggesting the activation of central hedonic systems *(174,175)*. The opioid antagonist naloxone blocked the enhancing effects of THC on electrical intracranial self-stimulation reward, at doses of naloxone that themselves have no effect on brain reward, which reveals the involvement of the endogenous opioid system in the rewarding effects of THC *(176)*. Interestingly, the facilitation of brain stimulation reward induced by most drugs of abuse is also reversed by naloxone administration *(177)*.

7.2. Place-Conditioning Paradigm

Conditioned place preference is a behavioral model that is currently used to measure the rewarding properties induced by the administration of a drug *(178,179)*. In this paradigm, the rewarding properties of a compound are associated with the particular characteristics of a given environment. A similar procedure can be used to explore the aversive properties of a drug. In this case, the animal will avoid staying in the compartment associated with a compound producing aversive/dysphoric effects. The administration of cannabinoid agonists currently produces aversive-like responses in the place conditioning paradigm *(21,180–183)*. Conditioned place aversion induced by cannabinoid agonists is abolished by the co-administration of SR 141716A, suggesting a selective involvement of CB1 cannabinoid receptors *(182)*.

Rewarding effects of cannabinoid agonists can also be revealed in this paradigm by using particular experimental conditions. Thus, THC produced conditioned place preference in rats when administered at lower doses than those used to induce place aversion and when animals were exposed to a 24-h washout period between the two THC-conditioning sessions *(184)*. THC also produces a clear conditioned place preference in mice using a long period of conditioning and avoiding the possible dysphoric consequences of the first drug exposure. Indeed, place preference was obtained with 1 mg/kg of THC when mice received a previous priming THC exposure in the home cage before the conditioning sessions *(185)*. Interestingly, doses of THC used in these studies were similar to those reported to facilitate intracranial self-stimulation in rats *(174)*.

Recent studies have used the place-conditioning paradigm in knockout mice to evaluate the involvement of the endogenous opioid system in the rewarding and aversive properties of cannabinoids. The motivational responses induced by THC have been investigated in knockout mice lacking μ-, δ-, or κ-opioid receptor *(144)*. The rewarding effects induced by 1 mg/kg of THC, avoiding the possible dysphoric consequences of the first drug exposure, were abolished in μ-opioid receptor knockout mice but were not modified in mice lacking δ- or κ-opioid receptor. The dysphoric effects induced by a high dose of THC (5 mg/kg) were not modified in δ-opioid receptor knockout mice, slightly attenuated in μ-knockout animals, and completely abolished in mice lacking κ-opioid receptor. These κ-opioid receptor knockout mice also showed place preference to THC (1 mg/kg) in animals that had not received any priming exposure, revealing the crucial role of this opioid receptor in THC-induced dysphoria. Results obtained with knockout mice deficient in the prodynorphin gene are in agreement with this hypothesis. Thus, the conditioned place aversion induced by a high dose of THC (5 mg/kg) was completely abolished in these prodynorphin knockout mice *(68)*. Interestingly, this interaction between the cannabinoid and opioid systems also seems to be bidirectional. Indeed, the rewarding effects induced by morphine in the conditioned place preference paradigm were blocked in knockout mice deficient in CB1 cannabinoid receptors *(186)*. Furthermore, the selective CB1 cannabinoid antagonist SR 141716A blocks the acquisition of morphine conditioned place preference, as well as the rewarding effects of other drugs of abuse, such as cocaine, and natural stimuli such as food *(182)*.

7.3. Self-Administration Studies

The procedure by which animals are permitted to intravenously self-administered drugs has provided a reliable method to evaluate directly the reinforcing properties of a psychoactive compound, and is probably the clearest indication in animals of its addictive potential in humans *(187)*. Numerous studies have shown that THC is unable to induce a self-administration behavior in any of the animal species studied *(148,149, 188–192)*. Animals that have already learned to self-administer other drugs of abuse did not self-infuse THC *(148,191,192)*. THC pharmacokinetic properties seem to be crucial for the behavioral responses on the self-administration paradigm. Indeed, WIN 55,212-2, a synthetic cannabinoid agonist that has a shorter half-life than THC, was intravenously self-administered in mice in a concentration-dependent manner, according to a two-phase "bell-shaped" curve *(193)*. Another synthetic cannabinoid agonist, CP 55,940 has been reported to sustain intracerebroventricular self-administration in rats *(194)*. A recent study has also revealed intravenous THC self-administration behavior, for doses much lower than those used previously, by monkeys that have a previous history of cocaine self-administration *(195)*. SR 141716A completely blocked self-administration induced by WIN 55,212-2 *(192)*, CP 55,940 *(194)*, and THC *(195)* indicating a selective involvement of CB1 cannabinoid receptors in the reinforcing properties of these cannabinoids.

Bidirectional interactions between cannabinoid and opioid systems have also been reported by using self-administration techniques. Indeed, morphine-induced intravenous self-administration was abolished in knockout mice lacking the CB1 cannabinoid receptors *(44)*. In agreement with this result, blockade of CB1 cannabinoid receptors by SR 141716A administration partially reversed heroin-induced intracerebroventricular self-administration *(194)*. These results indicate that the endogenous cannabinoid system is involved in opioid reinforcement. Interestingly, the endogenous opioid system also participates in the reinforcing properties induced by cannabinoids. Indeed, the opioid antagonist naloxone partially blocked self-administration induced by CP 55,940 *(194)*. THC self-administration behavior was also attenuated by the administration of the opioid antagonist naltrexone *(196)*.

7.4. Biochemical and Electrophysiological Studies

The mesolimbic dopaminergic system has been proposed as a common neuronal substrate for the rewarding properties of drugs of abuse *(5)*. Indeed, prototypical drugs of abuse, including opioids, psychostimulants, alcohol, and nicotine, increase the discharge rate of mesolimbic dopamine neurons *(197–199)*. This activation is associated with increased dopamine output in innervated projection structures, such as the nucleus accumbens *(200)*. In vivo microdialysis studies have revealed that acute administration of THC and other cannabinoid agonists also increases extracellular efflux of dopamine and its metabolites in nucleus accumbens of freely moving rats *(174,201)*. In agreement with these biochemical results, THC and other cannabinoid agonists produced a dose-dependent increase in the spontaneous firing rate of ventral tegmental area dopamine neurons *(202,203)*. The biochemical *(201)* and electrophysiological effects *(202,203)* of cannabinoids on in vivo dopamine transmission in the nucleus accumbens were prevented by the administration of SR 141716A, indicating the selective involvement of CB1 cannabinoid receptors.The endogenous opioid system has also been reported to be involved in the effects of cannabinoids on mesolimbic dopamine activity. Thus, administration of opioid antagonists was reported to block the increase in extracellular efflux of dopamine in the nucleus accumbens induced by the acute administration of cannabinoid agonists *(176,201)*. However, other studies have reported that naloxone did not modify THC-induced increase of the firing rate of dopaminergic neurons projecting to the nucleus accumbens *(203,204)* and neostriatum *(202)*.

8. Summary and Conclusions

The involvement of the endogenous opioid system in cannabinoid-induced antinociception, tolerance, and dependence has been evaluated by using pharmacological approaches and more recently by the use of new molecular tools. Pharmacological studies have suggested that the endogenous opioid system could be involved in cannabinoid antinociception. κ-Opioid antagonists were particularly effective to attenuate these cannabinoid antinociceptive responses. However, molecular studies have revealed that the suppression of a single opioid receptor did not modify THC antinociception. The different results obtained when using these two experimental approaches could be due to a crossreactivity with several opioid receptors when using opioid antagonists. In agreement with this hypothesis, molecular studies have demonstrated that opioid peptides derivatives from preproenkephalin and prodynorphin, which are not selective agonists of any opioid receptor, participate in cannabinoid antinociception. Furthermore, biochemical studies have revealed that cannabinoid administration increases the release of endogenous enkephalins and dynorphins in different brain areas.

The endogenous opioid system has been also reported to be involved in several adaptive and motivational responses induced by repeated cannabinoid administration that are related to their addictive properties. Pharmacological studies have suggested the involvement of the opioid system, and particularly κ-opioid receptors, in the development of tolerance to cannabinoid antinociception. Molecular studies using knockout mice have revealed that the suppression of a single opioid receptor has not significant consequences on the development of cannabinoid tolerance, whereas peptide derivatives from preproenkephalin gene participate in the development of tolerance to antinociceptive effects. Bidirectional interactions between cannabinoid and opioid dependence were demonstrated by using pharmacological approaches and have been now clarified by using knockout mice deficient in opioid receptors and peptide precursors. Thus, endogenous opioid peptides derivative from preproenkephalin, but not from prodynorphin, are important for the expression of cannabinoid abstinence by acting on μ- and probably other opioid receptors. The suppression of a single opioid receptor did not modify cannabinoid withdrawal, although the severity of cannabinoid abstinence was reduced in μ knockout mice when using very high doses of THC. Cannabinoid and opioid withdrawal share several biochemical common features, such as upregulation of the cyclic AMP pathway, elevation in extracellular levels of corticotropin-releasing factor in the mesolimbic system, and inhibition of mesolimbic dopamine activity. However, different brain structures have been reported to be involved in the expression of the withdrawal syndrome to these two drugs of abuse. The cerebellum seems to play an important role in the somatic manifestations of cannabinoid withdrawal, whereas other brainstem structures, such as the locus coeruleus, are responsible for the somatic signs of opioid abstinence.

Several behavioral models, such as intracranial self-stimulation, conditioned place preference, and intravenous self-administration, have revealed the rewarding properties of cannabinoid agonists. Cannabinoid rewarding properties, similarly to other prototypical drugs of abuse, seem to be related to the activation of the mesolimbic dopaminergic system. The endogenous opioid system has been also reported to play a crucial role in cannabinoid rewarding effects. Thus, opioid antagonists are able to block cannabinoid responses on intracranial self-stimulation, conditioned place preference, and intravenous self-administration, as well as the biochemical effects of THC on mesolimbic dopamine transmission. Furthermore, the rewarding properties of THC in the place-conditioning paradigm were abolished in knockout mice deficient in μ-opioid receptors, and the dysphoric effects induced by high doses of THC in this paradigm were suppressed in knockout mice deficient in κ-opioid receptors and prodynorphin gene.

In conclusion, pharmacological and molecular studies have demonstrated that the endogenous opioid system plays a key role in cannabinoid-induced antinociception, tolerance, physical dependence, and rewarding effects. The exact involvement of each component of the endogenous opioid system in some of these cannabinoid responses remains to be further elucidated.

References

1. Hollister, L. E. (1986) Health aspects of cannabis. *Pharmacol. Rev.* **38,** 1–17.
2. Haney, M., Ward, A. S., Corner, S. D., Foltin, R. W., and Fischman M. W. (1999). Abstinence symptoms following oral THC administration to humans. Psychopharmacology 141, 385–394.
3. Haney, M., Ward, A. S., Corner, S. D., Foltin, R. W., and Fischman M. W. (1999) Abstinence symptoms following smoked marijuana in human. *Psychopharmacology* **141,** 395–404.
4. Schulteis, G., Gold, L. H., and Koob, G. F. (1997) Preclinical behavioral models for addressing unmet needs in opiate addiction. *Neuroscience* **9,** 94–109.
5. Koob, G. (1992) Drugs of abuse: anatomy, pharmacology and function of reward pathways. *Trends Pharmacol. Sci.* **13,** 170–177.
6. Kieffer, B. L. (1995). Recent advances in molecular recognition and signal transduction of active peptides: receptors for opioid peptides. *Cell. Mol. Neurobiol.* **15,** 615–635.
7. Devane, W. A., Dysarz, F. A., Johnson, M. R., Melvin, L. S., and Howllett, A. C. (1988) Determination and characterization of cannabinoid receptor in rat brain. *Mol. Pharmacol.* **34,** 605–613.
8. Munro, S., Thomas, K. L., and Abu-Shaar, M. (1993) Molecular characterization of a peripheral receptor for cannabinoids. *Nature* **365,** 61–65.
9. Navarro, M., Chowen, J., Carrera, M. R. A., Del Arco, I., Villanua, M. A., Martin, Y., Roberts, A. J., Koob, G. F., and Rodriguez de Fonseca, F. (1998) CB1 cannabinoid receptor antagonist-induced opiate withdrawal in morphine-dependent rats. *Neuroreport* **9,** 3397–3402.
10. Tsou, K., Brown, S., Sañudo-Peña, M. C., Mackie, K., and Walker, J. M. (1998) Inmunohistochemical distribution of cannabinoid CB_1 receptors in the rat central nervous system. *Neuroscience* **83,** 393–411.
11. Wang H., Gracy K. N., and Pickel, V. M. (1999) Mu-opioid and NMDA-type glutamate receptors are often colocalized in spiny neurons within patches of the caudate putamen nucleus. *J. Comp. Neurol.* **412,** 132–146.

12. Hofmann A. G. and Herkenham, M. (2000) Localization of cannabinoid CB_1 receptor mRNA in neuronal subpopulations of rat striatum: a double-label in situ hibridization study. *Synapse* **37,** 71–80.
13. Rodríguez J. J., Mackie K., and Pickel, V. M. (2001) Ultrastructural localization of the CB_1 cannabinoid receptor in μ-opioid receptor patches of the rat caudate putamen nucleus. *J. Neurosci.* **21,** 823–833.
14. Howlett, A. C. and Fleming, R. M. (1984) Cannabinoid inhibition of adenylyl cyclase. Pharmacology of the response in neuroblastoma cell membranes. *Mol. Pharmacol.* **27,** 429–439.
15. Sharma, S. K., Nirenberg, M., and Klee, W. A. (1975) Morphine receptors as regulators of adenylate cyclase activity. *Proc. Natl. Acad. Sci. USA* **72,** 590–594.
16. Bouaboula, M., Poinot-chazel, C., Bourrie, B., Canat, X., Calandra, B., Rinaldi-Carmona, M., Le Fur, G., and Caselas, P. (1995) Activation of mitogen-activated protein kinases stimulation of the central cannabinoid receptor CB1. *Biochem. J.* **312,** 637–641.
17. Fukuda, K., Kato, S., Morikawa, H., Shoda, T., and Mori, K. (1996) Functional coupling of the delta-, mu-, and kappa-opioid receptors to mitogen-activated protein kinase and arachidonate release in Chinese hamster ovary cells. *J. Neurochem.* **67,** 1309–1316.
18. Morita, K. and North, R.A. (1982) Opiate activation of potassium conductance in myenteric neurons: inhibition by calcium ion. *Brain Res.* **242,** 145–150.
19. Hescheler, J., Rosentahl, W., Trautwein, W., and Schutz, G. (1987) The GTP-binding protein Go regulates neuronal calcium channels. *Nature* **325,** 445–447.
20. Felder, C. C., Veluz, J. S., Williams, H. L., Briley, E. M., and Matsuda, L. A. (1992) Cannabinoid agonists stimulate both receptor- and nonreceptor mediated signal transduction pathways in cells transfected with and expressing cannabinoid receptor clones. *Mol. Pharmacol.* **42,** 838–845.
21. Manzanares J., Corchero J., Romero J., Fernández-Ruiz J. J., Ramos J. A., and Fuentes J. A. (1999) Pharmacological and biochemical interactions between opioids and cannabinoids. *Trends Pharmacol. Sci.* **20,** 287–294.
22. Pertwee, R. G. (2001) Cannabinoid receptors and pain. *Prog. Neurobiol.* **63,** 569–611.
23. Buxbaum, D. M. (1972) Analgesic activity Δ^9-tetrahydrocannabinol in the rat and mouse. *Psychopharmacology* **25,** 275–280.
24. Hutcheson, D. M., Tzavara, E. T., Smadja, C., Valjent, E., Roques, B. P., Hanoune, J., and Maldonado, R. (1998) Behavioral and biochemical evidence for signs of abstinence in mice chronically treated with delta-9-tetrahydrocannabinol. *Br. J. Pharmacol.* **125,** 1567–1577.
25. Martin, B. R. (1985) Characterization of the antinociceptive activity of intravenously administered delta-9-tetrahidrocannabinol in mice, in: *Marihuan ´84* (Harvey, D. J., ed.), IRL Press, Oxford, UK, pp. 685–692.
26. Smith, F. I. (1998) The enhancement of morphine antinociception in mice by delta-9-tetrahydrocannabinol. *Pharmacol. Biochem. Behav.* **60,** 559–566.
27. Guilbert, P. E. (1981) A comparison of THC, nantradol, nabilone and morphine in the chronic spinal dog. *J. Clin. Pharmacol.* **21**(Suppl. I), 311S–319S.
28. Milne, G. M, Koe, B. K., and Johnson, M. R. (1979) Stereospecific and potent analgesic activity for nantradol. A structurally novel, cannabinoid-related analgesic, in *Problems of Drug Dependence,* vol. 27, (Harris, L. S., ed.), National Institute of Drug Abuse Research Monograph, NIDA, Rockville, MD, pp. 84–92.
29. Welch, S. P., Dunlow, L. D., Patrick, G. S., and Razdan, R. K. (1995) Characterization of anandamide- and fluoroanandamide-induced antinociception and crosstolerance to delta 9-THC after intrathecal administration to mice: blockade of delta 9-THC-induced antinociception. *J. Pharmacol. Exp. Ther.* **273,** 1235–1244.
30. Bicher, H. I. and Mechoulam, R. (1968) Pharmacological effects of two active constituents of marijuana. *Arch. Int. Pharmacodyn.* **172,** 24–31.

31. Weissman, A. and Milne, L. S. (1982) Cannabimimetic activity from CP-47,497, a derivative of 3-phenylcyclohexanol. *J Pharmacol. Exp. Ther.* **223,** 516–522.
32. Kaymakcalan, S., Türker, R. K. and Türker M. N. (1974) Analgesic effects of Δ^9-tetrahydrocannabinol in the dog. *Psychopharmacologia* **35,** 123–128.
33. Chapman, V. (1999) The cannabinoid CB_1 receptor antagonist, SR141716A, selectively facilitates nociceptive responses of dorsal horn neurones in the rat. *Br. J. Pharmacol.* **127,** 1765–1767.
34. Mazzari, S., Canella, R., Petrelli, L., Marcolongo, G., and Leon, A. (1996) N-(2-hydroxyethyl)hexadecanamide is orally active in reducing edema formation and inflammatory hyperalgesia by down-modulating mast cell activation. *Eur. J. Pharmacol.* **300,** 227–236.
35. Li, J., Daughters, R. S., Bullis, C., Bengiamin, R., Stucky, M. W., Brennan, J., and Simone, D. A. (1999) The cannabinoid receptor agonist WIN 55,212-2 mesylate blocks the development of hyperalgesia produced by capsaicin in rats. *Pain* **81,** 25–33.
36. Moss, D. E. and Johnson, R. L. (1980) Tonic analgesic effects of Δ^9 -tetrahydrocannabinol as measured with the formalin test. *Eur. J. Pharmacol.* **61,** 313–315.
37. Calignano, A., La Rana, G., Giuffrida, A., and Piomelli, D. (1998) Control of pain initiation by endogenous cannabinoids. *Nature* **394,** 277–281.
38. Jaggar, S. I., Hasnie, F. S., Sellaturay, S., and Rice, A. S. (1998) The anti-hyperalgesic actions of the cannabinoid anandamide and the putative CB2 receptor agonist palmitoylethanolamide in visceral and somatic inflammatory pain. *Pain* **76,** 189–199.
39. Martin, W. J., Coffin, P. O., Attias, E., Balinsky, M., Tsou, K., and Walker J. M. (1999) Anatomical basis for cannabinoid-induced antinociception as revealed by intracerebral microinjections. *Brain Res.* **822,** 237–242.
40. Herzberg, U., Eliav, E., Bennett, G. J., and Kopin, I. J. (1997) The analgesic effect of R(+)-Win 55,212-2 mesylate, a high affinity cannabinoid agonist, in a rat model of neurophatic pain. *Neurosci. Lett.* **221,** 157–160.
41. Mao, J., Price, D. D., and Mayer, D. J. (1995) Mechanisms of hyperalgesia and opiate tolerance: a current view of their possible interations. *Pain* **62,** 259–274.
42. Mao, J., Price, D. D., Lu, J., Keniston, L., and Mayer, D. J. (2000) Two distinctive antinociceptive systems in rats with pathological pain. *Neurosci. Lett.* **280,** 13–16.
43. Edsall, S. A., Knapp, R. J., Vanderah, T. W., Roeske, W. R., Consroe, P., and Yamamura, H. I.(1996) Antisense oligodeoxinucleotide treatment to the brain cannabinoid receptor inhibits antinoception. *Neuroreport* **7,** 593–596.
44. Ledent, C., Valverde, O., Cossu, G., Petitet, F., Aubert, J. F., Beslot, F., Bohme, G. A., Imperato, A., Pedrazzini, T., et al. (1999) Unresponsiveness to cannabinoids and reduced addictive effects of opiates in CB1 receptor knockout mice. *Science* **283,** 15–19.
45. Zimmer, A., Zimmer, A. M., Hohmann, A. G., Herkenham M., and Bonner, T. I. (1999) Increased mortality, hypoactivity, and hypoalgesia in cannabinioid CB_1 receptor knockout mice. *Proc. Natl. Acad. Sci. USA* **96,** 5780–578.
46. Smith, P. B., Welch, S. P., and Martin, B. R. (1994) Interactions between Δ^9-tetrahydrocannabinol and kappa opioids in mice. *J. Pharmacol. Exp. Ther.* **268,** 1381–1387.
47. Welch, S. P. (1993) Blockade of cannabinoid-induced antinociception by norbinaltorphimine but not N,N-diallyl-tyrosine-Aib-pheny-lalamine-leucine, ICI 174,864 or naloxone in mice. *J. Pharmacol. Exp. Ther.* **265,** 633–640.
48. Pugh, G., Mason, D. J., Combs, V., and Welch, S. P. (1997) Involvement of dynorphin B in the antinociceptive effects of the cannabinoid CP55,940 in the spinal cord. *J. Pharmacol. Exp. Ther.* **281,** 730–737.
49. Welch, S. P, Huffman, J. W., and Lowe, J. (1998) Differential blockade of the antinociceptive effects of centrally administered cannabinoids by SR141716A. *J. Pharmacol. Exp. Ther.* **286,** 1301–1308.

50. Raffa, R. B., Stone, D. J., and Hipp, S. J. (1999) Differential cholera-toxin sensitivity of supraspinal antinociception induced by the cannabinoid agonists Δ^9-THC, WIN 55,212-2 and anandamide in mice. *Neurosci. Lett.* **263,** 29–32.
51. Strangman, N. M., Patrick, S. L., Hohmann, A. G., Tsou, K., and Walker, J. M. (1998) Evidence for a role of endogenous cannabinoids in the modulation of acute and tonic pain sensitivity. *Brain Res.* **813,** 323–328.
52. Hanus, I., Breuer, A., Tchilibon, S., Shiloah, S., Godenberg, D., Horowitz, M., Peterwee, R. G., Ross, R. A., et al. (1999) HU-308: a specific agonist for CB_2, a peripheral cannabinoid receptor. *Proc. Natl. Acad. Sci. USA* **96,** 14,228–14,223.
53. Ko, M. C. and Woods, J. H. (1999) Local administration of Δ^9-tetrahydrocannabinol attenuates capsaicin-induced thermal nociception in rhesus monkeys: a peripheral cannabinoid action. *Psychopharmacology* **143,** 322-329.
54. Bhargava, H. N. and Matwyshyn, G. A. (1980) Influence of thyrotropin releasing hormone and histidyl-proline diketopieperazine on spontaneous locomotor activity and analgesia induced by Δ^9- tetrahydrocannabinol in mouse. *Eur. J. Pharmacol.* **68,** 147–154.
55. Meng, I. D., Manning, B. H., Martin, W. J., and Fields, H. L.(1998) An analgesia circuit activated by cannabinoids. *Nature* **395,** 381–383.
56. Welch, S. P. and Stevens, D. L. (1992) Antinociceptive activity of intrathecally administered cannabinoids alone, and in combination with morphine, in mice. *J. Pharmacol. Exp. Ther.* **262,** 10–18.
57. Vivian, J. A., Kishioka, S., Butelman, E. R., Broadbear, J., Lee, K. O., and Woods, J. H. (1998) Analgesic, respiratory and heart rate effects of cannabinoid and opioid agonists in rhesus monkeys: agonist effects of SR141716A. *J. Pharmacol. Exp. Ther.* **286,** 697–703.
58. Reche, I., Fuentes, J. A., and Ruiz-Gayo, M. (1996) A role for central cannabinoid and opioid systems in peripheral Δ^9-tetrahydrocannabinol-induced anlgesia in mice. *Eur. J. Pharmacol.* **301,** 75–81.
59. Wilson, R. S. and May, E. L. (1975) Analgesic properties of the tetrahydrocannabinols, their metabolites and analogs. *J. Med. Chem.* **18,** 700–703.
60. Reche, I., Ruiz-Gayo, M., and Fuentes, J. A. (1998) Inhibition of opioid-degrading enzymes potentiates Δ^9-tetrahydrocannabinol-induced antinociception in mice. *Neuropharmacology* **37,** 215–222.
61. Pugh, G., Smith, P. B., Dombrowski, D. S., and Welch, S. P. (1996) The role of endogenous opioids in enhancing the antinociception produced by the combination of delta 9-tetrahydrocannabinol and morphine in the spinal cord. *J. Pharmacol. Exp. Ther.* **279,** 608–616.
62. Mason, D. J., Lowe, J., and Welch, S. P. (1999) Cannabinoid modulation of dynorphin A: correlation to cannabinoid-induced antinociception. *Eur. J. Pharmacol.* **378,** 273–248.
63. Pugh, G., Abood, M. E., and Welch, S. P. (1995) Antisense oligodeoxynucleotides to the κ-1 receptor block the antinociceptive effects of Δ^9-THC in spinal cord. *Brain Res.* **689,** 157–158.
64. Rowen, D. W., Embrey, J. P., Moore, C. H., and Welch, S. P. (1998) Antisense oligodeoxynucleotides to the $kappa_1$ receptor enhance delta-9-THC-induced antinociceptive tolerance. *Pharmacol. Biochem. Behav.* **59,** 399–404.
65. Reche, I., Fuentes, J. A., and Ruiz-Gayo, M. (1996) Potentation of Δ^9-tetrahydrocannabinol-induced analgesia by morphine in mice: involvement of μ- and κ-opioid receptors. *Eur. J. Pharmacol.* **318,** 11–16.
66. Ghozland, S., Mathews, H. W. D., Simonin, F., Filliol, D., Kieffer, B. L., and Maldonado, R. (2002) Motivational effects of cannabinoids are mediated by mu- and κ-opioid receptors. *J. Neurosci.* **22,** 1146–1154.
67. Valverde, O., Maldonado, R., Valjent, E., Zimmer, A. M., and Zimmer, A. (2000) Cannabinoid withdrawal syndrome is reduced in preproenkephalin knock-out mice. *J. Neurosci.* **15,** 9284–9289.
68. Zimmer, A., Valjent, E., Koning, M., Zimmer, A. M., Clarke, S., Robledo, P., Chen, C. C., Hahn, H., Valverde, O., Hill, R. G., Kitchen, I., and Maldonado, R. (2001) Absence of delta-9-tetrahydrocannabinol dysphoric effects in dynorphin-deficient mice. *J. Neurosci.* **21,** 9499–9505.

69. Welch, S. P. and Eads, M. (1999) Synergistic interations of endogenous opioids and cannabinoid systems. *Brain Res.* **848,** 183–190.
70. Mathews, H. W. D., Maldonado, R., Simonin, F., Valverde, O., Slowe, S., Kitchen, I., Befort, K., Dierich, A., et al. (1996) Loss of morpine-induced analgesia, reward effect an withdrawal symptoms in mice lacking the mu-opioid-receptor gene. *Nature* **383,** 819–823.
71. Simonin, F., Valverde, O., Smadja, C., Slowe, S., Kitchen, I., Diereich, A., LeMeur, M., Roques, B. P., Maldonado, R., and Kieffer, B. (1998) Discruption of the kappa-opioid receptor gene in mice enhances sensibility to chemical visceral pain, impairs pharmacological actions of the selective kappa agonist U-50,488H and attenuates morphine withdrawal. *EMBO J.* **17,** 886–897.
72. Filliol, D., Ghozland, S., Chluba, J., Martin, M., Matthes, H. W. D., Simonin, F., Befort, K., Gaveriaux-Ruff, C., et al. (2000) Mice deficient for delta and mu-opioid receptors exhibit opposing alterations of emotional responses. *Nat. Genet.* **25,** 195–200.
73. Cichewicz, D. L., Martin, Z. L., Smith, F. L., and Welch, S. P. (1999) Enhancement of μ-opioid antinociception by oral Δ^9-tetrahydrocannabinol: dose-response analysis and receptor identification. *J. Pharmacol. Exp. Ther.* **289,** 859–867.
74. Richardson, J. D., Aanonsen, L., and Hargreaves, K. M. (1998) Hypoactivity of the spinal cannabinoid system results in NMDA-dependent hyperalgesia. *J. Neurosci.* **18,** 451–457.
75. Lichtman, A. H., Cook, S. A., and Martin, B. R. (1996) Investigation of brain sites mediating cannabinoid-induced antinociception in rats: evidence supporting periaqueductal gray involvement. *J. Pharmacol. Exp. Ther.* **276,** 585–593.
76. Martin, W. J., Tsou, K., and Walker, J. M. (1998) Cannabinoid receptor mediated inhibition of the rat tail-flick reflex after microinjeccion into the rostral ventromedial medulla. *Neurosci. Lett.* **242,** 33–36.
77. Martin, W. J., Hohmann, A. G., and Walker, J. M .(1996) Suppression of noxious stimulus-evoked activity in the ventral posterolateral nucleus of the thalamus by a cannabinoid agonist: correlation between electrophysiological and antinociceptive effects. *J. Neurosci.* **16,** 6601–6611.
78. Mailleux, P. and Vandergaeghen, J. J. (1992) Distribution of neuronal cannabinoid receptor in the adult rat brain: a comparative receptor binding radioautography and in situ hybridization histochemistry. *Neuroscience* **48,** 655–668.
79. Glass, M., Dragunow, M., and Faull, R. L. (1997) Cannabinoid receptors in the human brain: a detailed anatomical and quantitative autoradiographic study in the fetal, neonatal and adult human brain. *Neuroscience* **77,** 299–318.
80. Basbaum, A. I. and Fields, H. L. (1984) Endogenous pain control systems: brainstem spinal pathways and endorphin circuitry. *Ann. Rev. Neurosci.* **7,** 309–338.
81. Fields, H. L., Barbaro, N. M., and Heinricher, M. M. (1988) Brain stem neuronal circuitry underlying the antinociceptive action of opiates. *Prog. Brain Res.* **77,** 245–247.
82. Hohmann, A. G. and Herkenham, M. (1999) Localization of central cannabinoid CB1 receptor messenger RNA in neuron subpopulations of rat dorsal root ganglia: a double-label in situ hybridization study. *Neuroscience* **90,** 923–931.
83. Walker, J. M., Huang, S. M., Strangman, N. M., Tsou, K., and Sañudo-Peña, M. C. (1999) Pain modulation by release of the endogenous cannabinoid anandamide. *Proc. Natl. Acad. Sci. USA* **96,** 12,198–12,203.
84. Akil, H., Richardson, D. E., Hughes, J., and Barchas, J. D. (1978) Enkephalin-like material elevated in ventricular cerebrospinal fluid of pain patients after analgetic focal stimulation. *Science* **201,** 463–465.
85. Vaughan, C. W., MacGregor, I. S., and Christie, M. J. (1999) Cannabinoid receptor activation inhibits GABAergic neurotransmission in rostral ventromedial medulla neurons in vitro. *Br. J. Pharmacol.* **127,** 935–940.

86. Vaughan, C. W., Connor, M., Bagley, E. E., and Christie, M. J. (2000) Actions of cannabinoids on membrane properties and synaptic transmission in rat periaqueductal gray neurons in vitro. *Mol. Pharmacol.* **57,** 288–295.
87. Clarke, R. W. and Ford, T. W. (1987) The contribution of mu-, delta- and kappa-opioid receptors to the actions of endogenous opioids on spinal reflexes in the rabbit. *Br. J. Pharmacol.* **91,** 579–589.
88. Abbadie, C., Honoré, P., Fournié-Zaluski, M. C., Roques, B. P., and Besson, J. M. (1994) Effects of opioids and nonopioids on c-Fos immunoreactivity induced in rat lumbar spinal cord neurons by noxious heat stimulation. *Eur. J. Pharmacol.* **258,** 215–227.
89. Hohmann, A. G. and Herkenham, M. (1998) Regulation of cannabinoid and mu-opioid receptors in rat lumbar spinal cord following neonatal capsaicin treatment. *Neurosci. Lett.* **252,** 13–16.
90. Yaksh, T. L. (1993) The spinal actions of opioids, in *Handbook of Experimental Pharmacology. Opioids II* (Hertz, A., ed.), Springer-Verlag, Berlin, pp. 53–90.
91. Sañudo-Peña, M. C., Force, M., Tsou, K., Mclemore, G., Roberts, L., and Walker, J. M.(1995). Dopaminergic system does not play a major role in the precipitated cannabinoid withdrawal syndrome. *Acta Pharmacol. Sin.* **20,** 1121–1124.
92. Ross, R. A., Coutts, A. A., McFarlane, S. M., Irving, A. J., Pertwee, R. G., MacEvan, D. J., and Scott, R. H. (1999) Evidence for cannabinoid receptor-mediated inhibition of voltage-activated Ca^{2+} currents in neonatal rat cultured DRG neurones. *Br. J. Pharmacol.* **128,** 13P.
93. Minami, M., Maekawa, K., Yabuuchi, K., and Satoh, M. (1995) Double in situ hybridization study on coexistence of mu-, delta- and kappa-opioid receptors mRNAs preprotachykinin A mRNA in the rat dorsal root ganglia. *Brain Res. Mol. Brain Res.* **30,** 203–210.
94. Stein, C., Millan, M. J., Yassouridis, A., and Herz, A. (1988) Antinociceptive effects of mu- and kappa-agonists in inflammation are enhanced by a peripheral opioid receptor-specific mechanism. *Eur. J. Pharmacol.* **155,** 255–264.
95. Stein, C., Millan, M. J., Shippenberg, T. S., and Herz, A. (1988) Pheripheral effects of fentanyl upon nociception in inflammed tissues. *Neurosci. Lett.* **84,** 225–228.
96. Stein, C., Millan, M. J., Shippenberg, T. S., Peter, K., and Herz, A. (1989) Peripheral opioid receptors mediating antinociception in inflammation. Evidence for involvement of mu, delta and kappa receptors. *J. Pharmacol. Exp. Ther.* **248,** 1269–1275.
97. Stein, C., Hassan, A. H. S., Przewlocki, R., Gramsh, C., Peter, K., and Herz, A. (1990) Opioids and immunocytes interact with receptors on sensory nerves to inhibit nociception in inflammation. *Proc. Natl. Acad. Sci. USA* **87,** 5935–5939.
98. Parsons, C. G., Czlonkowski, A., Stein, C., and Herz, A. (1990) Peripheral opioid receptors mediating antinociception in inflammation. Activation by endogenous opioids and role of the pituitary-adrenal axis. *Pain* **41,** 81–93.
99. Maldonado, R., Valverde, O., Turcaud, S., Fournié-Zaluski, M. C., and Roques, B. P. (1994) Antinociceptive response induced by mixed inhibitors of enkephalin catabolism in peripheral inflammation. *Pain* **58,** 77–83.
100. Mason, J. R., Lowe, J., and Welch, S. P. (1999) A diminution of Δ^9-tetrahydrocannabinol modulation of dynorphin A-(1-17) in conjunction with tolerance development. *Eur. J. Pharmacol.* **381,** 105–111.
101. Houser, S. J., Eads, M., Embrey, J. P., and Welch, S. P. (2000) Dynorphyn B and spinal analgesia: induction of antinociception by the cannabinoids CP55,940, delta (9)-THC and anandamide. *Brain Res.* **857,** 337–342.
102. Valverde, O., Noble, F., Beslot, F., Daugé, V., Fournié-Zaluski, M. C., and Roques, B. P. (2001) Delta 9-tetrahydrocannabinol releases and facilitates the effects of endogenous enkephalins: reduction in morphine withdrawal syndrome wihtout change in rewarding effect. *Eur. J. Neurosci.* **13,** 1816–1824.

103. Mason, D. J., Lowe, J., and Welch, S. P. (1999) Cannabinoid modulation of dynorphin A: correlation to cannabinoid-induced antinociception. *Eur. J. Pharmacol.* **378,** 237–248.
104. Corchero J., Avila, M. A., Fuentes, J. A., and Manzanares, J. (1997) Delta-9-tetrahydrocannabinol increases prodynorphin and proenkephalin gene expression in the spinal cord of the rat. *Life Sci.* **61,** PL 39–PL 43.
105. Manzanares, J., Corchero, J., Romero, J., Fernandez-Ruiz, J. J., Ramos, J. A. and Fuentes, J. A. (1998) Chronic administration of cannabinoids regulates proenkephalin mRNA levels in selected regions of the rat. *Mol. Brain Res.* 55, 126–132.
106. Corchero, J., Fuentes, J. A., and Manzanares, J. (1997) Δ^9- tetrahydrocannabinol increases proopiomelanocortin gene expression in the arcuate nucleus of the rat hypothalamus. *Eur. J. Pharmacol.* **323,** 193–195.
107. Magour, S., Copher, H., and Fahndrich, C. (1977) Is tolerance to delta-9-THC cellular or metabolic? The subcellular distribution of delta-9-THC and its metabolites in brains of tolerant and nontolerant rats. *Psychopharmacology* **51,** 141–145.
108. Karler, R., Calder, L. D., Sangdee, P., and Turkanis, S. A. (1984) Interaction between Δ 9-tetrahydrocannabinol and kindling by electrical and chemical stimuli in mice. *Neuropharmacology* **23,** 1315–1320.
109. Abood, M. E., Sauss, C., Fan, F., Tilton, C. L., and Martin, B. R. (1993) Development of behavioral tolerance to delta 9-THC without alteration of cannabinoid receptor binding or mRNA levels in whole brain. *Pharmacol. Biochem. Behav.* **46,** 575–579.
110. Thompson, G., Fleischman, R., Rosenkrantz, H., and Braude, M. (1974) Oral and intravenous toxicity of delta 9-tetrahydrocannabinol in rhesus monkeys. *Toxicol. Appl. Pharmacol.* **27,** 648–665.
111. Pertwee, R. (1974) Tolerance to the effect of delta-9-tetrahydrocannabinol on corticosterone levels in mouse plasma produced by repeated administration of cannabis extract or delta-9-tetrahydrocannabinol. *Br. J. Pharmacol.* **51,** 391–397.
112. Kosersky, D. S., McMillan, D. E., and Harris, L. S. (1974) Delta-9-tetrahydrocannabinol and 11-hydroxy-Δ^9-tetrahydrocannabinol: behavioral effects and tolerance development. *J. Pharmacol. Exp. Ther.* **189,** 61–65.
113. Lamb, R. J., Jarbe, T. U., Makriyannis, A., Lin, S., and Goutopoulos, A. (2000) Effects of delta-9-tetrahydrocannabinol, (R)-methanandamide, SR 141716, and d-amphetamine before and during daily delta-9-tetrahydrocannabinol dosing. *Eur. J. Pharmacol.* **398,** 251–258.
114. Anderson, P. F., Jackson, D. M., and Chesher, G. B. (1974) Interaction of delta 9-tetrahydrocannabinol and cannabidiol on intestinal motility in mice. *J. Pharm. Pharmacol.* **26,** 136,137.
115. Birmighan, M. K. (1973) Reduction by delta-9-tetrahydrocannabinol in the blood pressure of hypertensive rats bearing regenerated adrenal glands. *Br. J. Pharmacol.* **48,** 169–171.
116. Adams, M. D., Chait, L. D., and Earnhardt, J. T. (1976) Tolerance to the cardiovascular effects of delta 9-tetrahydrocannabinol in the rat. *Br. J. Pharmacol.* **56,** 43–48.
117. Colasanti, B., Lindamood, C., and Craig, C. (1982) Effects of marihuana cannabinoids on seizure activity in cobalt-epileptic rats. *Pharmacol. Biochem. Behav.* **16,** 573–578.
118. Martin, B. R., Dewey, W. L., Harris, L. S., and Belckner, J. S. (1976) 3H-Δ-9-tetrahydrocannabinol tissue and subcellular distribution in the central nervous system and tissue distribution in peripheral organs of tolerant and nontolerant dogs. *J. Pharmacol. Exp. Ther.* **196,** 128–144.
119. Miczek, K. A. and Dihit, B. N. (1980) Behavioral and biochemical effects of chronic delta 9-tetrahydrocannabinol in rats. *Psychopharmacology* **67,** 195–202.
120. Abood, M. E. and Martin, B. R. (1992) Neurobiology of marijuana abuse. *Trends Pharmacol. Sci.* **13,** 201–207.
121. McMillan, D. E., Dewey, W. L., and Harris, L. S. (1971) Characteristics of tetrahydrocannabinol tolerance. *Ann. N.Y. Acad. Sci.* **191,** 83–99.
122. Bass, C. E. and Martin, B. R. (2000) Time course for the induction and maintenance of tolerance to delta-9-tetrahydrocannabinol in mice. *Drug Alcohol Depend.* **60,** 113–119.

123. Dewey, W. L., McMillan, D. E., Harris, L. S., and Turk, R. F. (1973) Distribution of radioactivity in brain of tolerant and nontolerant pigeons treated with ^{3}H-Δ^9-tetrahydrocannabinol. *Biochem. Pharmacol.* **22,** 399–405.
124. Oviedo, A., Glowa, J., and Herkenham, M. (1993) Chronic cannabinoid administration alters cannabinoid receptor binding in rat brain: a quantitative autoradiographic study. *Brain Res.* **616,** 293–302.
125. Rodriguez de Fonseca, F., Gorriti, M. A., Fernandez-Ruiz, J. J., Palomo, T., and Ramos, J. A. (1994) Downregulation of rat brain cannabinoid binding sites after chronic delta 9-tetrahydrocannabinol treatment. *Pharmacol. Biochem. Behav.* **47,** 33–40.
126. Fan, F., Tao, Q., Abood, M., and Martin, B. R. (1996) Cannabinoid receptor down-regulation with alteration of the inhibitory effect of CP-55,940 on adenylyl cyclase in the cerebellum of CP-55,940-tolerant mice. *Brain Res.* **706,** 13–20.
127. Rubino, T., Patrini, G., Massi, P., Fuzio, D., Vigano, D., Giognoni, G., and Parolaro, D. (1998) Cannabinoid-precipitated withdrawal: a time-course study of the behavioral aspect and its correlation with cannabinoid receptors and G protein expression. *J. Pharmacol. Exp. Ther.* **285,** 813–819.
128. Rubino, T., Vigano, D., Massi, P., and Parolaro, D. (2000) Changes in the cannabinoid receptor binding, G protein coupling, and cyclic AMP cascade in the CNS of rats tolerant to and dependent on the synthetic cannabinoid compound CP55,940. *J. Neurochem.* **75,** 2080–2086.
129. Rubino, T., Vigiano, D., Massi, P., Spinello, M., Zagato, E., Giagnoni, G., and Parolaro, D. (2000) Chronic delta-9-tetrahydrocannabinol treatment increases cAMP levels and cAMP-dependent protein kinase activity in some rat brain regions. *Neuropharmacology* **39,** 1331–1336.
130. Rubino, T., Massi, P., Patrini, G., Venier, I., Giagnoni, G., and Parolaro, D. (1994) Chronic CP-55,940 alters cannabinoid receptors mRNA in the rat brain: an in situ hybridization study. *Neuroreport* **5,** 2493–2496.
131. Romero, J., Berrendero, F., Manzanares, J., Pérez, A., Corchero, J., Fuentes, A., Fernández-Ruiz, J. J., and Ramos, J. A. (1998) Time-course of the cannabinoid receptor down-regulation in the adult rat brain caused by repeated exposure to delta 9-tetrahydrocannabinol. *Synapse* **30,** 298–308.
132. Rubino, T., Patrini, G., Parenti, G., Massi, P., and Parolaro, D. (1997) Chronic treatment with a synthetic cannabinoid CP55,940 alters G protein expression in the rat central nervous system. *Mol. Brain Res.* **44,** 191–197.
133. Sim, L. J., Hampson, R. E., Deadwyler, S. A., and Childers, S. R. (1996) Effects of chronic treatment with delta-9-tetrahydrocannabinol on cannabinoid-stimulated [^{35}S]GTPγS autoradigraphy in rat brain. *J. Neurosci.* **16,** 8057–8066.
134. Pertwee, R., Stevenson, L. A., and Griffin, G. (1993) Crosstolerance between delta-9-tetrahydrocannabinol and the cannabimimetic agents, CP 55,940, WIN 55,212-2 and anandamide. *Br. J. Pharmacol.* **110,** 1483–1490.
135. Fan, F., Compton, D. R., Ward, S., Melvin, L., and Martin B. R. (1994) Development of crosstolerance between delta-9-tetrahydrocanabinol, CP 55,940 and WIN 55,212. *J. Pharmacol. Exp. Ther.* **271,** 383–1390.
136. Kaymakcalan, S., and Deneau, G. A. (1972). Some pharmacologic properties of synthetic Δ^9-tetrahydrocannabinol (THC). *Acta Med. Turcica* **(Suppl. 1)**, 5–27.
137. Hine, B., Friedman, E., Torrelio, M., and Gershon, S. (1975) Tetrahydrocannabinol-attenuated abstinence and induced rotation in morphine-dependent rats: possible involvement of dopamine. *Neuropharmacology* **14,** 607–610.
138. Thorat, S. N. and Bhargava, H. N. (1994) Evidence for a bidirectional crosstolerance between morphine and Δ^9-tetrahydrocannabinol in mice. *Eur. J. Pharmacol.* **260,** 5–13.
139. Martin, B. R., Welch, S. P., and Abood, M. (1994) Progress toward understanding the cannabinoid receptor and its second messenger systems. *Adv. Pharmacol.* **25,** 341–397.

140. Bloom, A. S. and Dewey, W. L. (1978). A comparison of some pharmacological actions of morphine and Δ^9-tetrahydrocannabinol in the mouse. *Psychopharmacology* **57,** 243–248.
141. Melvin, L. S., Milne, G. M., Johnson, M. R., Subramaniam, B., Wilken, G. H., and Howlett, A. C. (1993) Structure-activity relationships for cannabinoid receptor-binding and analgesic activity: studies of bicyclic cannabinoid analogs. *Mol. Pharmacol.* **44,** 1008–1015.
142. Welch, S. P. (1997) Characterization of anandamide-induced tolerance: comparison to delta-9-tetrahydrocanabinol induced interactions with dynorphinergic systems. *Drug Alcohol Depend.* **45,** 39–45.
143. Ghozland, S., Aguado, F., Espinosa, J. F., Soriano, E., and Maldonado, R. (2002) *In vivo* chronic cannabinoid administration impairs spontaneous network activity of cerebellar granule neurons. *Eur. J. Neurosci.*, in press.
144. McMillan, D. E., Harris, L. S., Frankenheim, J. M., and Kennedy, J. S. (1970) L-Δ^9 -tetrahydrocannabinol in pigeons: tolerance to the behavioral effects. *Science* **169,** 501–503.
145. Dewey, W. L., Jenkins, J., O'Rourke, T., and Harris, L. S. (1972) The effects of chronic administration of *trans*-Δ^9-tetrahydrocannabinol on behavior and the cardiovascular system of dogs. *Arch. Int. Pharmacodyn. Ther.* **198,** 118–131.
146. Chesher, G. B. and Jackson, D. M. (1974) The effect of withdrawal from cannabis on pentylentetrazol convulsive threshold in mice. *Psychopharmacologia* **40,** 129–135.
147. Harris, R. T., Waters, W., and Mcendon, D. (1974) Evaluation of reinforcing capability of delta-9-tetrahydrocannabinol in rhesus monkeys. *Psychopharmacologia* **37,** 23–29.
148. Leite, J. L. and Carlini, E. A. (1974) Failure to obtain "cannabis directed behavior" and abstinence syndrome in rats chronically treated with cannabis sativa extracts. *Psychoparmacologia* **36,** 133–145.
149. Diana, M., Melis, M., Muntoni, A. L., and Gessa, G. L. (1998) Mesolimbic dopaminergic decline after cannabinoid withdrawal. *Proc. Natl. Acad. Sci. USA* **95,** 10,269–10,273.
150. Aceto, M. D., Scates, S. M., Lowe, J. A., and Martin, B. P. (2001) Spontaneous and precipitated withdrawal with a synthetic cannabinoid, WIN 55,212-2. *Eur. J. Pharmacol.* **416,** 75–81.
151. Tsou, K., Patrick, S. L., and Walker, J. M. (1995) Physical withdrawal in rats tolerant to delta 9-tetrahydrocannabinol precipitated by a cannabinoid receptor antagonist. *Eur. J. Pharmacol.* **280,** R13–R15.
152. Aceto, M. D., Scates, S. M., Lowe, J. A., and Martin, B. P. (1996) Dependence on delta 9-tetrahydrocannabinol: studies on precipitated and abrupt withdrawal. *J. Pharmacol. Exp. Ther.* **278,** 1290–1295.
153. Cook, S. A., Lowe, J. A., and Martin, B. R. (1998). CB1 receptor antagonist precipitates withdrawal in mice exposed to Δ^9-tetrahydrocannabinol. *J. Pharmacol. Exp. Ther.* **285,** 1150–1156.
154. Tzavara, E. T., Valjent, E., Firmo, C., Mas, M., Beslot, F., Defer, N., Roques, B. P., Hanoune, J., and Maldonado, R. (2000) Cannabinoid withdrawal is dependent upon PKA activation in the cerebelum. *Eur. J. Neurosci.* **12,** 1038–1046.
155. Kaymakcalan, S., Ayhan, I. H., and Tulunay, F. C. (1977) Naloxone-induced or postwithdrawal abstinence signs in Δ^9-tetrahydrocannabinol-tolerant rats. *Psychopharmacology* **55,** 243–249.
156. Romero, J., Fernández-Ruiz, J. J., Vela, G., Ruiz-Gayo, M., Fuentes, J. A., and Ramos, J. A. (1998) Autoradiographic analysis of cannabinoid receptor binding and cannabinoid-agonist-stimulated [^{35}S]GTPγS binding in morphine-dependent mice. *Drug. Alcohol Depend.* **50,** 241–249.
157. Lichtman, A. H., Sheikh, H. H., Loh, S. M., and Martin, B. R. (2001) Opioid and cannabinoid modulation of precipitated withdrawal in delta-9-tetrahydrocanabinol and morphine-dependent mice. *J. Pharmacol. Exp. Ther.* **298,** 1007–1014.
158. Hine, B., Friedman, E., Torrelio, M., and Gershon, S. (1975) Tetrahydrocannabinol-attenuated abstinence and induced rotation in morphine-dependent rats: possible involvement of dopamine. *Neuropharmacology* **14,** 607–610.
159. Hine, B., Friedman, E., Torrelio, M., and Gershon, S. (1975) Morphine-dependent rats: blockade of precipitated abstinence by tetrahydrocannabinol. *Science* **187,** 443–445.

160. Bhargava, H. N. (1976) Effect of some cannabinoids on naloxone-precipitated abstinence in morphine-dependent mice. *Psychopharmacology* **49,** 267–270.
161. Bhargava, H. N. and Way, E. L. (1976) Morphine tolerance and physical dependence: influence of cholinergic agonists and antagonists. *Eur. J. Pharmacol.* **36,** 79–88.
162. Vela, G., Ruiz-Gayo, M., and Fuentes, J. A. (1995) Anandamide decreases naloxone-precipitated withdrawal signs in mice chronically treated with morphine. *Neuropharmacology* **34,** 665–668.
163. Koob, G. F. (1996) Drug addiction: the yin and yang of hedonic homeostasis. *Neuron* **16,** 893–896.
164. Rossetti, Z. L., Hmaidan, Y., and Gessa, G. L. (1992) Marked inhibition of mesolimbic dopamine release: common feature of ethanol, morphine, cocaine and amphetamine abstinence in rats. *Eur. J. Pharmacol.* **221,** 227–234.
165. Rodriguez de Fonseca, F., Carrera, M. R. A., Navarro, M., Koob, G. F., and Weiss, F. (1997) Activation of corticotropin-releasing factor in the limbic system during cannabinoid withdrawal. *Science* **276,** 2050–2054.
166. Sharma, S. K., Klee, W. A., and Nierenberg, M. (1975) Dual regulation of adenylate cyclase accounts for narcotic dependence and tolerance. *Proc. Natl. Acad. Sci. USA* **72,** 3092–3096.
167. Nestler, E. J. and Tallman, J. F. (1988) Chronic morphine treatment increases cyclic AMP-dependent protein kinase activity in the rat locus coeruleus. *Mol. Pharmacol.* **33,** 127–132.
168. Aghajanian, G. K. (1978) Tolerance of locus coeruleus neurons to morphine and suppression of withdrawal response by clonidine. *Nature* **276,** 186–188.
169. Maldonado, R., Stinus, L., Gold, L. H., and Koob, G. F. (1992) Role of different brain structures in the expression of the physical morphine withdrawal syndrome. *J. Pharmacol. Exp. Ther.* **261,** 669–677.
170. Olds, M. E. and Travis, R. P. (1954) Possitive reinforcement produced by electrical stimulation of the septal area and other regions of rat brain. *J. Comp. Physiol. Psychol.* **47,** 419–427.
171. Esposito, R. U. and Kornetsky, C. (1978) Opioids and rewarding brain stimulation. *Neurosci. Biobehav. Rev.* **2,** 115–122.
172. Wise, R. A. (1980) Action of drugs of abuse on brain reward systems. *Pharmacol. Biochem. Behav.* **13 (Suppl. 1),** 213–223.
173. Gardner, E. L., Paredes, W., Smith, D., Donner, A., Milling, C., Cohen, D., and Morrison, D. (1988) Facilitation of brain stimulation reward by delta 9-tetrahydrocannabinol. *Psychopharmacology* **96,** 142–144.
174. Lepore, M., Liu, X., Savage, V., Matalon, D., and Gardner, E. L. (1996) Genetic differences in delta 9-tetrahydrocannabinol-induced facilitation of brain stimulation reward as measured by a rate-frequency curve-shift electrical brain stimulation paradigm in three different rat strains. *Life Sci.* **58,** PL365–PL372.
175. Chen, J., Paredes, W., Li, J., Smith, D., Lowinson, J., and Gardner, E. L. (1990) Δ^9-tetrahydrocannabinol produces naloxone-blockable enhancement of presynaptic basal dopamine efflux in nucleus accumbens of conscious, freely-moving rats as measured by intracerebral microdialysis. *Psychopharmacology* **102,** 156–162.
176. Gardner, E. L. and Lowinson, J. H. (1991) Marijuana's interaction with brain reward system: update 1991. *Pharmacol. Biochem. Behav.* **40,** 571–580.
177. Mucha, R. F., Van der Kooy, D., O'Shaughnessy, M., and Bucenieks, P. (1982) Drug reinforcement studied by use of place conditioning in rat. *Brain Res.* **243,** 91–105.
178. Schechter, M. D. and Calcagnetti, D. J. (1993) Trends in place preference conditioning with a cross-indexed bibliography: 1957–1991. *Neurosci. Biobehav. Rev.* **17,** 21–41.
179. Parker, L. A., and Gillies, T. (1995) THC-induced place and taste aversion in Lewis and Sprague-Dawley rats. *Behav. Neurosci.* **109,** 71–78.
180. McGregor, I. S., Issackidis, C., and Prior, G. (1996) Aversive effects of the synthetic cannabinoid CP 55,940 in rats. *Pharmacol. Biochem. Behav.* **53,** 657–664.

181. Sañudo-Peña, M. C., Tsou, K., Delay, E. R., Hohman, A. G., Force, M., and Walker, J. M. (1997) Endogenous cannabinoids as an aversive or counter-rewarding system in the rat. *Neurosci. Lett.* **223,** 125–128.
182. Chaperon, F., Soubrie, P., Puech, A. J., and Thiebot, M. H. (1998) Involvement of central cannabinoid (CB1) receptors in the establishment of place conditioning in rats. *Psychopharmacology* **135,** 324–332.
183. Mallet, P. E. and Beninger, R. J. (1998) Delta 9-tetrahydrocannabinol, but not the endogenous cannabinoid receptor ligand anandamide, produces conditioned place avoidance. *Life Sci.* **62,** 2431–2439.
184. Lepore, M., Vorel, S. R., Lowinson, J., and Gardner, E. L. (1995) Conditioned place preference induced by delta 9-tetrahydrocannabinol: comparison with cocaine, morphine, and food reward. *Life Sci.* **56,** 2073–2080.
185. Valjent, E. and Maldonado, R. (2000) A behavioural model to reveal place preference to delta 9-tetrahydrocannabinol in mice. *Psychopharmacology* **147,** 436–438.
186. Martin, M., Ledent, M., Parmentier, M., Maldonado, R., and Valverde, O. (2000) Cocaine but not morphine, induces conditioned place preference and sensitization to locomotor responses in CB1 knockout mice. *Eur. J. Neurosci.* **12,** 4038–4046.
187. Deneau, G. A., Yanagita, T., and Seevers, M. H. (1964) Self-administration of drugs by monkeys. *Bull. Drug Addict. Narc.* **11,** 3812.
188. Kaymakcalan, S. and Deneau, G. A. (1972). Some pharmacologic properties of synthetic Δ^9-tetrahydrocannabinol (THC). *Acta Med. Turcica* **(Suppl. 1)**, 5–27.
189. Pickens, R., Thompson, T., and Muchow, D. C. (1973) Cannabis and phencyclidine self-administration by animals. In: Goldberg, L., Hoffmeister, F. (eds.) *Bayer Sympos Psychic Dependence.* Berlin, Springe-Verlag, pp. 78–86.
190. Corcoran, M. E. and Amit, Z. (1974) Reclutance of rats to drink hashish suspensions: free choice and forced consumption, and the effects of hypothalamic stimulation. *Psycopharmacologia* **352,** 129–147.
191. Carney, J. M., Uwaydah, M. I., and Balster, R. L. (1977) Evaluatins of a suspension system for intravenous self-administration studies of water-insolubles compounds in the rhesus monkey. *Pharmacol. Biochem. Behav.* **7,** 357–364.
192. Mansbach, R. S., Nicholson, K. L., Martin, B. R., and Balster, R. L. (1994) Failure of Δ^9-tetrahydrocannabinol and CP 55,940 to maintain intravenous self-administration under a fixed-interval schedule in rhesus monkeys. *Behav. Pharmacol.* **5,** 210–225.
193. Martellotta, M. C., Cossu, G., Fattore, L., Gessa, G. L., and Fratta, W. (1998) Self-administration of the cannabinoid receptor agonist WIN 55,212-2 in drug-naive mice. *Neuroscience* **85,** 327–330.
194. Braida, D., Pozzi, M., Parolaro, D., and Sala, M. (2001) Intracerebral self-administration of the cannabinoid receptor agonist CP 55,940 in the rat: interaction with the opioid system. *Eur. J. Pharmacol.* **413,** 227–234.
195. Tanda, G., Munzar, P., and Goldberg, S. R. (2000) Self-administration behavior is maintained by the psychoactive ingredient of marijuana in squirrel monkeys. *Nat. Neurosci.* **3,** 1073–1074.
196. Goldberg, S. R., Munzar, P., Justinova, Z., and Tanda, G. (2001) Effects of naltrexone on intravenous self-administration of delta-9-tetrahydrocannabinol (THC) by squirrel monkeys under fixed-ratio and second-order schedules. *International Cannabis Research Society Meeting,* p. 102.
197. Matthews, R. T. and German, D. C. (1984) Electrophysiological evidence for excitation of rat ventral tegmental area dopamine neurons by morphine. *Neuroscience* **11,** 617–625.
198. Mereu, G., Fadda, F., and Gessa, G. L. (1984) Ethanol stimulates the firing rate of nigral dopaminergic neurons in unanesthetized rats. *Brain Res.* **292,** 63–69.
199. Mereu, G., Kong-Woo, P. Y., Boi, V., Gessa, G. L., Naes, L., and Westfall, T. C. (1987)

Preferential stimulation of ventral tegmental dopaminergic neurons by nicotine. *Eur. J. Pharmacol.* **141,** 395–399.

200. Di Chiara, G. and Imperato, A. (1988) Drug abused by humans preferentially increase synaptic dopamine concentrations in the mesolimbic system of freely moving rats. *Proc. Natl. Acad. Sci. USA* **94,** 5274–5278.
201. Tanda, G., Pontieri, F. E., and Di Chiara, G. (1997) Cannabinoid and heroin activation of mesolimbic dopamine transmission by a common μ1 opioid receptor mechanism. *Nature* **276,** 2048–2050.
202. Gessa, G. L., Melis, M., Muntoni, A. L., and Diana, M. (1998) Cannabinoids activate mesolimbic dopamine neurons by an action on cannabinoid CB1 receptors. *Eur. J. Pharmacol.* **341,** 39–44.
203. Gessa, G. L. and Diana, M. (2000) Different mechanisms for dopaminergic excitation induced by opiates and cannabinoids in the rat midbrain. *Prog. Neuropsychopharmacol. Biol. Psychiat.* **24,** 993–1006.
204. French, E. D. (1997) Δ 9-tetrahydrocannabinol excites rat VTA dopamine neurons through activation of cannabinoid CB1 but not opioid receptors. *Neurosci. Lett.* **226,** 159–162.

Part IV

Alcohol and Nicotine Addiction

13

Current Strategies for Identifying Genes for Alcohol Sensitivity

John C. Crabbe

1. Introduction

The familial occurrence of alcoholism has been known for many years. Many twin, adoption, and family studies now concur that this familial pattern is to a great extent conferred by genes transmitted to biological offspring *(1,2)*. Approximately 50–60% of individual differences in risk for alcoholism is genetic, and this proportion is approximately equal in men and women *(2)*. Thus, it is an easy task to predict that a close biological relative of an alcoholic is at higher risk for alcoholism. However, risk is not inherited—alleles at specific risk-promoting or -protective genes are inherited. To date, there are only two specific genes known to confer substantial protection against alcoholism, variants at the ALDH2*2 and ADH2*2 metabolic enzymes. The variant alleles lead to the accumulation of alcohol's metabolite, acetaldehyde, when susceptible individuals drink alcohol. This toxic compound produces nausea, flushing, dizziness, and other unpleasant effects, and slow alcohol metabolizers avoid excessive drinking *(3)*. Therefore, progress from assigning risk statistically to ascertaining whether specific individuals possess risk-promoting or -protective alleles will require the identification of the specific genes underlying risk.

The goal of this chapter is to review the current status in the search for such genes. There are three basic approaches to identifying important genes. First, one can target genes based on their presumed importance in influencing alcohol sensitivity. The targeted gene can be overexpressed, underexpressed, or disrupted to the extent that its function is ablated (a gene knockout). Second, one can seek genes that are identified as important because they are differentially expressed. Third, one can seek variations in the sequence of genes that are associated with alcohol sensitivity. The first approach can only be attempted using genetic animal models. Each approach will be reviewed, with an emphasis on the second and third.

Alcohol, like any drug, produces a wide spectrum of behavioral effects, and the genetic sensitivity to one effect is to a large degree distinct from that to another. For drugs of abuse, the spectrum of responses is typically conceptualized as representing the following: initial sensitivity to the drug in a naive organism; tolerance or sensitization of responses following chronic administration; dependence on the drug, as indicated by the occurrence of withdrawal signs when the drug is discontinued; and

From: *Molecular Biology of Drug Addiction*
Edited by: R. Maldonado © Humana Press Inc., Totowa, NJ

reinforcing effects of the drug, inferred from self-administration or from drug-seeking behavior. In addition, differences in drug metabolism are in part genetically mediated. Another chapter in this volume discusses ethanol reward extensively *(4)*. I will therefore concentrate my discussions on other aspects of drug sensitivity. Finally, I will concentrate on pharmacodynamic differences, that is, those not attributable to drug metabolism.

2. Gene Targeting

The literature on transgenic technologies to alter specific genes' functions in laboratory mice is extensive. Several chapters in this volume review studies exploring transgenics' responses to most classes of abused drugs. In their chapter, Cunningham and Phillips discuss more than two dozen studies testing behaviors related to alcohol reward in a like number of genetically engineered mutants *(4)*. The genes targeted range from neurotransmitter receptors to protein kinases to other enzymes. It is evident from their review that a substantial number of disrupted genes have been shown to affect alcohol self-administration or other traits relevant to reward. Many of the same mutants have been tested for alcohol sensitivity, tolerance, or withdrawal using a variety of behavioral assays, and many display altered responses when compared to their wild-type controls *(5)*.There is little to be added by listing here the results of these experiments for behavioral endpoints other than reward phenotypes, and nearly all the results for these other traits can be found in the references in their chapter *(4)*.

Cunningham and Phillips also present a lucid analysis of the strengths, weaknesses, and future promise of gene targeting technologies *(4)*. In particular, they note the need for stringent behavioral controls to enable straightforward interpretation of the results from such studies as relevant specifically to alcohol. One aspect of such control is the selection of the phenotype tested. As they note, virtually all studies of alcohol reward chose a single assay of "reinforcement," two-bottle preference drinking, to compare knockouts and wild types. However, they note that this paradigm (like all others) has several problems that hinder its interpretation. Multiple assays that target a putative behavioral domain (e.g., "reinforcement," "anxiety") should be studied before inferring a genetic effect on that domain. One example is studies of sensitivity to "alcohol-induced motor incoordination" conducted in mice lacking the serotonin 1B receptor. The 5-HT_{1B} null mutants were less sensitive to alcohol intoxication than wild types when tested on a balance beam or in an apparatus called the grid test. However, the genotypes did not differ in sensitivity when tested on a fixed-speed or an accelerating rotarod, the screen test, a static dowel, or when tested for grip strength or sensitivity to loss of righting reflex *(6)*. Even though these (and several other) tasks are used interchangeably in many laboratories, the effect of altering even one gene is not uniform across tests.

Good examples of how to probe a null mutant are on the increase in the literature. For example, Homanics reported no effect of deletion of the γ-aminobutyric acid A ($GABA_A$) receptor subunit 6 gene on alcohol withdrawal, and tested both acute withdrawal from a single injection and chronic withdrawal from a period of ethanol vapor inhalation *(7)*. Coste et al. found no effect of deletion of the corticotropin receptor 2 on three measures thought to reflect anxiety *(8)*. (Interestingly, two other groups who independently deleted the same receptor gene found different results; *see* ref. *9* for discussion).

Knockout technology, particularly when the altered gene construct can be expressed in brain with both temporal and anatomical precision, will certainly provide much additional information regarding the genetics of alcohol sensitivity, but new methods will need to be developed to realize this potential (for example, *see* ref. *10*).

3. Gene Expression

Until very recently, studies of gene expression related to alcohol were limited to animal studies and were driven by a candidate gene approach. Because such studies depended on laborious assays (e.g., Northern analysis) and were conducted one gene at a time, expression differences were sought only in neural systems where much prior evidence implicated changes as important for alcohol responses. Substantial evidence has been gathered over the years implicating gene expression in *N*-methyl-D-aspartic acid (NMDA), GABA, and glycine receptors, G-protein-coupled receptors, neurotransmitter transporters, second messenger systems, and transcription factors, including immediate early genes, among others. These studies have been reviewed *(11–13)*, and gene expression relevant to alcohol reward is reviewed by Cunningham and Phillips *(4)*.

Now that it is possible to manufacture gene chips, expression array profiling has become a very hot area in the biological sciences *(14)*. By affixing short-sequence probes of mRNA complementary to the DNA sequences of many thousands of genes to a glass chip and exposing the chip to a sample of DNA, those genes currently expressed can be captured by hybridization. This and related technologies allows studies with tissue from humans as well as nonhuman animal models. When expression patterns are compared between alcohol-treated groups and controls, for example, a number of genes appear to be differentially expressed. Only three such studies for alcohol-related gene expression have been reported to date. Lewohl and her colleagues compared postmortem frontal cortical tissue from human alcoholics vs controls. They examined expression patterns for >4000 genes and found that expression of 163 differed by more than 40% between alcoholics and controls. Several myelin-related genes were expressed less in alcoholic tissue, and they posit that this may suggest mechanisms underlying the loss of white matter in brains of alcoholics *(15)*. Thibault et al. used oligonucleotide arrays with 6000 human gene sequences to compare human neuroblastoma cell lines chronically exposed to ethanol with untreated lines. Forty-two genes showed either increased or decreased expression after 3 d of alcohol treatment. Of particular interest was the increase in expression of the dopamine hydroxylase (DBH) gene, which was accompanied by releasable norepinephrine in the cultures. Increased DBH expression was also seen in adrenal tissue 24 h after a single high-dose injection of ethanol to DBA/2J mice *(16)*.

Finally, the large differences in behavioral sensitivity to ethanol characterizing ILS and ISS mice (*see* Section 4.4.) were exploited in a expression profiling experiment with whole brain tissue *(17)*. These mice were selectively bred to display high or low sensitivity to ethanol-induced loss of righting reflex. Basal expression of 6000 genes or expressed sequence tags was studied, and 41 genes were differentially expressed.

Gene expression profiling will feature very prominently in advances in understanding genetic determinants for many complex traits, including responses to alcohol. The number of genes affected in such experiments is daunting. Much like the situation with gene targeting experiments, there will soon be so much provocative data that making sense of it will be the next problem. Already, research groups are struggling to establish the bioinformatics capacity for collecting and managing such data, which surely is the first step toward sensibly interpreting it. For a wide-ranging review of these methods and their interpretation, see ref. *(18)*.

4. Gene Sequence

4.1. Mapping Alcohol Response Genes

The third approach to finding specific genes depends on screening for allelic differences among individuals. Such approaches are essentially mapping genes by finding polymorphisms either in the gene itself or in a region of the chromosome that is physically closely linked to the gene. When the polymorphism (an altered sequence of bases in the DNA) is physically linked, it can be considered a genetic marker, and it escapes the Mendelian laws of segregation and independent assortment. The closer the linkage, the less likely that an individual will inherit the marker from one parent and the mapped gene from the other, which occurs if there is a recombination during meiosis in the region between the marker and the mapped gene.

Responses to alcohol are not all or none. When a population of individuals is studied, the majority tend to show an intermediate response while few show either extremely high or low sensitivity. That is, alcohol responses are quantitative, not qualitative, traits. This implies that multiple genes contribute to that portion of individual differences that is genetic. As mapping experiments seek such genes, they first establish rough chromosomal locations, and then gradually narrow the focus to smaller and smaller regions surrounding the gene. Hence, the search for alcohol response genes, as for genes affecting all complex traits, begins with the identification of *quantitative trait loci*, or QTLs.

There are numerous methods for performing QTL mapping experiments, and they are conceptually similar whether one is using humans or nonhuman animal models. Good descriptions of the various methods and their relative strengths and weaknesses can be found elsewhere, and technical description will be avoided in this chapter *(19–23)*. All methods depend on the existence of many genetic markers, scattered throughout the genome, and a densely populated map that locates each such marker on a specific chromosome relative to other markers. The utility of these markers (initially restriction fragment length polymorphisms, but now typically DNA microsatellites, or short repeated sequences of base pairs or triplets) for gene mapping was first realized in the late 1980s *(24)*. Now that the Human Genome Project has essentially finished the task of capturing each human gene and placing it on the map, the number of markers for mapping is greatly amplified. Because humans and mice share an ancient evolutionary ancestor, we also share large strings of DNA that have remained linked. This means that mapping QTLs in mice (and finding the responsible gene) tells us where to look in the human genome more than 80% of the time *(25)*. While the rough mouse genome map was completed only in 2002, the new generation of markers (single nucleotide polymorphisms, or SNPs) is rapidly filling in the remaining spaces.

4.2. Recombinant Inbred Mouse Strains and Confirming Populations

Mice have been used in QTL mapping studies for alcohol-related traits since the early 1990s *(26)*. Most laboratories have employed a multistage mapping strategy. One such popular strategy starts with recombinant inbred (RI) strains. For example, there are 26 RI strains that have been created from the intercross of two standard inbred strains, C57BL/6J (B6) and DBA/2J (D2). Each of these represents a more or less random reshuffling of the chromosomal segments (and, therefore, the genetic maps) derived from the B6 and D2 strains. Each BXD RI strain has been genotyped at more than 1500 genetic markers differing between B6 and D2 and the map of these markers assembled. By ascertaining the sensitivity to alcohol of each BXD RI strain and then comparing those values with the marker maps, it is possible to identify clusters of markers from one strain that are associated with high (or low) sensitivity. When such an association is seen, it suggests the presence of a QTL in that chromosomal region.

An early review of QTL mapping studies in RI strains *(27)* reported associations for 19 different responses to alcohol in approximately 50 different chromosomal regions. However, mapping could not be accomplished using only these 26 genotypes, because the resolving power of the analysis was insufficient. Many hundreds of phenotype–marker correlations are calculated in such an analysis, and after correcting for the multiple comparisons involved, many such possible linkages are actually false-positive associations. To achieve the statistical certainty generally accepted by the genetic mapping community *(28)*, it is necessary to study additional populations of animals. For many alcohol traits, such additional data have now been gathered and several QTLs have now been mapped with certainty. A more recent review of progress *(29)* reported that 15 QTLs had been reliably mapped in mouse and one in rat; the number now is more than 40. QTLs related to alcohol drinking and other reward phenotypes are reviewed by Cunningham and Phillips and will not be discussed here *(4)*. These QTLs have one interesting feature—there is occasional evidence for sex-specific QTLs for drinking. This is somewhat unusual, as sex differences have not featured as prominently in QTLs for the other alcohol phenotypes.

4.3. QTLs for Alcohol Withdrawal

The goal of any QTL mapping project is to progress from locus to gene. Two clear examples of such progress are now available for alcohol traits. The first is an ongoing project in the laboratory of Kari Buck. In collaboration with John Belknap, Buck is pursuing several QTLs that harbor genes influencing the severity of acute withdrawal from alcohol. When 4 g/kg ethanol is administered intraperitoneally to a mouse, it is sedated for 2–3 h. However, between 3 and 24 h after injection, an acute withdrawal reaction can be seen, where mice display handling-induced convulsions (HICs) that wax and then wane, peaking at about 7 h after injection *(30)*. This reaction, first noted in the early 1970s *(31)*, indicates a state of modest physical dependence on alcohol. The reaction becomes more pronounced when alcohol is administered chronically. Goldstein demonstrated through a short breeding project that the severity of alcohol withdrawal from vapor inhalation was heritable *(32)*, which we later confirmed *(33)*, and inbred strains of mice are known to differ markedly in acute *(34)* and chronic *(35)* ethanol withdrawal severity. The D2 strain had the highest acute (and chronic) withdrawal score of any strain tested, while the B6 strain was a low-withdrawal strain, so these strains are a good starting population for genetic mapping.

The BXD RI strains differ markedly in acute *(36)* and chronic *(37)* withdrawal HIC scores, and comparison of RI strain values with the BXD RI marker database identified several provisional QTLs for acute withdrawal *(36)*. We then tested 400 mice from the F2 cross between B6 and D2 inbred strains. Each individual F2 mouse possessed a unique pattern of markers due to recombinations during meiosis. By comparing the allelic status of the highest-scoring F2 mice with those from the lowest-scoring mice in the chromosomal regions nominated by the BXD RI analysis, we obtained further evidence for some of the provisional QTLs. Finally, starting with the F2 population, we selectively bred for four generations, mating high scorers together and low scorers together. As selection led to divergence in withdrawal severity between the two populations, the allele frequencies for the D2 (high-scoring) strain increased in the "high alcohol withdrawal" selected line for markers in some, but not all of the QTL regions. Together, this evidence allowed us to map three significant QTLs contributing to acute alcohol withdrawal on mouse chromosomes 1, 4, and 11 *(38)*. QTLs in the same three regions were also mapped using the same methods for acute pentobarbital withdrawal *(39)*.

Although each QTL was now known with certainty to harbor a gene or genes affecting withdrawal severity, the major problem to be overcome was the size of the confidence interval surrounding the map location. Even after combining data from three mapping populations, the confidence intervals were large, containing hundreds of genes, any one (or group) of which could actually be responsible. We chose, as other groups have also, to pursue fine mapping by developing *congenic strains (40)*. We started with B6D2F1 hybrid mice, backcrossed them to D2, and genotyped the backcross generation for four markers surrounding the QTL on chromosome 4, called *Alcw2*. We retained only those mice with B6 markers surrounding *Alcw2*. We then repeated this process for 10 generations to create a congenic strain. This strain was about 98% D2 genotype, and possessed DNA from the B6 strain only in the small region surrounding the QTL, about 35 centiMorgans (cM) in length. Nonetheless, this region still contained 600–700 genes.

When the congenics were tested for acute alcohol withdrawal vs the D2 background strain, the congenics had less severe withdrawal HICs, showing that the gene responsible for the QTL effect was captured in the B6 piece of DNA. Next, three further generations of backcrossing to D2 were pursued, and additional congenic strains were created. These *interval-specific congenic strains* (ISCS) possessed smaller regions of B6 DNA in the chromosome 4 QTL region. Testing of the five ISCS showed that four still retained the responsible gene, whereas one did not. The pattern of markers allowed us to exclude all but a very small region of chromosome 4, less than 1 cM. This region probably contains only 18 genes, and is captured in the ISCS5 strain. Combined with the previous mapping data, this QTL's gene significantly affects withdrawal with high probability (LOD = 8.8, $p = 2.2 \times 10^{-10}$). It similarly captures a gene affecting acute pentobarbital withdrawal severity *(40)*.

After sequencing 10 of the 18 genes or gene-related sequences in the QTL interval, only one with a substantial number of polymorphisms has yet been found. This gene, *Mpdz*, codes for a multiple PDZ-domain protein. We examined *Mpdz* in 15 standard inbred mouse strains (including B6 and D2) that were known to vary substantially in acute ethanol withdrawal *(34)*. Eight haplotypes yielding three protein variants were identified. There was a substantial correlation between the severity of withdrawal and the particular protein variant each strain possessed. *Mpdz* participates as a cellular scaffolding protein in the collocalization of receptors such as several serotonin receptor subtypes, a tyrosine kinase receptor, and a nerve growth factor receptor with their reaction partners. Thus, it is an attractive candidate gene for the QTL effect *(40)*.

4.4. QTLs for Alcohol-Induced Loss of Righting Reflex

The other long-standing gene mapping project in alcohol research that has made substantial progress toward gene identification has been studying the effects of a high dose of alcohol to cause mice to lose their ability to right when placed on their backs. The duration of suppression of this reflex has been used as an index of sensitivity to alcohol and numerous depressant drugs. Many years ago, mice were bred to be sensitive to this effect by breeding together those with a "long sleep" (LS) response and those with a "short sleep" (SS) response *(41)*. Although the mice are not actually sleeping, the LS and SS mouse lines after many generations of selective breeding differ markedly in duration of loss of righting reflex. LS and SS strains were crossed to create LSXSS RI strains, and these strains were phenotyped *(42)*. LS and SS strains were subsequently inbred (to form ILS and ISS strains, respectively), and a new set of ILSXISS RI strains has also been created.

Following logic parallel to that described in the previous section for the acute withdrawal mapping project, Tom Johnson, Beth Bennett and their colleagues have been pursuing several QTLs for sensitivity to the loss of righting reflex after ethanol. Early studies proposed several provisional QTLs and gradually proved the existence of four of them *(43–45)*. Numerous congenic and ISCS strains have been developed to isolate each QTL within smaller regions of the genome *(46,47)*, and the confidence interval for some is approaching the subcentiMorgan level (Bennett, personal communication).

A unique direction pursued in search of the loss of righting reflex QTLs is the search for gene coding variants within the QTL regions *(48)*. Using high-throughput comparative DNA sequencing technologies to compare DNA from ILS and ISS mice, >1.7 million bases of comparative DNA sequences were generated from 68 candidate genes located within the four QTLs. Eight central nervous system genes that together displayed 36 changes in protein coding were identified from this search.

4.5. Other Gene Identification Strategies

The congenic strains derived from QTL mapping studies are useful beyond the scope of the project for which they are prepared. For example, one congenic useful in pursuit of *Alcw1*, the acute withdrawal QTL on chromosome 1, was obtained from Thomas Ferraro, who constructed it to search for QTLs for kainate-induced seizures. A compendium of congenics available from various QTL mapping projects has been published *(49)*. In addition, a series of more than 60 congenics spanning the genome with D2 genotype introduced onto a B6 background is in progress *(50)*, and other specialized populations for mapping also exist *(51)*.

Mapping can of course be performed in highly polymorphic, genetically segregating populations, where more than two alleles are present for each gene. A QTL for alcohol-induced activity has been successfully mapped using such a strategy *(52)*, and other behavioral traits have been mapped in similar populations *(53,54)*.

A technique developed many years ago but recently repopularized is the generation of mutations thorough random mutagenesis. By treating male mice with a toxic chemical, N-ethyl-N-nitrosourea (ENU), many mutations are induced at random throughout the genome. The offspring of treated mice are then tested and phenotypic outliers sought on the assumption that they are extreme scorers because of an induced mutation. If such a mutation can be isolated by breeding, it can then be mapped and eventually identified. (It may seem a little odd to consider ENU mutagenesis under the rubric of mapping rather than as variant of candidate gene targeting, but it combines both approaches and the genes targeted are not identified *a priori*). Many ENU projects are in progress, and their utility for mapping complex trait genes is highly touted (*see* the special issue of *Mammalian Genome*, **7** [11], July, 2000). It has been argued that QTL mapping methods are inferior to ENU-based approaches *(55)*. However, neither approach has as yet identified a gene, although progress is continuing on both fronts. We have argued that it is more prudent to see what success is actually achieved by each approach before deciding which is a superior methodology *(56)*.

Another new technique depends on the proliferation of SNP markers. Using 15 standard inbred strains, Grupe et al. surveyed them for SNPs and identified 500 useful for mapping *(57)*. This mapping database was combined with preexisting information about nearly 3000 other SNPs among the strains. By examining published phenotypic data for the strains and associating that data with the mapping information, they identified several regions with potential QTLs. These included the three QTLs for acute alcohol withdrawal mapped by Buck et al. *(38)*, and QTLs for alcohol-preference drinking. The method currently does not allow assignment of statistical confidence to such predictions, but it may prove to be a useful alternative to the laborious process of genotyping individual mice as more phenotypic data on inbred strains are accumulated in the literature *(58)*.

5. Identifying Human Genes

As noted, studies of gene expression differences in humans are in their infancy. Gene mapping efforts in human populations uses variants of the methods described for mice. However, the problem of mapping in humans is markedly more difficult. Mouse populations are characterized by controlled breeding and the existence of inbred strains (each of which is essentially a collection of genetic clones). Humans breed with whomever they want, and are much more polymorphic than mice typically are. In addition, humans frequently form genetic subpopulations with characteristic gene frequencies. Comparison of the gene frequencies between a patient group and a "control" is apt to fall victim to stratification, meaning that controls may tend to be drawn from one population while alcoholics (for example) are drawn from another. The apparent gene frequency differences (or marker associations) are then likely not to be replicated in additional samples. Added to this is the fact that studies with humans tend to look for markers associated with a diagnosis. A diagnosis of alcoholism can be achieved with many different clusters of symptoms, so any group of alcoholics in a genetic study is likely to be diagnostically (and potentially etiologically) heterogeneous.

The statistical and methodological difficulties with human genetic mapping for complex traits are discussed more authoritatively elsewhere *(59–61)*.

Consequently, the search for markers associated with alcoholism, like similar searches for other complex traits, has been a succession of findings of linkage followed by a failure to replicate. Most such efforts have attempted to search intensively for markers in the vicinity of a candidate gene of interest (a serotonin receptor, for example). When considered collectively, the literature offers mixed evidence for several potential associations. This evidence has been reviewed elsewhere *(62–65)*.

One large association study has been ongoing for more than 10 yr. The Collaborative Study on the Genetics of Alcoholism (COGA) has accumulated information and genotypes from more than 10,000 subjects from hundreds of multigeneration families.

A few suggestive and even signficant linkages have been reported *(2,66–68)*. The best hope for such studies is currently thought to reside in shifting the focus to "endophenotypes." These are heritable traits associated with the real trait of interest. For example, alcoholics (and their close genetic relatives) have lower-amplitude sensory event-related potentials, called P3 or P300 waves, than controls. QTLs for P300 amplitude have been mapped in the COGA data set *(69)*. Mapping results with these and other phenotypes have been reviewed *(70)*.

As the mouse and human mapping projects continue to develop, it will be of increasing interest to compare results directly, given the evolutionary conservation of the genetic maps. Thus far, the most encouraging possibility of convergent QTL identification comes from the domain of alcohol withdrawal. The chromosome 11 QTL in mice falls in a region spanning genes for several $GABA_A$ receptor subtypes *(38)*. One variant form of the GABAα2 receptor subtype has been shown to be significantly associated with withdrawal severity across a number of inbred mouse strains *(71)*. Human QTL studies have also reported some evidence for an association of alcohol dependence with markers in the region of these subunit genes (72–74). This is far from proof that the same gene has been mapped in humans and mice. In neither species has the 2-subunit gene been proven to underlie the QTL association. Nonetheless, it offers hope that the efforts to find genes in mice and humans will be increasingly informative and facilitate progress in both species.

6. Conclusions

The search for alcohol-related genes has only been underway for about 10 yr. During that time, significant progress has been made, but no gene has yet been captured. This chapter has largely ignored the formidable complexities that must be addressed for further progress to occur. These include the highly specific nature of phenotypes under investigation, an important point because any gene-finding study is only as good as the phenotyping on which it rests, and the genes eventually found will certainly have a restricted range of phenotypic influence. Genes interact with other genes (epistasis), and both human and animal studies are faced with extracting genes' presence from this backgound of noise caused by other genes. Genes' effects are not necessarily the same even when identical genotypes are compared in multiple environmental settings *(75)*. Such gene *X* environment interactions are also of importance. Many of these issues are discussed elsewhere *(63)*. Nonetheless, the progress in 10 years has been marked, and the tools are improving daily *(76)*. Soon we will need to deal with the issue of what use to make of this knowledge of specific risk-promoting and -protective genes, and this must be an important future consideration for all geneticists.

Acknowledgments

Preparation of this chapter was supported by grants from the U.S. Department of Veterans Affairs, the National Institute on Alcohol Abuse and Alcoholism, and the National Institute on Drug Abuse.

References

1. Rose, R. J. (1995) Genes and human behavior. *Ann. Rev. Psychol.* **4,** 625–654.
2. Enoch, M. A. and Goldman, D. (2001) The genetics of alcoholism and alcohol abuse. *Curr. Psychiat. Rep.* **3,** 144–151.

3. Chen, C. C., Lu, R. B., Chen, Y. C., Wang, M. F., Chang, Y. C., Li, T. K., and Yin, S. J. (1999) Interaction between the functional polymorphisms of the alcohol-metabolism genes in protection against alcoholism. *Am. J. Hum. Genet.* **6,** 795–807.
4. Cunningham, C. L. and Phillips, T. J. (2002) Genetic basis of alcohol reward, in *Molecular Biology of Drug Addiction* (Maldonado, R., ed.),. Humana Press, Totowa, NJ, pp. 00–00.
5. Buck, K. J., Crabbe, J. C., and Belknap, J. K. (2000) Alcohol and other abused drugs, in *Genetic Influences on Neural and Behavioral Functions* (Pfaff, D. W., et al., eds.), CRC Press, Boca Raton, FL, pp. 159–183.
6. Boehm II, S. L., Schafer, G. L., Phillips, T. J., Browman, K. E., and Crabbe, J. C. (2000) Sensitivity to ethanol-induced motor incoordination in 5-HT_{1B} receptor null mutant mice is task-dependent: implications for behavioral assessment of genetically altered mice. *Behav. Neurosci.* **114,** 401–409.
7. Homanics, G. E., Le, N. Q., Kist, F., Mihalek, R., Hart, A. R., and Quinlan, J. J. (1998) Ethanol tolerance and withdrawal responses in $GABA_A$ receptor $\alpha 6$ subunit null allele mice and in inbred C57BL/6J and strain 129/SvJ mice. *Alcohol Clin.Exp.Res.* **22,** 259–265.
8. Coste, S. C., Kesterson, R. A., Heldwein, K. A., Stevens, S. L., Heard, A. D., Hollis, J. H., Murray, S. E., Hill, J. K., Pantely, G. A., Hohimer, A. R., Hatton, D. C., Phillips, T. J., Finn, D. A., Low, M. J., Rittenberg, M. B., Stenzel, P., and Stenzel-Poore, M. P. (2000) Abnormal adaptations to stress and impaired cardiovascular function in mice lacking corticotropin-releasing hormone receptor-2. *Nat. Genet.* **24,** 403–409.
9. Crabbe, J. C. (2001) Use of genetic analyses to refine phenotypes related to alcohol tolerance and dependence. *Alcohol Clin. Exp. Res.* **25,** 288–292.
10. Bond, C. T., Sprengel, R., Bissonnette, J. M., Kaufmann, W. A., Pribnow, D., Neelands, T., Storck, T., Baetscher, M., Jerecic, J., Maylie, J., Knaus, H. G., Seeburg, P. H., and Adelman, J. P. (2000) Respiration and parturition affected by conditional overexpression of the Ca^{2+}-activated K^+ channel subunit, SK3. *Science* **289,** 1942–1946.
11. Crabbe, J. C. (1997) Where does alcohol act in the brain? *Mol. Psychiatr.* **2,** 17–20.
12. Ryabinin, A. E. (2000) ITF mapping after drugs of abuse: pharmacological versus perceptional effects. *Acta Neurobiol. Exp. (Warsz.)* **60,** 547–555.
13. Reilly, M. T., Fehr, C., and Buck, K. J. (2001) Alcohol and gene expression in the central nervous system, in *Nutrient–Gene Interactions in Health and Disease* (Moussa-Moustaid, N., et al., eds.), CRC Press, Boca Raton, FL, pp. 131–162.
14. Watson, S. J. and Akil, H. (1999) Gene chips and arrays revealed: a primer on their power and their uses. *Biol. Psychiatr.* **45,** 533–543.
15. Lewohl, J. M., Wang, L., Miles, M. F., Zhang, L., Dodd, P. R., and Harris, R. A. (2000) Gene expression in human alcoholism: microarray analysis of frontal cortex. *Alcohol Clin. Exp. Res.* **24,** 1873–1882.
16. Thibault, C., Lai, C., Wilke, N., Duong, B., Olive, M. F., Rahman, S., Dong, H., Hodge, C. W., Lockhart, D. J., and Miles, M. F. (2000) Expression profiling of neural cells reveals specific patterns of ethanol-responsive gene expression. *Mol. Pharmacol.* **58,** 1593–1600.
17. Xu, Y., Ehringer, M., Yang, F., and Sikela, J. M. (2001) Comparison of global brain gene expression profiles between inbred long-sleep and inbred short-sleep mice by high-density gene array hybridization. *Alcohol Clin. Exp. Res.* **25,** 810–818.
18. Lockhart, D. J. and Winzeler, E. A. (2000) Genomics, gene expression and DNA arrays. *Nature* **405,** 827–836.
19. Palmer, A. A. and Phillips, T. J. (2002) Quantitative trait locus (QTL) mapping in mice, in *Methods in Alcohol-Related Neuroscience Research* (Liu, Y. and Lovinger, D., eds.), CRC Press, Boca Raton, FL, pp. 1–30.
20. Moore, K. J. and Nagle, D. L. (2000) Complex trait analysis in the mouse: The strengths, the limitations and the promise yet to come. *Annu. Rev. Genet.* **34,** 653–686.

21. Blizard, D. A. and Darvasi, A. (1999) Experimental strategies for mapping quantitative trait loci (QTL) analysis in laboratory animals. In *Handbook of Molecular-Genetic Techniques for Brain and Behavior Research* (W. E. Crusio and R. T. Gerlai, eds.), Elsevier, Amsterdam, pp. 82–99.
22. Belknap, J. K., Mitchell, S. R., O'Toole, L. A., Helms, M. L., and Crabbe, J. C. (1996) Type I and type II error rates for quantitative trait loci (QTL) mapping studies using recombinant inbred mouse strains. *Behav. Genet.* **26,** 149–160.
23. Almasy, L. and Blangero, J. (1999) Linkage strategies for mapping genes for complex traits in man, in *Handbook of Molecular-Genetic Tachniques for Brain and Behavior Research* (Crusio, W. E. and Gerlai, R. T., eds.), Elsevier, Amsterdam, pp. 100–112.
24. Lander, E. S. and Botstein, D. (1989) Mapping Mendelian factors underlying quantitative traits using RFLP linkage maps. *Genetics* **121,** 185–199.
25. Copeland, N. G., Jenkins, N. A., Gilbert, D. J., Eppig, J. T., Maltais, L. J., Miller, J. C., Dietrich, W. F., Weaver, A., Lincoln, S. E., and Steen, R. G. (1993) A genetic linkage map of the mouse: current applications and future prospects. *Science* **262,** 57–66.
26. Plomin, R., McClearn, G. E., Gora-Maslak, G., and Neiderhiser, J. M. (1991) Use of recombinant inbred strains to detect quantitative trait loci associated with behavior. *Behav. Genet.* **21,** 99–116.
27. Crabbe, J. C., Belknap, J. K., and Buck, K. J. (1994) Genetic animal models of alcohol and drug abuse. *Science* **264,** 1715–1723.
28. Lander, E. and Kruglyak, L. (1995) Genetic dissection of complex traits: guidelines for interpreting and reporting linkage results. *Nat. Gen.* **11,** 241–247.
29. Crabbe, J. C., Phillips, T. J., Buck, K. J., Cunningham, C. L., and Belknap, J. K. (1999) Identifying genes for alcohol and drug sensitivity: recent progress and future directions. *Trends Neurosci.* **22,** 173–179.
30. Crabbe, J. C., Merrill, C. D., and Belknap, J. K. (1991) Acute dependence on depressant drugs is determined by common genes in mice. *J. Pharmacol. Exp. Ther.* **257,** 663–667.
31. Goldstein, D. B. and Pal, N. (1971) Alcohol dependence produced in mice by inhalation of ethanol: grading the withdrawal reaction. *Science* **172,** 288–290.
32. Goldstein, D. B. (1973) Inherited differences in intensity of alcohol withdrawal reactions in mice. *Nature* **245,** 154–156.
33. Crabbe, J. C., Kosobud, A., Young, E. R., Tam, B. R., and McSwigan, J. D. (1985) Bidirectional selection for susceptibility to ethanol withdrawal seizures in *Mus musculus. Behav. Genet.* **15,** 521–536.
34. Metten, P. and Crabbe, J.C. (1994) Common genetic determinants of severity of acute withdrawal from ethanol, pentobarbital and diazepam in inbred mice. *Behav. Pharmacol.* **5,** 533–547.
35. Crabbe, J. C., Jr., Young, E. R., and Kosobud, A. (1983) Genetic correlations with ethanol withdrawal severity. *Pharmacol. Biochem. Behav.* **18 (Suppl. 1),** 541–547.
36. Belknap, J. K., Metten, P., Helms, M. L., O'Toole, L. A., Angeli-Gade, S., Crabbe, J. C., and Phillips, T. J. (1993) Quantitative trait loci (QTL) applications to substances of abuse: physical dependence studies with nitrous oxide and ethanol in BXD mice. *Behav. Genet.* **23,** 213–222.
37. Crabbe, J. C. (1998) Provisional mapping of quantitative trait loci for chronic ethanol withdrawal severity in BXD recombinant inbred mice. *J. Pharmacol. Exp. Ther.* **286,** 263–271.
38. Buck, K. J., Metten, P., Belknap, J. K., and Crabbe, J. C. (1997) Quantitative trait loci involved in genetic predisposition to acute alcohol withdrawal in mice. *J. Neurosci.* **17,** 3946–3955.
39. Buck, K., Metten, P., Belknap, J., and Crabbe, J. (1999) Quantitative trait loci affecting risk for pentobarbital withdrawal map near alcohol withdrawal loci on mouse chromosomes 1, 4, and 11. *Mamm. Genome* **10,** 431–437.

40. Fehr, C., Shirley, R. L., Belknap, J. K., Crabbe, J. C., and Buck, K. J. (2002) Congenic mapping of alcohol and pentobarbital withdrawal liability loci to a <1 centimorgam interval of murine chromosome 4: identification of *Mpdz* as a candidate gene. *J. Neurosci.* **22,** 3730–3738.
41. McClearn, G. E. and Kakihana, R. (1981) Selective breeding for ethanol sensitivity: short-sleep and long-sleep mice, in *Development of Animal Models as Pharmacogenetic Tools* (McClearn, G. E., et al., eds.), USDHHS:NIAAA, Rockville, MD, pp. 147–159.
42. DeFries, J. C., Wilson, J. R., Erwin, V. G., and Petersen, D. R. (1989) LS × SS recombinant inbred strains of mice: initial characterization. *Alcohol Clin. Exp. Res.* **13,** 196–200.
43. Markel, P. D., Fulker, D. W., Bennett, B., Corley, R. P., DeFries, J. C., Erwin, V. G., and Johnson, T. E. (1996) Quantitative trait loci for ethanol sensitivity in the LS × SS recombinant inbred strains: interval mapping. *Behav. Genet.* **26,** 447–458.
44. Markel, P. D., Bennett, B., Beeson, M., Gordon, L., and Johnson, T. E. (1997) Confirmation of quantitative trait loci for ethanol sensitivity in long- sleep and short-sleep mice. *Genome Res.* **7,** 92–99.
45. Bennett, B., Beeson, M., Gordon, L., and Johnson, T. E. (1997) Quick method for confirmation of quantitative trait loci. *Alcohol Clin. Exp. Res.* **21,** 767–772.
46. Bennett, B. and Johnson, T. E. (1998) Development of congenics for hypnotic sensitivity to ethanol by QTL- marker-assisted counter selection. *Mamm. Genome* **9,** 969–974.
47. Whatley, V. J., Johnson, T. E., and Erwin, V. G. (1999) Identification and confirmation of quantitative trait loci regulating alcohol consumption in congenic strains of mice. *Alcohol Clin. Exp. Res.* **23,** 1262–1271.
48. Ehringer, M. A., Thompson, J., Conroy, O., Xu, Y., Yang, F., Canniff, J., Beeson, M., Gordon, L., Bennett, B., Johnson, T. E., and Sikela, J. M. (2001) High-throughput sequence identification of gene coding variants within alcohol-related QTLs. *Mamm. Genome* **12,** 657–663.
49. Bennett, B. (2000) Congenic strains developed for alcohol- and drug-related phenotypes. *Pharmacol. Biochem. Behav.* **67,** 671–681.
50. Iakoubova, O. A., Olsson, C. L., Dains, K. M., Ross, D. A., Andalibi, A., Lau, K., Choi, J., Kalcheva, I., Cunanan, M., Louie, J., Nimon, V., Machrus, M., Bentley, L. G., Beauheim, C., Silvey, S., Cavalcoli, J., Lusis, A. J., and West, D. B. (2001) Genome-tagged mice (gtm): two sets of genome-wide congenic strains. *Genomics* **74,** 89–104.
51. Nadeau, J. H., Singer, J. B., Matin, A., and Lander, E. S. (2000): Analysing complex genetic traits with chromosome substitution strains. *Nat. Genet.* **24,** 221–225.
52. Demarest, K., Koyner, J., McCaughran, J. J., Cipp, L., and Hitzemann, R. (2001) Further characterization and high-resolution mapping of quantitative trait loci for ethanol-induced locomotor activity. *Behav. Genet.* **31,** 79–91.
53. Flint, J. and Mott, R. (2001) Finding the molecular basis of quantitative traits: Successes and pitfalls. *Nat. Rev. Genet.* **2,** 437–445.
54. Turri, M. G., Talbot, C. J., Radcliffe, R. A., Wehner, J. M., and Flint, J. (1999) High-resolution mapping of quantitative trait loci for emotionality in selected strains of mice. *Mamm. Genome* **10,** 1098–1101.
55. Nadeau, J. H. and Frankel, W. N. (2000) The roads from phenotypic variation to gene discovery: mutagenesis versus QTLs. *Nat. Genet.* **25,** 381–384.
56. Belknap, J. K., Hitzemann, R., Crabbe, J. C., Phillips, T. J., Buck, K. J., and Williams, R. W. (2001) QTL analysis and genome-wide mutagenesis in mice: complementary genetic approaches to the dissection of complex traits. *Behav. Genet.* **31,** 5–15.
57. Grupe, A., Germer, S., Usuka, J., Aud, D., Belknap, J. K., Klein, R. F., Ahluwalia, M. K., Higuchi, R., and Peltz, G. (2001) *In silico* mapping of complex disease-related traits in mice. *Science* **292,** 1915–1918.

58. Paigen, K. and Eppig, J. T. (2000) A mouse phenome project. *Mamm. Genome* **11,** 715–717.
59. Gelernter, J. (1999) Clinical molecular genetics, in *Neurobiology of Mental Illness* (Charney, D. S., et al., eds.), Oxford University Press, NY, pp. 108–120.
60. Long, A. D. and Langley, C. H. (1999) The power of association studies to detect the contribution of candidate genetic loci to variation in complex traits. *Genome Res.* **9,** 720–731.
61. Uhl, G. R. (1999) Molecular genetics of substance abuse vulnerability: a current approach. *Neuropsychopharmacology* **20,** 3–9.
62. Reich, T., Hinrichs, A., Culverhouse, R., and Bierut, L. (1999) Genetic studies of alcoholism and substance dependence. *Am. J. Hum. Genet.* **65,** 599–605.
63. Crabbe, J. C. (2002) Genetic contributions to addiction. *Ann. Rev. Psychol.* **53,** 435–462.
64. Lichtermann, D., Hranilovic, D., Trixler, M., Franke, P., Jernej, B., Delmo, C. D., Knapp, M., Schwab, S. G., Maier, W., and Wildenauer, D. B. (2000) Support for allelic association of a polymorphic site in the promoter region of the serotonin transporter gene with risk for alcohol dependence. *Am. J. Psychiatr.* **157,** 2045–2047.
65. LaForge, K. S., Yuferov, V., and Kreek, M. J. (2000) Opioid receptor and peptide gene polymorphisms: potential implications for addictions. *Eur. J. Pharmacol.* **410,** 2493–268.
66. Foroud, T., Bucholz, K. K., Edenberg, H. J., Goate, A., Neuman, R. J., Porjesz, B., et al. (1998) Linkage of an alcoholism related severity phenotype to chromosome 16. *Alcohol Clin. Exp. Res.* **22,** 2035–2042.
67. Foroud, T., Edenberg, H. J., Goate, A., Rice, J., Flury, L., Koller, D. L., Bierut, L. J., Conneally, P. M., Nurnberger, J. I., Bucholz, K. K., Li, T. K., Hesselbrock, V., Crowe, R., Schuckit, M., Porjesz, B., Begleiter, H., and Reich, T. (2000) Alcoholism susceptibility loci: confirmation studies in a replicate sample and further mapping. *Alcohol Clin. Exp. Res.* **24,** 933–945.
68. Schuckit, M. A., Edenberg, H. J., Kalmijn, J., Flury, L., Smith, T. L., Reich, T., Bierut, L., Goate, A. A., and Foroud, T. (2001) A genome-wide search for genes that relate to a low level of response to alcohol. *Alcohol Clin. Exp. Res.* **25,** 323–329.
69. Begleiter, H.,and Porjesz, B. (1999) What is inherited in the predisposition toward alcoholism? A proposed model. *Alcohol Clin. Exp. Res.* **23,** 1125–1135.
70. Schuckit, M. A. (2000) Biological phenotypes associated with individuals at high risk for developing alcohol-related disorder. Part 2. *Addict. Biol.* **5,** 23–36.
71. Hood, H. M. and Buck, K. J. (2000) Allelic variation in the $GABA_A$ receptor 2 subunit is associated with genetic susceptibility to ethanol-induced motor incoordination and hypothermia, conditioned taste aversion, and withdrawal in BXD/Ty recombinant inbred mice. *Alcohol Clin. Exp. Res.* **24,** 1327–1334.
72. Sander, T., Ball, D., Murray, R., Patel, J., Samochowiec, J., Winterer, G., Rommelspacher, H., Schmidt, L. G., and Loh, E. W. (1999) Association analysis of sequence variants of $GABA_A$ $\alpha 6$, $\alpha 2$, and $\gamma 2$ gene cluster and alcohol dependence. *Alcohol Clin. Exp. Res.* **23,** 427–431.
73. Loh, E. W., Higuchi, S., Matsushita, S., Murray, R., Chen, C. K., and Ball, D. (2000) Association analysis of the $GABA_A$ receptor subunit genes cluster on 5q33-34 and alcohol dependence in a Japanese population. *Mol. Psychiatr.* **5,** 301–307.
74. Iwata, N., Virkkunen, M., and Goldman, D. (2000) Identification of a naturally occurring Pro385-Ser385 substitution in the $GABA_A$ receptor $\alpha 6$ subunit gene in alcoholics and healthy volunteers. *Mol. Psychiatr.* **5,** 316–319.
75. Crabbe, J. C., Wahlsten, D., and Dudek, B. C. (1999) Genetics of mouse behavior: interactions with laboratory environment. *Science* **284,** 1670–1672.
76. Phillips, T. J., Belknap, J. K., Hitzemann, R. J., Buck, K. J., Cunningham, C. L., and Crabbe, J. C. (2002) Harnessing the mouse to unravel the genetics of human disease. *Gen. Brain Behav.* **1,** 14–26.

14

Genetic Basis of Ethanol Reward

Christopher L. Cunningham and Tamara J. Phillips

1. Introduction

Substantial evidence supports the suggestion that development of alcoholism is strongly influenced by family history and that genes underlie a significant portion of that influence *(1)*. Despite recognition of a genetic influence for many decades, it is only within the last 10 years that scientists have had the sophisticated research tools needed to begin the task of identifying specific genes involved in alcoholism. These tools have been provided by advances in molecular biology and by improvements in the animal genetic models and behavioral models used to study the processes that contribute to excessive ethanol consumption.

Use of animal models has enabled examination of a wide range of ethanol-related behavioral and neurobiological processes. Although many different behavioral processes have been considered, theorists place strong emphasis on ethanol's rewarding and aversive motivational effects. These effects are believed to be critical in determining whether individuals exposed to ethanol will continue to consume ethanol and increase their intake over time. Moreover, genetic differences in sensitivity to ethanol's rewarding and aversive effects are hypothesized to contribute importantly to development of the excessive drinking patterns characteristic of ethanol abuse and alcoholism *(2)*.

This chapter focuses on recent animal studies that address the contribution of genotype to ethanol's rewarding and aversive effects. We begin with a brief overview of the major behavioral models currently used to draw inferences about ethanol's motivational consequences. This overview is followed by more detailed consideration of genetic influences in these behavioral models using a variety of different strategies. Finally, we conclude with a summary of the current status of research in this area along with suggestions for possible future directions.

2. Models of Ethanol Reward and Aversion

Most studies examining genetic influences on ethanol reward have used behavioral procedures that fall into one of two categories: (a) self-administration models and (b) conditioning models. Self-administration models typically involve procedures in which animals have substantial control over their ethanol intake, including control over the amount (dose) and temporal pattern of intake. Within this category, it is useful to distinguish between relatively simple home-cage drinking procedures and procedures con-

From: *Molecular Biology of Drug Addiction*
Edited by: R. Maldonado © Humana Press Inc., Totowa, NJ

ducted in operant chambers that require an explicit "seeking" response (e.g., bar pressing) to obtain ethanol. In contrast, conditioning models are characterized by experimenter administration of a fixed drug dose in combination with exposure to a gustatory (e.g., taste conditioning) or exteroceptive (e.g., place conditioning) stimulus. The value attached to this drug-paired stimulus is typically assessed in a test conducted without ethanol. Each of these models is briefly reviewed in the following subsections, with emphasis on issues of potential importance to interpretation of genetic studies described later. More comprehensive discussions of these and related behavioral models can be found elsewhere *(3–6)*.

2.1. Home Cage Drinking

The oldest and most commonly used technique for studying genetic differences in ethanol reward is simply to place a drinking tube on the home cage and measure the amount/dose of ethanol consumed or the preference for ethanol relative to a concurrently available alternative solution such as water *(7)*. The principal advantages of this procedure are its simplicity and high face validity. Although it is recognized that many variables influence ethanol drinking in such procedures, rewarding postingestive pharmacological effects are assumed to play a significant role.

In home-cage procedures, ethanol is usually available continuously and intakes/preferences are measured over 24-h (or longer) periods of time. A potential problem with this approach is that 24-h measures may hide important variations in the temporal pattern of ethanol exposure. For example, a strain that drinks a large number of relatively small ethanol bouts each day could achieve the same total daily intake as a strain that drinks a small number of relatively large bouts. Although analysis of total daily intakes would suggest no strain difference, consideration of bout size suggests a difference in the pharmacological effects experienced by each strain. One solution to this problem, of course, is to measure drinking patterns within each day of continuous access (e.g., using a drinkometer: *[8–10]*). Alternatively, one can measure intakes over smaller periods of time. Another solution is to actually reduce the total amount of time per day that ethanol is available (e.g., 30–60 min), so that animals have the opportunity to engage in only one or a relatively small number of drinking bouts. Such "limited-access" procedures promote relatively high ethanol intakes in short time periods *(11,12)*, making it more likely that ethanol's pharmacological effects influence intake.

Although interpretation of genetic differences in home-cage ethanol drinking often focuses on hypothesized differences in sensitivity to postabsorptive rewarding pharmacological effects, ethanol drinking/preference can also be influenced by other factors such as taste or sensitivity to aversive orosensory or pharmacological effects. To address the role of taste, investigators have sometimes examined intakes/preferences for primary tastants (e.g., sweet, bitter) in the same strains. When strains differ in ethanol intake/preference, but not in intake/preference for the primary tastants, interpretations based on differences in sensitivity to postabsorptive pharmacological effects become more compelling. Testing animals in operant or conditioning procedures in which ethanol is administered by a nonoral route offers another strategy for eliminating interpretations based on taste differences. Conditioning procedures are also useful for addressing the possibility of genetic differences in sensitivity to aversive drug effects (*see* below).

2.2. Operant Self-Administration

In home-cage self-administration procedures, animals simply approach and consume ethanol. In contrast, operant procedures require animals to engage in an explicit "seeking" response (e.g., bar press or nose poke) to gain access to ethanol *(13,14)*. Although ethanol is usually consumed orally, operant models allow for the possibility of administering ethanol directly into the stomach, blood, or brain via surgically implanted cannulae *(15)*. Thus, in contrast to home-cage procedures, operant models allow one to separate ethanol "consumption" from ethanol "seeking" and, if desired, to eliminate the oral route entirely. The value of the latter feature is well illustrated by a recent study showing that two mouse strains known to differ dramatically in home-cage ethanol intake (DBA/2 and C57BL/6) showed little difference when nose poking produced intravenous ethanol injections in an operant procedure *(16)*. This finding supports the suggestion that aversive orosensory (preabsorptive) effects of orally administered ethanol contribute to the normally low home-cage intakes of DBA/2 mice *(17)*.

Interpretive issues raised in the study of home-cage oral self-administration generally also apply to operant studies using the oral route (e.g., concern over the temporal pattern of ethanol intake). In both models, investigators will sometimes add a sweetener or other flavor to the ethanol solution in an effort to increase overall intakes. Although these flavor additives may be "faded out" over time *(18)*, use of this strategy in genetic studies raises the possibility that strain differences are caused by differences in sensitivity to the added flavor rather than to postingestive effects of ethanol. Again, as suggested earlier, this hypothesis can be addressed by examining responding for the flavor additive in the absence of ethanol.

One difficulty in the interpretation of both home-cage and operant self-administration procedures is that increases (or decreases) in the target behavior produced by a genetic manipulation do not unambiguously reflect increases (or decreases) in ethanol reward. This ambiguity arises due to the inverted U-shaped relationship between ethanol intake and variables that presumably affect ethanol's reinforcing efficacy, such as dose or concentration *(19,20)*. Thus, a genetic manipulation that reduced ethanol reward (e.g., a null mutation) could either increase or decrease ethanol intake, depending on where control intakes fell on the concentration/dose–response function. Strategies for addressing this problem include examining a range of ethanol doses/concentrations and using conditioning procedures to provide converging evidence on effects of the genetic manipulation *(21–23)*. A unique feature of the operant model is its potential to separate genetic differences in appetitive processes underlying ethanol-seeking behavior from consummatory processes involved in the regulation of ethanol intake *(14)*. This feature could be important for determining whether different genes influence appetitive and consummatory processes. However, as recently noted *(24,25)*, much of the literature on operant self-administration of ethanol has not allowed a clear separation of these processes, due to frequent alternation between "seeking" and "consuming" within self-administration sessions. Although promising alternative procedures involving chain schedules have recently been introduced *(24,25)*, these schedules have not yet been used in the study of genetic differences.

2.3. Conditioning Models

Conditioning models are based on the premise, derived from Pavlovian conditioning, that stimuli paired with drug exposure acquire the ability to elicit affective states similar to those evoked by the drug itself *(26,27)*. Thus, approach or contact with a drug-associated conditioned stimulus (CS) is interpreted as evidence of a drug's rewarding effect, whereas avoidance or withdrawal from a drug-paired CS is viewed as indicating the drug's aversive effect. Early proponents of these techniques sometimes implied that conditioning models could substitute for self-administration models in the analysis of drug reward *(28)*. Although there is certainly overlap in the information provided by these models, theorists now recognize that conditioning models measure a learning process that is fundamentally distinct from that measured by self-administration models *(29)*. Consequently, genetic differences observed in self-administration models may not always match those obtained in conditioning models.

Place and taste conditioning are the two conditioning procedures that have been used most often to examine genetic differences in ethanol's rewarding and aversive effects. These procedures differ primarily in the nature of the CS paired with ethanol, and the response used to index learning. In place conditioning, the experimenter initially pairs drug with distinctive environmental cues (e.g., tactile, visual) and later measures approach toward or withdrawal from those cues in a choice test *(30)*. In taste conditioning, the experimenter pairs drug with novel taste or flavor cues (e.g., saccharin, sodium chloride) and measures changes in subsequent intake or preference for the flavored food or fluid *(31,32)*. Although avoidance of foods or fluids previously paired with abused drugs has often been attributed to presumed aversive properties of such drugs *(33,34)*, some theorists have argued that taste avoidance may actually be caused by rewarding drug effects *(35,36)*. On the basis of genetic correlational data from standard inbred mouse strains, however, it seems likely that taste avoidance induced by ethanol reflects an aversive drug effect *(37)*.

One of the main advantages of conditioning procedures is their sensitivity to both drug reward and aversion. In the case of ethanol, both the taste and place conditioning procedures have yielded evidence of conditioned preference and conditioned aversion *(4)*. Although a detailed discussion of variables producing these diverse outcomes is beyond the scope of this chapter, it appears that prior ethanol experience *(38)*, species *(39)*, and the temporal relationship between CS and ethanol *(40)* are among the variables determining whether preference or aversion is obtained.

Other advantages of conditioning models include the ability to test animals in a drug-free state and the fact that dose–effect curves are typically monophasic rather than biphasic. However, these models also have certain disadvantages that must be considered in the analysis of genetic differences. For example, initial biases (i.e., unconditioned preference or aversion) for the target CS may interfere with measurement of conditioned preference or aversion. Moreover, genetic differences in those unconditioned biases will complicate interpretation of genetic differences in conditioned effects. Interpretation of genetic differences in a target phenotype may be further complicated by genetic differences in other phenotypes that have an important influence on the target phenotype. For example, genetic differences in initial fluid intake may hinder interpretation of intake changes produced by taste conditioning *(41)*. Similarly, genetic differences in basal locomotor activity may affect detection of conditioned changes produced by place conditioning *(42)*.

3. Genetic Strategies and Findings

Several different genetic strategies have been applied to understanding the role of genes that influence ethanol's rewarding and aversive effects. These approaches include using traditional animal genetic models such as inbred strains and selectively bred lines, as well as more recently developed techniques that involve the mapping of quantitative trait loci (QTLs), the insertion or inactivation of targeted genes, or the measurement of changes in gene expression. Recent advances based on the application of these strategies to the study of ethanol reward and aversion are summarized below.

3.1. Standard Inbred Strains

The heading, "Standard Inbred Strains," is used to distinguish this panel of strains from specialized groups such as recombinant inbred strains. Standard inbred strains differ genetically from each other, unsystematically. They are unique, relative to a population of heterogeneous individuals, in that each individual is homozygous at every genetic locus. Further, each individual of an inbred strain is genetically identical to all other individuals of its strain. Therefore, differences among individuals must arise from epigenetic influences. The laboratory is ideal for holding environmental conditions such as temperature, housing density, and handling procedures constant. When several inbred strains are examined under controlled conditions, differences among individuals within each strain can be used to estimate epigenetic sources of trait variation, and differences among strains to estimate genetic influences.

Inbred strains have been widely used to demonstrate genetic influences on motivational traits, including those thought to reflect sensitivity to the rewarding and aversive effects of ethanol. Some of the earliest examples are studies that characterized strain differences for home-cage ethanol drinking *(7)*. Larger numbers of strains have been included in more recent examinations of this trait *(43)*. In addition, standard inbred strains have been used to demonstrate genetic regulation of ethanol-induced conditioned place preference (CPP) *(44)* and conditioned taste aversion (CTA) *(37)*. The many two-strain comparison studies, which also pepper the literature, suggest genetic regulation, but are inconclusive due to their lack of power for estimating heritability. Panels of 12 or more strains provide reasonable power for estimating genetic contribution, and for examining genetic correlations between traits. However, they have typically fallen short of providing specific genetic information with regard to chromosomal location of genes influencing these complex motivational traits.

As marker and sequence information expands for specific inbred strains, their utilization for gene mapping purposes is likely to grow. A recent publication by Grupe et al. *(45)* demonstrated that single nucleotide polymorphism data from standard inbred strains could be utilized to rapidly predict the chromosomal regions associated with trait variation. Furthermore, the Mouse Phenome Project *(46)*, initiated with the goal of promoting systematic characterization of many standard inbred mouse strains for a wide range of phenotypes of potential importance to biomedicine and biological science, may provide a rich phenotypic database, including ethanol reward traits, to which genetic information can be applied. In addition, the characteristics of inbred strains make them invaluable for QTL mapping (discussed below) via the production of recombinant inbred (RI) strains. They also provide a fixed genetic background for assessing effects of targeted mutations and transgene insertions, a characteristic that is proving to be critical in the interpretation of results in single-gene mutants *(47–54)*.

3.2. Selected Lines

Artificial selection is highly effective at producing extreme-scoring individuals, when some proportion of the trait variability within a population is genetically influenced. In this case, mating of extreme-scoring individuals results in offspring that are more extreme-scoring as a group, relative to the original population mean. Rodent lines selectively bred to study ethanol reward and sensitivity have been described elsewhere *(55–57)* and their use in the new genetic era discussed *(58,59)*. These selection programs have been bidirectional, producing genetic animal models with extreme sensitivity opposite those with extreme insensitivity for a presumably single trait. Like standard inbred strains, a typical use of selected lines has been to search for correlated traits to selection—traits that are pleiotropically affected by some of the same genes. For example, short-term selected lines bred for gene mapping purposes (discussed below) were used to demonstrate a genetic relationship between ethanol consumption and genetic susceptibility to exhibit withdrawal *(60)*.

In the current era, a time when mapping of complex trait genetics has become approachable, selected line utilization has been adjusted. The extreme nature of their selected traits makes them ideal for QTL mapping. For example, short-term selected lines were used to verify the locations of QTLs for home-cage ethanol preference drinking *(61)*; P and NP rats, long-term selected lines bred for high (P and HAD) and low (NP and LAD) home-cage ethanol preference drinking, have also been used to narrow the gene search to chromosomal regions *(62–64)*. Genotypic, rather than phenotypic, selection has also been used to verify a QTL for alcohol acceptance (consumption of ethanol in the absence of a water choice) in mice *(65)*. In this case, animals were chosen for breeding based on their genotype at chromosomal regions thought to harbor QTLs for alcohol acceptance.

The extreme traits of selected lines also make them suitable for examining the effects of genetic manipulations. Once a gene has been identified as a candidate for regulation of the selection trait or a genetically correlated trait, a logical approach might be the use of antisense oligodeoxynucleotides against that particular gene. This approach has been used successfully for non-alcohol-related traits. For example, injection of the antisense oligodeoxynucleotide against angiotensinogen significantly reduced the elevated systolic blood pressure, which is the hallmark of SHR rats *(66)*. Currently, problems with delivery of antisense molecules to the brain, with duration of action, and with difficulty in determining appropriate sites of injection, complicate the widespread utilization of this technology.

3.3. QTL Mapping

We have already mentioned QTL mapping, a means by which genes for several ethanol reward traits have been localized to segments of particular chromosomes. A chapter that serves as a primer on QTL mapping has recently been published *(67)*, and several other treatments of this topic also exist *(68–72)*. We review here the findings to date with regard to nominated and confirmed QTL locations for traits thought to reflect sensitivity to ethanol reward and aversion. We mention when similar regions have been implicated among studies. However, we recognize that the same gene or a different linked gene could be the source of similarity.

It is not surprising that ethanol consumption has been a focus of QTL mapping. Already mentioned are QTL studies using selected lines. Table 1 lists QTL studies in

Table 1
Nominated, Suggestive, and Confirmed QTLs for Ethanol Reward and Aversion Phenotypes in Mice

References	Trait	Genetic model	Nominated QTL locations	Suggestive and confirmed QTL locations
Home-cage two-bottle-choice drinking				
(73)	Home-cage drinking, 10% ethanol vs water	BXD RI	Chr 2, 49 cM Chr 3, 73 cM Chr 4, 59 cM Chr 7, 11 cM Chr 7, 57 cM Chr 9, 28 cM	
(74)	Home-cage drinking, 10% ethanol vs water	BXD RI	Chr 1, 33 cM Chr 1, 107 cM Chr 2, 53 cM Chr 6, 50 cM Chr 7, 13 cM Chr 7, 57 cM Chr 10, 67 cM Chr 11, 62 cM Chr 12, 53 cM Chr 15, 48 cM Chr 17, 4 cM	
(76)	Home-cage drinking, 10% ethanol vs water	(B6 × D2) × B6 backcross mice		Chr 2, 28 cM Chr 11, 45 cM
(75)	Home-cage drinking, 10% ethanol vs water	B6D2 F2 Mice		Chr 1, 76 cM Chr 2, 40 cM Chr 3, 45 cM Chr 4, 75 cM Chr 9, 25 cM Chr 10, 28 cM

(continued)

Table 1 (continued)

References	Trait	Genetic model	Nominated QTL locations	Suggestive and confirmed QTL locations
(84) also see *(60)*	Home-cage drinking, 10% ethanol vs water	BXD RI + B6D2 F2 + short-term selected lines		Chr 2, 49 cM Chr 3, 77 cM Chr 4, 56 cM Chr 7, 58 cM Chr 9, 29 cM Chr 15, 43 cM
(77)	Home-cage drinking, 10% ethanol vs water	(B6 × D2) × B6 backcross mice		Chr 1, 18 cM Chr 3, 80 cM
(83)	Home-cage drinking, 10% ethanol vs water	LSXSS RI	Chr 1, 78 cM Chr 2, 33 cM Chr 4, 6 cM Chr 5, 60 cM Chr 9, 29 cM Chr 9, 42 cM Chr 15, 39 cM X Chr , 4 cM	
(80)	Home-cage drinking, 10% ethanol vs water	AXB/BXA RI	Chr 2, 107 cM Chr 4, 9 cM Chr 5, 26 cM Chr 7, 44 cM Chr 7, 64 cM Chr 10, 21 cM Chr 11, 48 cM Chr 11, 70 cM Chr 16, 46 cM Chr 19, 24 cM	

(81)	Home-cage drinking, 12% ethanol vs water	B6.C recombinant QTL introgression strain mice		Chr 15, 53 cM
(82)	Home-cage drinking, 12% ethanol vs water	B6.C recombinant QTL introgression strain mice		Chr 1, 7 cM Chr 2, 106 cM
Ethanol-conditioned traits				
(42)	Conditioned place preference	BXD RI	Chr 4, 56 cM Chr 8, 13 cM Chr 9, 31 cM Chr 18, 43 cM Chr 19, 7 cM	
(41)	Conditioned taste aversion, 2 g/kg	BXD RI	Chr 1, 99 cM Chr 3, 21 cM Chr 3, 49 cM Chr 4, 85 cM Chr 11, 43 cM Chr 17, 17 cM	
(41)	Conditioned taste aversion, 4 g/kg	BXD RI	Chr 1, 85 cM Chr 2, 25 cM Chr 4, 88 cM Chr 6, 56 cM Chr 9, 49 cM	

mice for ethanol consumption that have nominated chromosomal regions or reported suggestive and confirmed linkages. The first published QTL study for home-cage ethanol drinking offered ethanol vs water to recombinant inbred strain (RI) mice generated from C57BL/6J (B6) and DBA/2J (D2) progenitors *(73)*. This panel of BXD RI strains was particularly appropriate for genetic analysis of this trait because of the well-known, extreme difference in ethanol preference and consumption between the B6 and D2 mouse strains *(7,43)*. QTLs nominated by the RI data were followed by studies intended to confirm or eliminate them from further consideration. These studies utilized short-term selected lines *(61)* and a genetically segregating B6 × D2 F2 cross *(22)*, and in combination, confirmed significant linkage of home-cage ethanol consumption with QTLs on chromosomes 2 and 9. At about the same time, similar studies were being conducted by another group, also in BXD RI mice *(74)*. This group ultimately confirmed QTLs for ethanol drinking on chromosomes 1, 4, and 9, using a B6 × D2 F2 intercross *(75)*.

A full genome scan for ethanol consumption QTLs in a (B6 × D2) × B6 backcross *(76)* also identified a QTL on chromosome 2, which was male-specific, at a location somewhat more proximal to the location suggested by the studies of *(22)* and *(75)*. A female-specific QTL mapped to chromosome 11 was not confirmed in a subsequent study *(77)*, but two new linkages to markers on chromosomes 1 (female-specific) and 3 (male-specific) were suggested. The QTL on chromosome 1 was similar in location to that identified by *(75)*, but it was not sex-specific in their F2 population. However, the QTL identified on chromosome 3 by *(75)* was male-specific. Finally, another confirmation study used congenic strains, created using a classical backcross procedure *(78)*, in which D2 alleles were introgressed onto a B6 background. Male B6 × D2 F1 mice were chosen for breeding based on low ethanol preference scores, and were crossed to female B6 mice *(79)*. A full genome scan was accomplished for multiple congenic lines that ultimately confirmed linkages of ethanol consumption with markers on chromosomes 1 and 2. Locations were similar to QTL locations identified by others.

Mice of genotypes other than those derived from B6 and D2 progenitors have also been used to map QTLs for home-cage ethanol drinking. A study in a set of RI strains derived from B6 and A/J progenitors (AXB/BXA RI) provided evidence for an ethanol drinking QTL on chromosome 2 *(80)*, but considerably distal to the location identified by others. Studies in B6.C Recombinant QTL Introgression strains, mice that carry approx 5% BALB/cJ (C) genome on a B6 background, have suggested linkage of ethanol consumption to markers on chromosomes 1, 2, and 15 *(81,82)*. The chromosome 2 QTL was in a location most similar to that suggested by *(80)*. Screening for voluntary ethanol consumption in LSXSS RI mice, strains derived from selected lines bred for extreme sensitivity and insensitivity to ethanol's sedative effects, nominated regions on chromosomes 1, 2, 4, 5, 9, 15, and X *(83)*. Several of these QTLs were in regions similar to those identified in other studies, particularly those on chromosomes 1, 2, 9, and 15.

It should be clear from these descriptions that several chromosomal regions have been implicated by many independent studies to harbor genes that influence voluntary ethanol consumption. This question has been addressed directly. Based on the eight QTL mapping studies described above that used B6/D2-derived populations of mice *(61,73–77,79,84)*, a meta-analysis was performed with the goal of combining the mapping information from these independent studies *(85)*. The QTLs that appeared consistently across studies were on chromosomes 2, 3, 4, and 9. The combined *P* values ranged from 1×10^{-7} to 1×10^{-15}, highly significant according to stringent criteria *(86)*. Two other QTLs, one on chromosome 1 and one on 11, were reported as appearing less consistently, but still reaching overall stringent statistical criteria ($P < 0.0001$). These results indicate the remarkable convergence across studies that were not procedurally identical and pinpoint the chromosomal regions that should be most avidly pursued in the identification of genes that influence ethanol consumption in mice.

Rats have also been used in the search for QTLs relevant to ethanol consumption. The first study that appeared in the literature utilized the F2 cross of the inbred P and NP rat lines, selectively bred for high and low levels of ethanol-preference drinking *(62)*. A QTL was identified on chromosome 4 with a LOD score sufficient to meet suggested statistical criteria for significant linkage. The gene for neuropeptide Y is located within the QTL region, which led to studies in knockout mice to examine the possible role of this peptide on ethanol consumption (reviewed below). However, QTL mapping in mice has not implicated the syntenic region (Chr 6, 26 cM). Subsequent work in P × NP F2 rats nominated two additional QTLs on chromosomes 3 and 8, both in regions syntenic to those previously nominated in mice; the rat chromosome 3 region is syntenic with a region on mouse chromosome 2, and the rat chromosome 8 regions is syntenic with mouse chromosome 9 *(63)*. One other rat study used an F2 cross of noninbred LAD and HAD rats, similarly selectively bred for high and low ethanol drinking levels *(64)*. Evidence was obtained for linkage of ethanol consumption to markers on chromosomes 5, 10, 12, and 16. Surprisingly, none of these QTLs overlap with those identified in the P × NP F2 cross. However, as the authors indicate, the genetic origins of the two sets of selected lines are different.

Only a few other traits relevant to ethanol reward have been subjected to QTL analyses. One of these traits is ethanol-induced CPP. Unlike rats, which tend to exhibit aversion to stimuli previously associated with ethanol *(32)*, ethanol-naïve mice develop preference for ethanol-associated cues *(87,88)*. A QTL study in BXD RI mice provided the strongest evidence for associations with ethanol-induced CPP on chromosomes 4, 8, 9, 18, and 19 *(42)*. The chromosome 4 region at about 56 cM and the chromosome 9 region at about 31 cM are within the confidence intervals of QTLs identified for ethanol consumption. Thus far, studies designed to confirm these nominated regions have not been published. Genetic correlations estimated from BXD RI means from this CPP study and the BXD drinking study of *(73)* showed a trend toward a positive genetic correlation ($r = 0.34$, $n = 17$, $.05 < p < 0.10$, one-tailed), offering weak support of a common genetic influence on these phenotypes *(22)*.

The other ethanol reward relevant trait subjected to QTL analysis is ethanol-induced CTA produced by pairing of a saccharin solution with injection of 2 or 4 g/kg ethanol *(41)*. For 2 g/kg of ethanol, regions of chromosomes 1, 3, 4, 7, 11, and 17 were implicated. For 4 g/kg of ethanol, regions on chromosomes 1, 2, 4, 6, and 9 were nominated. The nominated chromosome 2 QTL is in a region similar to that for ethanol consumption, whereas the nominated chromosome 9 QTL is more distal. Preliminary results suggest confirmation of the CTA QTL located on chromosome 1 *(70)*. Genetic correlations based on BXD RI strain means from this CTA study and strain means from our CPP study *(42)* or our drinking study *(73)* were not statistically significant, suggesting minimal genetic overlap for these traits. However, a recent study of ethanol-induced

CTA in standard inbred strains *(37)* reported a significant genetic correlation between CTA and home-cage ethanol preference *(43)*. More specifically, strains showing stronger ethanol-induced CTA tended to have a lower ethanol preference than strains showing weaker CTA ($r = 0.68$, $n = 13$, $p < 0.05$). Thus, in the standard inbred model, there is evidence of overlap in the genetic mechanisms influencing ethanol preference and ethanol-induced CTA.

Gene by gene analysis will not fully define the array of genetic influences on a complex trait. A developing focus in QTL studies is identification of epistatic interactions that influence trait variability. Pairwise combinations of markers are analyzed to estimate the proportion of trait variability attributable to their interaction. Such an analysis has been applied to the ethanol consumption trait and identified a significant interaction between loci on chromosomes 2 and 3 *(89)*. In other words, the influence on ethanol consumption of the QTL on chromosome 2 was modified by genotype at a chromosome 3 region. Specifically, those mice that were homozygous for B6 alleles in both the chromosome 2 and 3 regions consumed more ethanol than those that were homozygous B6 for chromosome 2, but heterozygous or homozygous D2 on chromosome 3. These data, and others *(90)*, illustrate the need to develop statistical models that routinely screen for interactive genetic influences.

3.4. Candidate Gene Approaches

As in most areas of contemporary neurobiology, investigators interested in the genetic basis of ethanol reward have exploited emerging genetic technologies that allow inactivation of a specific target gene or the insertion of a foreign gene. In most instances, these approaches have involved permanent manipulation of the genome at the embryonic stage and the creation of a new mouse strain (e.g., a "knockout" or "transgenic" strain). However, it is also possible to temporarily inactivate a target gene (using antisense oligonucleotides) or to insert a novel gene (using an adenoviral vector) into specific brain areas of individual animals. A detailed description of these technologies and their limitations is beyond the scope of this chapter. Interested readers can find overviews of these techniques and their application to the study of alcohol-related and other phenotypes elsewhere *(91–97)*. This section focuses on use of gene-targeting manipulations to address the role of specific candidate genes in ethanol reward. Presentation of these studies is organized below according to the neurobiological system thought to be affected by the genetic manipulation. A listing of these studies is also provided in Table 2.

3.4.1. Serotonin System

Given substantial evidence from animal studies linking the serotonin (5-HT) system to alcohol intake *(98)*, several investigators have focused on the role of 5-HT receptor genes in ethanol reward. Interest in the 5-HT_{1B} receptor gene in particular was encouraged by an initial report that 5-HT_{1B}−/− mice drank twice as much ethanol as 129/Sv-ter control mice across a range of ethanol concentrations in a two-bottle home-cage drinking procedure *(99)* . This difference did not appear to be related to taste or calories because there were no differences in food intake or consumption of sucrose, saccharin, or quinine. A subsequent pair of conditioning studies extended these observations by showing that deletion of the 5-HT_{1B} receptor reduced ethanol reward as indexed by

Table 2
Summary of Mouse Transgenic and Knockout Studies Involving Ethanol Reward Phenotypes

System	Ref.	Gene Target	Phenotype	Outcome
Serotonin	*(101)*	5-HT_{1B} receptor gene (knockout)	Two-bottle drinking (continuous access)	No difference from control
	(99)	5-HT_{1B} receptor gene (knockout)	Two-bottle drinking (continuous access)	KO mice drank more ethanol
	(3)	5-HT_{1B} receptor gene (knockout)	Two-bottle drinking (continuous access)	No difference from control
	(104)	5-HT_3 receptor gene (forebrain over-expression transgenic)	Two-bottle drinking (continuous access)	TG mice drank less ethanol
	(106)	MAOA gene (transgene-induced deficiency)	Two-bottle drinking (continuous access)	No difference from control
	(88)	5-HT_{1B} receptor gene (knockout)	Conditioned place preference	KO mice showed reduced CPP
	(88)	5-HT_{1B} receptor gene (knockout)	Conditioned taste aversion	No difference from control
	(103)	5-HT_{1B} receptor gene (knockout)	Operant (lever) self-administration	KO mice initially responded more for ethanol, but effect disappeared
Dopamine	*(21)*	Dopamine D2 receptor gene (knockout)	Conditioned place preference	KO mice showed reduced CPP
	(110)	Dopamine D1 receptor gene (knockout)	Two-bottle drinking (limited and continuous access)	KO mice drank less ethanol
	(110)	Dopamine D1 receptor gene (knockout)	One-bottle drinking (continuous access)	KO mice drank less ethanol
	(22)	Dopamine D2 receptor gene (knockout)	Two-bottle drinking (continuous access)	KO mice drank less ethanol
	(23)	Dopamine D2 receptor gene (knockout)	Operant (lever) self-administration	KO mice responded less for ethanol
	(114)	DARPP-32 gene (knockout)	Conditioned place preference	KO mice showed reduced CPP
	(114)	DARPP-32 gene (knockout)	Conditioned taste aversion	No difference from control
	(114)	DARPP-32 gene (knockout)	Operant (lever) self-administration	KO mice responded less for ethanol

(continued)

Table 2 (continued)

System	Ref.	Gene Target	Phenotype	Outcome
Opioid	*(119)*	POMC gene (β-endorphin knockout)	Intravenous self-administration	KO mice responded more for ethanol
	(118)	POMC gene (β-endorphin knockout)	Two-bottle drinking (continuous access)	KO mice drank more ethanol
	(117)	μ-Opioid receptor gene (knockout)	Conditioned place preference	KO mice showed reduced CPP
	(117)	μ-Opioid receptor gene (knockout)	Two-bottle drinking (continuous access)	KO mice drank less ethanol
	(116)	μ-Opioid receptor gene (knockout)	Operant (lever) self-administration	KO mice responded less for ethanol
	(116)	μ-Opioid receptor gene (knockout)	Operant (nose poke) self-administration	KO mice responded less for ethanol
	(116)	μ-Opioid receptor gene (knockout)	Two-bottle drinking (continuous access)	KO mice drank less ethanol
	(116)	μ-Opioid receptor gene (knockout)	One-bottle drinking (continuous access)	No difference from control
Protein kinase pathways	*(120)*	PKCε gene (knockout)	Two-bottle drinking (continuous access)	KO mice drank less ethanol
	(121)	PKCε gene (knockout)	Operant (lever) self-administration	KO mice responded less for ethanol
	(122)	PKA RIIβ gene (knockout)	Two-bottle drinking (continuous access)	KO mice drank more ethanol
	(122)	PKA RIβ gene (knockout)	Two-bottle drinking (continuous access)	No difference from control
	(122)	PKA Cβ1 gene (knockout)	Two-bottle drinking (continuous access)	No difference from control
	(123)	G_{nas} gene ($G_{s\alpha}$ knockout); heterozygotes on 3 different genetic backgrounds	Two-bottle drinking (continuous access)	Heterozygote mice drank less ethanol
	(123)	PKA-R(AB) gene (forebrain over-expression transgenic)	Two-bottle drinking (continuous access)	TG mice drank less ethanol
	(123)	$G_{s\alpha}$Q227L (forebrain over-expression transgenic)	Two-bottle drinking (continuous access)	No difference from control

(continued)

Table 2 (continued)

System	Ref.	Gene Target	Phenotype	Outcome
Neuropeptide Y	*(124)*	NPY gene (knockout on mixed genetic background)	Two-bottle drinking (continuous access)	KO mice drank more ethanol
	(124)	NPY gene (overexpression transgenic)	Two-bottle drinking (continuous access)	KO mice drank less ethanol
	(125)	NPY gene (knockout on 129/SvEv background)	Two-bottle drinking (continuous access)	KO mice drank more ethanol (at 20%, but not at lower concentrations)
	(125)	NPY Y5 receptor gene (knockout)	Two-bottle drinking (continuous access)	No difference from control
Other	*(131)*	Catalase gene (radiation-induced knockout)	Two-bottle drinking (continuous access)	KO mice drank more ethanol
	(127)	GIRK2 channels (knockout)	Two-bottle drinking (continuous access)	No difference from control in standard test; however, KO drank more ethanol when it was presented in favored location
	(129)	Angiotensinogen gene (knockout)	Two-bottle drinking (continuous access)	KO mice drank less ethanol
	(129)	Angiotensinogen gene (overexpression transgenic)	Two-bottle drinking (continuous access)	TG mice drank more ethanol
	(130)	PEPCK/βGH gene (overexpression transgenic)	Two-bottle drinking (continuous access)	Male TG mice drank more ethanol; female TG mice drank less ethanol
	(128)	Neutral endopeptidase (knockout)	Two-bottle drinking (continuous access)	KO mice drank more ethanol
	(126)	Dopamine β-hydroxylase gene (knockout)	Two-bottle drinking (continuous access)	KO mice drank less ethanol
	(126)	Dopamine β-hydroxylase gene (knockout)	Two-bottle drinking (continuous access)	KO mice slower to extinguish

[a]KO, knockout; TG, transgenic; outcomes expressed relative to wild-type control mice.

CPP, but not ethanol aversion as indexed by CTA *(100)*. However, later self-administration studies from the same laboratories and elsewhere have failed to show consistently greater ethanol intakes in knockout mice using either home-cage *(101,102)* or operant *(103)* procedures. The reasons underlying these later failures are not well understood. One possibility is that the null mutation's effect on the drinking phenotype was lost due to changes in the background genotype caused by genetic drift or the presence of alleles from multiple 129 substrains *(53)*.

The role of the serotonin system in ethanol reward has also received attention in two recent candidate gene studies that involved insertion of transgenes. In the first study, transgenic mice that overexpressed the 5-HT_3 receptor gene in forebrain areas showed significantly lower intake of 10% ethanol in a two-bottle home-cage procedure than B6SJL/F1 control mice *(104)*. Although conclusions are limited by use of only one ethanol concentration and lack of data on intake of other flavored solutions, this finding generally supports a modulatory influence of the 5-HT_3 receptor on ethanol reward.

The other transgenic study involved mice with a monoamine oxidase A (MAOA) deficiency produced by insertion of an interferon β (IFNβ) transgene into the MAOA gene. This enzyme deficiency produces a substantial elevation in whole-brain serotonin, as well as other amines *(105)*. Transgenic mice did not differ from C3H/HeJ control mice in intake of 10% ethanol or water in a home-cage two-bottle procedure *(106)*, suggesting no effect of MAOA deficiency on ethanol reward. However, conclusions from this study are limited by use of only one ethanol concentration and relatively short test durations (only 2 or 24 h after 24 h of water deprivation). Moreover, given that serotonin enhancement in these transgenic mice is known to diminish with age *(105)*, implications of this finding for understanding serotonin's role in ethanol reward are uncertain in the absence of brain serotonin measurements in mice tested for ethanol drinking. Nevertheless, despite absence of an effect on drinking, transgenic mice of the same age showed reductions in other measures of ethanol sensitivity, including ethanol-induced sleep time and hypothermia *(106)*.

3.4.2. Dopamine System

The dopamine (DA) system has been repeatedly implicated in the rewarding effects of most abused drugs *(107,108)*, including ethanol *(109)*. Thus, several candidate gene studies have targeted DA receptor genes, especially those encoding the D1 and D2 receptor subtypes. In an extensive study of home-cage ethanol drinking, mice lacking functional D1 receptors drank less ethanol and showed lower preference across a range of concentrations compared to heterozygous and wild-type littermate control mice in both limited and continuous-access two-bottle procedures *(110)*. D1 knockout mice also drank less ethanol (12%) when it was the only fluid available for 24 h. A companion series of experiments examined effects of several pharmacological pretreatments on intake of 12% ethanol in a two-bottle choice procedure. These experiments showed that treatment with selegiline (10 mg/kg), an MAOB inhibitor that increases synaptic DA levels, reduced ethanol intake in wild-type and heterozygous mice, but had no effect on D1 knockout mice. Similarly, treatment with a D1 receptor antagonist (SCH-23390, 1 mg/kg) reduced ethanol intake in wild-type and heterozygous mice, but not in D1 knockout mice. In contrast, injection of a D2 receptor antagonist (sulpiride, 50 mg/kg) nearly eliminated the already low ethanol intakes of D1 knockout mice at a dose that had relatively modest effects on drinking in control mice. While acknowledging involvement of both DA receptor subtypes, the authors concluded that their data supported a greater role of the D1 receptor subtype in ethanol reward. However, given the substantial growth retardation (20–30%) noted in this genetic model *(110,111)*, one must be cautious in interpreting these effects as specific to ethanol reward rather than a more general effect on feeding behavior *(112)*, especially since intakes of other palatable substances were not reported.

Several recent studies have used targeted mutation of the DA D2 receptor gene to provide information about its role in ethanol reward. An initial study of ethanol drinking in a home-cage continuous-access choice procedure showed significant reductions in ethanol intake and preference across several concentrations in D2 receptor knockouts compared to either heterozygous or wild-type control mice, which did not differ *(84)*. Alternative interpretations based on differences in taste sensitivity were eliminated by data showing no genotype effect on intake or preference for saccharin or quinine solutions presented without ethanol. The suggestion that reduced ethanol intake reflected a reduction in ethanol reward was supported by a subsequent conditioning study showing that D2 receptor knockout mice failed to develop ethanol-induced CPP *(21)*. Consistent with findings from the drinking study, heterozygous and wild-type mice did not differ in place preference, suggesting that one functional D2 receptor allele is sufficient to maintain ethanol reward in this procedure. The latter finding is of additional interest because place preference was maintained in heterozygotes despite a significant reduction in basal activity levels. Finally, an important role for the D2 receptor gene was indicated by a study showing that the null mutation significantly reduced responding for several ethanol concentrations in a continuous-access operant self-administration procedure *(23)*. However, D2 knockout mice also responded less for water, food, and saccharin, suggesting a relatively broad, nonspecific impairment of motivated responding.

The influence of the DA D2 receptor gene on ethanol drinking has received additional support in a very recent study that produced temporary overexpression of the receptor in nucleus accumbens of genetically heterogeneous rats by direct microinfusion of the gene via adenoviral vector *(115)*. More specifically, exposure to the gene-containing vector produced significant within-group reductions in 7% ethanol intake and preference in a continuous access home-cage choice procedure relative to intakes measured after exposure to a null vector, which had no effect on baseline drinking. Although interpretation of these findings would have been aided by examination of other palatable substances, these data raise the interesting possibility that an excess of DA D2 receptors reduces ethanol reward, a conclusion that appears at odds with earlier conclusions that ethanol reward is reduced by a lack of D2 receptors *(21,84)*. Such data suggest that one may not be able to extrapolate behavioral effects of the receptor's absence to predict effects of variation in receptor levels.

Although most candidate gene studies of DA system involvement in ethanol reward have focused on receptor gene manipulations, a recent series of experiments targeted the gene encoding DARPP-32, a phosphoprotein important for regulation of striatal dopaminergic systems *(116)*. DARPP-32 null mutants showed lower responding for ethanol and saccharin, but not for food or water in a continuous access operant self-administration procedure. Companion studies used conditioning procedures to support the conclusion of a deficit in ethanol reward as shown by absence of CPP, but no change in ethanol's aversive effect as measured by CTA. Thus, disruption of the DARPP-32 gene appeared to produce effects similar to those produced by targeted mutation of DA receptor genes.

3.4.3. Opioid System

Ethanol reward has also been linked to activity within the opioid system *(115)*, encouraging examination of mice lacking μ-opiate receptors. In two operant self-administration studies, μ-receptor knockout mice consistently responded less for ethanol than wild-type control mice *(116)*. μ-Knockout mice also consumed less ethanol in home-cage two-bottle tests, but only if they had prior ethanol self-administration experience in either the operant procedure or a home-cage single-bottle procedure. However, knockout mice also showed lower operant self-administration of water and sucrose, suggesting a broader involvement of the μ-receptor in ingestive behavior. A later report using a different μ-receptor knockout strain has generally confirmed and extended these observations by showing lower ethanol intakes in a home-cage drinking procedure and lack of ethanol-induced CPP in mice carrying the mutation *(117)*. Although these effects were statistically reliable only in females, findings from the previous study *(115)*, which involved only males, argue against the suggestion that effects of μ-receptor deletion are gender-specific.

In light of the prominent role that endogenous opioids play in current theorizing about mechanisms of ethanol reward *(118)*, it is not surprising that attention has also been given to gene manipulations that alter expression of the endogenous opioid peptide β-endorphin. In a home-cage two-bottle procedure, transgenic mice with β-endorphin deficiencies were found to drink more ethanol than sibling wild-type mice *(119)*. The suggestion that β-endorphin normally interferes with sensitivity to ethanol reward was supported in another study that showed greater intravenous self-administration of ethanol by β-endorphin-deficient mice than by wild-type control mice *(120)*.

3.4.4. Protein Kinase Signaling

Several recent studies of ethanol reward have targeted genes in protein kinase signaling pathways, especially those related to protein kinase C (PKC) and cAMP-dependent protein kinase (PKA). Mutant mice lacking the PKCε isozyme were found to self-administer less ethanol than wild-type sibling controls, both in an operant oral self-administration procedure *(120)* and in a home-cage two-bottle choice procedure *(121)*. However, PKCε-deficient mice did not differ in intake of food, water, saccharin, or quinine *(121)*, suggesting a selective effect on ethanol reward. A selective effect on home-cage ethanol drinking has also been reported in mice with disruption of the gene encoding the regulatory IIβ (RIIβ) subunit of PKA. In this case, however, the null mutation increased ethanol consumption across several concentrations, while having no effect on intakes of food, water, sucrose, or quinine *(122)*. The effect on ethanol intake appeared to be specific to disruption of the RIIβ subunit because mutations that produced a deficiency in either of two other PKA subunits (RIβ or Cβ1) had no effect on ethanol consumption.

The PKA signal transduction pathway has also been implicated in studies of mice with mutations in the gene encoding $G_{s\alpha}$, the α subunit of the stimulatory G protein of adenylyl cyclase (G_{nas}). These studies are especially noteworthy because effects of the mutation were examined on three different genetic backgrounds (C57BL/6J, 129/SvEv, and CD1 × 129/SvEv). In all three cases, ethanol intake or preference in a home-cage two-bottle procedure was lower in mice with heterozygous inactivation of the G_{nas} gene (G_{nas}−/+) compared to wild-type littermates *(123)*. Companion studies showed no effect of $G_{s\alpha}$ deficiency on intakes of food, sucrose, or quinine, suggesting a selective influence on ethanol reward. To provide additional information on the role of the PKA signaling pathway, these investigators also measured home-cage drinking in mice carrying one of two transgenes known to affect PKA activity. In both cases, expression of the transgene was limited to the hippocampus and other forebrain areas by using the promoter from the gene encoding CaMKIIα. In the first case, transgenic mice with reduced PKA activity caused by expression of R(AB) (an inhibitory form of the regulatory subunit of PKA) drank less ethanol than wild-type littermates. However, transgenic mice that showed increased adenylyl cyclase activity caused by expression of a constitutively active mutant form of $G_{s\alpha}$ (*G_{sa}Q227L*) drank ethanol at levels identical to controls. The authors suggested that the failure to see an increase in ethanol drinking in the latter case may have reflected a ceiling effect related to testing the transgene on a genetic background (C57BL/6J) that already drinks ethanol at relatively high levels. Overall, this series of studies provides good support for a role of the PKA pathway in ethanol reward.

3.4.5. Neuropeptide Y

As noted earlier, QTL mapping studies in rats derived from the selectively bred alcohol-preferring (P) and alcohol-nonpreferring (NP) lines have identified neuropeptide Y (NPY) as a possible candidate gene for ethanol self-administration *(62,63)*. More direct support for NPY's role has subsequently come from studies showing that targeted disruption of the NPY gene increases ethanol intake, whereas overexpression of the NPY gene reduces ethanol intake relative to wild-type littermate controls in a home-cage two-bottle procedure *(124)*. The increase produced by NPY deficiency did not appear to be related to taste or calories as shown by the lack of effect on intake of food, sucrose, or quinine. A later study extended these findings by examining effects of the NPY knockout on a different genetic background and by testing mice lacking the NPY Y5 receptor *(125)*. These experiments showed that the enhanced drinking phenotype was retained when the NPY mutation was shifted from a mixed C57BL/6J × 129/SvEv background to an inbred 129/SvEv background, but only at the highest ethanol concentration (20% v/v). However, targeted disruption of the Y5 receptor (129/SvEv background) had no effect on ethanol consumption or preference.

3.4.6. Other Candidate Genes

Ethanol reward-related behaviors have also been examined in several other candidate gene studies that do not readily fit into the above categories. Gene targets in these studies include dopamine β-hydroxylase, which affects norepinephrine synthesis *(126)*; the G-protein-coupled inwardly rectifying potassium channel GIRK2 *(127)*, the peptide-degrading enzyme neutral endopeptidase *(128)*, angiotensinogen *(129)*, bovine growth hormone *(130)*, and catalase *(131)*. Outcomes of these studies are briefly summarized in Table 2.

3.4.7. Summary

To date, about 25 different mouse transgenic or knockout models have been tested using one or more of the ethanol reward phenotypes (*see* Table 1). The most commonly studied phenotype is two-bottle home-cage drinking, which was tested in nearly two-thirds of the published comparisons. In almost all of these drinking studies, a continuous-access procedure was used and intakes or preferences were based on measurements averaged over 24 h. Thus, most of the information currently available from candidate gene studies is based on a single procedure that, as noted earlier, has several limitations.

Of particular interest is the finding that manipulation of candidate genes did not universally reduce ethanol intakes or fluid intakes more generally. For example, the outcomes from two-bottle studies were distributed such that about one-third of the studies showed an increase in ethanol intake or preference, about one-third showed a decrease, and about one-third showed no effect of the genetic manipulation. Such findings increase confidence that effects of gene insertion or deletion are related to the targeted genes themselves rather than nonspecific effects of genetic engineering. Moreover, as noted above, many of the drinking and operant studies examined the effect of the gene manipulation on intakes of food or other palatable substances, thereby addressing whether effects were specific to ethanol or reflected a more general influence on ingestive or motivated behavior.

Overall, however, there are several weaknesses in the candidate gene studies. For example, the influence of background genotype was addressed explicitly in only a few studies. In several cases, it was difficult to discern the exact nature of the background genotype. It was also sometimes difficult to determine whether wild-type and mutant mice were littermates derived from breeding of heterozygous parents or were bred separately from homozygous parents. Given the potentially important influence of background genotype on ability to detect effects of a targeted mutation, future studies should give greater attention to these issues. (For further discussion of this issue, *see* refs. *50–54* and *132*.)

Another weakness, which is not unique to the study of ethanol phenotypes, is that genetic alterations in these studies were generally present throughout development, raising the possibility that observed effects on target phenotypes reflect compensatory changes in other neurobiological systems *(133)*. Moreover, although a few of the transgene studies restricted expression to the forebrain, most of these studies placed no anatomical limits on the gene manipulation. Both of these concerns may be addressed in the near future, with greater use of genetic technologies that allow for creation of tissue-specific mutations and "conditional" mutations that can be introduced or removed in individual mice under experimenter control.

Finally, future studies involving genetic mutations should give greater attention to alternative interpretations that do not involve ethanol reward mechanisms. For example, only a few of the studies included in Table 1 considered the possibility that the gene manipulation produced its effect by altering ethanol pharmacokinetics *(84,120,125)*. Another alternative interpretation that must be considered, especially in the case of conditioning procedures (i.e., operant, taste, or place), is whether the gene manipulation has affected sensory-motor or learning abilities that are critical to acquisition and performance of behavior in those tasks.

Table 1 suggests that ethanol reward, or at least ethanol drinking, is likely influenced by a relatively large number of genes acting through several different neurobiological systems. In some cases, the influence of these genes was already strongly predicted on the basis of previous neuropharmacological studies directed at the protein products of these genes (e.g., dopamine and opioid receptors). In other cases, however, these studies have revealed novel information on genetic influences that have not been, or in some cases could not be, identified using conventional approaches. The challenge for future investigators is to find a way to incorporate this complex pattern of information into a conceptual framework that will provide an integrated explanation of the genetic and neurobiological mechanisms underlying ethanol reward.

3.5. Gene Expression

A burgeoning focus in alcohol research is gene expression profiling for specific alcohol-related traits. Gene expression profiling is the study of changes in gene expression for multiple genes at one time. Although studying changes in gene expression is by no means new to alcohol research *(134,135)*, advances in technology are beginning to alter drastically the way in which this research is conducted. If a candidate gene is the subject of study, expression may be compared by *in situ* hybridization, Northern analysis, Western analysis, or by other analogous methods. However, if the question is which genes are regulated by alcohol, the ability to screen thousands of genes or genetic sequences for expression changes, simultaneously, poses a major advantage. This approach has been termed microarray analysis *(136)* because gene expression changes are assessed using an array of unique mRNA sequences, which have been adhered to a solid support, such as a small glass microscope slide. Hybridization with a double-labeled probe provides a signal that permits automated detection of gene up- or downregulation. This approach is being applied to human alcoholism *(137)*, and is likely to make inroads into the question of genetic regulation of alcohol addiction. Expression profiling has been used in cell lines treated with ethanol and in genetic animal models of altered ethanol sensitivity *(138,139)*, but microarray analyses in animal models of ethanol reward are eagerly awaited.

The information that can be derived from the literature about ethanol reward and gene expression comes largely from studies of candidate genes and immediate early genes (IEGs). In a putative study of "craving," rats were denied the beer they had become accustomed to consuming in a distinctive environment. The expression of c-fos was increased in corticolimbic and brainstem regions of craving rats, relative to rats given free access to beer or to control rats with no beer experience *(140)*. Another study found an interaction between stress and ethanol consumption in B6 mice *(141)*. Significant changes in IEG expression were found in some brain regions of stressed ethanol/sucrose-consuming B6 mice, relative to stressed sucrose controls. For example, lower c-Fos expression was found in the hippocampus and higher expression in the nucleus accumbens. However, no significant effects of ethanol were found on the gene expression of nonstressed mice. In a follow-up study that achieved somewhat higher levels of ethanol consumption, c-Fos and FosB were increased in the nucleus accumbens, central nucleus of the amygdala, and Edinger-Westphal (EW) nucleus, in ethanol/sucrose-consuming mice, relative to mice offered sucrose alone or water *(142)*. This study also found a significant reduction in c-Fos expression in the dentate gyrus from alcohol-experienced mice. Most recently, this research group has reported increased c-Fos expression in the EW nucleus of B6 mice when first given the opportunity to consume ethanol, which increased with additional ethanol experience *(143)*. Furthermore, rats bred for high levels of voluntary alcohol drinking, Alko Alcohol (AA) rats, trained operantly to self-administer alcohol or alcohol in saccharin, have also exhibited increased c-Fos expression in the EW nucleus, relative to water or saccharin self-administering rats *(144)*. The EW nucleus has previously been associated with oculomotor functions *(145)*. These results and others, for example, suggesting a role for the EW nucleus in response to stress *(146)*, appeal for a reexamination of this nucleus and its connections as they relate to ethanol's motivational effects.

Candidate gene expression studies include examination of dopamine D2 receptor gene expression in alcohol preferring AA vs alcohol-avoiding ANA rats *(147)*. Using slot-blot to quantify mRNA, no significant differences were found between the rat lines in the dopaminergic nuclei examined. Another study used RT-PCR and found that ribosomal protein L18A mRNA and diacylglycerol kinase iota mRNA were differentially expressed between AA and ANA rats *(148)*. In this study, the goal was to search for novel genetic substrates for ethanol preference, and did not target specific genes. There is the possibility that these expression differences are unrelated to the ethanol-drinking phenotype, since selected lines would be expected to differ genetically for some trait-irrelevant genes due to genetic drift.

One study examined the expression specifically of genes coding for receptors and enzymes involved in GABAergic and glutamatergic neurotransmission. Compared were rats given the choice of drinking ethanol or water for 2 mo or over 9 mo, rats given ethanol as the sole source of fluid for over 9 mo, and rats offered only water *(149)*. mRNA content was evaluated in several brain regions chosen on the basis of known expression of the genes of interest. The most pronounced effects were seen in the parieto-occipital cortex. Here, reductions in mRNA levels for all eight of the $GABA_A$ receptor subunits examined were found in rats with free choice of ethanol for over 9 mo. mRNA levels for six of the $GABA_A$ subunits were also reduced in rats offered free choice of ethanol for 2 mo. Only mRNA for the $\alpha3$ subunit was reduced in the group of rats forced to consume ethanol. Similar results were found for some glutamate receptor subunits and enzymes associated with GABA and glutamate production, with respect to changes being associated primarily with the voluntary consumption groups. Data from this study must be carefully interpreted given differences among groups in daily ethanol consumption. However, it is interesting that larger changes were seen in the group of rats offered ethanol vs water for 2 mo than in those under the forced-consumption condition, since the later group consumed relatively more alcohol.

Given the focus of alcoholism research on genes associated with ethanol metabolism *(150–153)*, it is not surprising that the effects of ethanol consumption on the expression of *Adh-1*, *Ahd-2*, and *Cas-1*, three genes involved in ethanol metabolism, have been examined *(154)*. In this study, ethanol preference was characterized in B6, BALB/c, their F1, and six recombinant inbred strains derived from the B6 × BALB/c F2. Separate groups of mice of these strains were offered ethanol-containing liquid diet or isocaloric liquid diet as their only source of fluid and food. Using cDNA probes, liver mRNA for the genes of interest was quantified. Ethanol feeding resulted in increases in the level of mRNA for *Adh-1* and *Ahd-2*, across most genotypes. A smaller effect of ethanol was seen for *Cas-1*. A strong association between ethanol preference and change in *Ahd-2*-specific mRNA following ethanol feeding was found.

Gene expression analysis in the study of ethanol's motivational effects will likely expand as chip technology and bioinformatics continue to improve. One of the greatest advantages of the gene profiling method, relative to analysis of one gene at a time, is that coordinated changes in related systems can be detected. Detection of patterns of increased and decreased gene expression are likely to prove critical to our ultimate understanding of the genetic processes influencing alcohol addiction.

4. Current Status and Future Directions

Tremendous strides have been made over the last 10 years in the search for genes that influence ethanol reward-related phenotypes. For example, as illustrated above, QTL mapping studies in both rats and mice have identified and confirmed several candidate chromosomal regions that contain genes thought to influence ethanol intake or preference in the two-bottle home-cage procedure (*see* Table 1). Although the QTL approach has not yet yielded evidence supporting the involvement of any specific gene in ethanol reward, these studies have pointed to several interesting candidates that have been pursued using other strategies such as targeted gene disruption. Efforts to narrow the chromosomal regions containing these QTLs are ongoing, and it seems likely that this approach will lead to the identification of specific genes in the very near future.

Candidate gene studies using genetically engineered mouse strains (knockouts, transgenics) have also been quite successful in suggesting a role in ethanol reward for many genes across several different neurobiological systems, especially as indexed by the two-bottle home-cage drinking procedure (Table 2). The targets of these studies have included genes for receptors that were already strongly implicated in ethanol reward on the basis of neuropharmacological studies (e.g., D2 dopamine receptor, μ-opioid receptor), as well as signaling pathway genes and other genes whose influence on behavior has been more difficult to establish by conventional approaches (e.g., protein kinase Cε, Gsα).

The application of gene-expression profiling technology to the study of ethanol reward-related behaviors holds great promise for the future, although present efforts in this area are still at a very early stage of development. This situation is expected to change quickly as access to this technology increases and costs are reduced. It may prove fruitful in future studies to compare basal gene expression levels and ethanol-induced expression changes in rodent lines that have been selectively bred for sensitivity to a reward-related phenotype such as ethanol intake or preference. A selected line difference in the expression of a particular gene would strongly implicate that gene in producing the phenotypic difference, especially if expression differences were observed across multiple replications of the selection study *(155)*. One might also adopt the strategy of studying basal and ethanol-induced gene expression in large panels of standard inbred strains that have been well characterized for ethanol-reward phenotypes. Genetic correlational analysis could then be used to identify the subset of genes whose patterns of expression levels or changes were most strongly associated with ethanol's rewarding or aversive effects.

Although the field has advanced substantially through the application of all of these new genetic technologies, there are still several limitations in our current understanding. For example, most existing information is based on a single behavioral procedure: continuous-access, two-bottle home-cage drinking. Because behavior in this procedure is influenced by variables other than ethanol's postingestive pharmacological effects, it is critical that future genetic research give greater attention to examining ethanol's effects in other reward-relevant procedures, especially operant and Pavlovian conditioning. By using multiple behavioral models, the field will eventually develop a better understanding of the relationships among these models and, it is hoped, will converge on a common set of genes that mediate ethanol reward and aversion.

As correlational approaches such as QTL mapping and gene expression profiling yield an increasing number of candidate genes, gene manipulation strategies (i.e., targeted mutation, transgene insertion, antisense oligonucleotides, adenoviral vector insertion) will assume an even more important role in providing an experimental basis for eliminating "false positives." Strengths and weaknesses of the various genetic approaches have recently been reviewed *(59,156,157)*. An issue that has not yet been well addressed in the literature is the extent to which these techniques model natural genetic variation. For example, knowing that complete inactivation of a gene changes an ethanol-reward phenotype may not be very useful for understanding the normal function of that gene when the range of naturally occurring alleles does not include a "null" variant. Thus, it will be important for future studies and technologies to examine gene manipulations that more closely approximate natural genetic variation.

Future progress in this area of research will also be facilitated by continued appreciation of the fact that ethanol reward-related phenotypes are *polygenic*. That is, multiple genes determine these traits and each of these genes may have only a relatively small influence on total phenotypic variance. Moreover, as noted earlier, the impact of a given gene may depend critically on the form or level of expression of other genes (epistasis) and environmental conditions (gene × environment interaction). When attempting to integrate and interpret findings, greater attention should also be given to the principle of *pleiotropy*, which refers to the fact that a given gene may affect multiple traits. There is a strong tendency among alcohol researchers to place a very high value on genes whose influence appears alcohol-specific, even though it seems quite unlikely that natural selection pressure would have fostered retention of such genes. Rather, it seems likely that many of the genes that will eventually be found to influence ethanol reward and aversion were selected because of more general roles they play in determining an organism's sensitivity to motivational variables that affect survival (e.g., ability to distinguish between nutrients and toxins, ability to learn and remember based on previous experience).

These are very exciting times for scientists interested in the genetic basis of complex behavioral traits. The postgenomics era has placed great emphasis on the development of tools and conceptual frameworks for understanding the functions and interrelationships among genes, environment and behavior. Advances during the last 10 years have nearly overshadowed those in the preceding 50 years, and progress over the next 10 years is likely to be even more astonishing.

Acknowledgments

Preparation of this chapter was supported by NIAAA grants AA10760, AA07702, and AA07468, and a grant from the US Department of Veterans Affairs.

References

1. NIAAA. (2000) 10th Special Report to the U.S. Congress on Alcohol and Health. Public Health Service, U.S. Department of Health and Human Services, Washington, DC.

2. Tabakoff, B. and Hoffman, P. L. (1988) A neurobiological theory of alcoholism, in *Theories on Alcoholism* (Chaudron, C. D. and Wilkinson, D. A., eds.), Addiction Research Foundation, Toronto, Canada, pp. 29–72.
3. Crabbe, J. C. and Cunningham, C. L. (1999) Drug and alcohol dependence-related behaviors, in *Handbook of Molecular Techniques for Brain and Behavior Reasearch (Techniques in the Behavioral and Neutral Sciences)*, vol. 13, (Crusio, W. E. and Gerlai, R. T., eds.), Elsevier, Amsterdam, pp. 652–666.
4. Cunningham, C. L., Fidler, T. L., and Hill, K. G. (2000) Animal models of alcohol's motivational effects. *Alcohol Res. Health* **24,** 85–92.
5. Heyman, G. M. (2000) An economic approach to animal models of alcoholism. *Alcohol Res. Health* **24,** 132–139.
6. Spanagel, R. (2000). Recent animal models of alcoholism. *Alcohol Res. Health* **24,** 124–131.
7. McClearn, G. E. and Rodgers, D. A. (1959) Differences in alcohol preference among inbred strains of mice. *Quart. J. Stud. Alcohol* **20,** 691–695.
8. Dole, V. P., Ho, A., and Gentry, T. (1983) An improved technique for monitoring the drinking behavior of mice. *Physiol. Behav.* **30,** 971–974.
9. Gill, K., Mundl, W. J., Cabilio, S., and Amit, Z. (1989) A microcomputer controlled data acquisition system for research on feeding and drinking behavior in rats. *Physiol. Behav.* **45,** 741–746.
10. Stromberg, M. F., Mackler, S. A., Volpicelli, J. R., O'Brien, C. P., and Dewey, S. L. (2001) The effect of gamma-vinyl-GABA on the consumption of concurrently available oral cocaine and ethanol in the rat. *Pharmacol. Biochem. Behav.* **68,** 291–299.
11. Le, A. D., Corrigall, W. A., Harding, J. W., Juzytsch, W., and Li, T. K. (2000) Involvement of nicotinic receptors in alcohol self-administration. *Alcohol. Clin. Exp. Res.* **24,** 155–163.
12. Marcucella, H. (1989) Predicting the amount of ethanol consumed per bout from schedule of access to ethanol. *Animal Learn. Behav.* **17,** 101–112.
13. Meisch, R. A. (1977) Ethanol self-administration: infrahuman studies, in *Advances in Behavioral Pharmacology,* Vol. 1, (Thompson, T and Dews, P., eds.), Academic Press, New York, pp. 35–84.
14. Samson, H. H. and Hodge, C. W. (1996) Neurobehavioral regulation of ethanol intake, in *Pharmacological Effects of Ethanol on the Nervous System* (Deitrich, R. A. and Erwin, V. G., eds.). CRC Press, Boca Raton, FL, pp. 203–226.
15. Rodd-Henricks, Z. A., McKinzie, D. L., Crile, R. S., Murphy, J. M., and McBride, W. J. (2000) Regional heterogeneity for the intracranial self-administration of ethanol within the ventral tegmental area of female Wistar rats. *Psychopharmacology* **149,** 217–224.
16. Grahame, N. J., and Cunningham, C. L. (1997) Intravenous ethanol self-administration in C57BL/6J and DBA/2J mice. *Alcohol. Clin. Exp. Res.* **21,** 56–62.
17. Belknap, J. K., Belknap, N. D., Berg, J. H., and Coleman, R. (1977) Preabsorptive vs postabsorptive control of ethanol intake in C57BL/6J and DBA/2J mice. *Behav. Genet.* **7,** 413–425.
18. Samson, H. H. (1986) Initiation of ethanol reinforcement using a sucrose-substitution procedure in food- and water-sated rats. *Alcohol. Clin. Exp. Res.* **10,** 436–442.
19. Richter, C. P., and Campbell, K. H. (1940) Alcohol taste thresholds and concentrations of solution preferred by rats. *Science* **91,** 507–509.
20. Meisch, R. A., and Thompson, T. (1974) Ethanol intake as a function of concentration during food deprivation and satiation. *Pharmacol. Biochem. Behav.* **2,** 589–596.
21. Cunningham, C. L., Howard, M. A., Gill, S. J., Rubinstein, M., Low, M. J., and Grandy, D. K. (2000) Ethanol-conditioned place preference is reduced in dopamine D2 receptor-deficient mice. *Pharmacol. Biochem. Behav.* **67,** 693–699.

22. Phillips, T. J., Belknap, J. K., Buck, K. J., and Cunningham, C. L. (1998) Genes on mouse chromosomes 2 and 9 determine variation in ethanol consumption. *Mammal. Genome* **9,** 936–941.
23. Risinger, F. O., Freeman, P. A., Rubinstein, M., Low, M. J., and Grandy, D. K. (2000) Lack of operant ethanol self-administration in dopamine D2 receptor knockout mice. *Psychopharmacology* **152,** 343–350.
24. Samson, H. H., Slawecki, C. J., Sharpe, A. L., and Chappell, A. (1998) Appetitive and consummatory behaviors in the control of ethanol consumption: a measure of ethanol seeking behavior. *Alcohol. Clin. Exp. Res.* **22,** 1783–1787.
25. Samson, H. H., Sharpe, A. L., and Denning, C. (1999) Initiation of ethanol self-administration in the rat using sucrose substitution in a sipper-tube procedure. *Psychopharmacology* **147,** 274–279.
26. Cunningham, C. L. (1993) Pavlovian drug conditioning, in *Methods in Behavioral Pharmacology* (van Haaren, F., ed.), Elsevier, Amsterdam, pp. 349–381.
27. Cunningham, C. L. (1998) Drug conditioning and drug-seeking behavior, in *Learning and Behavior Therapy* (O'Donohue, W., ed.), Allyn and Bacon, Boston, pp. 518–544.
28. Katz, R. J. and Gormezano, G. (1979) A rapid and inexpensive technique for assessing the reinforcing effects of opiate drugs. *Pharmacol. Biochem. Behav.* **11,** 231–233.
29. Bardo, M. T. and Bevins, R. A. (2000) Conditioned place preference: what does it add to our preclinical understanding of drug reward? *Psychopharmacology* **153,** 31–43.
30. Tzschentke, T. M. (1998) Measuring reward with the conditioned place preference paradigm: a comprehensive review of drug effects, recent progress and new issues. *Prog. Neurobiol.* **56,** 613–672.
31. Goudie, A. J. (1987). Aversive stimulus properties of drugs: the conditioned taste aversion paradigm, in *Experimental Psychopharmacology: Concepts and Methods* (Greenshaw, A. J. and Dourish, C. T., eds.), Humana Press, Clifton, NJ, pp. 341–391.
32. Sherman, J. E., Jorenby, D. E., and Baker, T. B. (1988) Classical conditioning with alcohol: acquired preferences and aversions, tolerance, and urges/cravings, in *Theories on Alcoholism* (Chaudron, C. D. and Wilkinson, D. A., eds.), Toronto: Addiction Research Foundation, Canada, pp. 173–237.
33. Cappell, H., LeBlanc, A. E., and Endrenyi, L. (1973) Aversive conditioning by psychoactive drugs: effects of morphine, alcohol and chlordiazepoxide. *Psychopharmacologia* **29,** 239–246
34. Reicher, M. A. and Holman, E. W. (1977) Location preference and flavor aversion reinforced by amphetamine in rats. *Animal Learn. Behav.* **5,** 343–346.
35. Grigson, P. S. (1997) Conditioned taste aversions and drugs of abuse: a reinterpretation. *Behav. Neurosci.* **111,** 129–136.
36. Hunt, T. and Amit, Z. (1987). Conditioned taste aversion induced by self-administered drugs: paradox revisited. *Neurosci. Biobehav. Rev.* **11,** 107–130.
37. Broadbent, J., Muccino, K. J., and Cunningham, C. L. (2002) Ethanol-induced conditioned taste aversion in fifteen inbred mouse strains. *Behav. Neurosci.*, in press.
38. Holloway, F. A., King, D. A., Bedingfield, J. B., and Gauvin, D. V. (1992) Role of context in ethanol tolerance and subsequent hedonic effects. *Alcohol* **9,** 109–116.
39. Cunningham, C. L., Niehus, J. S., and Noble, D. (1993) Species difference in sensitivity to ethanol's hedonic effects. *Alcohol* **10,** 97–102.
40. Cunningham, C. L., Okorn, D. M., and Howard, C. E. (1997) Interstimulus interval determines whether ethanol produces conditioned place preference or aversion in mice. *Animal Learn. Behav.* **25,** 31–42.
41. Risinger, F. O. and Cunningham, C. L. (1998) Ethanol-induced conditioned taste aversion in BXD recombinant inbred mice. *Alcohol. Clin. Exp. Res.* **22,** 1234–1244.
42. Cunningham, C. L. (1995) Localization of genes influencing ethanol-induced conditioned place preference and locomotor activity in BXD recombinant inbred mice. *Psychopharmacology* **120,** 28–41.

43. Belknap, J. K., Crabbe, J. C., and Young, E. R. (1993) Voluntary consumption of ethanol in 15 inbred mouse strains. *Psychopharmacology* **112,** 503–510.
44. Cunningham, C. L., Okorn, D. M., and Howard, C. E. (1996) Ethanol-induced conditioned place preference and activation in 15 inbred mouse strains. *Alcohol. Clin. Exp. Res.* **20,** 59A.
45. Grupe, A., Germer, S., Usuka, J., Aud, D., Belknap, J. K., Klein, R. F., Ahluwalia, M. K., Higuchi, R., and Peltz, G. (2001) In silico mapping of complex disease-related traits in mice. *Science* **292,** 1915–1918.
46. Paigen, K. and Eppig, J. T. (2000) A mouse phenome project. *Mammal. Genome* **11,** 715–717.
47. Crusio, W. E. (1996) Gene-targeting studies: new methods, old problems. *Trends Neurosci.* **19,** 186–187.
48. Lathe, R. (1996) Mice, gene targeting and behaviour: more than just genetic background. *Trends Neurosci.* **19,** 183–186.
49. Gerlai, R. (1996) Gene-targeting studies of mammalian behavior: is it the mutation or the background genotype? *Trends Neurosci.* **19,** 177–181.
50. Banbury Conference on Genetic Background in Mice. (1997) Mutant mice and neuroscience: recommendations concerning genetic background. *Neuron* **19,** 755–759.
51. Kelly, M. A., Rubinstein, M., Phillips, T. J., Lessov, C. N., Burkhart-Kasch, S., Zhang, G., Bunzow, J. R., Fang, Y., Gerhardt, G. A., Grandy, D. K., and Low, M. J. (1998) Locomotor activity in D2 dopamine receptor-deficient mice is determined by gene dosage, genetic background, and developmental adaptations. *J. Neurosci.* **18,** 3470–3479.
52. Low, M. J., Kelly, M. A., Rubinstein, M., and Grandy, D. K. (1998) Single genes and complex phenotypes. *Mol. Pharmacol.* **3,** 375–377.
53. Phillips, T. J., Hen, R., and Crabbe, J. C. (1999) Complications associated with genetic background effects in research using knockout mice. *Psychopharmacology* **147,** 5–7.
54. Gingrich, J. A. and Hen, R. (2000) The broken mouse: the role of development, plasticity and environment in the interpretation of phenotypic changes in knockout mice. *Curr. Opin. Neurobiol.* **10,** 146–152.
55. Crabbe, J. C., Belknap, J. K., and Buck, K. J. (1994) Genetic animal models of alcohol and drug abuse. *Science* **264,** 1715–1723.
56. Lumeng, L., Murphy, J. M., McBride, W. J., and Li, T. K. (1995) Genetic influences on alcohol preference in animals, in *The Genetics of Alcoholism* (Begleiter, H. and Kissin, B., eds.), Oxford University Press, New York, pp. 165–201.
57. Phillips, T. J. and Crabbe, J. C. (1991) Behavioral studies of genetic differences in alcohol action, in *The Genetic Basis of Alcohol and Drug Actions* (Crabbe, J. C. and Harris, R. A., eds.), Plenum Press, New York, pp. 25–104.
58. Grahame, N. J. (2000) Selected lines and inbred strains. Tools in the hunt for the genes involved in alcoholism. *Alcohol Res. Health* **24,** 159–163.
59. Phillips, T. J., Belknap, J. K., Hitzemann, R., Buck, K. J., Cunningham, C. L., and Crabbe, J. C. (2002) Harnessing the mouse to unravel the genetics of human disease. *Genes Brain Behav.* **1,** 14–26.
60. Metten, P., Phillips, T. J., Crabbe, J. C., Tarantino, L. M., McClearn, G. E., Plomin, R., Erwin, V. G., and Belknap, J. K. (1998) High genetic susceptibility to ethanol withdrawal predicts low ethanol consumption. *Mammal. Genome* **9,** 983–990.
61. Belknap, J. K., Richards, S. P., O'Toole, L. A., Helms, M. L., and Phillips, T. J. (1997) Short-term selective breeding as a tool for QTL mapping: Ethanol preference drinking in mice. *Behav. Genet.* **27,** 55–66.
62. Carr, L. G., Foroud, T., Bice, P., Gobbett, T., Ivashina, J., Edenberg, H., Lumeng, L., and Li, T. K. (1998) A quantitative trait locus for alcohol consumption in selectively bred rat lines. *Alcohol. Clin. Exp. Res.* **22,** 884–887.

63. Bice, P., Foroud, T., Bo, R., Castelluccio, P., Lumeng, L., Li, T. K., and Carr, L. G. (1998) Genomic screen for QTLs underlying alcohol consumption in the P and NP rat lines. *Mammal. Genome* **9,** 949–955.
64. Foroud, T., Bice, P., Castelluccio, P., Bo, R., Miller, L., Ritchotte, A., Lumeng, L., Li, T. K., and Carr, L. G. (2000) Identification of quantitative trait loci influencing alcohol consumption in the high alcohol drinking and low alcohol drinking rat lines. *Behav. Genet.* **30,** 131–140.
65. McClearn, G. E., Tarantino, L. M., Rodriguez, L. A., Jones, B. C., Blizard, D. A., and Plomin, R. (1997) Genotypic selection provides experimental confirmation for an alcohol consumption quantitative trait locus in mouse. *Mol. Psychiatr.* **2,** 486–489.
66. Sugano, M., Tsuchida, K., Sawada, S., and Makino, N. (2000) Reduction of plasma angiotensin II to normal levels by antisense oligodeoxynucleotides against liver angiotensinogen cannot completely attenuate vascular remodeling in spontaneously hypertensive rats. *J. Hypertens.* **18,** 725–731.
67. Palmer, A. A. and Phillips, T. J. (2002) Quantitative trait locus mapping in mice, in *Methods for Alcohol Related Neuroscience Research.* (Liu, Y. and Lovinger, D., eds.), CRC Press, Boca Raton, FL, pp. 1–30.
68. Rikke, B. A. and Johnson, T. E. (1998) Towards the cloning of genes underlying murine QTLs. *Mammal. Genome* **9,** 963–968.
69. Zeng, Z. B., Kao, C. H., and Basten, C. J. (1999) Estimating the genetic architecture of quantitative traits. *Genet. Res.* **74,** 279–289.
70. Crabbe, J. C., Phillips, T. J., Buck, K. J., Cunningham, C. L., and Belknap, J. K. (1999) Identifying genes for alcohol and drug sensitivity: recent progress and future directions. *Trends Neurosci.* **22,** 173–179.
71. Grisel, J. E. (2000) Quantitative trait locus analysis. *Alcohol. Res. Health* **24,** 169–174.
72. Flint, J. and Mott, R. (2001) Finding the molecular basis of quantitative traits: successes and pitfalls. *Nat. Rev. Genet.* **2,** 427–445.
73. Phillips, T. J., Crabbe, J. C., Metten, P., and Belknap, J. K. (1994) Localization of genes affecting alcohol drinking in mice. *Alcohol. Clin. Exp. Res.* **18,** 931–941.
74. Rodriguez, L. A., Plomin, R., Blizard, D. A., Jones, B. C., and McClearn, G. E. (1995) Alcohol acceptance, preference, and sensitivity in mice. II. Quantitative trait loci mapping analysis using BXD recombinant inbred strains. *Alcohol. Clin. Exp. Res.* **19,** 367–373.
75. Tarantino, L. M., McClearn, G. E., Rodriguez, L. A., and Plomin, R. (1998).Confirmation of quantitative trait loci for alcohol preference in mice. *Alcohol. Clin. Exp. Res.* **22,** 1099–1105.
76. Melo, J. A., Shendure, J., Pociask, K., and Silver, L. M. (1996) Identification of sex-specific quantitative trait loci controlling alcohol preference in C57BL/ 6 mice. *Nat. Genet.* **13,** 147–153.
77. Peirce, J. L., Derr, R., Shendure, J., Kolata, T., and Silver, L. M. (1998) A major influence of sex-specific loci on alcohol preference in C57BL/6 and DBA/2 inbred mice. *Mammal. Genome* **9,** 942–948.
78. Dudek, B. C. and Tritto, T. (1995) Classical and neoclassical approaches to the genetic analysis of alcohol-related phenotypes. *Alcohol. Clin. Exp. Res.* **19,** 802–810.
79. Whatley, V. J., Johnson, T. E., and Erwin, V. G. (1999) Identification and confirmation of quantitative trait loci regulating alcohol consumption in congenic strains of mice. *Alcohol. Clin. Exp. Res.* **23,** 1262–1271.
80. Gill, K., Desauiniers, N., Desjardins, P., and Lake, K. (1998) Alcohol preference in AXB/ BXA recombinant inbred mice: gender differences and gender-specific quantitative trait loci. *Mammal. Genome* **9,** 929–935.
81. Vadasz, C., Saito, M., Balla, A., Kiraly, I., Vadasz, C., 2nd, Gyetvai, B., Mikics, E., Pierson, D., Brown, D., and Nelson, J. C. (2000) Mapping of quantitative trait loci for ethanol preference in quasi- congenic strains. *Alcohol* **20,** 161–171.
82. Vadasz, C., Saito, M., Gyetvai, B., Mikics, E., and Vadasz, C., II. (2000) Scanning of five chromosomes for alcohol consumption loci. *Alcohol* **22,** 25–34.

83. Gehle, V. M. and Erwin, V. G. (1998) Common quantitative trait loci for alcohol-related behaviors and CNS neurotensin measures: voluntary ethanol consumption. *Alcohol. Clin. Exp. Res.* **22,** 401–408.
84. Phillips, T. J., Brown, K. J., Burkhart-Kasch, S., Wenger, C. D., Kelly, M. A., Rubinstein, M., Grandy, D. K., and Low, M. J. (1998) Alcohol preference and sensitivity are markedly reduced in mice lacking dopamine D2 receptors. *Nat. Neurosci.* **1,** 610–615.
85. Belknap, J. K. and Atkins, A. L. (2001). The replicability of QTLs for murine alcohol preference drinking behavior across eight independent studies. *Mammal. Genome*, **12,** 893–899.
86. Lander, E. and Kruglyak, L. (1995) Genetic dissection of complex traits: guidelines for interpreting and reporting linkage results. *Nat. Genet.* **11,** 241–247.
87. Cunningham, C. L. and Prather, L. K. (1992) Conditioning trial duration affects ethanol-induced conditioned place preference in mice. *Animal Learn. Behav.* **20,** 187–194.
88. Risinger, F. O. and Oakes, R. A. (1996) Dose- and conditioning-trial dependent ethanol-induced conditioned place preference in Swiss-Webster mice. *Pharmacol. Biochem. Behav.* **55,** 117–123.
89. Fernandez, J. R., Tarantino, L. M., Hofer, S. M., Vogler, G. P., and McClearn, G. E. (2000) Epistatic quantitative trait loci for alcohol preference in mice. *Behav. Genet.* **30,** 431–437.
90. Hood, H. M., Belknap, J. K., Crabbe, J. C., and Buck, K. J. (2001) Genomewide search for epistasis in a complex trait: pentobarbital withdrawal convulsions in mice. *Behav. Genet.* **31,** 93–100.
91. Beal, M. F. (2001) Experimental models of Parkinson's disease. *Nat. Rev. Neurosci.* **2,** 325–334.
92. Garver, E., Tu, G. C., Cao, Q. N., Aini, M., Zhou, F., and Israel, Y. (2001) Eliciting the low-activity aldehyde dehydrogenase asian phenotype by an antisense mechanism results in an aversion to ethanol. *J. Exp. Med.* **194,** 571–580.
93. Homanics, G. E. and Hiller-Sturmhofel, S. (1997) New genetic technologies in alcohol research. *Alcohol Health Res. World* **21,** 298–309.
94. Homanics, G. E., Quinlan, J. J., Mihalek, R. M., and Firestone, L. L. (1998) Alcohol and anesthetic mechanisms in genetically engineered mice. *Front. Biosci.* **3,** 548–558.
95. Landel, C. P. (1991) The production of transgenic mice by embryo microinjection. *Genet. Anal. Tech. Appl.* **8,** 83–94.
96. Pravenec, M., Landa, V., Zidek, V., Musilova, A., Kren, V., Kazdova, L., Aitman, T. J., Glazier, A. M., Ibrahimi, A., Abumrad, N. A., Qi, N., Wang, J. M., St. Lezin, E. M., and Kurtz, T. W. (2001) Transgenic rescue of defective Cd36 ameliorates insulin resistance in spontaneously hypertensive rats. *Nat. Genet.* **27,** 156–158.
97. Wehner, J. M. and Bowers, B. J. (1995) Use of transgenics, null mutants, and antisense approaches to study ethanol's actions. *Alcohol. Clin. Exp. Res.* **19,** 811–820.
98. LeMarquand, D., Pihl, R. O., and Benkelfat, C. (1994) Serotonin and alcohol intake, abuse, and dependence: findings of animal studies. *Biol. Psychiatr.* **36,** 395–421.
99. Crabbe, J. C., Phillips, T. J., Feller, D. J., Hen, R., Wenger, C. D., Lessov, C. N., and Schafer, G. L. (1996) Elevated alcohol consumption in null mutant mice lacking 5-HT1B serotonin receptors. *Nat. Genet.* **14,** 98–101.
100. Risinger, F. O., Bormann, N. M., and Oakes, R. A. (1996) Reduced sensitivity to ethanol reward, but not ethanol aversion in mice lacking 5-HT1b receptors. *Alcohol. Clin. Exp. Res.* **20,** 1401–1405.
101. Bouwknecht, J. A., Hijzen, T. H., van der Gugten, J., Maes, R. A., Hen, R., and Olivier, B. (2000) Ethanol intake is not elevated in male 5-HT(1B) receptor knockout mice. *Eur. J. Pharmacol.* **403,** 95–98.
102. Crabbe, J. C., Wahlsten, D., and Dudek, B. C. (1999) Genetics of mouse behavior: interactions with laboratory environment. *Science* **284,** 1670–1672.
103. Risinger, F. O., Doan, A. M., and Vickrey, A. C. (1999) Oral operant ethanol self-administration in 5-HT1b knockout mice. *Behav. Brain Res.* **102,** 211–215.

104. Engel, S. R., Lyons, C. R., and Allan, A. M. (1998) 5-HT_3 receptor over-expression decreases ethanol self administration in transgenic mice. *Psychopharmacology* **140,** 243–248.
105. Cases, O., Seif, I., Grimsby, J., Gaspar, P., Chen, K., Pournin, S., Muller, U., Aguet, M., Babinet, C., Shih, J. C., and et al. (1995) Aggressive behavior and altered amounts of brain serotonin and norepinephrine in mice lacking MAOA. *Science* **268,** 1763–1766.
106. Popova, N. K., Vishnivetskaya, G. B., Ivanova, E. A., Skrinskaya, J. A., and Seif, I. (2000) Altered behavior and alcohol tolerance in transgenic mice lacking MAO A: a comparison with effects of MAO A inhibitor clorgyline. *Pharmacol. Biochem. Behav.* **67,** 719–727.
107. Di Chiara, G. (1998) A motivational learning hypothesis of the role of mesolimbic dopamine in compulsive drug use. *J. Psychopharmacol.* **12,** 54–67.
108. Spanagel, R. and Weiss, F. (1999) The dopamine hypothesis of reward: past and current status. *Trends Neurosci.* **22,** 521–527.
109. Koob, G. F., Roberts, A. J., Schulteis, G., Parsons, L. H., Heyser, C. J., Hyytiä, P., Merlo-Pich, E., and Weiss, F. (1998) Neurocircuitry targets in ethanol reward and dependence. *Alcohol. Clin. Exp. Res.* **22,** 3–9.
110. El-Ghundi, M., George, S. R., Drago, J., Fletcher, P. J., Fan, T., Nguyen, T., Liu, C., Sibley, D. R., Westphal, H., and O'Dowd, B. F. (1998) Disruption of dopamine D1 receptor gene expression attenuates alcohol-seeking behavior. *Eur. J. Pharmacol.* **353,** 149–158.
111. Drago, J., Gerfen, C. R., Lachowicz, J. E., Steiner, H., Hollon, T. R., Love, P. E., Ooi, G. T., Grinberg, A., Lee, E. J., Huang, S. P., and et al. (1994) Altered striatal function in a mutant mouse lacking D1A dopamine receptors. *Proc. Natl. Acad. Sci. USA* **91,** 12,564–12,568.
112. Robinson, S. W., Dinulescu, D. M., and Cone, R. D. (2000) Genetic models of obesity and energy balance in the mouse. *Ann. Rev. Genet.* **34,** 687–745.
113. Thanos, P. K., Volkow, N. D., Freimuth, P., Umegaki, H., Ikari, H., Roth, G., Ingram, D. K., and Hitzemann, R. (2001) Overexpression of dopamine D2 receptors reduces alcohol self-administration. *J. Neurochem.* **78,** 1094–1103.
114. Risinger, F. O., Freeman, P. A., Greengard, P., and Fienberg, A. A. (2001) Motivational effects of ethanol in DARPP-32 knock-out mice. *J. Neurosci.* **21,** 340–348.
115. Herz, A. (1997) Endogenous opioid systems and alcohol addiction. *Psychopharmacology* **129,** 99–111.
116. Roberts, A. J., McDonald, J. S., Heyser, C. J., Kieffer, B. L., Matthes, H. W., Koob, G. F., and Gold, L. H. (2000) Mu-Opioid receptor knockout mice do not self-administer alcohol. *J. Pharmacol. Exp. Ther.* **293,** 1002–1008.
117. Hall, F. S., Sora, I., and Uhl, G. R. (2001) Ethanol consumption and reward are decreased in mu-opiate receptor knockout mice. *Psychopharmacology* **154,** 43–49.
118. Grisel, J. E., Mogil, J. S., Grahame, N. J., Rubinstein, M., Belknap, J. K., Crabbe, J. C., and Low, M. J. (1999) Ethanol oral self-administration is increased in mutant mice with decreased beta-endorphin expression. *Brain Res.* **835,** 62–67.
119. Grahame, N. J., Low, M. J., and Cunningham, C. L. (1998) Intravenous self-administration of ethanol in beta-endorphin-deficient mice. *Alcohol. Clin. Exp. Res.* **22,** 1093–1098.
120. Hodge, C. W., Mehmert, K. K., Kelley, S. P., McMahon, T., Haywood, A., Olive, M. F., Wang, D., Sanchez-Perez, A. M., and Messing, R. O. (1999) Supersensitivity to allosteric GABA(A) receptor modulators and alcohol in mice lacking PKCepsilon. *Nat. Neurosci.* **2,** 997–1002.
121. Olive, M. F., Mehmert, K. K., Messing, R. O., and Hodge, C. W. (2000) Reduced operant ethanol self-administration and in vivo mesolimbic dopamine responses to ethanol in PKCε-deficient mice. *Eur. J. Neurosci.* **12,** 4131–4140.
122. Thiele, T. E., Willis, B., Stadler, J., Reynolds, J. G., Bernstein, I. L., and McKnight, G. S. (2000) High ethanol consupmtion and low sensitivity to ethanol-induced sedation in protein kinase A-mutant mice. *J. Neurosci.* **20,** RC75.
123. Wand, G., Levine, M., Zweifel, L., Schwindinger, W., and Abel, T. (2001) The cAMP-protein kinase A signal transduction pathway modulates ethanol consumption and sedative effects of ethanol. *J. Neurosci.* **21,** 5297–5303.

124. Thiele, T. E., Marsh, D. J., Ste Marie, L., Bernstein, I. L., and Palmiter, R. D. (1998) Ethanol consumption and resistance are inversely related to neuropeptide Y levels. *Nature* **396,** 366–369.
125. Thiele, T. E., Miura, G. I., Marsh, D. J., Bernstein, I. L., and Palmiter, R. D. (2000) Neurobiological responses to ethanol in mutant mice lacking neuropeptide Y or the Y5 receptor. *Pharmacol. Biochem. Behav.* **67,** 683–691.
126. Weinshenker, D., Rust, N. C., Miller, N. S., and Palmiter, R. D. (2000) Ethanol-associated behaviors of mice lacking norepinephrine. *J. Neurosci.* **20,** 3157–3164.
127. Blednov, Y. A., Stoffel, M., Chang, S. R., and Harris, R. A. (2001) Potassium channels as targets for ethanol: studies of G-protein-coupled inwardly rectifying potassium channel 2 (GIRK2) null mutant mice. *J. Pharmacol. Exp. Ther.* **298,** 521–530.
128. Siems, W., Maul, B., Krause, W., Gerard, C., Hauser, K. F., Hersh, L. B., Fischer, H. S., Zernig, G., and Saria, A. (2000) Neutral endopeptidase and alcohol consumption, experiments in neutral endopeptidase-deficient mice. *Eur. J. Pharmacol.* **397,** 327–334.
129. Maul, B., Siems, W. E., Hoehe, M. R., Grecksch, G., Bader, M., and Walther, T. (2001) Alcohol consumption is controlled by angiotensin II. *FASEB J.* **15,** 1640–1642.
130. Meliska, C. J., Bartke, A., Vandergriff, J. L., and Jensen, R. A. (1995) Ethanol and nicotine consumption and preference in transgenic mice overexpressing the bovine growth hormone gene. *Pharmacol. Biochem. Behav.* **50,** 563–570.
131. Aragon, C. M. and Amit, Z. (1993) Differences in ethanol-induced behaviors in normal and acatalasemic mice: systematic examination using a biobehavioral approach. *Pharmacol. Biochem. Behav.* **44,** 547–554.
132. Lariviere, W. R., Chesler, E. J., and Mogil, J. S. (2001) Transgenic studies of pain and analgesia: mutation or background genotype? *J. Pharmacol. Exp. Ther.* **297,** 467–473.
133. Wang, Q., Hummler, E., Maillard, M., Nussberger, J., Rossier, B. C. K. I., Brunner, H. R., and Burnier, M. (2001) Compensatory up-regulation of angiotensin II subtype 1 receptors in alpha ENaC knockout heterozygous mice. *Kidney Int.* **59,** 2216–2221.
134. Miles, M. F., Barhite, S., Sganga, M., and Elliott, M. (1993) Phosducin-like protein: an ethanol-responsive potential modulator of guanine nucleotide-binding protein function. *Proc. Natl. Acad. Sci. USA* **90,** 10,831–10,835.
135. Miles, M. F., Diaz, J. E., and DeGuzman, V. (1992) Ethanol-responsive gene expression in neural cell cultures. *Biochim. Biophys. Acta* **1138,** 268–274.
136. Lander, E. S. (1999) Array of hope. *Nat. Genet.* **21(Suppl. I),** 3–4.
137. Lewohl, J. M., Wang, L., Miles, M. F., Zhang, L., Dodd, P. R., and Harris, R. A. (2000). Gene expression in human alcoholism: microarray analysis of frontal cortex. *Alcohol. Clin. Exp. Res.* **24,** 1873–1882.
138. Thibault, C., Lai, C., Wilke, N., Duong, B., Olive, M. F., Rahman, S., Dong, H., Hodge, C. W., Lockhart, D. J., and Miles, M. F. (2000) Expression profiling of neural cells reveals specific patterns of ethanol-responsive gene expression. *Mol. Pharmacol.* **58,** 1593–1600.
139. Xu, Y., Ehringer, M., Yang, F., and Sikela, J. M. (2001) Comparison of global brain gene expression profiles between inbred long-sleep and inbred short-sleep mice by high-density gene array hybridization. *Alcohol. Clin. Exp. Res.* **25,** 810–818.
140. Topple, A. N., Hunt, G. E., and McGregor, I. S. (1998) Possible neural substrates of beer-craving in rats. *Neurosci. Lett.* **252,** 99–102.
141. Ryabinin, A. E., Wang, Y.-M., Freeman, P., and Risinger, F. O. (1999) Selective effects of alcohol drinking on restraint-induced expression of immediate early genes in mouse brain. *Alcohol. Clin. Exp. Res.* **23,** 1272–1280.
142. Bachtell, R. K., Wang, Y.-M., Freeman, P., Risinger, F. O., and Ryabinin, A. E. (1999) Alcohol drinking produces brain region-selective changes in expression of inducible transcription factors. *Brain Res.* **847,** 157–165.

143. Ryabinin, A. E., Bachtell, R. K., Freeman, P., and Risinger, F. O. (2001) ITF expression in mouse brain during acquisition of alcohol self- administration. *Brain Res.* **890,** 192–195.
144. Weitemier, A., Woerner, A., Bäckström, P., Hyytiá, P., and Ryabinin, A. E. (2001) Expression of c-Fos in Alko Alcohol rats responding for ethanol in an operant paradigm. *Alcohol. Clin. Exp. Res.* **25,** 704–710.
145. Buttner-Ennever, J. A., Horn, A. K., Scherberger, H., and D'ascanio, P. (2001) Motoneurons of twitch and nontwitch extraocular muscle fibers in the abducens, trochlear, and oculomotor nuclei of monkeys. *J. Comp. Neurol.* **438,** 318–335.
146. Weninger, S. C., Peters, L. L., and Majzoub, J. A. (2000) Urocortin expression in the Edinger-Westphal nucleus is up-regulated by stress and corticotropin-releasing hormone deficiency. *Endocrinology* **141,** 256–263.
147. Syvalahti, E. K., Pohjalainen, T., Korpi, E. R., Palvimaki, E. P., Ovaska, T., Kuoppamaki, M., and Hietala, J. (1994) Dopamine D2 receptor gene expression in rat lines selected for differences in voluntary alcohol consumption. *Alcohol. Clin. Exp. Res.* **18,** 1029–1031.
148. Sommer, W., Arlinde, C., Caberlotto, L., Thorsell, A., Hyytia, P., and Heilig, M. (2001) Differential expression of diacylglycerol kinase iota and L18A mRNAs in the brains of alcohol-preferring AA and alcohol-avoiding ANA rats. *Mol. Psychiatr.* **6,** 103–108.
149. Eravci, M., Schulz, O., Grospietsch, T., Pinna, G., Brodel, O., Meinhold, H., and Baumgartner, A. (2000) Gene expression of receptors and enzymes involved in GABAergic and glutamatergic neurotransmission in the CNS of rats behaviourally dependent on ethanol. *Br. J. Pharmacol.* **131,** 423–432.
150. Agarwal, D. P. and Goedde, H. W. (1992) Pharmacogenetics of alcohol metabolism and alcoholism. *Pharmacogenetics* **2,** 48–62.
151. Crabb, D. W., Dipple, K. M., and Thomasson, H. R. (1993) Alcohol sensitivity, alcohol metabolism, risk of alcoholism, and the role of alcohol and aldehyde dehydrogenase genotypes. *J. Lab. Clin. Med.* **122,** 234–240.
152. Enomoto, N., Takase, S., Yasuhara, M., and Takada, A. (1991) Acetaldehyde metabolism in different aldehyde dehydrogenase-2 genotypes. *Alcohol. Clin. Exp. Res.* **15,** 141–144.
153. Higuchi, S. (1994) Polymorphisms of ethanol metabolizing enzyme genes and alcoholism. *Alcoh. Alcohol. Suppl.* **2,** 29–34.
154. Tagliabracci, C. E. and Singh, S. M. (1996) Genetic regulation of gene-specific mRNA by ethanol in vivo and its possible role in ethanol preference in a cross with RI lines in mice. *Biochem. Genet.* **34,** 219–238.
155. Crabbe, J. C., Phillips, T. J., Kosobud, A., and Belknap, J. K. (1990) Estimation of genetic correlation: Interpretation of experiments using selectively bred and inbred animals. *Alcohol. Clin. Exp. Res.* **14,** 141–151.
156. Belknap, J. K., Hitzemann, R., Crabbe, J. C., Phillips, T. J., Buck, K. J., and Williams, R. W. (2001) QTL analysis and genomewide mutagenesis in mice: complementary genetic approaches to the dissection of complex traits. *Behav. Genet.* **31,** 5–15.
157. Nadeau, J. H. and Frankel, W. N. (2000) The roads from phenotypic variation to gene discovery: mutagenesis versus QTLs. *Nat. Genet.* **25,** 381–384.

15

Behavioral and Molecular Aspects of Alcohol Craving and Relapse

Rainer Spanagel

Alcohol dependence and addiction, here equated with alcoholism, is a clinically well-defined disorder in which normal behavioral control is lost by the individual, leading to particularly severe consequences. Several key aspects of alcoholism have been modeled in experimental animals. These animal models mimic various behavioral aspects seen in human alcoholics, such as loss of control over drinking, tolerance, physical dependence, craving, and relapse *(1,2)*.

There are opposing views in the field of drug abuse research regarding the term "craving"—whether it describes a physiological, subjective, or behavioral state, and if it is necessary at all to explain addictive behavior or it is an epiphenomenon that is not necessary for the production of continued drug use in addicts. An Expert Committee gathered by the United Nations International Drug Control Programme (UNDCP) and the World Health Organization (WHO) agreed on the definition of craving as "the desire to experience the effect(s) of a previously experienced psychoactive substance." Markou et al. *(3)* conceptualized craving within the framework of incentive motivational theories of behavior and modified the definition of craving as "incentive motivation to self-administer a psychoactive substance." Such an operational definition of craving has the advantage of making the phenomenon of craving accessible to experimental investigation and making it measurable. On the basis of this definition, animal models of alcohol craving and relapse have been developed.

1. Animal Models of Craving and Relapse

1.1. Long-Term Alcohol Self-Administration with Repeated Deprivation Phases

A promising animal model to study craving and relapse behavior in heterogenous Wistar rats or in alcohol-preferring P and HAD-rat lines is a long-term alcohol self-administration procedure with repeated deprivation phases *(4–7)*. Alcohol-experienced animals show a transient increase in alcohol consumption and alcohol preference after a period of forced abstinence (alcohol deprivation), which is termed the *alcohol deprivation effect*. The alcohol deprivation effect reflects an incentive motivation to take alcohol—that is, alcohol craving—and can also be seen as relapse-like drinking behavior. It can be observed in long-term alcohol-drinking rats that have developed a behavioral alcohol dependence *(7–9)*, as well as in nondependent rats *(6,10)*, both under

From: *Molecular Biology of Drug Addiction*
Edited by: R. Maldonado © Humana Press Inc., Totowa, NJ

home-cage drinking and under operant self-administration conditions; and in monkeys *(11)* and humans *(12)*.

Interestingly, the alcohol deprivation effect is prolonged and enhanced in alcohol-preferring P and HAD-rat lines after repeated deprivation phases *(4,13)* and changes its characteristics with repeated deprivation phases *(7,14)*. Thus, the alcohol deprivation effect in long-term alcohol self-administering rats that had experienced repeated deprivation phases has interesting characteristics: during an alcohol deprivation effect these animals consume large amounts of highly concentrated alcohol solutions, even at unusual times. Pronounced changes in the diurnal rhythm of drinking activity were observed in long-term alcohol-drinking rats that had repeated deprivation phases *(7)*. Tested in a fully automated electronic drinkometer device, age-matched control animals showed normal drinking activity: drinking activity during the active night phase was high, whereas drinking activity during the inactive light phase was very low, reaching zero for some hours. In contrast, in long-term alcohol-drinking rats during the alcohol deprivation effect, the pattern of drinking activity changed completely. In particular, during the inactive phase, most of the animals still showed high drinking activity. Moreover, some animals were found that even demonstrated level drinking—that is, drinking activity during the dark and light phases no longer differed *(7)*. Such drinking activity is far beyond normal controlled behavioral mechanisms seen in the appropriate control animals.

Furthermore, alcohol drinking behavior during an alcohol deprivation effect cannot be modified by taste adulteration with quinine or the additional choice of a highly palatable sucrose solution *(7,14)*. These findings suggest that the observed alcohol drinking behavior is pharmacologically and not nutritionally motivated. In conclusion, alcohol drinking during the alcohol deprivation effect, especially following repeated deprivation phases, seems to consist of an uncontrolled incentive motivation to self-administer the drug. This statement is fully compatible with the operational definition of craving by Markou et al. *(3)*. An incentive motivation to drink a highly concentrated alcohol solution following deprivation is further supported by the introduction of various progressive ratio tasks. Under those operant conditions, animals have to work more and more in order to receive a reinforcer. Here the breaking point (the number of consecutive lever responses in order to receive one reinforcer) for alcohol responding is significantly higher following deprivation compared to baseline responding *(15)*. However, it is important to note that the measurement of an alcohol deprivation effect assesses only a behavioral outcome and cannot tell us anything about a subjective state associated with an incentive motivation to drink alcohol. Nevertheless, the fact that clinically effective antirelapse drugs also reduce the alcohol deprivation effect *(16–18)* lends predictive value to this animal model for the development of new and better drugs for the treatment of alcoholism. In addition, alcohol self-administration with repeated deprivation phases can be used as an animal model to further study the neurobiological and molecular basis of craving and relapse.

1.2. The Reinstatement Model

The reinstatement model is also used for the measurement of craving and relapse behavior *(19)*. In this paradigm, the animal is trained to self-administer a drug and is then subjected to extinction—that is, the animal is tested under conditions of nonreinforcement until operant responding appears to be extinguished. When the animal reaches some criterion of unresponsiveness, various stimuli are presented. A stimulus is said to reinstate the drug-seeking behavior if it causes renewed responding—lever pressing—without any further response-contingent drug reward. At least five conditions can reinstate responding: (a) drug priming, that is, the injection of a small dose of the drug; (b) stress; (c) conditioned stimuli; (d) withdrawal; and (e) electrical brain stimulation.

Although reinstatement of intravenous self-administration of psychostimulants and opioids has been established for many years, only a few attempts have been undertaken to transfer this paradigm into the alcohol field. In 1995 the first alcohol reinstatement study in rats was reported by Chiamulera and co-workers *(20)*. In this study, rats acquirred operant responding for alcohol over several months. After stable lever pressing was obtained between subsequent sessions, the rats were tested in extinction, meaning that animals received water instead of alcohol following lever pressing. After 8–10 extinction sessions, reexposure to a small quantity of ethanol was able to reinstate previously extinguished alcohol-seeking behavior. These results are consistent with the widely reported description of the "first drink" phenomenon: ingestion of a small quantity of alcohol may induce in abstinent alcoholics a strong subjective state of craving and then relapse to drug-taking behavior *(21)*. The "priming effect" due to alcohol preload may be evident even after years of abstinence from the drug *(22)*. Only very recently has the alcohol reinstatement paradigm been followed by other research groups. It could be demonstrated that intermittent foot shock stress can also reinstate previously extinguished responding for alcohol *(23)*. Furthermore, it has been shown that alcohol-associated olfactory cues and other cues can reinstate extinguished alcohol-seeking behavior *(24,25)*. In conclusion, reinstatement of alcohol-seeking behavior has similar characteristics in comparison to other drugs of abuse and can be used to study the neurobiological and molecular bases of craving and relapse.

1.3. Ethanol-Induced Behavioral Sensitization

Repeated administration of psychoactive drugs can have neuroadaptive consequences that lead either to a decrease (tolerance or desensitization) or an increase (sensitization) of their behavioral effects. Sensitization processes have been implicated in the development of compulsive drug use. One of the most prominent theories on the significance of drug-induced sensitization, proposed by Robinson and Berridge *(26)*, posits that compulsive drug-seeking behavior is a result of a progressive hypersensitivity of neural systems that mediate "incentive salience," resulting in a transformation of ordinary "wanting" into excessive craving *(26)*. Although this theory is extremely difficult to test, drug-induced sensitization processes may give some indirect insights into behavioral and neurobiological processes of craving.

In contrast to most other drugs of abuse, ethanol-induced behavioral sensitization is difficult to demonstrate. Sensitization to the locomotor stimulant effects of low doses of ethanol has been documented only in some mouse genotypes, such as the inbred strain DBA/2J *(27,28)*. While DBA/2J mice are known to develop sensitization to ethanol, they have also been characterized as alcohol avoiders based on alcohol self-administration studies, raising questions regarding the relevance of examining sensitization in these mice. However, recent evidence indicates that DBA/2J mice are sensitive to the motivational effects of ethanol in that they show ethanol-induced place preference *(29)* and intravenous self-administration of ethanol *(30)*, suggesting a causal link between ethanol-induced sensitization and reinforcement processes. This suggestion is further supported by the finding that voluntary ethanol consumption induces sensitization to ethanol´s locomotor activating effects in a high-alcohol-preferring mouse line *(31)*. On the other hand, sensitized DBA/2J mice did not show enhanced alcohol intake in a subsequent drinking experiment, and prior voluntary intake did not alter ethanol-induced behavioral sensitisation, showing that ethanol-induced behavioral sensitization may not be linked directly to ethanol-induced reinforcement processes in DBA/2J mice *(32)*. In summary, in some mouse strains repeated intermittent injections of ethanol can induce behavioral sensitization to this drug, and there is some evidence that ethanol-induced behavioral sensitization is linked to reinforcement processes. In rats it is even more difficult to demonstrate ethanol-induced sensitization; however, a recent study showed that in Sprague-Dawley rats that were divided into either high or low responders to novelty, high responders to novelty exhibit a sensitized locomotor response to a very low challenge dose (0.25 g/kg) *(33)*. Thus, depending on the mouse or rat strain, ethanol-induced behavioral sensitization can be induced by repeated intermittent injections. This provides the possibility of studying neurobiological and molecular mechanisms of ethanol-induced sensitization processes. However, findings from those studies have only limited significance to extend our understanding of neurobiological and molecular mechanisms underlying alcohol craving.

2. Neurobiological and Molecular Bases of Alcohol Craving and Relapse

2.1. General Considerations

The neurobiological and molecular bases of alcohol craving and relapse are still not well understood; however, preclinical as well as clinical data strongly imply that craving and relapse for alcohol (and other drugs of abuse) can be induced through different mechanisms *(34)*. A first pathway may induce alcohol craving and relapse due to the mood-enhancing, positive reinforcing effects of alcohol consumption *(35)*. This pathway seems to involve opioidergic and dopaminergic systems in the ventral striatum *(36,37)*. The role of the dopaminergic system may lie in the direction of attention toward reward-indicating stimuli *(38)*, whereas the induction of euphoria and positive mood states may be mediated by opioidergic systems *(36)*. Associative learning may, in turn, transform positive mood states and previously neutral environmental stimuli into alcohol-associated cues that acquire positive motivational salience and induce reward craving *(26)*.

A second and potentially independent pathway may induce alcohol craving and relapse by negative motivational states, including conditioned withdrawal and stress *(35,39)*. This pathway seems to involve the glutamatergic system and the corticotropin-releasing hormone (CRH)-system *(40,41)*. Chronic alcohol intake leads to compensatory changes within these systems. During withdrawal and abstinence, increased glutamatergic excitatory neurotransmission as well as increased CRH release *(42)* lead to a state of hyperexcitability that becomes manifest as craving, anxiety, seizures, and autonomic dysregulation *(43)*. Moreover, cues associated with prior alcohol intake that are not followed by actual drug consumption may induce conditioned withdrawal *(21)*.

2.2. Findings from an Animal Model with Long-Term Alcohol Self-Administration with Repeated Deprivation Phases

2.2.1. Involvement of Opioidergic Systems in the Alcohol Deprivation Effect

A first pathway that may induce alcohol craving and relapse due to the mood-enhancing, positive-reinforcing effects of alcohol consumption and that seems to involve opioidergic systems was studied in long-term voluntary alcohol-drinking rats that had repeated deprivation phases. In particular, the role of opioid receptors on the alcohol deprivation effect was studied. Three different types of opioid receptors have been identified: μ-, δ-, and κ-opioid receptors. Opioid receptor blockade, either by the unspecific antagonists naloxone and naltrexone *(44–46)* or by selective μ-opioid receptor antagonists *(47)*, reduces ethanol consumption. Furthermore, μ-opioid receptor knockout mice do not self-administer alcohol either under operant or under home-cage drinking conditions *(48)*. These findings lead to the conclusion that at least a part of the rewarding effect of ethanol is mediated by the activation of μ-opioid receptors, which in turn reinforces ethanol intake. Hölter and Spanagel *(18)* determined under which treatment conditions naltrexone is effective in reducing the alcohol deprivation effect in long term alcohol experienced rats. The effects of chronic naltrexone treatment via osmotic minipumps and repeated intermittent naltrexone injections were studied. Chronic naltrexone treatment did not reduce the alcohol deprivation effect but enhanced alcohol preference. In contrast to chronic treatment, intermittent injections of naltrexone attenuated the alcohol deprivation effect. These opposing effects can be explained on a pharmacological level: chronic opiate receptor blockade leads to an up-regulation of opioid receptors, rendering the endogenous opioid system more sensitive to the effects of alcohol, whereas intermittent naltrexone injections at moderate doses do not induce functionally relevant opioid receptor changes. These findings emphasize the importance of the treatment regimen for obtaining a suppressant effect on alcohol drinking and relapse behavior. Thus, a treatment regimen with a low dose and frequency of administration that prevents drug accumulation is presumably necessary to maintain the reductive effect of opioid receptor blockade on alcohol relapse drinking. In view of these findings it seems questionable whether the use of naltrexone depots (some clinical trials have been already initiated) would be an appropriate treatment regimen.

The role of δ-opioid receptors in ethanol reinforcement is less clear. Some of the published reports that used selective δ-opioid receptor antagonists showed an attenuation of voluntary alcohol drinking in laboratory animals *(49–51)*, whereas others show no effect of this receptor subtype in ethanol consumption *(47,52–53)*. In contrast, δ-opioid receptor knockout mice showed increased ethanol consumption *(54)*. The selective δ-opioid receptor antagonist naltrindole had no effect on the alcohol deprivation effect in voluntary long-term alcohol-drinking rats (Spanagel, unpublished data), demonstrating that it is unlikely that δ-opioid receptors play a critical role in alcohol craving and relapse behavior.

The role of the dynorphin/κ-opioid receptor system in ethanol reinforcement is even less clear. κ-Opioid receptor stimulation, in contrast to μ- and δ-opioid receptor stimulation, has aversive motivational consequences in animals *(55,56)*. To further elucidate the role of the dynorphin/κ-opioid receptor system in ethanol reinforcement and relapse behavior, Hölter et al. *(57)* investigated the effects of the highly selective κ-opioid receptor agonist CI-977 and of the long-acting selective κ-opioid receptor antagonist nor-BNI on the alcohol deprivation effect in long-term ethanol-experienced rats. The results of this study show that chronic treatment with the selective κ-opioid receptor agonist CI-977 strongly increased the alcohol deprivation effect, whereas the long-acting selective κ-opioid receptor antagonist nor-BNI had no effect on the alcohol deprivation effect in long-term ethanol-experienced rats. In conclusion, the κ-opioid receptor system can play a role in ethanol reinforcement in cases of increased receptor activation, as it was experimentally induced in this study by chronic treatment with a selective κ-opioid receptor agonist. The subsequent increase in ethanol intake might be an attempt to counteract the aversive motivational consequences of this treatment. Previous findings indicate that increased dynorphin activity might occur during the early phases of alcohol withdrawal, but disappear later on *(58)*. Hence, an increased endogenous κ-opioid receptor activation might enhance the probability of relapse during the early phases of alcohol withdrawal. However, since the κ-opioid receptor antagonist nor-BNI had no effect on the alcohol deprivation effect, endogenous κ-opioid receptor stimulation does not seem to be involved in relapse-like drinking after protracted abstinence.

2.2.2. Involvement of the Mesolimbic Dopaminergic System in the Alcohol Deprivation Effect

Several lines of evidence indicate that ethanol activates the mesolimbic dopaminergic system. Alcohol injected intravenously increased firing of dopamine neurons in the ventral tegmental area *(59)* and acute administration of alcohol results in preferential release of dopamine from the nucleus accumbens shell region *(60)*. Dopamine D1 and D2 antagonists, administered either systemically or locally into the nucleus accumbens, decrease home-cage drinking and operant responding for alcohol *(61)*. Furthermore, alcohol intake and preference are markedly reduced in mice lacking dopamine D2 receptors *(62)*, and these mice show reduced ethanol-induced conditioned place preference *(63)*. Examination of dopaminergic neuron function in the nucleus accumbens associated with the alcohol deprivation effect revealed that relapse behavior was accompanied by enhanced dopamine release *(64)*, which is in line with preliminary findings in voluntary long-term drinking rats that had repeated deprivation phases. These animals showed augmented dopamine release during the onset of an alcohol deprivation effect compared to age-matched control animals that were alcohol naïve (Spanagel, unpublished data). The dopamine D2 receptor agonist lisuride was tested in voluntary long-term alcohol-drinking rats. Lisuride treatment significantly increased alcohol intake, indicating a “pro-craving/pro-relapse” effect of this compound *(65)*. The dopamine D2 receptor antagonist flupenthixol was also tested. A “pro-craving/pro-relapse” effect of flupenthixol treatment similar to that of lisuride treatment was observed in voluntary long-term drinking rats *(66)*. These studies show that dopamine D1 and D2 receptors are important in reinforcement processes involved in the acquisition of alcohol drinking, but it seems that these receptors play no critical role in alcohol craving and relapse and if at all can induce a “pro-craving/pro-relapse” effect due to pharmacological stimulation. In another study, voluntary long-term ethanol consumption led to specific changes in dopamine D3 receptor gene expression, whereas ethanol drinking did not alter the mRNA expression of dopamine D1, D2, D4, or D5 receptors *(67)*. However, chronic bilateral infusion of endcapped phosphorothioate dopamine D3 receptor antisense oligonucleotides into the nucleus accumbens shell region had no effect on the alcohol deprivation effect, although this treatment led to a selective reduction of this dopamine receptor subtype that was accompanied by suppressed food-reinforced behavior *(68)*. In summary, mesolimbic dopamine release and the activation of its receptors play an important role in the acquisition of alcohol-seeking behavior; however, whether dopamine receptor activation during craving and relapse plays a critical role is uncertain.

2.2.3. Involvement of the Glutamatergic System in the Alcohol Deprivation Effect

Numerous microdialysis studies revealed that, depending on the brain region and the rat strain, low doses of ethanol can increase extracellular glutamate levels whereas high intoxicating doses can decrease glutamate levels *(43)*. The mechanism behind the inhibitory effect of high intoxicating doses of ethanol on glutamate release is not clear. Although multiple mechanisms have been implicated in this action, considering the general inhibitory influence of γ-aminobutyric acid (GABA) on glutamatergic neurotransmission, it may be suggested that the inhibitory effect of ethanol on glutamate is due to an initial increase of GABA release, which in turn inhibits the release of glutamate. Whatever the mechanism might be, the inhibitory effect of high intoxicating doses of ethanol leads to several adaptive responses within the glutamatergic system following its chronic administration. Indeed, extracellular glutamate levels are enhanced during withdrawal *(43)*, and these changes underly the occurrence of some physical signs of withdrawal such as seizures. However, a recent study by Dahchour and De Witte *(69)* demonstrated that, following repeated alcohol withdrawal episodes, no enhancement of glutamate levels in the hippocampus can be observed despite the fact that the severity of ethanol-induced withdrawal seizures, as well as the duration of seizures, is exacerbated in rats that had a prior history of repeated alcohol withdrawal episodes *(70)*. These data suggest that the phenomenon of "kindling" observed following repeated alcohol withdrawal phases *(71)* may not be due simply to a further increase in glutamate, but that other mechanisms, that is, changes in glutamate receptors, may be implicated in the occurrence of seizures *(71)*. It also seems to be unlikely that enhanced glutamatergic neurotransmission in the nucleus accumbens and other brain sites following long-term abstinence is associated with craving and relapse. However, it has recently been demonstrated that ethanol-conditioned stimuli can induce an increase in extracellular glutamate levels in the amygdala *(72)*. Therefore, one might speculate that conditioned responses to extracellular glutamate may participate in environmental cues-induced conditioned craving for ethanol and relapse behavior. As pointed out later, however, changes in glutamate receptors seem to be more likely involved in craving and relapse behavior.

2.2.3.1. The Role of NMDA Receptors in the Alcohol Deprivation Effect

Electrophysiological and neurochemical studies show that ethanol at behaviorally relevant concentrations (>5 m*M*) inhibits *N*-methyl D-aspartic acid (NMDA) receptors expressed in neurons and in recombinant expression systems *(73–76)*. The fact that ethanol-induced inhibition of NMDA receptors differs across brain sites *(77)* and that NMDA receptor subunits are differentially distributed throughout the central nervous system suggest that differences in the subunit composition are important determinants of ethanol sensitivity. Despite intensive research, the site of action of ethanol on the NMDA receptor remains unknown, but, there is good hope that an "alcohol receptor" on the extracellular domaine of the receptor complex will be identified soon *(78)*. Following long-term alcohol intake, adaptive responses such as changes in the number and affinity of synaptic glutamate receptors and glutamate transporters will occur to counterbalance for the acute inhibitory effect of ethanol on NMDA receptor function and glutamate release. Initial binding studies detected increases in NMDA receptor density after chronic ethanol treatment *(79–81)*, whereas others have not *(82,83)*. A very extensive study was recently performed by Rudolph and collaborators *(84)* in order to determine if the variation observed in past NMDA receptor-binding studies might be related to genetic differences between strains and/or the chronic ethanol protocol used. The outcome of this comprehensive study and other unpublished reports indicate that robust increases in NMDA receptor binding do not occur, and suggest that NMDA receptor supersensitivity observed following chronic ethanol administration is not due simply to changes in the density of NMDA receptors. In fact, because of the complicated pharmacology, allosterism, and subunit heterogeneity of the NMDA receptor complex, it seems insufficient to use classical binding parameters. Therefore, following chronic alcohol treatment more information can be obtained by *in situ* hybridization studies or competitive RT-PCR studies aiming for different splice variants of the NR1 subunit and/or different NR2 subunits *(85)*. In addition, highly sensitive and specific antibodies are available to analyze relative amounts of NR subunits. Using these tools it became clear that chronic alcohol administration leads to specific changes in NMDA receptor subunit composition in various brain sites, which results in enhanced NMDA receptor function *(86–90)*. Furthermore, following long-term alcohol self-administration, enhanced NMDA receptor function *(91)* and long-lasting changes in NMDA receptor composition have been found *(92)*. All these studies support the hypothesis that changes in NMDA receptor subunit composition represent neuronal adaptive responses to alcohol-induced inhibition of the NMDA receptor and that these changes are long-lasting and can contribute to withdrawal responses as well as to states of hyperexcitability that can occur long after cessation of alcohol intake and can eventually contribute to alcohol craving and relapse *(34)*.

The role of NMDA receptors in modulating alcohol self-administration has only recently been examined. The functional NMDA receptor antagonist acamprosate *(93)* was effective in a series of preclinical studies *(10,14,17,34)* in reducing alcohol consumption and relapse. Competitive NMDA receptor antagonists attenuated operant responding for ethanol without affecting baseline levels of water self-administration *(94)*; however, the selectivity of such an effect was questioned by demonstrating that the competitive NMDA receptor antagonist CPPene decreased both ethanol and saccharin self-administration *(95)*. Noncompetitive NMDA receptor antagonists such as PCP and memantine were also tested, and both compounds reduced alcohol-reinforced behavior. However, both compounds have also modified other types of operant behavior in control experiments *(95,96)*. In a more recent study, chronic treatment with a novel low-affinity, noncompetitive NMDA receptor antagonist, MRZ 2/579 (neramexane)—the cyclohexan MRZ 2/579 has very similar characteristics to memantine—has been evaluated in nonoperant and operant models of alcohol drinking behavior *(97)*. Nonoperant two-bottle-choice ethanol drinking was not affected by a 6-d sc infusion of MRZ 2/579 (9.6 mg/d). Only intermittent injections of the drug (5 mg/kg) but not constant infusion led to significant decreases in operant responding for ethanol.

These promising results with low-affinity noncompetitive NMDA receptor antagonists led to further studies on alcohol craving and relapse behavior. In order to study alcohol craving and relapse, a new model of long-term alcohol self-administration with repeated deprivation phases was used *(2,7)*. Low-affinity noncompetitive NMDA receptor antagonists such as memantine or the cyclohexan MRZ 2/579 were tested in this model. Chronic administration of memantine or MRZ 2/579 via osmotic minipumps selectively reduced relapse behavior *(98,99)*. In alcohol discrimination experiments, memantine as well as MRZ 2/579 dose-dependently generalized to the ethanol cue *(99,100)* suggesting that low-affinity noncompetitive NMDA receptor antagonists might exert their reducing effect on alcohol intake and relapse behavior by generalizing for some of the stimulus properties of ethanol.

2.2.3.2. The Role of AMPA and Kainate Receptors in the Alcohol Deprivation Effect

Acute and chronic effects of ethanol on other ionotropic glutamate receptors such as AMPA (alpha amino 3 hydroxy 5 methyl 4 isoxazole propionate) and kainate (KA) receptors have also been studied. It has been found that ethanol at physiologically relevant concentrations (≤50 m*M*) can suppress both AMPA *(101)* and KA receptor-associated currents *(102)*. However, little is known about adaptations in AMPA/KA receptors after chronic ethanol treatment. In a single study on the role of AMPA/KA receptors on the reinforcing properties of ethanol, Stephens and Brown *(103)* have found that neither a competitive AMPA/KA antagonist, NBQX, nor the preferential noncompetitive AMPA antagonist, GYKI 52466, exerted any selective effects on ethanol-reinforced behavior in rats trained to self-administer ethanol, sucrose, or saccharin on a progressive ratio schedule of reinforcement. GYKI 52466 had no effect on operant responding, but increased spontaneous locomotor activity. In contrast, NBQX significantly decreased operant responding for ethanol, sucrose, and saccharin, but only at the doses that suppressed spontaneous locomotor activity. The authors concluded that non-NMDA ionotropic glutamate receptors may not play any specific role in the positive-reinforcing properties of ethanol *(103)*. This conclusion is supported by a recent finding in AMPA receptor knockout mice: these mice did not differ in alcohol intake and preference from wild-type animals (M. S. Cowen, personal communication). The AMPA antagonist GYKI 52466 was also studied in voluntary long-term alcohol drinking rats following repeated deprivation phases. This compound did not affect the alcohol deprivation effect, showing that AMPA receptors do not play an important role in alcohol craving and relapse behavior (Spanagel, unpublished data).

2.2.4. Involvement of the CRH System in the Alcohol Deprivation Effect

Stress is consistently viewed as an important factor in alcohol drinking in laboratory animals *(104)*. Alcohol is postulated to reduce physiological effects induced by stress. As a result, alcohol ingestion itself or alcohol-induced relief from stress becomes reinforcing. CRH regulates endocrine responses to stress *(105)* and mediates stress-related behavioral responses via extrahypothalamic sites *(106)*. The CRH signal is transmitted by two types of CRH receptors, termed CRH1 receptor (CRHR1) and CRH2 receptor (CRHR2), which differ in their pharmacology and expression pattern in the brain

(107,108). Dysregulation in the CRH system has been attributed to a variety of stress-related psychiatric disorders including alcoholism *(109)*. The detection of altered levels of endogenous CRH in distinct brain areas of selectively bred, ethanol-preferring animals *(110,111)* is evidence for a possible relationship between the activity of the CRH system and voluntary alcohol consumption. Since CRH acting at CRHR1 plays a key role in mediating the central stress response, CRHR1 activation might also affect alcohol-drinking behavior. Therefore, the effects of the novel nonpeptide CRHR1-antagonist R121919 *(112)* on the alcohol deprivation effect in long-term alcohol-experienced rats was studied. This compound did not influence relapse-like drinking (Hölter, personal communication). The conclusion from this study is that CRHR1 activation is not involved in the expression of an alcohol deprivation effect; however, it is known that following repeated alcohol deprivation phases a subsequent alcohol deprivation effect is not associated with behavioral or endocrinological stress responses, offering an explanation as to why selective CRHR1 blockade does not affect the alcohol deprivation effect. Interestingly, mice lacking a functional CRHR1 *(113)* markedly increased their alcohol intake in response to repeated stress and deprivation phases, although acquisition of alcohol drinking under resting conditions was not influenced by the dysfunctional CRH system, as alcohol consumption of knockout mice did not differ from wild-type mice during this phase *(114)*.

2.3. Findings from Reinstatement Studies

2.3.1. Involvement of Opioid Receptors in Reinstatement of Alcohol-Seeking Behavior

As already mentioned, one pathway that may induce alcohol craving and relapse due to the mood-enhancing, positive-reinforcing effects of alcohol consumption *(35)* involves the opioidergic systems *(36)*. The motivational effects of positive mood states and external alcohol-associated cues on alcohol craving and relapse (= alcohol-associated positive mood states) may be blocked by opioid receptor antagonists.

Reinstatement studies of alcohol-seeking behavior have shown that naltrexone attenuates the reinstatement of responding elicited by conditioned cues *(25)*. In this study reinstatement was tested with conditioned auditory stimuli, olfactory stimuli, and a combination of both stimuli. Auditory stimuli alone did not lead to reinstatement of responding, whereas olfactory stimuli and in particular the combination of both stimuli led to a strong reinstatement of responding. Naltrexone (0.2 g/kg) selectively decreased responding elicited by the ethanol-associated olfactory cues; specifically, naltrexone reversed the increases in responding induced by the ethanol-associated cues but did not alter responding in the presence of water-associated cues. This finding supports a role for opioid receptor activation in the motivational effects of ethanol-related environmental stimuli and further suggests that craving and relapse elicited by conditioned stimuli can be prevented by opioid receptor blockade. In another study by Lê et al. *(115)* reinstatement of alcohol-seeking behavior was elicited by either priming injections of ethanol (0.25–1.0 g/kg) or exposure to intermittent foot shock. Naltrexone pretreatment (0.2–0.4 g/kg) blocked alcohol-induced, but not stress-induced, reinstatement. In another study, naltexone (1–3 mg/kg) potently and dose-dependently inhibited reinstatement of alcohol-seeking behavior induced by noncontingent deliveries of the liquid dipper filled with 8% ethanol solution, which is comparable to an ethanol priming injection *(24)*. Together these studies show that naltrexone suppresses reinstatement of ethanol-responding induced by either ethanol priming or conditioned cues. In contrast, naltrexone is ineffective when reinstatment is initiated by stress. These finding are in agreement with the hypothesis that activation of opioid receptors underly the induction of "alcohol-associated positive mood states" and of cue-induced "alcohol-associated positive mood states."

2.3.2. Involvement of NMDA Receptors in Reinstatement of Alcohol-Seeking Behavior

Acamprosate acts as a functional NMDA receptor antagonist *(93)*. Together with other noncompetitive, low-affinity NMDA receptor antagonists, the therapeutic effectiveness of these compounds may be exerted by decreasing stress- and cue-induced conditioned withdrawal and withdrawal relief craving (= alcohol-associated negative mood states) *(34)*.

It remains to be established whether acamprosate or noncompetitive NMDA receptor antagonists might alter reinstatement of ethanol seeking after extinction induced by stress and cue-induced conditioned withdrawal, but, preliminary findings may indicate that the noncompetitive NMDA receptor antagonist MRZ 2/579 nonselectively inhibits reinstatement of ethanol- and water-seeking behavior induced by conditioned stimuli in the rat (Bienkowski, personal communication). In a very recent study, Vosler et al. *(116)* used a discriminative two-lever test to examine the ability of dizocilpine to reinstate ethanol-seeking behavior. In this study, one group of rats was trained to lever-press for ethanol and another group to lever-press for sucrose. After extinction, rats were injected with ethanol (0.5 g/kg). The ethanol group showed reinstatement of lever responding, whereas the sucrose group showed minimal responding following ethanol priming. In contrast, dizocilpine increased responding in both groups, suggesting a loss of discriminative control.

2.3.3. Involvement of CRHR1-Receptors in Reinstatement of Alcohol-Seeking Behavior

Most recently, the role of CRHR1 blockade in reinstatement of alcohol seeking induced by intermittent foot-shock stress was studied in rats *(117)*. Rats were given alcohol in a two-bottle free-choice procedure (water vs alcohol) for 30 d and were then trained for 1 h/d to press a lever for alcohol for 1 mo in operant conditioning boxes. After stable drug intake was obtained, lever pressing for alcohol was extinguished by terminating alcohol delivery. Subsequently, reinstatement of alcohol seeking was determined after exposure to intermittent foot shock in different groups of rats that were pretreated with selective CRHR1 antagonists. The CRHR1 antagonists D-Phe-CRF or CP-154,526 attenuated foot-shock-induced reinstatement of alcohol seeking. On the other hand, the removal of circulating corticosterone by adrenalectomy had no effect on foot-shock stress-induced reinstatement of alcohol-seeking behavior. These data suggest that CRHR1 activation contributes to foot-shock stress-induced reinstatement of alcohol seeking via its actions on extrahypothalamic sites.

2.4. Findings from Sensitization Studies

Very little systematic work has been performed so far to elucidate the neurobiological and molecular mechanisms underlying ethanol-induced behavioral sensitization.

Several studies have shown that the noncompetitive NMDA receptor antagonist dizocilpine prevents the development of drug-induced sensitization processes, and recently, Broadbent and Weitemier *(118)* showed that dizocilpine also prevents the development of sensitization to ethanol in DBA/2J mice. The inhibtory effect of dizocilpine on the development of behavioral sensitizatzion to ethanol could be explained in terms of state-dependent effects. Thus a sensitized response to ethanol can occur only under the influence of dizocilpine. However, this possibility was ruled out by further experiments demonstrating that dizocilpine treatment inhibits the development of sensitization, independent of a state-dependent effect.

3. Conclusions and Further Perspectives

Studies in voluntary long-term alcohol-drinking rats with repeated deprivation phases and reinstatement experiments clearly show that there are at least two different neurobiological pathways that induce alcohol craving and relapse. The first pathway involves the opioidergic system (in particular, μ-opioid receptors) and probably the mesolimbic dopaminergic system and may induce alcohol craving and relapse due to the mood-enhancing, positive-reinforcing effects of alcohol consumption. Associative learning may, in turn, transform positive mood states during alcohol drinking and previously neutral stimuli into alcohol-associated cues that acquire positive motivational salience and induce craving and relapse. These conditioned cues also depend on the activation of the opioidergic system and probably on the dopaminergic system. A second pathway involves the glutamatergic system (in particular, NMDA receptors) and the CRH system and may induce alcohol craving and relapse by negative motivational states including withdrawal and stress. Associative learning may, in turn, transform negative mood states during alcohol withdrawal and previously neutral stimuli into alcohol withdrawal-associated cues that produce an aversive state and induce craving and relapse. These conditioned cues might also depend on the activation of the glutamatergic system, since it has recently been shown that the NMDA receptor modulator acamprosate inhibits conditioned abstinence behavior in rats. Clearly, more systematic work is needed to define fully these different neurobiological/molecular pathways involved in alcohol craving and relapse behavior. Nevertheless, the findings presented here already have important implications for relapse prevention in humans. Since the neurobiological pathways of craving and relapse inducing mood states are different, pharmacological treatments that are normally aimed at a selective biological substrate should be specific to certain motivational states and stimuli, that is, effective in reducing relapse induced by a given mood state and stimuli but unable to counteract relapse induced by another mood state and stimuli. Consequently, relapse treatments, to be successful, should be adapted to the factor that has a greater chance to induce relapse in a given subject. Developing a new approach that takes into account relapse-specific mood states and stimuli can certainly be considered a breakthrough in the management and prevention of relapse in alcoholism.

References

1. Li, T. K. (2000) Clinical perspectives for the study of craving and relapse in animal models. *Addiction* **95,** 55–60.
2. Spanagel, R. (2000) Recent animal models of alcoholism. *Alcohol Res. Health* **24,** 124–131.

3. Markou, A., Weiss, F., Gold, L. H., Caine, S. B., Schulteis, G. and Koob, G. F. (1993) Animal models of drug craving. *Psychopharmacology* **112,** 163–182.
4. Rodd-Henricks, Z. A., McKinzie, D. L., Murphy, J. M., McBride, W. J., Lumeng, L., and Li, T. K. (2000) The expression of an alcohol deprivation effect in the high-alcohol-drinking replicate rat lines is dependent on repeated deprivations. *Alcohol. Clin. Exp. Res.* **24,** 747–753.
5. Rodd-Henricks, Z. A., McKinzie, D. L., Shaikh, S. R., Murphy, J. M., McBride, W. J., Lumeng, L. and Li, T. K. (2000) Alcohol deprivation effect is prolonged in the alcohol preferring (P) rat after repeated deprivations. *Alcohol. Clin. Exp. Res.* **24,** 8–16.
6. Sinclair, J. D. and Li, T. K. (1989) Long and short alcohol deprivation: effects on AA and P alcohol-preferring rats. *Alcohol* **6,** 505–509.
7. Spanagel, R. and Hölter, S. M. (1999) Long-term alcohol self-administration with repeated alcohol deprivation phases: an animal model of alcoholism? *Alcoholism* **34,** 231–243.
8. Hölter, S. M., Engelmann, M., Kirschke, C., Liebsch, G., Landgraf, R., and Spanagel, R. (1998) Long-term ethanol self-administration with repeated ethanol deprivation episodes changes ethanol drinking pattern and increases anxiety-related behavior during ethanol deprivation in rats. *Behav. Pharmacol.* **9,** 41–48.
9. Hölter, S. M., Linthorst, A. C. E., Reul, J. M. H. M., and Spanagel, R. (2000) Withdrawal symptoms in a long-term model of voluntary alcohol drinking in wistar rats. *Pharmacol. Biochem. Behav.* **66,** 143–151.
10. Heyser, C. J., Schulteis G., and Koob G. F. (1997) Increased ethanol self-administration after a period of imposed ethanol deprivation in rats trained in a limited access paradigm. *Alcohol. Clin. Exp. Res.* **21,** 784–791.
11. Sinclair, J. D. (1971) The alcohol-deprivation effect in monkeys. *Psychonom. Sci.* **25,** 21–25.
12. Burish, T. G., Maisto, S. A., Cooper, A. M., and Sobell, M. B. (1981) Effects of voluntary short-term abstinence from alcohol on subsequent drinking patterns of college students. *J. Stud. Alcohol.* **42,** 1013–1020.
13. Rodd-Henricks, Z. A., Bell, R. L., Kuc, K. A., Murphy, J. M., McBride, W. J., Lumeng, L., and Li, T. K. (2001) Effects of concurrent access to multiple ethanol concentrations and repeated deprivations on alcohol intake of alcohol-preferring rats. *Alcohol. Clin. Exp. Res.* **25,** 1140–1150.
14. Spanagel, R., Hölter S. M., Allingham, K., Landgraf, R. and Zieglgänsberger, W. (1996) Acamprosate and alcohol: I. Effects on alcohol intake following alcohol deprivation in the rat. *Eur. J. Pharmacol.* **305,** 39–44.
15. Spanagel, R. and Hölter, S.M. (2000) Pharmacological validation of a new animal model of alcoholism. *J. Neural. Transm.* **107,** 669–680.
16. Heyser, C. J., Schulteis, G., Durbin, P., and Koob, G.F. (1998) Chronic acamprosate eliminates the alcohol deprivation effect while having limited effects on baseline responding for ethanol in rats. *Neuropsychopharmacology* **18,** 125–133.
17. Hölter, S. M., Landgraf, R., Zieglgänsberger, W., and Spanagel, R. (1997) Time course of acamprosate action on operant self-administration following ethanol deprivation. *Alcohol. Clin. Exp. Res.* **21,** 862–869.
18. Hölter, S. M. and Spanagel, R. (1999) Effects of opiate antagonist treatment on the alcohol deprivation effect in long-term ethanol-experienced rats. *Psychopharmacology* **145,** 360–369.
19. Stewart, J., and De Wit, H. (1987) Reinstatement of drug-seeking behavior as a method of assessing incentive motivational properties of drugs, in *Methods of Assessing the Reinforcing Properties of Abused Drugs* (Bozarth, M. A., ed.), Springer, New York, pp. 211–227.
20. Chiamulera, C., Valerio, E., and Tessari, M. (1995) Resumption of ethanol-seeking behavior in rats. *Behav. Pharmacol.* **6,** 32–39.
21. Ludwig, A. M., Wikler, A., and Stark L. H. (1974) The first drink: psychobiological aspects of craving. *Arch. Gen. Psych.* **30,** 539–547.

22. Besancon, F. (1993) Time to alcohol dependence after abstinence and first drink. *Addiction* **88,** 1647–1650.
23. Lê, A. D., Quan, B., Juzytch, W., Fletcher, P. J., Joharchi, N., and Shaham, Y. (1998) Reinstatement of alcohol-seeking by priming injections of alcohol and exposure to stress in rats. *Psychopharmacology* **135,** 169–174.
24. Bienkowski, P., Koros, E., Kostowski, W., and Bogucka-Bonikowska, A. (2000) Reinstatement of ethanol seeking in rats: behavioral analysis. *Pharmacol. Biochem. Behav.* **66,** 123–128.
25. Katner, S. N., Magalong, J. G., and Weiss, F. (1999) Reinstatement of alcohol-seeking behavior by drug-associated discriminative stimuli after prolonged extinction in the rat. *Neuropsychopharmacology* **20,** 471–479.
26. Robinson, T. E. and Berridge, K. C. (1993) The neural basis of drug craving: an incentive-sensitization theory of addiction. *Brain Res. Brain Res. Rev.* **18,** 247–291.
27. Broadbent, J., Grahame, N. J., and Cunningham, C. L. (1995) Haloperidol prevents ethanol-stimulated locomotor activity but fails to block sensitization. *Psychopharmacology* **120,** 475–482.
28. Cunningham, C. L. (1995) Localization of genes influencing ethanol-induced conditioned place preference and locomotor activity in BXD recombinant inbred mice. *Psychopharmacology* **120,** 28–41.
29. Cunningham, C. L., Niehus, D. R., Malott, D. H., and Prather, L. K. (1992) Genetic differences and activating effectsof morphine and ethanol. *Psychopharmacology* **107,** 385–393.
30. Grahame, N. J. and Cunningham, C. L. (1997) Intravenous ethanol self-administration in C57BL/6J and DBA/2J mice. *Alcohol. Clin. Exp. Res.* **21,** 56–62.
31. Grahame, N. J., Rodd-Henricks, K., Li, T. K., and Lumeng, L. (2000) Ethanol locomotor sensitization, but not tolerance correlates with selection for alcohol preference in high- and low-alcohol preferring mice. *Psychopharmacology* **151,** 252–260.
32. Lessov, C. N., Palmer, A. A., Quick, E. A., and Phillips, T. J. (2001) Voluntary ethanol drinking in C57BL/6J and DBA/2J mice before and after sensitization to the locomotor stimulant effects of ethanol. *Psychopharmacology* **155,** 91–99.
33. Hoshaw, B. A. and Lewis, M. J. (2001) Behavioral sensitization to ethanol in rats: evidence from the Sprague-Dawley strain. *Pharmacol. Biochem. Behav.* **68,** 685–690.
34. Spanagel, R. and Zieglgänsberger, W. (1997) Anti-craving compounds for ethanol: new pharmacological tools to study addictive processes. *Trends Pharmacol. Sci.* **18,** 54–59.
35. Koob, G. F. and Le Moal, M. (1997) Drug abuse: hedonic homeostatic dysregulation. *Science* **278,** 52–58.
36. Herz, A. (1997) Endogenous opioid systems and alcohol addiction. *Psychopharmacology* **129,** 99–111.
37. Spanagel, R. and Weiss, F. (1999) The dopamine hypothesis of reward: past and current status. *Trends Neurosci.* **22,** 521–527.
38. Schultz, W., Dayan, P., and Montague, P.R. (1997) A neural substrate of prediction and reward. *Science* **275,** 1593–1599.
39. Littleton, J. (1995) Acamprosate in alcohol dependence: how does it work? *Addiction* **90,** 1179–1188.
40. Tsai, G. E. and Coyle J. T. (1998) The role of glutamatergic neurotransmission in the pathophysiology of alcoholism. *Annu. Rev. Med.* **49,** 173–184.
41. Weiss, F., Ciccocioppo, R., Parsons, L. H., Katner, S., Liu, X., Zorrilla, E .P., Valdez, G.R., Ben-Shahar, O., Angeletti, S., and Richter, R. R. (2001) Compulsive drug-seeking behavior and relapse. Neuroadaptation, stress, and conditioning factors. *Ann. N.Y. Acad. Sci.* **937,** 1–26.
42. Pich, E. M., Lorang, M., Yeganeh, M., Rodriguez de Fonseca, F., Raber, J., Koob, G. F., and Weiss, F. (1995) Increase of extracellular corticotropin-releasing factor-like immunoreactivity levels in the amygdala of awake rats during restraint stress and ethanol withdrawal as measured by microdialysis. *J. Neurosci.* **15,** 5439–5447.
43. Spanagel, R. and Bienkowski, P. (2001) Glutamatergic mechanisms involved in alcohol dependence and addiction, in *Therapeutic Potential of Ionotropic Glutamate Receptor*

Antagonists and Modulators (Lodge, D., Danysz, W., and Parsons, C. G., eds.), F.P. Publishing Co., Johnson City, IN, in press.
44. Altshuler, H. L., Phillips, P. E., and Feinhandler, D. A. (1980) Alteration of ethanol self-administration by naltrexone. *Life Sci.* **26,** 679–688.
45. Hyytiä, P. and Sinclair, J. D. (1993) Responding for oral ethanol after naloxone treatment by alcohol-preferring AA rats. *Alcohol. Clin. Exp. Res.* **17,** 631–636.
46. Samson, H. H. and Doyle T. F. (1985) Oral ethanol self-administration in the rat: effect of naloxone. *Pharmacol. Biochem. Behav.* **22,** 91–99.
47. Hyytiä, P. (1993) Involvement of mu-opioid receptors in alcohol drinking by alcohol- preferring AA rats. *Pharmacol. Biochem. Behav.* **45,** 697–701.
48. Roberts, A. J., McDonald, J. S., Heyser, C. J., Kieffer, B. L., Matthes, H. W., Koob, G. F., and Gold, L. H. (2000) Mu-opioid receptor knockout mice do not self-administer alcohol. *J. Pharmacol. Exp. Ther.* **293,** 1002–1008.
49. Froehlich, J. C., Zweifel, M., Harts, J., Lumeng, L., and Li, T. K. (1991) Importance of delta opioid receptors in maintaining high alcohol drinking. *Psychopharmacology* **103,** 467–472.
50. Krishnan-Sarin, S., Jing, S. L., Kurtz D. L., Zweifel, M., Portoghese, P. S., and Li, T. K. (1995) The delta opioid receptor antagonist naltrindole attenuates both alcohol and saccharin intake in rats selectively bred for alcohol preference. *Psychopharmacology* **120,** 177–185.
51. June, H. L., McCune, S. R., Zink, R. W., Portoghese, P. S., Li, T. K., and Froehlich, J. C. (1999) The δ-opioid receptor antagonist naltriben reduces motivated responding for ethanol. *Psychopharmacology* **147,** 81–99.
52. Honkanen, A., Vilamo, L., Wegelius, K., Sarviharju, M., Hyytiä, P., and Korpi, E. R. (1996) Alcohol drinking is reduced by a μ 1- but not by a δ-opioid receptor antagonist in alcohol-preferring rats. *Eur. J. Pharmacol.* **304,** 7–13.
53. Middaugh, L. D., Kelley, B. M., Groseclose, C. H., and Cuison, E. R. Jr. (2000) Delta-opioid and 5-HT3 receptor antagonist effects on ethanol reward and discrimination in C57BL/6mice. *Pharmacol. Biochem. Behav.* **65,** 145–154.
54. Roberts, A. J., Gold, L. H., Polis, I., McDonald J. S., Filliol, D., Kieffer, B. L., and Koob, G. F. (2001) Increased ethanol self-administration in delta-opioid receptor knockout mice. *Alcohol. Clin. Exp. Res.* **25,** 1249–1256.
55. Mucha, R. F. and Herz, A. (1985) Motivational properties of kappa and mu opioid receptor agonists studied with place and taste preference conditioning. *Psychopharmacology* **86,** 274–280.
56. Spanagel, R., Herz, A., and Shippenberg, T. S. (1992) Opposing tonically active endogenous opioid systems modulate the mesolimbic dopaminergic pathway. *Proc. Natl. Acad. Sci. USA* **89,** 2046–2050.
57. Hölter, S. M., Henniger, M. S. H., Lipowski, A. W., and Spanagel, R. (2000) Kappa-opioid receptors and relapse-like drinking in long-term ethanol-experienced rats. *Psychopharmacology* **153,** 93–102.
58. Przewlocka, B., Turchan, J., Lason, W., and Przewlocki, R. (1997) Ethanol withdrawal enhances the prodynorphin system activity in the rat nucleus accumbens. *Neurosci. Lett.* **238,** 13–16.
59. Gessa, G. L., Muntoni, F., Collu, M., Vargiu, L., and Mereu, G. (1985) Low doses of ethanol activate dopaminergic neurons in the ventral tegmental area. *Brain Res.* **25,** 201–203.
60. Di Chiara, G. and Imperato, A. (1988) Drugs abused by humans preferentially increase synaptic dopamine concentrations in the mesolimbic system of freely moving rats. *Proc. Natl. Acad. Sci. USA* **85,** 5274–5278.
61. Hodge, C. W., Samson, H. H., and Chappelle, A. M. (1997) Alcohol self-administration: further examination of the role of dopamine receptors in the nucleus accumbens. *Alcohol. Clin. Exp. Res.* **21,** 1083–1091.
62. Risinger, F. O., Freeman, P. A., Rubinstein, M., Low, M. J., and Grandy, D. K. (2000) Lack of

operant ethanol self-administration in dopamine D2 receptor knockout mice. *Psychopharmacology* **152,** 343–350.

63. Cunningham, C. L., Howard, M. A., Gill, S. J., Rubinstein, M., Low, M. J., and Grandy, D. K. (2000) Ethanol-conditioned place preference is reduced in dopamine D2 receptor-deficient mice. *Pharmacol. Biochem. Behav.* **67,** 693–699.
64. Nestby, P., Vanderschuren, L. J., De Vries, T. J., Mulder, A. H., Wardeh, G., Hogenboom, F., and Schoffelmeer, A. N. (1999) Unrestricted free-choice ethanol self-administration in rats causes long-term neuroadaptations in the nucleus accumbens and caudate putamen. *Psychopharmacology* **141,** 307–314.
65. May, T., Wolf, U., and Wolffgramm, J. (1995) Striatal dopamine receptors and adenylyl cyclase activity in a rat model of alcohol addiction: effects of ethanol and lisuride treatment. *J. Pharmacol. Exp. Ther.* **275,** 1195–1203.
66. Wolffgramm, J., Galli, G., Thimm, F., and Heyne, A. (2000) Animal models of addiction: models for therapeutic strategies? *J. Neural. Transm.* **107,** 649–668.
67. Eravci, M., Grosspietsch, T., Pinna, G., Schulz, O., Kley, S., Bachmann, M., Wolffgramm, J., Gotz, E., Heyne, A., Meinhold, H., and Baumgartner, A. (1997) Dopamine receptor gene expression in an animal model of "behavioral dependence" on ethanol. *Mol. Brain. Res.* **50,** 221–229.
68. Spanagel, R., Probst C., Mash, D. C., and Skutella T. (1999) How to test antisense oligonucleotides in animals. In *Manual of antisense methodology* (S. Endres, ed.), pp. 145–165.
69. Dahchour, A. and de Witte, P. (1999) Effect of repeated ethanol withdrawal on glutamate microdialysate in the hippocampus. *Alcohol. Clin. Exp. Res.* **23,** 1698–1703.
70. Veatch, L. M.,and Gonzalez, L. P. (1996) Repeated ethanol withdrawal produces site-dependent increases in EEG spiking. *Alcohol. Clin. Exp. Res.* **20,** 262–267.
71. Gonzalez, L. P., Veatch, L. M., Ticku, M. K., and Becker, H. C. (2001) Alcohol withdrawal kindling: mechanisms and implications for treatment. *Alcohol. Clin. Exp. Res.* **25,** 197–201.
72. Quertemont, E., de Neuville, J., and De Witte, P. (1998) Changes in the amygdala amino acid microdialysate after conditioning with a cue associated with ethanol. *Psychopharmacology* **139,** 71–78.
73. Göthert, M. and Fink, K. (1989) Inhibition of N-methyl-D-aspartate (NMDA)- and L-glutamate-induced noradrenaline and acetylcholine release in the rat brain by ethanol. *Naunyn Schmiedebergs Arch. Pharmacol.* **340,** 516–521.
74. Hoffman, P. L., Rabe, C. S., Moses, F., and Tabakoff, B. (1989) N-methyl-D -aspartate receptors and ethanol: inhibition of calcium flux and cyclic GMP production. *J. Neurochem.* **52,** 1937–1940.
75. Lovinger, D. M., White, G., and Weight, F. F. (1989) Ethanol inhibits NMDA-activated ion current in hippocampal neurons. *Science* **243,** 1721–1724.
76. Lovinger, D. M., White, G., and Weight, F. F. (1990) NMDA receptor-mediated synaptic excitation selectively inhibited by ethanol in hippocampal slice from adult rat. *J. Neurosci.* **10,** 1372–1379.
77. Narahashi, T., Kuriyama, K., Illes, P., Wirkner, K., Fischer, W., Muhlberg, K., Scheibler, P., Allgaier, C., Minami, K., Lovinger, D., Lallemand, F., Ward, R. J., De Witte, P., Itatsu, T., Takei, Y., Oide, H., Hirose, M., Wang, X. E., Watanabe, S., Tateyama, M., Ochi, R., and Sato, N. (2001) Neuroreceptors and ion channels as targets of alcohol. *Alcohol. Clin. Exp. Res.* **25,** 182–188.
78. Peoples, R. W. and Stewart, R. R. (2000) Alcohols inhibit N-methyl-D-aspartate receptors via a site exposed to the extracellular environment. *Neuropharmacology* **39,** 1681–1691.
79. Gulya, K., Grant, K. A., Valverius, P, Hoffman, P. L., and Tabakoff, B (1991) Brain regional specificity and time-course of changes in the NMDA receptor-ionophore complex during ethanol withdrawal. *Brain Res.* **547,** 129–134.

80. Hu, X. J., and Ticku, M. K. (1995) Chronic ethanol treatment upregulates the NMDA receptor function and binding in mammalian cortical neurons. *Brain Res. Mol. Brain Res.* **30,** 347–356.
81. Snell, L. D., Tabakoff, B., and Hoffman, P. L. (1993) Radioligand binding to the N-methyl-D-aspartate receptor/ionophore complex: alterations by ethanol in vitro and by chronic in vivo ethanol ingestion. *Brain Res.* **602,** 91–98.
82. Carter, L. A., Belknap, J. K., Crabbe, J. C., and Janowsky, A. (1995) Allosteric regulation of the N-methyl-D-aspartate receptor-linked ion channel complex and effects of ethanol in ethanol-withdrawal seizure-prone and -resistant mice. *J. Neurochem.* **64,** 213–219.
83. Tremwel, M. F., Anderson, K. J., and Hunter, B. E. (1994) Stability of [3H]MK-801 binding sites following chronic ethanol consumption. *Alcohol. Clin. Exp. Res.* **18,** 1004–1008.
84. Rudolph, J. G., Walker, D. W., Iimuro, Y., Thurman, R. G., and Crews, F. T. (1997) NMDA receptor binding in adult rat brain after several chronic ethanol treatment protocols. *Alcohol. Clin. Exp. Res.* **21,** 1508–1519.
85. Winkler, A., Mahal, B., Zieglgänsberger, W., and Spanagel, R (1999) Accurate quantification of the mRNA of NMDAR1 splice variants measured by competitive RT-PCR. *Brain Res. Protoc.* **4,** 69–81.
86. Dodd, P. R., Beckmann, A. M., Davidson, M. S., and Wilce P. A. *(2000)* Glutamate-mediated transmission, alcohol, and alcoholism. *Neurochem. Int.* **37,** 509–533.
87. Hardy, P.A., Chen, W., and Wilce, P.A. (1999) Chronic ethanol exposure and withdrawal influence NMDA receptor subunit and splice variant mRNA expression in the rat cerebral cortex. *Brain Res.* **819,** 33–39.
88. Kalluri, H. S., Mehta, A. K., and Ticku, M. K. (1998) Up-regulation of NMDA receptor subunits in rat brain following chronic ethanol treatment. *Mol. Brain Res.* **58,** 221–224.
89. Trevisan, L., Fitzgerald, L. W., Brose, N., Gasic, G. P, Heinemann, S. F., Duman, R. S., and Nestler, E. J. (1994) Chronic ingestion of ethanol up-regulates NMDAR1 receptor subunit immunoreactivity in rat hippocampus. *J. Neurochem.* **62,** 1635–1638.
90. Winkler, A., Mahal, B., Kiianmaa, K., Zieglgänsberger, W., and Spanagel, R. (1999) Effects of chronic alcohol treatment on the expression of different splice variants in the brain of AA and ANA lines of rats. *Mol. Brain Res.* **72,** 166–175.
91. Darstein, M., Albrecht, C., Lopez-Francos, L., Knorle, R., Hölter, S. M., Spanagel, R., and Feuerstein, T. J. (1998) Release and accumulation of neurotransmitters in the rat brain: acute effects of ethanol in vitro and effects of long-term voluntary ethanol intake. *Alcohol. Clin. Exp. Res.* **22,** 704–709.
92. Putzke, J., Wolf, G., Zieglgänsberger, W., and Spanagel, R. (2001) Chronic ethanol ingestion leads to long-lasting changes in NMDA receptor subunit composition in rat hippocampus. Submitted.
93. Rammes, G., Mahal, B., Putzke, J., Parsons, C., Spielmanns, P., Pestel, E., Spanagel, R., Zieglgansberger, W., and Schadrack, J. (2001) The anti-craving compound acamprosate acts as a weak NMDA-receptor antagonist, but modulates NMDA-receptor subunit expression similar to memantine and MK-801. *Neuropharmacology* **40,** 749–760.
94. Rassnick, S., Pulvirenti, L., and Koob, G. F. (1992) Oral ethanol self-administration in rats is reduced by the administration of dopamine and glutamate receptor antagonists into the nucleus accumbens. *Psychopharmacology* **109,** 92–98.
95. Shelton, K. L. and Balster, R. L. (1997) Effects of gamma-aminobutyric acid agonists and N-methyl-D-aspartate antagonists on a multiple schedule of ethanol and saccharin self-administration in rats. *J. Pharmacol. Exp. Ther.* **280,** 1250–1260.
96. Piasecki, J., Koros, E., Dyr, W., Kostowski, W., Danysz, W., and Bienkowski, P. (1998) Ethanol-reinforced behavior in the rat: effects of uncompetitive NMDA receptor antagonist, memantine. *Eur. J. Pharmacol.* **354,** 135–143.
97. Bienkowski P., Krzascik, P., Koros, E., Kostowski, W., Scinska, A., and Danysz, W. (2001)

Effects of a novel uncompetitive NMDA receptor antagonist, MRZ 2/579 on ethanol self-administration and ethanol withdrawal seizures in the rat. *Eur. J. Pharmacol.* **413,** 81–89.

98. Hölter, S. M., Danysz, W., and Spanagel, R. (1996). Evidence for alcohol anti-craving properties of memantine. *Eur. J. Pharmacol.* **314,** 1–2.
99. Hölter, S. M., Danysz, W., and Spanagel, R. (2000) The noncompetitive NMDA receptor antagonist MRZ 2/579 suppresses the alcohol deprivation effect in long-term alcohol drinking rats and substitutes the alcohol cue in a discrimination task. *J. Pharmacol. Exp. Ther.* **292,** 545–552.
100. Hundt, W., Danysz, W., Hölter, S. M., and Spanagel, R. (1998) Ethanol and N-methyl-D-aspartate receptor complex interactions: a detailed drug discrimination study in the rat. *Psychopharmacology* **135,** 44–51.
101. Akinshola, B. E., Stewart, R. R., Karvonen, L., Taylor, R. E., and Liesi, P. (2001) Involvement of non-NMDA receptors in the rescue of weaver cerebellar granule neurons and sensitivity to ethanol of cerebellar AMPA receptors in oocytes. *Mol. Brain Res.* **10,** 8–17.
102. Weiner, J. L., Dunwiddie, T. V., and. Valenzuela, C. F. (1999) Ethanol inhibition of synaptically evoked kainate responses in rat hippocampal CA3 pyramidal neurons. *Mol. Pharmacol.* **56,** 85–90.
103. Stephens, D. N. and Brown, G. (1999) Disruption of operant oral self-administration of ethanol, sucrose, and saccharin by the AMPA/kainate antagonist, NBQX, but not the AMPA antagonist, GYKI 52466. *Alcohol. Clin. Exp. Res.* **23,** 1914–1920.
104. Pohorecky, L. A. (1990) Interaction of ethanol and stress: research with experimental animals—an update. *Alcohol Alcohol.* **25,** 263–276
105. Vale, W., Spiess, J., Rivier, C., and Rivier, J. (1981) Characterization of a 41-residue ovine hypothalamic peptide that stimulates secretion of corticotropin and beta-endorphin. *Science* **213,** 1394–1397.
106. Heinrichs, S. C., Menzaghi, F., Merlo Pich, E., Britton, K. T., and Koob, G. F. (1995) The role of CRF in behavioral aspects of stress. *Ann. N .Y. Acad. Sci.* **771,** 92–104.
107. Chalmers, D. T., Lovenberg, T. W., and De Souza, E. B. (1995) Localization of novel corticotropin-releasing factor receptor (CRF2) mRNA expression to specific subcortical nuclei in rat brain: comparison with CRF1 receptor mRNA expression. *J. Neurosci.* **15,** 6340–6350.
108. De Souza, E. B. (1995) Corticotropin-releasing factor receptors: physiology, pharmacology, biochemistry and role in central nervous system and immune disorders. *Psychoneuroendocrinology* **20,** 789–819.
109. Sarnyai, Z., Shamam, Y., and Heinrichs, S. C. (2001) The role of corticotropin-releasing factor in drug addiction. *Pharmacol. Rev.* **53,** 209–243.
110. Ehlers, C. L., Chaplin, R. I., Wall, T. L., Lumeng, L., Li, T. K., Owens, M. J., and Nemeroff, C. B. (1992) Corticotropin releasing factor (CRF): studies in alcohol preferring and non-preferring rats. *Psychopharmacology* **106,** 359–364.
111. Richter, R. M., Zorilla, E. P., Basso, A. M., Koob, G. F., and Weiss, F. (2000) Altered amygdalar CRF release and increased anxiety-like behavior in Sardinian alcohol-preferring rats: a microdialysis and behavioral study. *Alcohol Clin. Exp. Res.* **24,** 1765–1772.
112. Keck, M. E., Welt, T., Wigger, A., Renner, U., Engelmann, M., Holsboer, F., and Landgraf, R. (2001) The anxiolytic effect of the CRH(1) receptor antagonist R121919 depends on innate emotionality in rats. *Eur. J. Neurosci.* **13,** 373–380.
113. Timpel, P., Spanagel, R., Sillaber I., Kresse, A., Reul, J. M. H. M., Stalla, J., Planquet, V. Stekler, T., Holsboer, F., and Wurst, W. (1998) Impaired stress response and reduced anxiety in mice lacking a functional corticotropin-releasing hormone receptor. *Nat. Genet.* **19,** 162–166.
114. Sillaber, I., Rammes, R., Zimmermann, S., Mahal, B., Zieglgänsberger, W., Wurst, W., et al. (2002) Enhanced and delayed stress-induced alcohol drinking in mice lacking functional CRH1 receptors. *Science* **296,** 931–933.

115. Lê, A. D., Poulos, C. X., Harding, S., Watchus, J., Juzytsch, W., and Shaham, Y. (1999) Effects of naltrexone and fluoxetine on alcohol self-administration and reinstatement of alcohol seeking induced by priming injections of alcohol and exposure to stress. *Neuropsychopharmacology* **21,** 435–444.
116. Vosler, P. S., Bombace, J. C., and Kosten, T. A. (2001) A discriminative two-lever test of dizocilpine's ability to reinstate ethanol-seeking behavior. *Life Sci.* **69,** 591–598.
117. Lê, A. D., Harding, S., Juzytsch, W., Watchus, J., Shalev, U., and Shaham, Y. (2000) The role of corticotrophin-releasing factor in stress-induced relapse to alcohol-seeking behavior in rats. *Psychopharmacology* **150,** 317–324.
118. Broadbent, J. and Weitemier A. Z. (1999) Dizocilpine (MK-801) prevents the development of sensitization to ethanol in DBA/2J mice. *Alcohol Alcohol.* **34,** 283–288.

16

Molecular and Behavioral Aspects of Nicotine Dependence and Reward

Emilio Merlo Pich, Christian Heidbreder, Manolo Mugnaini, and Vincenzo Teneggi

1. Introduction

Tobacco produces dependence or addiction in humans. Dependence can be defined as a maladaptive pattern of substance use or drug taking over an extended time period. The American Psychiatric Association, in its DSM-IV manual for psychiatric diagnosis *(1)*, suggest an operative diagnosis of tobacco dependence when three or more of the following seven symptoms or signs are identified in a subject:

1. Persistent desire and unsuccessful attempts to quit
2. Use of large amounts of drug and for a longer period than intended
3. Continued use in the face of medical, familial, and social problems
4. Important social, familial, and recreational activities given up or reduced because of drug craving
5. Expenditure of a great deal of time and activity in relation to drugs
6. Tolerance
7. Physical dependence (withdrawal)

Interestingly, most of the symptoms can be related either to the loss of control over the drug use habit (i.e., *1–5*), or to the physiological adaptation to the drug (i.e., *6–7*). Since only three symptoms are required for the diagnosis, dependence can be present without the signs of tolerance and withdrawal, once considered necessary prerequisites of dependence and addiction (2). In fact, in the 1964 Report of the U.S. Surgeon General, chronic exposure to nicotine was recognized as not producing physical dependence, and was regarded as "habit forming" rather than addicting *(3)*. However, in the 1988 Report of the U.S. Surgeon General the definition was changed, recognizing nicotine as addictive, indicating that tolerance and withdrawal syndrome are neither necessary nor sufficient criteria for diagnosis *(4)*. This definition was included in the DSM-IIIR, and then in DSM-IV. Psychological dependence, that is, liability detectable by behavioral changes, became a key qualifier of addiction. In this review the terms *dependence* and *addiction* are used interchangeably, following other authors who equate the two terms *(5,6)*.

1.1. Nicotine Is the Main Tobacco Constituent with Addictive Properties

Among tobacco's 3500 different constituents, nicotine has been considered the most likely responsible for tobacco's addictive effects *(7)*. When smoked, tobacco releases

From: *Molecular Biology of Drug Addiction*
Edited by: R. Maldonado © Humana Press Inc., Totowa, NJ

nicotine, which readily enters the blood circulatory system and rapidly accumulates in the brain, exerting its neuroactive action. To date, the neurobiology of nicotine dependence is only partially understood; the knowledge about the underlying molecular mechanisms so far accumulated is based mostly on findings obtained in animal models of nicotine dependence and, to a lesser extent, on observations in humans. According to a well-accepted psychopharmacological model *(8)*, substance use or drug-seeking behavior is controlled by four main processes: the positive reinforcing effects, the aversive effects, the discriminative effects, and the stimulus-conditioned effects of the drug. In more explicit terms, an individual consumes nicotine because it:

1. Produces pleasure (positive emotions)
2. Reduces the aversion due to abstinence (negative emotions)
3. Signals discrimination about its presence (assuming motivational value)
4. Promotes, as a conditioned stimulus, the unconditioned learned component of the drug-taking response

In the real world all these processes are interacting in determining the behavior of individuals smoking tobacco, and have been modeled according to the different weight given to certain processed *(9,10)*. The common factor to all these models is the presence of compulsive drug-taking.

1.2. Drug-Taking Is the Key Behavioral Trait in Smokers to Be Modeled in Animals for Understanding the Nicotine Reward Component in Nicotine Dependence

This review is focused mainly on in vivo drug-taking paradigms as a way to study the neurobiology of dependence. In naturalistic conditions, smokers titrate their optimal nicotine dose by taking different numbers of puffs per unit of time using inhalation devices (cigarettes, cigars, etc.). They consume several cigarettes per day, often clustering them around specific events, for example, at home after work, or after lunch. Similarly, in an experimental setting, smokers can be trained to self-administer nicotine intravenously and to titrate their optimal dose per session or daily *(11)*. The determinants leading to the way nicotine is abused, that is, smoking, chewing, or intravenous nicotine self-administration are described elsewhere. The most commonly accepted reason for taking an addictive drug is because of its positive reinforcing effects. A drug serves as positive reinforcer when its use increases the probability of eliciting a response on which it is contingent (i.e., smoke inhalation maintaining cigarette use in smokers; nicotine infusion maintaining lever pressing in rats trained in a Skinner box).

In this review we selected studies where molecular and neural mechanisms were investigated in subjects in which the effects of nicotine were measurable as reinforcing properties, that is, the disposition of an individual to take the drug compulsively and experience the drug effects again, is monitored, quantified, and expressed as increased probability of responding. This definition is consistent with research paradigms used for addictive drugs in general *(12)*. In our attempt to reduce the confounding factors intrinsic to several animal models of dependence, that is, the inclusion of equivocal or indirect measurements of such disposition, only operant behavior paradigms in which nicotine serves as positive reinforcer were selected. These paradigms, in particular intravenous (iv) drug self-administration, offer both acceptable theoretical support for the modeling of the positive reinforcing properties of addictive drugs and technical robustness in view of molecular investigations *(12–14)*.

1.3. The Reward System as Neural Substrates of the Reinforcing Properties of Nicotine

Most of our knowledge about the neural mechanism of reinforcement stems from the experiments of Olds and Milner *(15)* using intracranial self-stimulation (ICSS) procedures. ICSS can be triggered when the electrodes are targeting several discrete brain regions, spanning from the ventral tegmental area (VTA) to the lateral hypothalamus, via the medial forebrain bundle, to the nucleus accumbens and prefrontal cortex. The ensemble of these structures has been loosely defined as reward system *(16)*. At first, nicotine (0.2–0.4 mg/kg) was shown to enhance the rate of responding for ICSS in the rat VTA or medial prefrontal cortex without changing the current threshold *(17)*. Years later, Huston-Lyon et al. *(18)*, using an auto-titration procedure, showed that nicotine indeed lowered the current threshold for ICSS in a dose-dependent fashion, demonstrating the reinforcing efficacy of low-dose nicotine. More recently, nicotine was shown to produce a leftward shift of the ICSS curve relating the frequency of currents and the response rate in rats *(19)*. These data indicate that nicotine is acting directly on the reward system, affecting its efficiency in positive terms.

Other lines of investigation, using neurochemical-electrolytic brain lesions, local microinjections of antagonist drugs, or local measurements of neurotransmitter levels with microdialysis probes, have shed some light on the relevance of certain structures in mediating the rewarding effects of addictive drugs *(9,13,20)*. According to recent neurobiological models *(12,20)*, it is surmised that: (a) the complex processing involving cognitive and associative mechanisms associated with drug-taking, such as subjective attribution, cue assessment, and craving, is assumed to depend on the prefrontal and associative cortex, as well as the amygdala and hippocampus; (b) the mesocorticolimbic dopamine (DA) system, projecting from the ventral tegmental area to the nucleus accumbens and medial prefrontal cortex, is implicated in the control of instrumental behaviors and their outcomes; (c) the dorsal striatum, which receives projections from the DA neurons of the substantia nigra, is suggested to participate in habit formation.

Finally, chronic exposure to pharmacological doses of addictive drugs, by acting on some components of the reward system, in particular the mesocorticolimbic DA system, are believed to produce adaptive modifications of neural functioning. These changes result in several behavioral changes, some not related to changes in reward-dependent behavior—for example, locomotor sensitization, place preference, latent inhibition, increased DA release to addictive drug challenge, increased sensitivity to stress exposure, reduced palatable food consumption, and maintenance of drug self-administration *(21)*. It is interesting to note that most of these behaviors are observed neither in rats chronically self-administering nicotine nor in human smokers. In addition, it is not automatic to suggest that exposure to nicotine, independent of the paradigm of administration, produces dependence in individuals. In fact, continuous vs pulsed administration, or passive vs active self-administration, may differ importantly in defining the total dose effects, the neurobiological changes in the substrates, and the tolerance or sensitization profile of the drug *(10)*.

2. The Behavioral Model: Intravenous Nicotine Self-Administration as Model of Drug-Taking

2.1. The Role of Nicotine Reward

The authors are aware that if a drug sustains drug-taking behavior, that is, nicotine self-administration or smoking, this does not tell how it does so. Many factors have been proposed to contribute to the overall positive reinforcing efficacy of the drug, that is, nicotine reward, and these have been reviewed extensively *(8,10,12,22)*. Accordingly, a drug may:

1. Reinforce the stimulus–response habit directly
2. Modulate internal affective states and producing euphoria or reducing anxiety
3. Modulate other reinforcer effects for example, enhancing social or sexual reinforcers
4. Modulate attentional or perceptual mechanisms

It has been suggested that the first and second are distinctive mechanisms by which the drug acts directly on the neural substrate and mediate the reinforcing properties, whereas for the third and fourth the drug exerts its effects indirectly *(12)*.

In drug self-administration the subject starts by learning about the response-outcome contingency (incentive, stimulus–reward learning, goal-oriented behavior), and then the action gradually becomes a habit (stimulus–response learning). Certain addictive drugs may influence self-administration behavior by acting on stimulus-reward learning, leading to the acquisition of a drug-reinforced habit over time *(23,24)*. In this case environmental stimuli can become associated with the effects of the drugs, acquiring secondary reinforcing properties (conditioned reinforcer). Interestingly, using so called second-order schedules, conditioned reinforcers may take control over operant responding, indicating persistence of drug-taking induced by conditioned cues in the absence of the drug. It has been argued that, at least for human studies, substance use (i.e., smoking) is habitual, and relapse is most often triggered by exposure to eliciting cues, including environmental conditioned stimuli described above, or a priming stimulus with a single dose of the same drug *(8,25)*. In the latter case, the drug acts as a discriminative stimulus, producing interoceptive effects that contribute to the initiation of drug-taking behavior. Interestingly, low doses of nicotine can induce relapse of nicotine self-administration in rats following extinction *(26,27)*. White (22) maintains that both forms of stimulus–reward and stimulus–response leaning coexist in parallel in the brain of subjects who self-administer addictive drugs, and this appears particularly true for nicotine *(10)*. Eliciting cues, both environmental and interoceptive-discriminant, play an important role in determining the subjective feeling of craving for the drug. These conditioned processes have to be taken into account when addressing experiments on the molecular correlates of self-administration procedures. However, it is believed that they play important roles in sustaining drug self-administration only if the drug is able to maintain its activating properties on the neural substrate of the reward system over time.

2.2. The Other Side of the Coin: Relevance of Nicotine Withdrawal and the Negative Reinforcing Effects of Nicotine in Self-Administration Paradigms

Difficulties in giving up the habit of smoking are often reported to be also related to the nicotine withdrawal syndrome, that is, increase nervousness, frustration, anger, and desire to smoke *(7,28)*. Nicotine withdrawal syndrome can also be seen in rats passively infused with high doses of nicotine and then acutely treated with mecamylamine *(29,30)*. Stress, depression, and other negative affective states, and peripheral signs induced by nicotine withdrawal, have been suggested to be the basis of nicotine craving, and may constitute an important component of the maintenance of drug-taking behavior *(21,31)*. In smokers, the time to the first cigarette in the morning after awakening is considered a main feature in the Fargenström Questionaire for Smoking Dependence *(32)*. The value of such an indicator is based on the overnight abstinence from smoking due to sleep, which should induce a state of physiological unbalance (i.e., withdrawal) that will drive a motivation for drug-seeking behavior. This phenomenon has been identified as a sort of psychological withdrawal syndrome *(9,33)*. Such a syndrome can be measured objectively in rats by increases of ICSS threshold after the interruption of passive chronic infusions with nicotine, suggesting a disruption of the reward system *(30)*. However, relief from withdrawal symptoms has been regarded as possibly related to negative reinforcement, and it may belong to a separate category when compared with the positive reinforcement properties of a drug. Interestingly, in certain paradigms nicotine exerts aversive effects, and it was suggested that nicotine might serve as a negative reinforcer *(8,34)*. For example, Henningfield and Goldberg showed that three smokers who failed to self-administer nicotine intravenously learned an operant paradigm to avoid the scheduled injection of nicotine, suggesting aversion *(11)*. However, to date, no evidence of a role of negative reinforcement has been observed in animals trained to nicotine self-administration.

In addition, a recent elegant experiment *(35)* indicates that rats trained to stable levels of intravenous nicotine self-administration do not show signs of psychological withdrawal in a Social Interaction Test at 24 and 72 h after the last daily self-administration session. The Social Interaction Test, a very sensitive procedure to measure emotional negative states, showed that all rats were significantly more anxious than saline-exposed control rats at any time intervals following the self-administration sessions up to 72 h, suggesting a mild, generalized condition of negative emotional state. The lack of worsening of this condition over time following the last self-administration session suggests that psychological withdrawal is not involved in motivating the animal to start the following session 24 h afterward. Therefore, a prevalence of the positive reward component over the psychological withdrawal component is suggested in determining drug-seeking behavior in nicotine self-administering rats. In this regard, nicotine probably differs from other addictive drugs *(9,33)*.

2.3. Methodological Aspects of Intravenous Nicotine Self-Administration in Experimental Animals

Initially, nicotine self-administration was successfully developed in subhuman primates *(34,35)*. Some early works also described different protocols in rats *(36–40)*. It was only during late 1980s that Corrigalll and Coen (1989) developed reliable schedules of iv nicotine self-administration in rats *(41)*, which were replicated and used by other groups *(42–44)*. Under these specific methodological conditions, rats acquire stable self-administration. Critical schedule parameters are:

1. The short infusion time (1 s vs 5–6 s usually set for other drugs)

2. The limitation of aversive effects due to nicotine overdosing/overload (e.g., limited daily access to self-administration session, a "timeout" period after each infusion, checking nicotine solutions pH)
3. The fixed-ratio schedule between responding and nicotine infusion delivery

Overdosing, high intake levels, and any other exposure to potential aversive effects of nicotine need to be controlled into the schedule of reinforcement and in the protocol *(43–45)*. Most of the published studies described schedule ranging between 1 and 3 h duration, with methodological conditions essentially similar to those described by Corrigall and Coen in 1989. However, Valentine et al. (1997) showed that rats could also be exposed to unlimited access to nicotine self-administration (24 h/d), with responding stable for weeks *(46)*.

Male Long-Evans, Sprague Dawley, Wistar, Holtzman, but not Lewis or Fisher-344 rat strains have been shown to acquire nicotine self-administration *(43)*. Female Sprague-Dawley also acquire nicotine self-administration, but with rates of responding different from male subjects, suggesting a gender-dependent variability to the reinforcing effects of nicotine *(44)*. Training to lever-press for food reinforcement, cocaine infusion, or previous nicotine pretreatment facilitated the acquisition of iv nicotine self-administration *(37,42,43)*, even if training histories did not appear necessary *(44)*. Diet restriction is a preferred procedure, since it allows better control of rat body weight, an important factor to keep constant infusion volumes (infusion unit, mL/kg body weight) throughout experimental sessions *(36,41)*.

Nicotine is iv self-injected by rats in a dose-related fashion but within a narrow dose range, nicotine 0.03 mg/kg/infusion unit dose being the most widely used dose for training and baseline performance. In contrast to other drug self-administration protocols, such as for cocaine *(47)*, it is not possible to identify a dose–response curve "descending limb" for nicotine self-administration, where one increase of drug unit dose corresponds to a proportional decrease of the rate of responding. Only a steep dose–response curve, with a given threshold and a plateau of response at various active doses, has been reported for nicotine self-administration. Further increases in nicotine doses are associated with motor depression and a decrease of responding *(41,42)*.

Environmental factors influencing smoking behavior have also been shown to modulate nicotine self-administration in rodents *(48)*. Preexposure to a single injection of low-dose nicotine or to environmental stimuli previously paired with nicotine self-administration is able to reinstate nicotine-seeking behavior, even after weeks of abstinence *(26,27)*.

Recently, iv nicotine self-administration has also been developed in mice *(49,50)*, and effects obtained with different strains of mice or with nicotinic ligands were described. The development and characterization of these protocols offer the opportunity to use transgenic mice to study the effects of gene mutation on the reinforcing properties of nicotine *(51,52)*.

3. Molecular Substrates of Nicotine Reward

3.1. The Role on Neuronal Nicotinic Acetylcholine Receptors (nAChR) in Nicotine Self-Administration

When nicotine enters the brain, it binds to neuronal nicotinic acetylcholine receptors (nAChR), heterogenously expressed in central nervous system neurons *(53)*. Much like their cousins in the neuromuscular junction, brain nAChR are probably pentameric complexes arranged around a central pore that is permeable to K^+, Na^+, and Ca^{2+} *(54)*. In the mammalian brain, eight α subunits (α2–α7 and α9–α10) and three β subunits (β2–β4) are differently combined to define two principal subfamilies of receptors: the α-bungarotoxin-sensitive subfamily, consisting of homopentametic α7 nAChR subunits, and the hetero-oligomeric subfamily, made of different combinations of the other α subunits and the β(2–4) subunits *(53,55)*. Experiments performed in mutant mice lacking the α7-subunit nAChR receptors indicate that practically all the high-affinity α-bungarotoxin binding found in the mouse brain is due to α7-subunit-containing nAChR *(56)*. Similar experiments performed in mutant mice lacking the β2 *(51,57)* and α4 subunits *(58)* indicate that almost all of the high-affinity ^{3}H-nicotine-binding sites of the mouse brain are represented by an α4β2 nAChR. However, persistence of high-affinity binding sites for other nicotinic ligands with high affinity for α4β2 receptors, such as ^{3}H-cytisine or ^{3}H-epibatidine, were still found in restricted brain regions of both mutant mice, in particular the medial habenula and the interpeduncular nucleus, structures known to selectively express α2β4 and α3β4 nAChR *(59)*.

3.2. Chronic Nicotine Administration Produces AchR Upregulation

Recent reports consistently indicate that ^{3}H-nicotine binding, but not ^{125}I-α-bungarotoxin binding, is upregulated in human postmortem prefrontal cortex, hippocampus, entorhinal cortex, and, to a lesser extent, striatum of smokers when compared with age-matched controls *(60–62)*. Interestingly, nicotine binding is normalized in ex-smokers, confirming the reversibility of the upregulation of ^{3}H-nicotine-binding sites observed in rodent brain following chronic passive nicotine administration *(63,64)*. These results indicate that smoking and nicotine exposure produces upregulation of nAChR, most likely α4β2, but does not directly support receptor hypersensitivity, that is, increased effects of nicotine on postsynaptic target cells. Recent unpublished results obtained in rats self-administering nicotine (0.03 mg/kg/injection) for at least 3 wk (Tessari and Mugnaini, personal communication) indicate that ^{3}H-nicotine-binding sites are upregulated in the parieto-frontal cortex, but not the VTA, suggesting a potential relevant role of the cerebral cortex in maintenance of self-administration. This interpretation is highly speculative, since it is not clear yet if this receptor upregulation stands for functional hypersensitivity (similar to denervation hypersensitivity) or reduced membrane turnover of inactivated receptors. Pharmacological and electrophysiological experiments show that prolonged exposures to nicotine produces desensitization of nAChR into a high-affinity binding state *(53,65,66)*, suggesting the latter case. Finally, recent experiments in a cell line carrying heterologous α3β2 showed that a fraction of the upregulated receptor is still active *(67)*, suggesting that a similar phenomenon may apply to *in vivo* receptor upregulations.

3.3. nAChR as Primary Site of the Reinforcing Effects of Nicotine

The involvement of nAChR in the reinforcing properties of nicotine is also supported by evidence that the nonselective antagonist mecamylamine reduces nicotine self-administration in rats (*41*; Tessari and Chiamulera, unpublished data). Experiments performed on mutant mice lacking the β2 subunit and previously trained for cocaine self-administration showed a marked attenuation of nicotine self-administration when cocaine was substituted with nicotine *(51)*. The β2 mutant mice were also able to lever-press for food, and the selective attenuation of lever pressing contingent to nicotine infusions indicates the key role of β2-contaning nAChR in sustaining nicotine self-administration. Again, this effect was clearly mediated by the mesolimbic DA system *(51)*. In the next section the role of the mesocorticolimbic DA system will be discussed extensively.

3.4 The Mesocorticolimbic DA System: Pharmacological Neuroadaptation to Nicotine Exposure

The mesocorticolimbic DA system has been the object of intense research in the field of drug addiction because:

1. It is a relevant component of the reward system *(16)*.
2. It is involved in stimulus-reward learning and in the incentive motivational effects of drugs *(19,20)*.
3. When exposed to chronic treatment with addictive drugs, it is the site of neuroadaptive changes, such as sensitization or dysregulation associated with negative emotional states *(19,21,68,69)*.

Midbrain DA neurons contain mRNA for all the principal nAChR subunits *(70)*. However, it is not clear if all of them form functional receptors. High-affinity binding for ^{125}I-α-bungarotoxin and ^{3}H-nicotine suggests that α7 and β2-containing nAChR proteins, respectively, are actually expressed. Antibodies against α4 have been recently used to confirm the presence of this subunit as translated protein, strongly supporting the presence of a functioning α4β2 nAChR *(71)*.

Pharmacological doses of nicotine are known to stimulate DA neurons of the VTA preferentially *(72,73)* vs those of the substantia nigra *(74)*, and most likely by a direct action. Pioneering studies by Fuxe and collaborators using the Falk-Hillarp and related techniques showed that exposure to nicotine or smoked tobacco increases the rate of disappearance of fluorescent DA from mesolimbic DA terminals in rats *(75,76)*. These data suggest that nicotine enhances the impulse flow and release of DA in nucleus accumbens. Neurochemical experiments using *in vivo* microdialysis show that acute nicotine induces DA release in the terminal fields of midbrain DA neurons, in particular in the shell of the nucleus accumbens *(77,78,79)*. Nicotine effects in the nucleus accumbens are antagonized by mecamylamine microinjected into the VTA, but not into the nucleus accumbens *(80)*. This result suggests that, in vivo, nAChRs located in the VTA region are of relevance in controlling DA overflow. Interestingly, acute nicotine at doses of 50–100 μg/kg also increases the spontaneous burst firing of extracellularly recorded DA VTA neurons, most likely via NMDA receptor-dependent activation *(81)*, and burst firing of VTA neurons has been associated with DA release *(82)*. The mechanisms seem to be dependent on nAChR containing α4 or α3, but not α7 *(72)*.

Repeated daily exposures to systemic nicotine result in sensitization of its stimulatory effects on DA overflow in the nucleus accumbens and prefrontal cortex *(78)* in a manner that recalls the effects of chronic exposures to cocaine or amphetamine *(68)*. On the other hand, sensitized responses to acute systemic challenges with nicotine are antagonized by mecamylamine, and can be induced by local microinjection into the VTA *(78)*. These results point to a key role of nAChR located in the VTA region to induce sensitization to nicotine effects following chronic nicotine treatments. Interestingly, when nicotine is constantly infused using subcutaneous osmotic minipumps, and not injected, it results in desensitization and tolerance of nicotine effects on DA release *(83,84)*. Recent data indicates that repeated nicotine administration (0.4–0.5 mg/kg) can result in sensitization of DA release in the core of the nucleus accumbens and DA release tolerance in the shell *(85)*, but not in the medial prefrontal cortex *(84,85)*, suggesting a complexity due to differences in neuroanatomical compartments. All in all, these data suggest that repeated passive exposure to nicotine can enhance DA neurotransmission in the mesocorticolimbic DA system. However, to date, there is no evidence that this effect is also produced in rats trained to nicotine self-administration, casting doubt about the relevance of DA release sensitization for the reinforcing effects of nicotine.

In contrast to the central role of VTA neurons in the sensitization of DA release to nicotine, recent microdialysis experiments performed by locally perfusing nicotine into the terminal fields of midbrain DA neurons showed a dose-dependent increase of mecamylamine-sensitive DA release produced by acute nicotine. Chronic nicotine treatment further enhances DA release in response to nicotine when locally infused via microdialysis probes into the nucleus accumbens, striatum, and prefrontal cortex in rat *(86)*. Experiments performed in vitro on synaptosomes confirm that DA is released via $\alpha4\beta2$ and $\alpha3\beta2$ nAChR-dependent Na^{+-} and Ca^{2+}-sensitive channel-operated mechanisms *(87,88)*. Interestingly, when DA release was studied in striatal slices *ex mortem* in rats injected daily with nicotine, enhancement of nicotine-induced DA overflow response was measured *(89)*, whereas desensitization was seen in synaptosomes obtained from rats continuously infused with nicotine *(66,88)*. Therefore, nAChR expressed at the synaptic terminals of midbrain DA neurons is likely to contribute to the development of mesocorticolimbic DA system sensitization to nicotine effects only under specific conditions.

3.4.1. Effects of Nicotine on the DA Systems Mediated by α7-Containing AChR

Attenuation of systemic nicotine- and food-induced DA overflow in the nucleus accumbens was also produced by microinjections of the α7-subunit-selective antagonist methyllycaconetine (MLA) into the VTA *(90)*. These α7-dependent effects are most likely mediated by presynaptic nAChR located on glutamatergic neurons projecting to the VTA *(91)*. Electrophysiological experiments on brain slices containing VTA DA neurons of rats previously continuously infused with nicotine for several days show desensitization of α7-dependent currents, suggesting that chronic nicotine attenuates α7-mediated synaptic transmission *(73)*. However, in vivo studies indicate that MLA was unable to block the discriminative properties of nicotine in rats *(92)* nor was nicotine self-administration, suggesting that α7 is not involved in the discriminative effects of nicotine, most likely mediated by hippocampal or prefrontal cortex rather than the DA systems *(90)*.

3.4.2. Effects of Chronic Nicotine on Thyrosine Hydroxylase (TH) in Mesolimbic DA Neurons

In vivo acute nicotine stimulates the activity of tyrosine hydroxylase (TH), the rate-limiting enzyme in the synthetic pathway of DA, in DAergic terminals of the nucleus accumbens, whereas tolerance to this effect is observed following repeated nicotine administration *(93)*. Conversely, acute nicotine does not change the levels of TH immunoreactivity in the VTA and substantia nigra, whereas increases are observed following repeated nicotine administration *(94)*. No data are available on the effects of chronic nicotine self-administration on TH levels and activity.

3.5. Involvement of the Mesocorticolimbic DA System in the Positive Reinforcement Effects of Nicotine

The involvement of mesocorticolimbic DA system in the reinforcing effects of nicotine is supported by several experimental data.

1. Selective 6-hydroxydopamine (OHDA) lesions of the nucleus accumbens reduce nicotine self-administration *(95)*.
2. Low doses of systemic administration of neuroleptics attenuate nicotine self-administration in rats *(96)*.
3. In humans, neuroleptics increase nicotine plasma levels in smokers *(97)*, suggesting that the individuals smoke more to counteract the effects of the drug and experience rewarding effects.
4. Postmortem DA and DA metabolite levels were found to beincreased in the striatum of smokers when compared with nonsmoker controls *(61)*.
5. The indirect DA agonist bupropion has been used successfully to reduce the relapse rate in smokers who have manifested the intention to quit smoking *(98)*.

In rats, nicotine self-administration is attenuated by VTA microinfusions of the α4β2-preferential competitive antagonist dihydro-β-erythroidine, but not by the muscarinic antagonist atropine *(99)*. This procedure perturbates the nicotinic-dependent cholinergic input from the pedunculopontine tegmental nucleus (PPT) into the VTA, probably by reducing nAChR-dependent facilitation of DA neuronal firing. Thus, the rewarding effects of nicotine self-administration are most likely mediated by nAChR, and not by muscarinic transmission *(14)*.

The lack of enhancing effects of nicotine on firing in VTA neurons and on DA release in nucleus accumbens and striatum observed in the β2 mutant mice *(51)* strongly suggests a role for β2-containing nAChR in mediating the reinforcing properties of nicotine through the activation of the mesolimbic DA system. In support of a direct effect of nAChR on DA release, no changes of D1–D4 DA receptor levels and affinity nor of DA transporter were reported in mutant mice.

3.6. Evidence of Neuroadaptive Changes in the Terminal Fields of the Mesocorticolimbic DA System Associated with the DA-Dependent Positive Reinforcing Effects of Nicotine: Transcriptional Regulation

Chronic exposure to addictive drugs indirectly stimulates transcription of specific genes by increasing intracellular cyclic adenosine monophosphate (cAMP), which in turn results in activation of multifunctional protein kinases and phosphorylation of several cellular proteins, including transcription factors, in target neurons of the mesocorticolimbic DA system *(69,100,101)*. A limited but growing number of scientific reports indicate that nicotine also produces some adaptive changes in the target neurons of DA terminal fields, namely, the nucleus accumbens, striatum, and prefrontal cortex, most likely via DA release and D1 receptor mediation. One of the most well studied effects of D1 receptor activation in target neurons is the transcriptional regulation of the immediate early gene (IEG) *c-fos*, and medium-late genes of the Fos-related antigen family *(100,102)*. Expression of *c-fos* has been used largely to provide transcriptional activation brain maps induced by various stimuli, ranging from natural to pharmacological *(102)*. Acute nicotine passive administration induces expression of *c-fos* that is not restricted to the mesocorticolimbic DA system terminal fields, but includes several other rat brain regions, in particular the superior colliculus, the interpeduncular nucleus, the raphe, but not the midbrain DA neurons *(101,103,104,105)*. This activation is inhibited by mecamylamine. In the striatum, a dose-responsiveness was observed, with the lower dose (0.4 mg/kg) inducing *c-fos* in the medial and central portions. In the nucleus accumbens, the response was smaller than in the striatum, with prevalence in the shell. The acute administration of epibatidine, an nAChR agonist with preferential affinity for the $\alpha 4\beta 2$ form, induced c-fos IR in the prefrontal cortex, medial striatum, nucleus accumbens, amygdala, and superior colliculus *(106)*. In the nucleus accumbens and striatum, but not the prefrontal cortex, this effect was inhibited by pretreatment with D1–D4 receptor antagonists, but not D2 antagonist *(107)*. This result is in agreement with the nicotine-induced DA overflow measured in the terminal fields of the mesocorticolimbic DA system, and the D1 mediation of reinforcing properties of DA released in the nucleus accumbens by addictive drugs *(13)*. Finally, repeated nicotine administration in rats results in preferential Fos increases in visuomotor and limbic structures, including basal ganglia and prefrontal cortex *(105,108)*.

3.6.1. Activation of c-fos *and Fos-Related Antigen (FRA) Transcription Factor Expression in the Rat Brain by Nicotine Self-Administration*

Nicotine self-administration increases Fos immunoreactivity in most of the brain regions that are activated by passive nicotine treatment *(109)*, in particular the striatum, the shell and core of the nucleus accumbens, the lateral septum, the prefrontal cortex, and the cingulate cortex. Also, the retrosplenial cortex, the piriform cortex, the periventricular thalamic nucleus, and the superior colliculus showed increase of Fos immunoreactivity. In contrast to acute nicotine, nicotine self-administration did not increase Fos immunoreactivity in the hypothalamus, locus coeruleus, amygdala, and dentate gyrus *(109)*. Passive chronic nicotine administration produced patterns similar to those obtained with acute nicotine, showing little tolerance to the challenging dose of nicotine, and a more general activation of the core of the nucleus accumbens when compared to the selective increase in the ventral shell observed in nicotine self-administration *(105,109)*. These differences are important, since brain maps of immediate early gene expression simultaneously represent the direct pharmacological effects of nicotine on neural networks and the activation associated with the internal dispositional state that controls nicotine taking, which includes the reinforcing effects of nicotine *(110)*.

A more extensive study of the transcriptional regulation of the c-fos-related antigen family (FRA) expressed in the terminal fields of the mesocorticolimbic DA pathway was performed in rats trained for nicotine and cocaine self-administration. FRA heterodimers constitute the AP1 complex that transcriptionally regulate a large number of plasticity-related genes *(111)*. FRAs, and in particular the 35-kDa component recently identified as Δ-Fos, are medium–late onset genes and their products persist for several days in the nucleus of the target neurons *(100)*. Therefore, once induced by nicotine or cocaine, FRA-containing AP1 complexes may regulate plasticity-related genes for a long period. In one experiment, rats were trained to self-administering cocaine or nicotine *(110)*. Increased expression levels of FRA immunoreactivity were found in the anterior cingulate cortex, prefrontal cortex, nucleus accumbens, medial striatum, but not amygdala, of rats from both groups. In a second experiment the binding of AP1 complex to neuronal DNA was measured in brain tissues dissected out immediately before the last self-administration session. The results showed a persistent increase of AP1 complexes in the nucleus accumbens in cocaine and nicotine self-administering rats, but not in the prefrontal cortex or striatum. This differential effect points to a primary involvement of target neurons in the nucleus accumbens as cellular substrate for the maintenance of drug self-administration and/or the stimulus–reward learning produced by these two drugs. Interestingly, the lack of increase of AP1 complex in the prefrontal cortex suggests that long-term changes in target neurons of this DA terminal field are of modest relevance in drug self-administration maintenance.

3.6.2. cDNA Microarrays Measurement of Transcriptional Profiles in the Terminal Fields of DA Systems After Chronic Nicotine Treatment

cDNA microarray is one among several large-scale biology techniques proposed during the last few years to address the possibility of parallel massive measurement of changes in gene expression products of living cells or tissue following pharmacological manipulations *(112)*.

Only preliminary reports and one full article on nicotine treatments in rodents have been published so far *(113)*. In the latter study, Konu et al. *(113)* used cDNA microarrays containing 1117 genes and expressed sequence tags (ESTs) to assess the transcriptional response to chronic nicotine treatment in rats, by comparing 4 brain regions, the prefrontal cortex, nucleus accumbens, VTA, and amygdala. The results indicate 94 genes whose expression was altered above threshold. Results from principal component analysis and pairwise correlations suggested that brain regions studied were similar in terms of their absolute expression levels (with amygdala showing the lowest activity), but differed in the composition of transcriptional profiles in response to chronic nicotine *(113)*. Accordingly, prefrontal cortex and nucleus accumbens were significantly more similar to each other than to either VTA or amygdala, supporting the value of the concept of terminal field of the mesocorticolimbic DA system as relatively homogeneous neurobiological substrates for nicotine pharmacological properties.

Several genes involved in cellular signaling, structure/cytoskelectal maintenance, metabolism, cell cycle, and transcriptional regulation were found to be affected by chronic nicotine. In particular, the MAP-kinase, the phosphatidylinositol, and EGFR signaling pathways showed quite consistent changes and were proposed as possible targets in response to nicotine administration.

These results, as exciting as they appear, need a cautionary note for an appropriate interpretation. In fact, technical limitations can importantly question the reproducibility of obtaining the same up- or downregulation for the same gene using different large-scale biological approaches. In addition, given the relatively small percentage of the genome featured into the cDNA microarray (approx 2%), these conclusions can be considered preliminary in all senses. When quasi-exhaustive cDNA microarrays containing >90% of the genome are used, it may well be that other pathways will be most affected, shifting the overall interpretation of comparative relevance.

Independent of any criticisms, these types of experiments are opening the way to a novel "global" approach to understanding the neurobiological substrate of nicotine dependence, where all the cellular biochemical pathways are assessed in parallel, and large-scale computational models will be used to make inferences about results.

3.7. Notes on Neuroadaptation to Nicotine Exposure of Neurotransmitter Systems Other Than DA

Are all the positive reinforcing effects of nicotine mediated only by mesocorticolimbic DA? Some discrepancies in the experiments cited above, and evidence of neuroadaptation in other neurotransmitter systems, may open some alternatives. Two examples follow.

1. A cDNA microarray experiment reveals effects on biochemical pathways whose changes have not been related to DA effects (e.g., EGRF) *(113)*.
2. In rats, cocaine self-administration completely downregulates *c-fos* expression in the nucleus accumbens, whereas nicotine does not, suggesting a D1 receptor-independent mechanism for activating transcription of *c-fos (108,110)*.

In fact, here, as in the anterior cingulate cortex, nicotine may act directly either on postsynaptic nAChR located in cortical or accumbens interneurons, or in pre-synaptic nAChR located on nordadrenaline, glutamate, or GABAergic terminals, exerting important modulatory roles.

Below we list some other examples of possible involvement as substrate for the addictive properties of nicotine, organized along the major neurotransmitter systems (*see* also ref. *33*).

3.7.1. Noradrenaline (NE)

Nicotine stimulates the in vitro release of NE from slices or synaptosomes from hippocampus and cerebral cortex via $\alpha3$- or $\alpha4$-containing nAChR *(114,115)*. In rats, acute nicotine administration increases TH mRNA in the locus coeruleus, whereas chronic nicotine administration increases TH levels in the locus coeruleus and telencephalic terminal fields (cerebral cortex, hippocampus, etc.) *(94)*. Acute nicotine also increases NE release in the hippocampus as measured by microdialysis *(116)*. This release is antagonized by injections of mecamylamine into the locus coeruleus, but not into the hippocampus *(117)*, while intrahippocampal α-BG or MLA antagonizes the NE release *(118)*. These data suggest somatic mediation of NE release by $\alpha4\beta2$ nAChR located near the locus coeruleus, and distal mediation of NE release by presynaptic $\alpha7$ nAChR in hippocampus. Finally, chronic nicotine exposure enhances the nicotine-induced NE release *(117)*. Hippocampus has been involved in mediating some discriminative effects of nicotine *(119)*, supporting a potential role of NE in this component of drug self-administration. In addition, NE substrates can be involved in mediating the antismoking effects of bupropion, a mild DA- and NE-uptake blocker *(98,120)*. However, no NE antagonist or selective NE neurochemical lesions are known to affect nicotine self-administration in rats.

3.7.2. Serotonin (5-HT)

Serotonin has been suggested to participate to the control of motivated consummaroty behavior *(121)*, and evidence indicates some involvement in cocaine self-administration, in particular via 5-HT_{1B} receptors *(122)*. However, a role for serotonin in mediating the reinforcing effects of nicotine seems to be excluded by the lack of effects of several serotonergic drugs in affecting nicotine self-administration *(123,124,125)*. These results are at variance with other preclinical observations showing motivational effects of 5-HT_3 antagonists, for example in the rat place preference paradigm *(126)*.

These results were not confirmed in human smokers *(125)*. In addition, when smokers were treated with antidepressant seroronin-uptake inhibitors such as fluoxetine, no therapeutic effects were shown *(127)*. There is a general agreement that antidepressants are effective only in the subpopulation in which depression and smoking are comorbidities. Moreover, data on the effects of chronic nicotine on the molecular expression patterns of serotonin-related enzymes or receptor in CNS are missing, leaving the picture rather incomplete.

3.7.3. Glutamate

Glutamate is probably the most common excitatory neurotransmitter of the brain, and glutamatergic neurons are integral part of the reward system *(16,33)*. Several electrophysiological effects of nicotine are mediated by a direct effect on glutamatergic neurons. For example, in anesthetized rats, nicotine increases the firing of locus coeruleus dose-dependently *(128)* via a glutamate-dependent mechanism. More interesting, nicotine stimulates presynaptic α7 nicotinic receptors within the VTA localized on glutamatergic afferents from the medial prefrontal cortex, producing an increase in glutamate concentrations that stimulates the NMDA receptors expressed by DA-containing neurons in the VTA *(91)*. According to Svensson and collaborators, the resulting enhanced burst firing of VTA neurons would enhance DA release in the nerve terminal regions. In this view, glutamate-containing neurons are instrumental for the building of sensitized DA functions that underlies the condition of nicotine dependence.

Glutamate has been involved as one of the mediators of sensitization to nicotine. Co-administration of the NMDA receptor antagonist MK801 or D-CPPene during the nicotine pretreatment phase attenuates the development of locomotor and intra-accumbens DA release sensitisation *(78,129)*, suggesting a potential role in determining nicotine dependence.

The effects of glutamatergic drugs on nicotine self-administration have been rarely addressed *(14,33)*. So far, there is no evidence of direct effects of systemically administered glutamate antagonists.

3.7.4. Opioid Peptides

Chronic nicotine increases increases β-endorphin in the hypothalamus *(130)*, and Met-enkephaline in the striatum and nucleus accumbens only 24 h after last nicotine administration, and returns to basal level 7 d afterward *(131,132)*. This time-dependency partially explains the lack of effects reported in other studies (e.g., ref. *133*). Increase of Pro-dynorphin following chronic nicotine was reported in nucleus accumbens *(133)*. The role of opiates in nicotine dependence is not fully understood. Naloxone does not affect intravenous nicotine self-administration in rats *(91)*, nor does it modify DA release in nucleus accumbens in rats chronically infused with nicotine *(85)*. However, naloxone precipitates a withdrawal syndrome in rats chronically infused with nicotine *(95)*, and reduces cigarette consumption in smokers *(134)*, suggesting a motivational role as part of the central stress system *(21)*. Initial reports of craving reduction in smokers with naloxone *(135)* were not confirmed in other studies with naloxone or naltrexone *(134,136)*. In rats, Corrigal and collaborators reported that the δ-receptor agonist DAMGO, injected into the VTA, shows modest effects on intravenous nicotine self-administration *(137)*. Overall, these data indicate a minor involvement of opiate peptides in the positive reinforcing aspects of nicotine dependence.

3.7.5. Gamma Aminobutyric Acid (GABA)

GABAergic neurotransmission is a recognized component of the reward pathway *(16)*. Administration of a single dose of nicotine increases GABA release from the nerve terminals and synaptosomes of several brain structures, including the VTA-substantia nigra and hippocampus *(138,139)*, via β2-containing nAChR *(140)*. In VTA-substantia nigra, nicotine effects depend on viable presynaptic D1 neurotransmission, suggesting a presynaptic modulatory effect on the feedback loop from striatum-nucleus accumbens to VTA *(138)*. Corrigal and collaborators showed that intra-VTA microinjections of baclofen, a $GABA_B$ receptor agonist, attenuates intravenous nicotine self-administration in rats *(137)*, supporting a direct role of GABAergic neurotrasmission in the reinforcing properties of nicotine. The antimotivational role of $GABA_B$ receptor agonism is also supported by the baclofen-induced complete blockade of γ-hydroxybutyric acid self-administration in mice *(141)*. However, clinical data in 24-h-abstinent smokers administered 20 mg of baclofen showed no effects on number of cigarettes smoked or craving score during the following 3-h fee-smoking period *(142)*. Mild sedative-like effects (increased "relaxing"), and changes in the sensory perception of smoked cigarettes (increased "harsh" and decrease "liking") were also reported, suggesting a role for facilitation of smoking cessation. Recent evidence suggests that this role is shared by another pro-GABAergic drug, γ vinyl-GABA (Vigabatrin; *143*). Administration of 75–150 mg/kg of γ-vinyl-GABA to rats completely antagonizes the nicotine-induced DA release in nucleus accumbens and the development of nicotine-induced conditioned place preference. Presently, nicotine–GABA interactions are a topic for intense research, and a likely target for future antismoking therapy *(144)*.

4. Conclusion

The available data are insufficient to propose an acceptable model of the molecular and cellular substrates of the positive reinforcing properties of nicotine. Possible important information can be derived by a better knowledge of the genetic risk of smoking, in particular if the genetic differences are affecting the reinforcing properties of nicotine. The evidence so far accumulated does not support this hypothesis directly. For instance, significant association between two markers of the tryptophan hydroxylase gene (TPH C218A and C779A) studied in 780 genotyped subjects and the age of smoking initiation, but not nicotine dependence progression, was recently found *(145)*. In another study, association between polymorphism of the dopamine β-hydroxylase gene and smoking cessation was also found *(146)*. Both studies are suggestive of potential genetically determined differences in the reward system, but only indirectly, since they address clinical readouts that are not directly measuring reinforcing effects. Hypothetically, a genetically determined mesolimbic DA reward system that can be easily sensitized by nicotine may predispose to smoking because of its enhanced reward-related signaling associated with drug-taking. This genetically determined defective mesolimbic functioning has been sometimes indicated as reward deficiency syndrome *(147)*. Genetic differences are also known to be important in determining addictive drug effects in rodents, in particular nicotine *(148,149)*, including strain differences in the capacity to acquire and maintain operant responding, in particular intravenous drug self-administration *(150,151)*.

In spite of these limitations, a working hypothesis is presented here with the aim of conservatively summarizing the essential features of the data discussed so far. Nicotine exerts its positive reinforcing effects by acting via the α4β2 nAChR located on the somatodendritic membranes of the DA cells and possibly its sensitising effects via α7 nAChR located on glutamate terminals. Direct neuroadaptive effects of nicotine on a supposedly highly sensitive reward system, in particular the mesolimbic DA system, are believed to influence the stimulus–reward learning and to be the major player in sustaining nicotine self-administration. However, environmental cues previously positively associated with nicotine taking and, to a lesser extent, changes of the internal affective state toward anxiety–depression, are thought to participate in triggering drug-taking behavior. These dispositional states are suggested to be produced by some of the neural networks related to the reward system, as indicated by the diffuse changes of gene expression patterns in the terminal fields of the mesocorticolimbic DA system, in particular the nucleus accumbens and prefrontal cortex. Further research matching molecular investigation to behavioral paradigms, selectively addressing each phenomenological component of the behavioral processes underling nicotine dependence and its genotypic characterization, will provide the information necessary to describe the relevant molecular players in these important disorders.

Acknowledgments

The authors thank Michela Tessari, Bernd Bunneman, Lucia Carboni, and Cristiano Chiamulera for their expert contributions in discussing some of the topics presented in this chapter.

References

1. American Psychiatric Association. (1994) *Diagnostic and Statistical Manual of Mental Disorders*, DSM-IV. American Psychiatric Association, Washington, DC.
2. Jarvik, M. E. (1995) Commentary. *Psychopharmacology* **117,**18–20.
3. US Department of Health, Education and Welfare. (1964) *Smoking and Health*. Report of the advisory committee to the Surgeon General of the Public Health Service, Maryland.
4. US Department of Health and Human Services. (1988) *The Health Consequences of Smoking, Nicotine Addiction*. A Report of the Surgeon General, Office of Smoking and Health, Bethesda, Maryland.
5. O'Brien, C. P. (1996) Drug addiction and drug abuse, in *The Pharmacological Basis of Therapeutics* (Goodman, L. S. and Gilman, P. R., eds.), McGraw-Hill, New York, pp. 276–282.
6. Stolerman, I. P. and Jarvis, M. J. (1995) The scientific case that nicotine is addictive. *Psychopharmacology* **117,** 2–10.
7. Domino, E. F. (1997) Tobacco smoking and nicotine neuropsychopharmacology, some future research directions. *Neuropsychopharmacology* **18,** 456–468.
8. Stolerman, I. P. (1992) Drug of abuse, behavioral principles, methods and term. *TIPS* **13,** 170–176.
9. Koob, G. F. and Le Moal, M. (1997) Drug abuse, hedonic homeostatic dysregulation. *Science* **278,** 52–58.
10. Di Chiara, G. (2000) Behavioral pharmacology and neurobiology of nicotine reward and dependence, in *Handbook of Experimental Pharmacology*, vol. 144, (Clementi, F., Fornasari, D., and Gotti, C., eds.), Neural Nicotinic Receptors, Springer-Verlag, Berlin, Heidelberg, pp. 603–750.
11. Hennigfield, J. E. and Goldberg, S. R. (1983) Control of behavior by intravenous injection of nicotine in human subjects. *Pharmacol. Physiol. Behav.* **19,**1021–1026.

12. Altman, J., Everitt, B. J., Glautier, S., Markou, A., Nutt, D., Oretti, R., Phillips, G. D., and Robbins T. W. (1996) The biological, social, and clinical bases of drug addiction, commentary and debate. *Psychopharmacology* **125,** 285–345.
13. Koob, G. F. (1992) Drug of abuse, anatomy, pharmacology and function of reward pathways. *TIPS* **13,** 177–184.
14. Corrigall, W. A. (1999) Nicotine self-administration in animal as a dependence model. *Nicotine & Tobacco Res.* **1,** 11–20.
15. Olds, J. and Milner, P. M. (1954) Positive reinforcement produced by electrical stimulation of septal area and other regions of the rat brain. *J. Comp. Pshysiol. Psychol.* **47,** 419–427.
16. Wise, R. A. (1996) Neurobiology of addiction. *Curr. Opin. Neurobiol.* **6,** 243–251.
17. Schaefer, G. J. and Michael, R. P. (1986) Task-specific effects of nicotine in rats, intracranial self-stimulation and locomotor activity. *Neuropharmacology* **25,** 125–131.
18. Huston-Lyon, D. and Kornetski, C. (1992) Effects of nicotine on the threshold for rewarding brain stimulation in rats. *Pharmacol. Biochem. Behav.* **41,** 755–759.
19. Bauco, P. and Wise, R. A. (1994) Potentiation of lateral hypothalamic and midline mesencephalic brain stimulation reinforcement by nicotine, examination of repeated treatment. *J. Pharmacol. Exp. Ther.* **271,** 294–301.
20. Robbins, T. W.,and Everitt, B. J. (1996) Neurobehavioral mechanisms of reward and motivation. *Curr. Opin. Neurobiol.* **6,** 228–256.
21. Kreek, M. J. and Koob, G. F. (1998) Drug dependence, stress and dysregulation of brain reward pathways. *Drug & Alcohol Dependence* **51,** 23–47.
22. White, N. M. (1996) Addictive drugs as reinforcers, multiple partial actions on memory system. *Addiction* **91,** 921–946.
23. Cador, M., Robbins, T. W., and Everitt, B. J. (1989) Involvement of amygdala in stimulus-reward association, interaction with the ventral striatum. *Neuroscience* **30,** 77–86.
24. White, N. M. and Hiroi, N. (1993) Amphetamine conditioned cue preference and the neurobiology of drug seeking. *Semin. Neurosci.* **5,** 329–336.
25. Tiffani, S. T. (1990) A cognitive model of drug urges and drug-use behavior, role of automatic and nonautomatic processes. *Psychol. Rev.* **97,** 147–168.
26. Chiamulera, C., Borgo, C., Falchetto, S., Valerio, E., and Tessari, M. (1996) Nicotine reinstatement of nicotine self-administration after long term extinction. *Psychopharmacology* **127,** 102–107.
27. Shaham, Y., Adamson, L. K., Grocki, S., and Corrigall, W. A. (1997) Reinstatement and spontaneous recovery of nicotine seeking in rats. *Psychopharmacology* **130,** 396–403.
28. Snyder, F. R., Davis, F. C., and Henningfield, J. E. (1989) The tobacco withdrawal syndrome, performance decrements assessed on computerised test battery. *Drug Alcohol Depend.* **23,** 259–266.
29. Malin, D. H., Lake, J. R., Carter, V. A., Cunningham, J. S., and Wilson, O. B. (1993) Naloxone precipitates nicotien abstinence syndrome in the rat. *Psychopharmacology* **112,** 339–342.
30. Epping-Jordan, M. P., Watkins, S. S., Koob, G. F., and Markou, A. (1998) Dramatic decreases in brain reward function during nicotine withdrawal. *Nature* **393,** 76–79.
31. Markou, A., Kosten, T. R., and Koob, G. F. (1998) Neurobiological similarities in depression and drug dependence: a self–medication hypothesis. *Neuropsychopharmacology* **18,** 135–174.
32. Fagernstrom, K. O. (1978) Measuring degree of physical dependence to tobacco smoking with reference to individualization of treatment. *Add. Behav.* **3,** 235–241.
33. Watkins, S. S., Koob, G. F., and Markou A. (2000) Neural mechanisms underlying nicotine addiction, acute positive reinforcement and withdrawal. *Nicotine & Tobacco Res.* **2,**19–37.
34. Spealman, R. D. (1983) Maintenance of behavior by postponement of scheduled injections of nicotine in squirrel monkeys. *J. Pharmacol. Exp. Ther.* **227,** 154–159.

35. Irvine, E. E., Bagnalasta, M., Marcon, C., Motta, C., Tessari, M., File, S. E., and Chiamulera, C. (2001) Nicotine self-administration and withdrawal, modulation of anxiety in the social interaction test in rats. *Psychopharmacologia* **153,** 315–320.
36. Singer, G., Simpson, F., and Lang, W. J. (1978) Schedule induced self-injections of nicotine with recovered body weight. *Pharmacol. Biochem. Behav.* **9,** 387–389.
37. Latiff, A. A., Smith, L. A., and Lang, W. J. (1980) Effects of changing dosage and urinary pH in rats self-administering nicotine on a food delivery schedule. *Pharmacol. Biochem. Behav.* **13,** 209–213.
38. Smith, L.A. and Lang, W.J. (1980) Changes occuring in self-administration of nicotine by rats over a 28-day period. *Pharmacol. Biochem. Behav.* **13,** 215–220.
39. Goldberg, S. R., Spealman, R. D., and Goldberg D. M. (1981) Persistent behavior at high rates maintained by intravenous self-administration of nicotine. *Science* **214,** 573–575.
40. Cox, B. M., Goldstein, A., and Nelson, W. T. (1984) Nicotine self–administration in rats. *Br. J. Pharmacol.* **83,** 49–55.
41. Corrigall, W. A. and Coen, K. M. (1989) Nicotine maintains robust self–administration in rats on a limited-access schedule. *Psychopharmacology* **99,** 473– 478.
42. Tessari, M., Valerio, E., Chiamulera, C., and Beardsley, P. M. (1995) Nicotine reinforcement in rats with histories of cocaine self-administration. *Psychopharmacology* **121,** 282–283.
43. Shoaib, M., Schindler, C. W., and Goldberg, S. R. (1997) Nicotine self-administration in rats, strain, and nicotine pre-exposure effects on acquisition. *Psychopharmacology* **129,** 35–43.
44. Donny, E. C., Caggiula, A. R., Mielke, M. M., Jacobs, K. S., Rose, C., and Sved, A. F. (1998) Acquisition of nicotine self-administration in rats, the effects of dose, feeding schedule, and drug contingency. *Psychopharmacology* **136,** 83–90.
45. Dworkin, S. I., Vrana, S. L., Broadbent, J., and Robinson, J. H. (1993) Comparing the reinforcing effects of nicotine, caffeine, methylphenidate and cocaine. *Med. Chem. Res.* **2,** 593–602.
46. Valentine, J. D., Hokanson, J. S., Matta, S. G., and Sharp, B. M. (1997) Self–administration in rats allowed unlimited access to nicotine. *Psychopharmacology* **133,** 300–304.
47. Koob, G. F. (1995) Animal models of drug addiction, in *Pschopharmacology, The Fourth Generation of Progress.* (Bloom, F. E. and Kupfler, D. J., eds.), Raven Press, New York, pp. 759–772.
48. Goldberg, S. R. and Gardner, M. L. (1981) Second-order schedules, extended sequences of behavior controlled by brief environmental stimuli associated with drug self-administration. *NIDA Res. Monogr. Ser.* **37,** 241–270.
49. Rasmussen, T. and Swedberg, M. D. B. (1998) Reinforcing effects of nicotine compounds, intravenous self-administration in drug-naïve mice. *Pharmacol. Biochem. Behav.* **60,** 567–573.
50. Stolerman, I. P., Naylor, C., Elmer, G. I., and Goldberg, S. R. (1999) Discrimination and self-administration of nicotine by inbred strains of mice. *Psychopharmacology* **141,** 297–306.
51. Picciotto, M. R., Zoli, M., Rimondini, R., Lena, C., Marubio, L., Merlo Pich, E., Fuxe, K., and Changeux, J. P. (1998) Acetylcholine receptors containing the β2 subunit are involved in the reinforcing properties of nicotine. *Nature* **391,** 173–177.
52. Merlo-Pich, E. and Epping-Jordan, M. P. (1998) Transgenic mice in drug dependence research. *Ann. Med.* **30,** 390–396.
53. Changeux, J.-P., Bertrand, D., Corringer, P. J., Dehaene, S., Edelstein, S., Lena, C., Le Novere, N., Marubio, L., Picciotto, M., and Zoli, M. (1998) Brain nicotinic receptors, structure and regulation, role in learning and reinforcement. *Brain Res. Rev.* **26,** 198–216.
54. Colquhoun, L. M. and Patrick, J. W. (1997) Pharmacology of neuronal nicotinic acetylcholine receptor subtypes. *Adv. Pharmacol.* **39,**191–220.
55. Clarke, P. B. S., Schwartz, R. D., Paul, S. M., Pert, C. B., and Pert, A. (1985) Nicotine binding in rat brain, autoradiography comparison of ^{3}H-acetylcholine, ^{3}H-nicotine and ^{125}I-α-bungarotoxine. *J. Neurosci.* **5,** 1307–1315.

56. Orr-Uteger, A., Galdner, F. M., Saeki, M., Lorenzo, I., Goldberg, L., De Blasi, M., Dani, J. A., Patrick, J. K., and Beaudet, A. L. (1997) Mice deficent of the α7 neuronal nicotinic acethylcoline receptor lack α–bungarotoxin binding sites and hippocampal fast nicotinic currents. *J. Neurosci.* **17,** 9165–9171.
57. Zoli, M., Lena, C., Picciotto, M., and Changeux, J.-P. (1998) Identification of four classes of brain nicotinic receptors using β2 mutant mice. *J. Neurosci.* **18,** 4461–4472.
58. Marubio, L. M., Arroyo-Jimenez, M. M., Cordero-Erausquin, M., Lena, C., Le Novere, N., de Kerchove d'Exaert, A., Huchet, M., Damaj, M. I., and Changeux, J.-P. (1999) Reduced antinociception in mice lacking neuronal nicotinic receptor subunits. *Nature* **398,** 805–809.
59. Mulle, C., Vidal, C., Benoit, P., and Changeux, J.-P. (1991) Exsitence of different subtypes of nicotinic acetylcholine receptors in the rat habenulo–interpeduncular system. *J. Neurosci.* **11,** 2588–2597.
60. Benwell, M. E. M., Balfour, D. J. K., and Anderson, J. M. (1988) Evidence that tobacco smoking increases the density of (-)3H-nicotine binding site in human brain. *J. Neurochem.* **50,** 1243–1247.
61. Breese, C. R., Marks, M. J., Logel, J., Adams, C. E., Sullivan, B., Collins, A. C., and Leonard, S. (1997) Effects of smoking history on ^{3}H–nicotine binding in human post–mortem brain. *J. Pharmacol. Exp. Ther.* **255,** 187–196.
62. Court, J. A., Lloyd, S., Thomas, N., Piggot, M. A., Marshall, E. F., Morris, C. M., Lamb, H., Perry, R. H., Johnson, M., and Perry, E. K. (1998) Dopamine and nicotinic receptor binding and the level of dopamine and homovanillic acid in human brain related to tobacco use. *Neuroscience* **87,** 63–78.
63. Flores, C. M., Rogers, S. W., Pabreza, L. A., Wolfe, B. B., and Kellar, K. J. (1992) A subtype of nicotinic cholinergic receptor in rat brain is composed of α4 and β2 subunits and is up–regulated by chronic nicotine treatment. *Mol. Pharmacol.* **41,** 31–37.
64. Flores, C. M., Davila-Garcia, M. I., Ulrich, Y. M., and Kellar, K. J. (1997) Differential regulation of neuronal nicotinic receptor binding sites following chronic nicotine administration. *J. Neurochem.* **69,** 2216–2219.
65. Marks, M. J., Pauly, J. R., Gross, S. D., Deneris, E. S., Hermans-Borgmeyer, I., and Collins, A. C. (1992) Nicotine binding and nicotinic receptor subunit RNA after chronic nicotine treatment. *J. Neurosci.* **12,** 2765–2784.
66. Marks, M. J., Grady, S. R., Yang, J. M., Lippiello, P. M., and Collins A. C. (1983) Desensitisation of nicotine-stimulated ^{86}RB efflux from mouse brain synaptosome. *J. Neurochem.* **63,** 2125–2135.
67. Wang, F., Nelson, M. E., Kuryatov, A., Olale, F., Cooper, J., Keyser, K., and Lindstrom, J. (1998) Chronic nicotine treatment up–regulated human α3β4 but not acetylcholine receptors stably transfected in human embryonic kidney cells. *J. Biol. Chem.* **273,** 28731–28732.
68. Robinson, T. E. and Berridge, K. C. (1993) The neural basis of drug craving, an incentive-sensitisation theory of addiction. *Brain Res. Rev.* **18,** 247–291.
69. Nestler, E. J. (1996) Under siege, the brain on opiates. *Neuron* **16,** 897–900.
70. LeNovere, N., Zoli, M., and Changeux, J. P. (1996) Neuronal nicotinic receptors alpha-6 subunit mRNA is selectively concentrated in catecholaminergic nuclei of the rat brain. *Eur. J. Neurosci.* **8,** 2428–2433.
71. Arroyo-Jimenez, M. M., Bourgeois, J.-P., Marubio, L. M., Le Sourd, A. M., Ottersenn, O. P., Rinvik, P., Faire, A., and Changeux, J.-P., (1999) Ultrastructural localization of the α4-subunit of the neuronal acetylcholine nicotinic receptor in the rat substantia nigra. *J. Neurosci.* **19,** 6475–6487.
72. Calabresi, P., Lacey, M. G., and North, R. A. (1989) Nicotine excitation of rat ventral tegmental neurons in vitro studied by intracellular recording. *Br. J. Pharmacol.* **98,** 135–140.

73. Pidoplichko, V. I., DeBiasi, M., Williams, J. T., and Dani J. D. (1997) Nicotine activates and desensitizes midbrain dopamine neurons. *Nature* **390,** 401–404.
74. Mereu, G., Yoon, K. W. P., Boi, V., Gessa, G. L., Naes, L., and Wesfall, T. C. (1987) Preferential stimulation of ventral tegmental areas dopaminergic neurons by nicotine. *Eur. J. Pharmacol.* **141,** 395–399.
75. Anderson, K., Fuxe, K., and Agnati, L. F. (1981) Effects of single injection of nicotine on the ascending dopaminergic pathways in the rats. Evidence for increase dopamine turnover in the mesostriatal and mesolimbic dopamine neurones. *Acta Physiol. Scand.* **112,** 345–347.
76. Fuxe, K., Andersson, K., Harfstrand, A., and Agnati, L. F. (1986) Increase dopamine utilisation in certain limbic dopamine terminal populations after short period of intermittent exposure to male rats to cigarette smoke. *J. Neural. Transm.* **67,** 15–29.
77. Imperato, A., Mulas, A., and Di Chiara, G. (1986) Nicotine preferentially stimulates dopamine release in the limbic system of freely moving rats. *Eur. J. Pharmacol.* **132,** 337–338.
78. Balfour, D. J. K., Benwell, M. E. M., Birrel, C. E., Kelly, R. J., and Al-Aloul M. (1998) Sensitisation of measoaccumbens dopamine response to nicotine. *Pharmacol. Biochem. Behav.* **59,** 1021–1030.
79. Pontieri, F. E., Tanda, G., Orzi, F., and Di Chiara, G. (1997) Effects of nicotine on the nucleus accumbens and similarities to those of addictive drugs. *Nature* **382,** 255–257.
80. Nisell, M., Nomikos, G. G., and Svensson T. H. (1994) Systemic nicotine-induced dopamine release in the rat nucleus accumbens is regulated by nicotinic receptors in the ventral tegmental area. *Synapse* **16,** 36–44.
81. Chergui, K., Charlety, P. J., Akaoka, H., Saunier, C. F., Brunet, J. L., Buda, M., Svensson, T. H., and Chovet, G. (1993) Tonic activation of NMDA receptors causes spontaneous burst discharge of rat midbrain dopamine neurons in vivo. *Eur. J. Neurosci.* **5,** 137–144.
82. Gonon, F. G. (1988) Nonlinear relationship between impulse flow and dopamine released by rats midbrain dopaminergic neurons as studied by in vivo electrochemistry. *Neuroscience* **24,** 19–28.
83. Benwell, M. E. M., Balfour, D. J. K., and Birrel, C. E. (1995) Desensitisation of nicotine-induced dopamine response during constant infusion with nicotine. *Br. J. Pharmacol.* **114,** 211–217.
84. Hildebrand, B. E., Nomikos, G. G., Hertel, P., Schilstrom, B., Swensson, T. H. (1997) Reduced dopamine output in nucleus accumbens but not in the medial prefrontal cortex in rats displaying a mecamylamine–precipitated withdrawal syndrome. *Brain Res.* **779,** 214–225.
85. Carboni, E., Bortone, L., Giua, C., and Di Chiara, G. (2000) Dissociation of physical abstinence sign from changes in extracellular dopamine in the nucleus accumbens and in the prefronatal cortex of nicotine dependent rats. *Drug Alcohol Dep.* **58,** 93–102.
86. Marshall, D. L., Redfern, P. H., and Wonnacott, S. (1997) Presynaptic nicotinic modulation of dopamine release in the three ascending pathways studied by in vivo microdalysis, comparison of naive and chronic nicotine-treated rats. *J. Neurochem.* **68,** 1511–1519.
87. Wonnacott, S., Soliakov, L., Wilke, G., Redfern, P., and Marshall, D. (1996) Presynaptic nicotine acetylcholine receptors in the brain. *Drug Develop. Res.* **38,** 149–159.
88. Grady, S. R., Marks, M. J., and Collins, A. C. (1994) Desensitisation of nicotine-stimulated ^{3}H-dopamine release from mouse striatal synaptosomes. *J. Neurochem.* **62,** 1390–1398.
89. Yu, Y. J. and Wecker, L. (1994) Chronic nicotine administration differentially affects neurotransmitter release from rat striatal slices. *J. Neurochem,* **63,** 186–194.
90. Schilstrom, B., Svensson, H. M., Svensson, T. H., and Nomikos, G. G. (1998) Nicotine and food induced dopamine release in the nucleus accumbens of the rat, putative role of $\alpha 7$ nicotinic receptors in the ventral tegmental area. *Neuroscience* **85,** 1005–1009.
91. Nomikos, G. G., Schilstrom, B., Hildebrand, B. E., Panagis, G., Grenhoff, J., and Svensson, T. H. (2000) Role of alpha7 nicotinic receptors in nicotine dependence and implications for psychiatric illness. *Behav. Brain Res.* **113,** 97–103.

92. Brioni, J. D., Kim, D. J., and O'Neill, A. B. (1996) Nicotine cue, lack of the effect of α7 nicotinic receptor antagonist methyllycaconitine. *Eur. J. Pharmacol.* **301,** 1–5.
93. Carr, L. A., Rowell, P. P., and Pierce, W. M., Jr. (1989) Effects of subchronic nicotine administration on central dopaminergic mechanisms in the rat. *Neurochem. Res.* **14,** 511–515.
94. Smith, K. M., Mitchell, S. N., and Joseph, M. H. (1991) Effects of chronic and subchronic nicotine on tyrosine-hydroxylase activity in noradrenergic and dopaminergic neurones in the rat brain. *J. Neurochem.* **57,** 1750–1756.
95. Corrigall, W. A. and Coen, K. M. (1991) Selective dopamine antagonists reduce nicotine self-administration. *Psychopharmacology* **104,** 171–176.
96. Corrigal, W. A., Franklin, K. B. J., Coen, K. M., and Clarke, P. B. S. (1992) The mesolimbic dopamine system is implicated in the reinforcing effects of nicotine. *Psychopharmacology* **107,** 285–289.
97. Dawe, S., Giarada, C., Russel, M. A. H., and Gray, J. A. (1995) Nicotine intake in smokers increases following single dose of haloperidol. *Psychopharmacology* **117,** 110–116.
98. Hurt, R. D., Sachs, D. P. L., and Glover, E. D. (1997) A comparison of sustained-release bupropion and placebo for smoking cessation. *N. Engl. J. Med.* **337,** 1195–1202.
99. Corrigall, W. A., Coen, K. M., and Adamson, K. L. (1994) Self-administration of nicotine activates the mesolimbic nicotine system through the ventral tegmental area. *Brain Res.* **653,** 278–284.
100. Hope, B. T., Nye, H. E., Kelz, M. B., Self, D. W., Iadarola, M. J., Nakabeppu, Y., Duman, R. S., and Nestler, E. J. (1994) Induction of a long-lating AP-1 complex composed of altered Fos-like proteins in brain by chronic cocaine and other chronic treatment. *Neuron* **13,** 1235–1244.
101. Harlan, R. E. and Garcia, M. M. (1998) Drug of abuse and immediate–early genes in the forebrain. *Mol. Neurobiol.* **16,** 221–267.
102. Hughes, P. and Dragunow, M. (1995) Induction of immediate–early genes and the control of neurotransmitter-regulated gene expression within the nervous system. *Pharmacol. Rev.* **47,** 133–178.
103. Ren, T. and Segar, S. M. (1992) Induction of c-fos immunostaining in the rat brain after the systemic administration of nicotine. *Brain Res. Bull.* **29,** 589–597.
104. Pang, Y., Kiba, H., and Jayaraman, A. (1993) Acute nicotine injections induced c-fos mostly in non-dopaminegic cells of the midbrain of the rat. *Mol. Brain Res.* **20,** 162–170.
105. Mathieu-Kia, A. M., Pages, C., and Besson, M. J. (1998) Inducibility of c-Fos protein in visuomotor system and limbic structures after acute and repeated administration of nicotine in the rat. *Synapse* **29,** 343–354.
106. Watanabe, K. I., Hashimoto, K., Nishimura, T., Tsunashima, K. I., and Minabe, Y. (1998) Expression of Fos protein in rat brain following administration of a nicotinic acetylcholine receptor agonist epibatidine. *Brain Res.* **797,** 135–142.
107. Kiba, H. and Jayaraman, A. (1994) Nicotine induced c-fos expression in the striatum is mediated mostly by dopaminergic D1 receptor and is dependent on NMDA stimulation. *Mol. Brain Res.* **23,** 1–13.
108. Nisell, M., Nomikos, G. G., Chergui, K., Grillner, P., Svesson, T. H. (1997) Chronic nicotine enhances basal and nicotine-induced Fos immunoreactivity preferentially in the medial prefrontal cortex of the rat. *Neuropsychopharmacology* **17,**151–161.
109. Pagliusi, S. R., Tessari, M., DeVevey, S., Chiamulera, C., and Merlo Pich, E. (1996) The reinforcing properties of nicotine are associated with a specific patterning of c-fos expression in the rat brain. *Eur. J. Neurosci.* **8,** 2247–2256.
110. Merlo Pich, E., Pagliusi, S. R., Tessari, M., Talabot-Ayer, D., Hooft van Huijsduijnen, R., and Chiamulera, C. (1997) Common neural substrates for the addictive properties of nicotine and cocaine. *Science* **275,** 83–86.
111. Dobranzki, P., Nouguchi, T., Kovary, K., Rizzo, C., Lazo, P. S., and Bravo, R. (1991) Both products of the fosB gene, fosB and its short form, FosB/SF, are transcriptional activator in fibroblasts. *Mol Cell Biol.* **11,** 2063–2069.

112. Schena, M. (1995), Quantitative monitoring of gene expression patterns with a complementary DNA microarray. *Science* **270**,467–470.
113. Konu, O., Kane, J. K., Barrett, T., Vawter, M. P., Chang, R. Y., Ma, J. Z., Donovan, D. M., Sharp, B., Becker, K. G., and Li, M. D. (2001) Region-specific transcriptional response to chronic nicotine in rat brain. *Brain Res.* **909,** 194–203.
114. Sacaan, A. L., Dunlop, J. L., and Loyd, G. K. (1995) Pharmacological characterisation of neuronal acetylcholine gated ion channel-receptors mediated hippocampal norepinephrine and striatal dopamine release from rat brain slices. *J. Pharmacol. Exp. Ther.* **274,** 224–230.
115. Clarke, P. B. S. and Reuben, M. (1996) Release of [3H]-noradrenaline form rat hippocampal synaptosomes by nicotine, mediation from different nicotinic receptor subtypes from striatal [3H]-dopamine release. Br. J. Pharmacol. **117,** 595–606.
116. Brazel, M. P., Mitchell, S. N., and Gray, J. A. (1991) Effects of acute administration of nicotine on the in vivo release of noradrenaline in the hippocampus of freely moving rats, dose response and antagonism studies. *Neuropharmacology* **30,** 823–833.
117. Mitchell, S. N., Smith, K. M., Joseph, M. H., and Gray, G. A. (1993) Increases of thyrosine hydroxylase messenger RNA in the locus coeruleus after a single dose of nicotine are followed by time-dependent increases in enzyme activities and noradrenaline release. *Neuroscience* **56,** 989–997.
118. Fu, Y., Matta, S. G., and Sharp, B. M. (1999) Local α-bungarotoxine-sensitive nicotinic receptors modulate hippocampal norepinephrine release by systemic nicotine. *J. Pharmacol. Exp. Ther.* **289,**133–139.
119. Rosencrans, J. A. and Meltzer, L. T. (1981) Central sites and mechanisms of action of nicotine. *Neurosci. Biobehav. Rev.* **5,** 497–501.
120. Holmes, K. J. and Spencer, C. M. (2000) Bupropion, a review of its use in the management of smoking cessation. *Drugs* **59,**1007–1024.
121. Amit, Z., Smith, B. R., and Gill, K. (1991) Serotonin uptake inhibitor, effects of motivated consummatory behavior. *J. Clin. Psychiatr.* **52**(Suppl.),55–61.
122. Parsons, L. H., Weiss, F., and Koob, G. F. (1998) Serotonin 1B receptor stimulation enhances cocaine reinforcement. *J. Neurosci.* **18,** 10078–89.
123. Corrigall, W. A. and Coen, K. M. (1994) Nicotine self–administration and locomotor activity are not modified by the 5-HT3 antagonist ICS205-930 and MDL722222. *Pharmacol. Biochem. Behav.* **49,** 67–71.
124. Sannerud, C. A., Prada, J., Goldberg, D. M., and Goldberg, S. M. (1994) The effects of sertraline on nicotine self-administrationand food-maintained responding in squirrel monkeys. *Eur. J. Pharmacol.* **271,** 461–469.
125. Zacny, J. P., Apfelbaum, J. L., Lichtor, J. L., and Zacaroza, J. G. (1993) Effects of 5-hydroxythriptamine-3 receptor antagonist, ondansetron, on cigarette smoking, smoke exposure, and mood in humans. *Pharmacol. Biochem. Behav.* **44,** 387–391.
126. Carboni, E., Aquas, E., Leone, P., and Di Chiara, G. (1989) 5-HT3 receptor antagonist ondansetron blocks morphine- and nicotine- but not amphetamine- induced reward. *Psychopharmacology* **97,** 175–178.
127. Blondal, T., Gudmundsson, L. J., Tomasson, K., Jonsdottir, D., Hilmarsdottir, H., Kristjansson, F., Nilsson, F., and Bjornsdottir, U.S. (1999) The effects of fluoxetine combined with nicotine inhalers in smoking cessation, a randomized trial. *Addiction* **94,**1007–1015.
128. Engberg, G. (1989) Nicotine induced excitation of locus coeruleus neurons is mediated via release of excitatory amino acids. *Life Sci.* **44,** 1535–1540.
129. Shoaib, M., Benwell, M. E. M., Akbar, M. T., Stolerman, I. P., and Balfour, D. J. K. (1994) Behavioral and neurochemical adaptation to nicotine in rats: influence of NMDA antagonists. *Br. J. Pharamcol.* **111,** 1073–1080.

130. Rosenceans, J. A., Hendry, S. G., and Hong, J. S. (1985) Biphasic effects of chronic nicotine treatment on hypothalamic immunoreactive β-endorphins in the mouse. *Pharmacol. Biochem. Behav.* **23,** 141–143.
131. Houdi, A. A., Dasgupta, R., and Kindly, M. S. (1998) Effects of nicotine use and withdrawal on brain preproenkephalin A mRNA. *Brain Res.* **799,** 257–263.
132. Dhatt, R. K., Gudehithlu, K. P., Wemlinger, T. A., Tewjani, G. A., Neff, N. H., and Hdjiconstantinou, M. (1995) Preproenkephalin mRNA and methionine-enkephalin contant are increased in mouse striatum after treatment with nicotine. *J. Neurochm.* **64,** 1878–1883.
133. Mathieu-Kia, A. M. and Besson, M. J. (1998) Repeated administration of cocaine, nicotine and ethanol, effects on preprodynorphin, preprotachynin A, and preproenkephalin nRNA expression in the dorsal and ventral striatum of the rat. *Mol. Brain Res.* **54,** 141–151.
134. Gorelik, D. A., Rose, J. E., and Jarvik, M. E. (1989) Effects of naloxone on cigarette smoking. *J. Subst. Abuse* **1,** 153–159.
135. Karras, A. and Kane, J. M. (1980) Naloxone reduces cigarette smoking. *Life Sci.* **27,** 1541–1545.
136. Nemeth-Coslett, R. and Griffith, R.R. (1986) Naloxone does not affect cigarette smoking. *Psychopharmacology* **89,** 261–264.
137. Corrigal, W. A., Coen, K. M., Adamson, K. L., Chow, B. L., and Zhang, J. (2000) Response of nicotine self-administration in the rat to manipulations of mu-opioid and gamma-aminobutyric acid receptors in the ventral tegmental area. *Psychopharmacologia* **149,** 107–114.
138. Kayadjanian, N., Retaux, S., Menetrey, A., and Besson, M. J. (1994) Stimulation by nicotine of the spontaneous release of [3H]gamma-aminobutyric acid in the substantia nigra and in the globus pallidus of the rat. *Brain Res.* **649,** 129–135.
139. Alkondon, M., Pereira, E. F., Barbosa C. T., and Albunquerque, E. X. (1997) Neuronal nicotinic acetylcholine receptor activation modulates amma-aminobutyric acid release from CA1 neurons of rat hippocampal slices. *J. Pharmacol. Exp. Ther.* **283,** 1396–1411.
140. Lu, Y., Grady, S., Marks, M. J., Picciotto, M., Changeux, J. P., and Colins, A. C. (1998) Pharmacological characterization of nicotinic receptor-stimulated GABA release from mouse brain synaptosomes. *J. Pharmacol. Exp. Ther.* **287,** 648–657.
141. Fattore, L., Cossu, G., Martellotta, M. C., Deiana, S., and Fratta, W. (2001) Baclofen antagonises intravenous self-administration of gammahydroxybutyric acid in mice. *Neuroreport* **12,** 2243–2246.
142. Cousins, M. S., Stamat, H. M., and de Wit, H. (2001) Effects of a single dose of baclofen on self-reported subjective effects and tobacco smoking. *Nicotine & Tobacco Res.* **3,** 23–129.
143. Dewey, S. L., Brodie, J. D., Gerasimov, M., Horan, B., and Ashby, C. R., Jr. (1999) A pharmacologic strategy for the treatment of nicotine addiction. *Synapse* **31,** 76–86.
144. Wickelgren, I. (1998) Drug may suppress the craving for nicotine. *Science* **282,** 1797–1799.
145. Sullivan, P. F., Jiang, Y., Neale, M. C., Kendler, K. S., and Straub, R. E. (2001) Association of the tryptophan hydroxylase gene with smoking initiation but not progression to nicotine dependence. *Am. J. Med. Gen.* **105,** 479–484.
146. McKinney, E. F., Walton, R. T., Yudkin, P., Fuller, A., Haldar, N. A., Man, D., Murphy, M., Welsh, K. I., and Marshall, S. E. (2000) Association between polymorphisms in dopamine metabolic enzymes and tobacco consumption in smokers. *Pharmacogenetics* **10,** 483–491.
147. Comings, D. E. and Blum, D. E. (2000) Reward deficiency syndrome: genetic aspects of behavioral disorders. *Prog. Brain Res.* **126,** 325–341.
148. Stitzel, J. A., Lu, Y., Jimenez, M., Tritto, T., and Collins, A. C. (2000) Genetic and pharmacological strategies identify a behavioral function of neuronal nicotinic receptors. *Behav. Brain Res.* **113,** 57–64.
149. Mohammed, A. H. (2000) Genetic dissection of nicotine-related behavior: a review of animal studies. *Behav. Brain Res.* **113,** 35–41.

150. Heyser, C. J., McDonald, J. S., Beauchamp, V., Koob, G. F., and Gold, L. H. (1997) The effects of cocaine on operant responding for food in several strains of mice. *Psychopharmacology* **132,** 202–208.
151. Deroche, V., Caine, S. B., Heyser, C. J., Polis, I., Koob, G. F., and Gold, L. H. (1997) Differences in the liability to self–administer intravenous cocaine between C57BL/6 x SJL and BALB/cByJ mice. *Pharmacol. Biochem Behav.* **57,** 429–435.

Index

129 strain, 278
2-AG, 173–174, 176–177, 179, 183–184, 186, 201, 213
2-Arachidonoyl glycerol. *See* 2-AG
2-Arachidonyl glyceryl ether. *See* Noladin ether
2-Palmitoyl glycerol, 176, 183
^{35}S-GTPγS, 48–49, 180, 208–209, 227
3MT, 146
5 HT1A. *See* Serotonin receptor
5-HT1B. *See* Serotonin receptor
5-HT2. *See* Serotonin receptor
5-HT3. *See* Serotonin receptor
A
α2-Agonist, 61, 67
A5 noradrenergic area, 225
Aβ-fibers, 225
Abstinence. *See* Withdrawal
Acetylcholine receptor, 53, 320–321, 329–330
  muscarinic acetylcholine receptor, 53
  nicotinic acetylcholine receptors, 320–321, 329–330
Activating transcription factor 1. *See* ATF1
Adenosine receptor, 53, 61
Adenylate cyclase. *See* Adenylyl cyclase
Adenylyl cyclase, 39–41, 45, 50, 53, 221
  type I, 39–41, 50
  type II, 50, 53
  type IV, 50,
  type V, 50,
  type VI, 50,
  type VII, 50
  type VIII, 39–41
Ad-fibers, 225
Adh-1 gene, 284
ADH2, 249
Adrenalectomy, 133–152
Adrenaline, 86
Ahd-2 gene, 284
Alcohol, 10–11, 92, 142, 185, 232, 249–314
Alcoholic. *See* Alcoholism
Alcoholism, 249–294
Alcw1, 255
Alcw2, 254
ALDH2, 249
Aldosterone, 140
AM404, 181
Aminopeptidase N, 62–66
AMPA. *See* Glutamatergic receptor
Amphetamine, 10, 79–169, 322
Amygdala, 15, 38, 108–110, 112, 225, 230, 283, 317, 325–326
Analgesia, 3–13, 30, 32, 38, 65, 68, 177, 180, 222–228
Anandamide membrane transporter, 174, 176, 184
Anandamide, 173–175, 178, 180, 182, 184, 201, 205, 213, 225
Angiotensinogen, 277, 281
Antinociception. *See* Analgesia
Antisense oligonucleotides, 14
Anxiety, 92, 250
AP1. *See* Immediate early genes
AP2. *See* Immediate early genes
Apomorphine, 87, 88 181
Apoptosis signal regulating kinase. *See* ASK
ARC *See* Immediate early genes
ASK, 48
ATF1, 27, 29, 40
Atropine, 324

From: *Molecular Biology of Drug Addiction*
Edited by: R. Maldonado © Humana Press Inc., Totowa, NJ

AXB/BXA strain, 270
β-Arrestin, 8–9, 12–14, 17, 46, 48, 53, 209
β-arrestin1, 48
β-arrestin2, 8–9, 12–14, 17, 46, 48
γ-Aminobutyric acid. *See* GABA
B
Baclofen, 329
Basal ganglia, 179, 207
Basic leucine zipper. *See* bZIP
Benzodiazepine, 61
BDDF. *See* Neurotrophic factor
Bovine growth hormone, 281
Brain-derived neurotrophic factor. *See* BDNF
Bungarotoxin, 321
Bupropion, 80
Butorphanol, 49
BXD RI strain, 253–254, 269–271
bZIP, 27–28, 30
C
C57Bl/6J strain, 253, 265, 269–272, 274, 281, 283
C6 cells 50
C779A, 329
Ca^{++} channel. *See* Ion channel
Ca^{++}-calmodulin kinase, 40, 46–47, 281
Calcitonin gene-related peptide, 225
cAMP response element-binding protein. *See* CREB
cAMP response element modulation protein. *See* CREM
cAMP, 4, 27–44, 49, 52–53, 90–91, 107, 122, 201, 210, 214, 233, 324
Cannabinoid receptor, 6, 9, 10, 12, 15–18,173–197, 200–214, 221–223, 227, 232
CB1, 6, 9, 10, 12, 15–18, 173–174, 177–178, 180–182, 184, 201–214, 221–223, 227, 232
CB2, 201, 211, 213, 222
Cannabinoid, 12, 14, 173–245
Cannabis, 173–245
Cas-1 gene, 284
Catalase gene, 277, 281
Catalepsy, 207
Caudateputamen. *See* Striatum
CB1. *See* Cannabinoid receptor
CB2. *See* Cannabinoid receptor
CBP, 40
Cdk5, 97
Cerebellum, 52, 179, 184, 207, 210, 234
c-fos. *See* Immediate early genes
CGRP, 7, 13, 15, 18, 225
Cholecystokinin, 108
Chromosome 1, 269–271
Chromosome 2, 269–273
Chromosome 3, 269
Chromosome 4, 254, 269–271, 273
Chromosome 5, 270
Chromosome 6, 269, 271
Chromosome 7, 269–270
Chromosome 9, 269–273
Chromosome 10, 269–270
Chromosome 11, 269–272
Chromosome 12, 269
Chromosome 15, 269–270
Chromosome 16, 270
Chromosome 17, 269, 271
Chromosome 18, 271
Chromosome 19, 270
CI-977, 227, 300, 325
Citalopram, 80
c-Jun amino-terminal kinase. *See* JNK
Clonidine, 67–68
Cocaine, 10–11, 19, 33–34, 42, 79–169, 185, 320, 322, 327
Cognitive function, 181–182
Collaborative Study on the Genetics of Alcoholism, 256–257
COMT, 145–146
Conditional mutant mice, 31–34
Conditioned Place Paradigm, 4, 9–13, 16–19, 68, 70, , 89, 137, 185, 221 231–232, 264–267, 275–277, 279–280, 282, 298, 317
aversion 9, 17–18, 231, 300
preference, 4, 10–13, 16, 18–19, 68, 70, 89, 137, 185, 221, 231–232, 279–280, 298, 317
Corticosterone. *See* Glucocorticoid hormones

Cortisol. *See* Glucocorticoid hormones
CP-55,940, 201, 203–204, 206–208, 211, 226, 232
Craving, 66, 70, 161, 107, 185, 283, 295–313
Cre/loxP, 30, 33
CREB, 8–9, 13, 18, 27–44, 122
CREB-binding protein. *See* CBP
CREB^{loxP}, 28, 30, 33
CREM, 27, 29, 40
Cre-recombinase, 30, 33
CRF. *See* CRH
CRH, 141, 211, 298, 303–306
CTOP, 122
CTT9A, 329
Cyclin-dependent kinase 5. *See* Cdk5
Cytisine, 321
D
DADLE, 49
DAMGO, 16, 49, 52, 114, 328
DARPP-32, 96, 98, 275, 279
DAT, 6, 9, 11–13, 16–18, 79–84, 92, 95, 98, 107
DBA/2J strain, 253–254, 265, 269–272, 274, 297–298, 306
Delta. *See* Opioid receptor
Deltorphin, 9
Dentate gyrus, 325
Dependence, 3–75, 199–245, 316–317
Desensitization, 45, 208–209, 214, 297, 323
Desipramine, 80
Dexamethasone, 140
Δ*fos*B, 96–99, 164
Dihydro-β-erythroidine, 324
DOPAC, 145
Dopamine and cyclic adenosine 3',5'mono-phosphate-regulated phosphoprotein, 32kDa. *See* DARPP-32
Dopamine receptor, 6, 9, 11–12, 15–17, 34, 53, 67–68, 70, 81–82, 85, 88–99, 107, 110–111, 116–117, 121–123, 144, 146, 162, 164, 180, 275, 278–279, 284, 300, 324, 327, 329
  D1, 34, 53, 67–68, 70, 82, 88, 89–98, 107, 111, 116, 121–123, 144, 162, 275, 278–279, 300, 324, 327, 329
  D2, 6, 9, 11–12, 15–17, 81, 85, 88, 89–93, 95, 107, 110, 116–117, 146, 164, 180, 275, 278–279, 284, 300, 324
  D3, 82, 85, 88, 89, 92–93, 95, 98–99, 111, 300, 324
  D4, 88– 89, 92–95, 98, 300, 324
  D5, 88–89, 91, 95, 300
Dopamine, 16, 37, 42, 67–68, 70, 79–159, 181, 230, 232–233, 275, 283, 298, 300
Dopamine-β-hydroxylase, 251, 329
Dopamine transporter. *See* DAT
Down-regulation, 45, 49–50, 85, 207–208, 211, 214, 283, 326
DPDPE, 114
Drinking/preference, 264, 275–277, 282, 285
DSM IV, 315
DSM-IIIR, 315
Dynorphin, 42, 88, 89, 93, 108, 110–111, 120, 224, 226, 299–300
Dysphoria, 38
E
Edinger-Westphal nucleus, 283–284
Endocannabinoid, 173–197
Endomorphin, 110, 115, 122
  endomorphin 1, 122
  endomorphin 2, 122
Enkephalin catabolism, 61–75
Enkephalin 61–75, 88–89, 95, 108, 110, 120, 226, 328
Epibatidine, 321
Erk, 48
Ethanol. *See* Alcohol
Etorphine, 49
Excitotoxicity, 183
β-Endorphin, 49, 120, 280, 328
F
Fatty acid amide hydrolase (FAAH), 173–176, 184, 201–202, 205, 213
Feeding, 182
Fluoxetine, 80
Flupenthixol, 300
*Fos*B. *See* Immediate early genes
Fos-related antigen (FRAs), 97, 325

G

GABA, 41–42, 82, 96, 108–114, 117, 147, 173, 175, 222, 225, 251, 284, 301, 327, 329

GABA receptor, 208, 250, 257, 284, 329

  $GABA_A$, 250, 257, 284

  $GABA_B$, 208, 329

Gap junctions, 178

GBR 12909, 115

GBR 12935, 80

GDNF. *See* Neurotrophic factor

GIRK2. *See* Ion channel, K+ channel

Glial cell line-derived neurotrophic factor. *See* GDNF

Globus pallidus, 208

Glucocorticoid hormones, 133–152, 305

  corticosterone, 133–134, 136, 138–142, 145–146, 305

  cortisol, 133

Glucocorticoid receptor, 6, 9, 12, 15, 18, , 133, 135, 140, 143, 144–147, 163, 166, 250, 303–305

Glutamate, 109, 113, 162, 164, 175, 183, 284, 298, 301–303, 327–328

Glutamatergic receptor, 13, 15–16, 18, 52, 162, 165, 176, 178–179, 183, 223, 251, 301–303, 305–306, 328

  AMPA, 13, 15, 18, 165, 303

  kainate 303

  NMDA, 13, 16, 52, 162, 176, 178–179, 183, 223, 251, 301–303, 305–306, 328

Glycine, 251

G protein, 8–9, 11, 14, 28, 38–39, 42, 45, 48, 50–53, 66, 87, 89, 94, 96, 98, 107, 163, 175–176, 201, 208–209, 227, 276, 280–281

  G_{ia}, 14, 28, 38–39, 42, 45, 51, 89, 107, 163, 175–176, 201, 208–209, 227

  $G_{nas\alpha}$, 276, 281

  $G_{o\alpha}$, 28, 38–39, 42, 45, 89, 163, 175–176, 201, 208–209

  $G_{olf\alpha}$, 94, 96, 98, 107

  $G_{s\alpha}$, 50–53, 87, 94, 107, 176, 227, 276, 280

  $G_{z\alpha}$, 8–9, 11, 14

  $G_{\beta\gamma}$, 50, 52–53

G-protein-coupled receptor kinase (GRK), 46, 48, 53, 209

  GRK2, 46, 48, 53

  GRK3, 46, 48, 53

  GRK5, 46, 48

  GRK6, 46, 48

H

Hallucinogenic drug, 84

Haloperidol, 88, 91

Hashish, 221

Heroin, 7, 12, 62, 70

Herpes simplex virus vector, 33

Hippocampus, 47, 108–110, 123, 179, 181, 183–184, 207, 317, 327

Histamine, 86

Homer1a. *See* Immediate early genes

Homer1bc. *See* Immediate early genes

HU210, 178, 201, 203–204

HU211, 184

Hypothalamus, 15, 52, 110, 184, 226, 317, 325, 328

Hypothalamus-pititary-adrenal axis, 133–159

I

ICI-174,864, 223

IL6, 7, 12, 14, 19

Immediate early genes, 50, 53, 65, 83, 89, 91, 93, 96–98, 122 143, 163, 165, 211, 225, 283, 324–325–327

  AP1, 50, 96, 122, 325–326

  AP2, 50

  ARC, 163

  c-fos, 53, 65, 83, 89, 91, 93, 143, 163, 225, 283, 324–325, 327

  Fos B, 96–98, 283

  homer1a, 163

  homer1bc, 165

  junB, 91

  narp, 163

  zif268, 89, 91

Incentive salience, 297

Interferon β, 278

Interval-specific congenic strains, 254

Intracranial self-stimulation. *See* Self-stimulation
Ion channel, 38–39, 45–46, 51–52, 96, 146–147, 183, 201, 222, 277, 281, 321, 323
  K$^+$ channel, 38–39, 45, 51, 96, 146–147, 201, 222, 277, 281, 321
  Na$^+$ channel, 39, 96, 147, 321, 323
  Ca^{++} channel, 45–46, 52, 96, 146–147, 183, 201, 222, 321, 323
Ischemia, 183
ISCS strain, 255
J
JNK, 48
JunB. *See* Immediate early genes
K
K$^+$ channel. *See* Ion channel
Kainate. *See* Glutamatergic receptor
Kappa. *See* Opioid receptor
Ketanserin, 178
Ketoconazole, 136, 138, 140, 141
Knockout, 3–36, 79–106, 177, 179–180, 182, 206, 212–214, 224, 228–229, 231–233, 273–279, 282, 285, 299
L
L-DOPA, 81
Learning. *See* Cognitive function
Lisuride, 300
Locus coeruleus, 29, 32–33, 38–40, 42, 51–52, 64, 230, 325, 327
LSXSS strain, 270
M
Mammalian Genome, 256
MAO, 145–146, 275, 278
MAP kinase, 40, 45, 47, 51, 222, 326
Marijuana, 199–221
Mazindol, 80
MDL72222, 178
Mecamylamine, 319, 322, 327
Medulla oblongata, 52
  nucleus tractus solitari, 52
  nucleus prepositus hypoglossal, 52
Memantine, 302–303
Memory. *See* Cognitive function
Mesocorticolimbic system, 6–7, 15, 37–38, 41–43, 51–52, 66–68, 70, 96, 99, 107–119, 123, 141–148, 162–166, 181, 185, 226, 232–233, 283, 300, 317, 322–324, 325–326, 328–330
  nucleus accumbens, 6–7, 15, 33, 37–38, 41–43, 51, 67–68, 70, 96, 107–119, 123, 141–148, 164–165, 185, 226, 232, 283, 300, 317, 322–323, 325, 328, 330
    shell, 111, 142–144, 147, 300
    core, 142–144
  ventral tegmental area, 37–38, 41–42, 52, 66, 99, 108–115, 117, 123, 141–148, 185, 233, 317, 321–324, 328–329
  prefrontal cortex, 108–110, 113, 116 162–165, 181, 185, 317, 322–323, 325–326, 330
Methadone, 61–62, 68
Methanandamide, 208
Methyllycaconetine, 323
Methylphenidate, 80
Metyrapone, 134, 136, 139
Microdialysis, 67
Mineralocorticoid receptor, 133–135, 140, 143, 147
Mitogen-activated protein kinase. *See* MAP kinase
Mitogen and stress-activated protein kinase. *See* MSK
MK-801, 52, 328
Morphine, 3–26, 29–34, 49, 63–65, 68, 227, 229–230
Motor function, 180
Mouse Phenome Project, 267
Mpdz gene, 254
MRZ 2/579, 302–303, 305
MSK, 40
Mu. *See* Opioid receptor
Muscarinic acetylcholine receptor. *See* Acetylcholine receptor
Myenteric plexus, 38
N
N18TG2 cells, 175
N4TG cells, 49
Na$^+$ channel. *See* Ion channel

Nafadotride, 88
Naloxone, 31, 223, 229–230, 299, 328
Naltrexone, 121, 300, 304–305
Naltrindole, 68, 122
Narp. *See* Immediate early genes
NBQX, 303
Neramexane. *See* MRZ 2/579
Nerve growth factor. *See* Neurotrophic factor, NGF
NET, 79–80, 84–86, 98
N-ethylmaleimide-sensitive fusion protein, 48
N-ethyl-*N*-nitrosurea, 255
Neurokinin B, 108
Neuropeptide Y, 108, 276–277, 281
Neuroprotection, 183–184
Neurotensin, 88–89, 93, 108
Neurotrophic factor, 97–99
  BDNF, 97–99
  GDNF, 97–99
  NGF, 97
Neutral endopeptidase, 62–66, 277, 281
NDF. *See* Neurotrophic factor
NG108-15 cells, 49–50, 53
Nicotine, 10, 19, 142, 232, 315–338
Nicotinic acetylcholine receptors. *See* Acetylcholine receptors
Nitric oxide, 52
NK1 receptor, 6, 9, 11, 16–18
NMDA. *See* Glutamatergic receptor
N-methyl-D-aspartate. *See* NDMA
Noladin ether, 173–174
Nomifensine, 87
Norepinephrine transporter. *See* NET
Norepinephrine, 327
Norbinaltorphimine, 117, 119–120, 300
NSF, 48
Nucleus accumbens. *See* Mesocorticolimbic system

O

OFQ/N, 6, 9, 12–13, 18, 110, 112, 120
Oleamide, 176, 178
Olfactory bulb, 52
Olfactory nuclei, 52
Olfactory tubercle, 207
Opiate. *See* Opioid
Opioid, 3–75, 99, 108–115, 117, 123, 107–132, 141–148, 173, 185, 221–245, 275–276, 279–280, 298–300, 304–305, 328
Opioid receptor, 5, 9, 10, 13–14, 17, 38–39, 45–50, 62–63, 70, 110–114, 116, 120–121, 123, 221–224, 228, 231, 233–234, 276, 280, 284, 299–300, 306, 328
  mu, 5, 9, 10, 13, 38–39, 45–50, 52, 62–63, 70, 110–114, 116, 121, 123, 221–224, 228, 231, 233–234, 276, 280, 284, 299, 306
  delta, 5, 9, 10, 14, 17, 45–47, 49–50, 62–63, 70, 110–112, 114, 121, 123, 221, 223–224, 228, 231, 299, 328
  kappa, 5, 9, 10, 17, 42, 45–46, 62, 110–114, 116–117, 120, 123, 221, 223–224, 228, 231, 234, 299–300
  phosphorylation, 46
ORL1 receptor, 6, 9, 12–14, 18, 110–113

P

p38-MAPK, 179
P70/S6 kinase, 40
PAG, 38, 66, 180, 225, 230
Pain, 3–13, 45, 65, 180
Paroxetine, 80
Pedunculopontine tegmental nucleus, 324
PEPCK/βGH gene, 277
Periaqueductal gray matter. *See* PAG
Phospholipase A, 53
Phospholipase C, 45, 47, 51, 53
PKA, 38–41, 47, 50–51, 91, 96, 214, 230, 276, 280
PKB, 40
PKC, 8–9, 16, 46–47, 51, 117, 276, 280, 285
  PKCγ, 8–9, 16
  PKCε, 276, 280, 285
Pleiotropy, principle of , 286
PL017, 49
PP1, 96
Prefrontal cortex. *See* Mesocorticolimbic system
Prepositus hypoglossal nucleus. *See* Medulla oblongata
Preproenkephalin. *See* Proenkephalin
Prodynorphin, 62, 111–112, 115, 122, 224,

228, 233, 328
Proenkephalin, 14, 17, 62, 111, 115–116, 224, 228, 233
Proopiomelanocortin, 62, 115, 275–276
Protein kinase A. *See* PKA
Protein kinase B. *See* PKB
Protein kinase C. *See* PKC
Protein phosphatase 1. *See* PP1
Psychostimulant, 79–169, 232, 297

Q

QTLs, 252, 255–257, 263–294

R

R121919, 304
Raf, 48, 51
Rap, 51
Ras, 51
RB101, 63–70
Reboxetine, 80, 84
Recombinant inbred strains, 267
Reinforcement, 4, 16–17, 19, 34, 80, 83–85, 107, 121, 148, 185, 232, 250, 300, 316–317, 320–322
Reinstatement, 296, 304–305
Relapse, 43, 70, 107, 137, 141, 295–313
Reserpine,86
Reward, 4, 15–17, 29, 33–34, 42–43, 68, 70–71, 92, 137, 142–143, 148, 221, 230–233, 250, 263–294, 315–338
Rostral ventromedial medulla, 225
Rp-8Br-cAMP, 210, 230
RU 38486, 140, 143, 146
RU 39305, 143

S

SCH23390, 67–68, 88, 278
Second order schedule, 318
Self-administration, 6–7, 15, 19, 66, 68, 83, 89, 134, 136–137, 140–142, 147, 161, 213, 221, 230, 250, 263, 265, 278–279, 284, 295–297, 316–320, 324–325, 327, 330
Self-stimulation, 221–230–231, 317, 319
Sensitization, 10, 13, 19, 37, 53, 83, 90, 97, 107, 117, 120–121, 135, 138, 148, 161–169, 297–298, 317, 322–323, 328
Septum, 207
Serotonin, 82, 90, 173, 274–275, 327
Serotonin receptor, 83, 178, 274–275, 278, 327
  5-HT1A, 83
  5-HT1B, 274–275, 278, 327
  5-HT2, 178, 182
  5-HT3, 178, 275, 278, 327
Serotonin selective reuptake inhibitor, 80
Serotonin transporter. *See* SERT
SERT, 79–80, 83–84, 90, 98
SH-SY5Y cells, 49
SKF 82958, 144
SKF 38393, 88
Sleep, 176, 182
Somatostatin, 225
Sp-8Br-cAMP, 210
Spinal cord, 38
  dorsal horn, 38
Spiperone, 88
SR141716A, 177, 178, 181–183, 185, 204–205, 208–214, 228–232
SRC, 48
Standard inbred strains, 267
Striatum, 81–83, 93–97, 115, 164, 206–207, 210, 222, 226, 233, 317, 321, 325–326
SubcentiMorgan level, 255
Substance P, 12–13, 82, 89, 91–92, 108, 225
Substantia nigra, 88, 208, 317
Superactivation. *See* Up-regulation
Superior colliculus, 225, 325

T

Taste conditioning, 264, 266–267, 273–277, 280
Tetrabenazine, 86
Thalamus, 109, 225, 325
THC, 15, 175–176, 181–183, 185, 199–245
Tobacco, 315–338
Tolerance, 10, 13, 17, 19, 37–40, 45–60, 65–66, 83, 184, 199–200, 206, 221–245, 250, 297, 323
TPH C218A, 329
Tractus solitari nucleus. *See* Medulla oblongata
TrkB receptor, 97, 99, 165
Two-bottle paradigm. *See* Drinking preference

Tyrosine hydroxylase, 39, 41, 81, 92, 145, 147, 163, 323–324
Δ[9]-Tetrahydrocannabinol. *See* THC

U

U50,488, 46, 227
U69,593, 114, 116, 118
United Nations International Drug Control Programme, 295
Up-regulation, 38–42, 45, 92, 211, 214, 321, 326

V

Vanilloid receptor, 179
Ventral tegmental area. *See* Mesocorticolimbic system
Vesicular monoamine transporter. *See* VMAT.
Vigabatrin, 329
VMAT, 86–87
  VMAT 1, 86–87
  VMAT 2, 86–87,

W

WIN55,212-1, 179
WIN55,212-2, 179–180, 183, 185, 201, 203–205, 208, 211, 213, 223, 228, 232
Withdrawal, 16–19, 29–30, 37–43, 67–68, 162, 200, 202–205, 209, 250, 253–254, 297, 300–303, 318–319
World Health Organization, 295

Z

Zif268. *See* Immediate early genes